ELECTRONIC INSTRUMENTS AND MEASUREMENTS

ELECTRONIC INSTRUMENTS AND MEASUREMENTS

Second Edition

Larry D. Jones
Oklahoma State University

A. Foster Chin
Tulsa Junior College

PRENTICE HALL, Englewood Cliffs, New Jersey 07632

Library of Congress Cataloging-in-Publication Data

Jones, Larry D.
Electronic instruments and measurements / Larry D. Jones, A. Foster Chin.--2nd ed.
p. cm.
Includes bibliographical references.
ISBN 0-13-248469-2
1. Electronic instruments. 2. Electronic measurements. I. Chin, A. Foster. II. Title.
TK7878.4.J66 1990b 89-48280
621.381'54--dc20 CIP

Editorial/production supervision **Lillian Glennon**
Manufacturing buyer: **Gina Brennan**

Printed in the United States of America

10 □ 9 □ 8 □ 7 □ 6 □ 5 □ 4 □ 3 □ 2 □ 1

ISBN 0-13-248469-2

Prentice-Hall International (UK) Limited, *London*
Prentice-Hall of Australia Pty. Limited, *Sydney*
Prentice-Hall Canada Inc., *Toronto*
Prentice-Hall Hispanoamericano, S.A., *Mexico*
Prentice-Hall of India Private Limited, *New Delhi*
Prentice-Hall of Japan, Inc., *Tokyo*
Simon & Schuster Asia Pte. Ltd., *Singapore*
Editors Prentice-Hall do Brasil, Ltds., *Rio de Janeiro*

Contents

Preface

The ability to make measurements is vital to an understanding of the physical world in which we live. Some people say that measurement provides us with an understanding of physical phenomena and that instruments are tools for measurement.

A great many of the instruments used for obtaining measurements in various kinds of physical systems are electronic in nature, and electronic instrumentation has increased at an extraordinary rate during the last twenty years. Automation in many areas of industry and the accompanying need for more precise measurements have been largely responsible for the increased need for electronic indicating, recording, and controlling instruments. Integrated circuit technology has made possible the proliferation of instruments to meet these needs.

Obtaining accurate, reliable, and cost-effective measurements goes beyond the instrument to the user. Proper selection and use of instruments and interpretation of measurement data are the responsibility of the user. A person's ability to make an intelligent selection and to use the instrument properly is greatly increased by an understanding of the basic theory of operation and the capabilities and limitations of the test instrument.

Laboratory practice has a very important place in any science-related course. The laboratory experiments contained in the text will strengthen the student's knowledge of electronic test instruments by experimentally analyzing circuits used in these instruments and will help develop proficiency in the intelligent use of instruments of this kind.

We discuss both service and laboratory-quality instruments, but the major emphasis is on laboratory-quality instruments. Among the illustrations of instruments are a number of photographs. A case can be made against the use of photographs of actual commercial instruments in textbooks because it can cause a book to become obsolete very rapidly. We believe, however, that the selective use of photographs is beneficial to students in the type of course for which this text is intended.

The chances are remote that any textbook will coincide exactly with a particular course outline. Since each of our chapters is essentially self-supporting, the order in which they are arranged can be altered as required.

This book was written with one thought in mind: textbooks are mainly for students. Therefore, we have worked diligently to make the material as readable as possible. We hope that we have succeeded.

The work was a cooperative effort, and the order of authorship has no significance. We accept equal responsibility for the text's strengths and weaknesses.

We extend our appreciation to the editorial staff of Prentice Hall, to the many companies that provided photographs, data, and diagrams of their instruments, and to the reviewers of the manuscript for their constructive suggestions. Typing was skillfully and cheerfully done by Kay Porter and CeCe Henry. Finally, we thank our families for their patience and encouragement.

Larry Jones
A. Foster Chin

CHAPTER 1 Introduction to Instrumentation

1-1 Instructional Objectives

This introductory chapter discusses electrical units, various classes of standards, functions and characteristics of instruments, and different types of errors and methods of error analyses. After completing Chapter 1 you should be able to

1. List three functions of instruments.
2. List three advantages of electronic instruments over electrical instruments.
3. Define terms related to the quality of instruments.
4. List and describe four categories of standards.
5. List and describe the three elements of electronic instruments.
6. Define terms related to error and error analysis.
7. Perform basic statistic analysis calculations.

1-2 INTRODUCTION

Over the last century and a half, there have been many contributions to the art of measuring electrical quantities. During most of this period, the principal effort was aimed at perfecting a deflection-type **instrument** with a scale and movable pointer. The angle of deflection of the pointer is a function of, and is therefore analogous to, the value of the electrical quantity being measured. The term **analog instrument** was coined to identify deflection-type instruments and to distinguish these from totally different instruments, which display in *decimal* (digital) form the value of the quantity being measured. These newer instruments are called **digital instruments**.

1-3 FUNCTIONS AND CHARACTERISTICS OF INSTRUMENTS

Many instruments serve common purposes in supplying information about some variable quantity that is to be measured. Besides providing a visual *indication* of the quantity being measured, the instrument sometimes furnishes a permanent *record*. In addition, some instruments are used to *control* a quantity. Therefore, we can say that instruments serve three basic functions: *indicating*, *recording*, and *controlling*. A particular instrument may serve any one or all three of these functions simultaneously. *General-purpose* electrical and electronic test instruments primarily provide *indicating* and *recording* functions. The instrumentation used in *industrial-process* situations frequently provides a *control* function. The entire system may then be called a *control* or *automated system*.

There are many ways to measure the value of different quantities. Some physical quantities are best measured by purely mechanical means such as using a manometer gauge to measure gas pressure. Other quantities are measured by methods that are primarily electrical such as measuring solution conductivity with a current meter. Other measurements are made with electronic instruments that contain an amplifying circuit to increase the amplitude of the quantity being measured.

Although some electronic instruments are more expensive than simple electrical instruments, they offer some significant advantages for **measurement** purposes. The use of an electronic amplifier results in an instrument with a *high sensitivity* rating, capable of measuring very small (low-amplitude) signals. The greater sensitivity of the instrument also *increases* its *input impedance*, thereby reducing loading effects when measurements are made. A third advantage of electronic instruments is their ability to *monitor remote signals*.

1-4 ELECTRICAL UNITS

Because electrical and electronic instruments measure electrical quantities, this introductory chapter should include a discussion of electrical units. As with any quantitative science, a system of units is required before one can make a quantitative evaluation of parameters measured.

Six electrical quantities are of concern when we make electrical measurements. In order to treat them quantitatively, we need to establish and define a system of units. The six quantities are (1) electric charge, Q, (2) electric current, I, (3) electromotive force or potential difference, V, (4) resistance, R, (5) inductance, L, and (6) capacitance, C. The fundamental quantities of the MKSA or SI system of units and a few of the electrical quantities that can be derived from these units are listed in Table 1-1.

TABLE 1-1
Six Fundamental and Some Derived Quantities of the SI System

Quantity	Symbol	Unit	Unit Abbreviation	Dimensions
Fundamental				
Length	l	meter	m	l
Mass	m	kilogram	kg	m
Time	t	second	s	t
Temperature	T	degree Kelvin	°K	T
Luminous intensity		candela	cd	
Electric current	I	ampere	A	i
Derived				
Electromotive force	V	volt	V	$l^2mt^{-3}i^{-1}$
Quantity of charge	Q	coulomb	C	it
Electrical resistance	R	ohm	Ω	$ml^2t^{-2}i^{-2}$
Capacitance	C	farad	F	$l^{-2}m^{-1}t^4i^2$
Inductance	L	henry	H	$l^2mt^{-2}i^{-2}$

1-5 MEASUREMENT STANDARDS

Whether we wish to use an electrical or an electronic instrument, all instruments are calibrated at the time of manufacture against a measurement **standard**. Standards are defined in four categories.

The *international standards* are defined by international agreement. These standards are maintained at the International Bureau of Weight and Measures in Paris and are periodically evaluated and checked by absolute measurements in terms of the fundamental units of physics. They represent certain units of measurement to the closest possible accuracy attainable by the science and technology of measurement.

The *primary standards* are maintained at national standards laboratories in different countries. The National Bureau of Standards (NBS) in Washington, D.C., is responsible for maintaining the primary standards in North America. Primary standards are not available for use outside the national laboratories. Their principal function is the calibration and verification of secondary standards.

The *secondary standards* are the basic reference standards used by measurement and calibration laboratories in the industry to which they belong. Each industrial laboratory is completely responsible for its own secondary standards. Each laboratory periodically sends its secondary standards to the national standards laboratory for calibration. After calibration the secondary standards are returned to the industrial laboratory with a certification of measuring accuracy in terms of a primary standard.

The *working standards* are the principal tools of a measurements laboratory. They are used to check and calibrate the instruments used in the laboratory or to make comparison measurements in industrial application.

1-6 ERROR IN MEASUREMENT

Measurement is the process of comparing an unknown quantity with an accepted standard quantity. It involves connecting a measuring instrument into the system under consideration and observing the resulting response on the instrument. The measurement thus obtained is a quantitative measure of the so-called true value. Since it is very difficult to define the true value adequately, the term **expected value** is used throughout this text. Any measurement is affected by many variables; therefore, the results rarely reflect the expected value. For example, connecting a measuring instrument into the circuit under consideration always disturbs (changes) the circuit, causing the measurement to differ from the expected value.

Some factors that affect measurements are related to the measuring instruments themselves. Other factors are related to the person using the instrument. The degree to which a measurement conforms to the expected value is expressed in terms of the **error** of the measurement. Error may be expressed either as *absolute* or as a *percent of error*. *Absolute error* may be defined as the difference between the expected value of the variable and the measured value of the variable, or

$$e = Y_n - X_n \tag{1-1}$$

where

e = absolute error
Y_n = expected value
X_n = measured value

If we wish to express the error as a percentage we can say that

$$\text{Percent error} = \frac{\text{Absolute error}}{\text{Expected value}}(100)$$

or

$$\text{Percent error} = \frac{e}{Y_n}(100)$$

Substituting for e yields

$$\text{Percent error} = \left|\frac{Y_n - X_n}{Y_n}\right|(100) \tag{1-2}$$

It is frequently more desirable to express measurements in terms of **relative accuracy** rather than error, or

$$A = 1 - \left|\frac{Y_n - X_n}{Y_n}\right| \tag{1-3}$$

where A is the relative accuracy.

Accuracy expressed as percent accuracy, a, is

$$a = 100\% - \text{Percent error} = A \times 100 \qquad (1\text{-}4)$$

EXAMPLE 1-1

The expected value of the voltage across a resistor is 50 V; however, measurement yields a value of 49 V. Calculate

(a) The absolute error.
(b) The percent of error.
(c) The relative accuracy.
(d) The percent of accuracy.

Solution

(a) $e = Y_n - X_n$

$$= 50\text{ V} - 49\text{ V} = 1\text{ V} \qquad (1\text{-}1)$$

(b) $$\text{Percent error} = \frac{50\text{ V} - 49\text{ V}}{50\text{ V}} \times 100\%$$

$$= \frac{1}{50} \times 100\% = 2\%$$

(c) $$A = 1 - \frac{50\text{ V} - 49\text{ V}}{50\text{ V}} = 1 - \frac{1}{50}$$

$$= 1 - 0.2 = 0.98$$

(d) $$a = 100\% - \text{Percent error} = 100\% - 2\% = 98\%$$

$$= A \times 100 = 0.98 \times 100 = 98\%$$

If a measurement is accurate, it must also be *precise*; that is, accuracy implies precision. The reverse, however, is not necessarily true; that is, precision does not necessarily imply accuracy. The **precision** of a measurement is a quantitative, or numerical, indication of the *closeness* with which a repeated set of measurements of the same variable agrees with the *average* of the set of measurements. Precision can be expressed in a mathematical sense, or quantitatively, as

$$\text{Precision} = 1 - \left| \frac{X_n - \bar{X}_n}{\bar{X}_n} \right| \qquad (1\text{-}5)$$

where

X_n = the value of the nth measurement
$\bar{X}_n$ = the average of the set of n measurements

EXAMPLE 1-2

The following set of ten measurements was recorded in the laboratory. Calculate the precision of the fourth measurement.

Measurement Number	Measurement Value X_n (volts)
1	98
2	102
3	101
4	97
5	100
6	103
7	98
8	106
9	107
10	99

Solution

The average value for the set of measurements is equal to the sum of the ten measurements divided by 10 (refer to Eq. 1 in Experiment 1), which is 101.1. The precision of the fourth measurement is

$$\text{Precision} = 1 - \left|\frac{X_n - \bar{X}_n}{\bar{X}_n}\right|$$

$$= 1 - \left|\frac{97 - 101.1}{101.1}\right| = 1 - 0.04 = 0.96 \qquad (1\text{-}5)$$

An indication of the precision of a measurement is obtained from the number of significant figures to which the result is expressed. Significant figures convey information regarding the magnitude and preciseness of a quantity; additional significant digits represent a more precise measurement.

When making measurements or calculations, we retain only significant figures. Significant figures are the figures, including zeros and estimated figures, that have been obtained from measuring instruments known to be trustworthy. The position of the decimal point does not affect the number of significant figures. If a zero is used merely to locate the decimal point, it is not a significant figure. However, if it actually represents a digit read with an instrument, or estimated, then it is a significant figure. When making calculations, we should drop nonsignificant figures. This avoids false conclusions since too many figures imply greater accuracy than the figures actually represent.

The following rules should be used regarding significant figures when making calculations.

1. When performing addition and subtraction, do not carry the result beyond the first column that contains a doubtful figure. As a general rule, all figures in columns to the right of the last column in which all figures are significant should be dropped.

2. When performing multiplication and division, retain only as many significant figures as the least precise quantity contains.
3. When dropping nonsignificant figures, do not change the last figure to be retained if the figures dropped equal less than one-half. The last figure retained should increase by 1 if the figures dropped have a value equal to, or greater than, one-half. The following example illustrates these rules.

EXAMPLE 1-3

(a) The voltage drops across two resistors in a series circuit are measured as

$$V_1 = 6.31\ \text{V}$$
$$V_2 = 8.736\ \text{V}$$

The applied voltage is the sum of the voltage drops. Often it is advisable to add values without rounding off the individual numbers, and then round off the sum to the correct number of significant figures. If we follow this practice, the applied voltage is given as

$$E = 6.31\ \text{V} + 8.736\ \text{V}$$
$$= 15.046\ \text{V}$$

Rounded to the same precision as the least precise voltage drop, the supply voltage is given as

$$E = 15.05\ \text{V}$$

(b) The current in the same series circuit is measured as 0.0148 A. Using this value of current and the voltage drops, we can compute the value of each resistor as

$$R_1 = \frac{V_1}{I} = \frac{6.31\ \text{V}}{0.0148\ \text{A}} = 426\ \Omega$$

and

$$R_2 = \frac{V_2}{I} = \frac{8.736\ \text{V}}{0.0148\ \text{A}} = 590.3\ \Omega$$

Both values are expressed with the same number of significant figures as the least accurate value used in the computation and are rounded according to rule 3.

(c) The power dissipated by each resistor is the product of the voltage drop across the resistor and the current through the resistor and is given as

$$P_1 = V_1 I$$
$$= (6.31\ \text{V})(0.0148\ \text{A}) = 0.093\ \text{W}$$

and

$$P_2 = V_2 I$$

$$= (8.736\ \text{V})(0.0148\ \text{A}) = 0.1293\ \text{W}$$

Both values for power dissipation are expressed with the same number of significant figures as the least accurate value used in the computations and are rounded according to rule 3.

The accuracy and precision of measurements depend not only on the quality of the measuring instrument, but also on the person using the instrument. However, regardless of the quality of the instrument or the care exercised by the user, some error is always present in measurements of physical quantities.

Error, which has been described quantitatively, may be defined as the deviation of a reading or set of readings from the expected value of the measured variable. Errors are generally categorized under the following three major headings.

1-6.1 Gross Errors

Gross errors are generally the fault of the person using the instruments and are due to such things as incorrect reading of instruments, incorrect recording of experimental data, or incorrect use of instruments.

1-6.2 Systematic Errors

Systematic errors are due to problems with instruments, environmental effects, or observational errors. These errors recur if several measurements are made of the same quantity under the same conditions.

1. *Instrument errors.* Instrument errors may be due to friction in the bearings of the meter movement, incorrect spring tension, improper calibration, or faulty instruments. Instrument error can be reduced by proper maintenance, use, and handling of instruments.
2. *Environmental errors.* Environmental conditions in which instruments are used may cause errors. Subjecting instruments to harsh environments such as high temperature, pressure, or humidity, or strong electrostatic or electromagnetic fields, may have detrimental effects, thereby causing error.
3. *Observational errors.* Observational errors are those errors introduced by the observer. The two most common observational errors are probably the parallax error introduced in reading a meter scale and the error of estimation when obtaining a reading from a meter scale.

1-6.3 Random Errors

Random errors are those that remain after the gross and systematic errors have been substantially reduced, or at least accounted for. Random errors are generally the accumulation of a large number of small effects and may be of real concern only in measurements requiring a high degree of accuracy. Such errors can only be analyzed statistically.

Several other terms that have been discussed or described quantitatively need to be formally defined. These terms and their definitions are listed below.

- *Instrument*. A device or mechanism used to determine the present value of a quantity under observation.
- *Measurement* The art (or process) of determining the amount, quantity, degree, or capacity by comparison (direct or indirect) with accepted standards of the system of units employed.
- *Expected value*. The design value, that is, "the most probable value" that calculations indicate one should expect to measure.
- *Accuracy*. The degree of exactness of a measurement compared to the expected value, or the most probable value, of the variable being measured.
- *Resolution*. The smallest change in a measured variable to which an instrument will respond.
- *Precision*.[1] A measure of the consistency or repeatability of measurements.

EXAMPLE 1-4

The following table of values represents a meter output in terms of the angular displacement of the needle, expressed in degrees, for a series of identical input currents. Determine the *worst-case* precision of the readings.

I_{in} (μA)	Output Displacement (degrees)
10	20.10
10	20.00
10	20.20
10	19.80
10	19.70
10	20.00
10	20.30
10	20.10

[1]This definition for precision may be modified to define the precision of measuring instrument (as opposed to the precision of a *measurement* made with the instrument). Precision, as it applies to the instrument, is the *consistency* of the instrument output for a given value of input. In the case of an instrument designed around a basic deflection-type meter movement, the input to the instrument is current whereas the output is the angular displacement of the needle.

Solution

The average output, determined by adding the output values and dividing by eight, is equal to 20.02 degrees. The fifth reading differs from the average by the greatest amount; therefore, the worst-case precision is related to and found from this reading.

$$\text{Precision} = 1 - \left|\frac{X_n - \bar{X}_n}{\bar{X}_n}\right|$$

$$= 1 - \left|\frac{19.70 - 20.02}{20.02}\right| = 1 - 0.016 = 0.984 \qquad (1\text{-}5)$$

1-7 STATISTICAL ANALYSIS OF ERROR IN MEASUREMENT

When we measure any physical quantity, our measurements are affected by a multitude of factors. For example, if we measure the resistance of a piece of wire, we find that several factors will influence the resistance value we obtain. Some of these factors are very important whereas others are rather insignificant. The factors to be considered include the type and purity of the wire material, its temperature, the length and cross-sectional area, current distribution through the wire, manufacturing practices such as heat treating, and tension placed on the wire during measurement.

Ideally, the degree to which any individual parameter influences the measurement should be known so that the measurement can be understood and explained. If a measurement is repeated several times, the readings may differ because of the experimenter's failure or inability to keep all the above factors constant.

Let us assume that the *influences* of any parameters (beyond our control) act in a *random* manner. Let us also assume that small *variations* in parameters, which we attempt to hold constant, also act in a *random* manner. If all these variations are purely random, then there is *equal* probability that a variation may be equally positive or negative. This fact allows us to make a statistical determination of the quality of our experimental data. Our evaluation is based on a statistical analysis to allow us to obtain such information as the *mean* value, *average deviation*, and *standard deviation* of our data. Such information allows us to make quantitative judgments of the variations, or errors, in our data.

An *average* is a value typical (or most representative) of a set of data. Since such typical values tend to lie *centrally* within the data when arranged according to magnitude, averages are also called *measures of central tendency*.

The most frequently used average is the **arithmetic mean**, which is the sum of a set of numbers divided by the total number of pieces of data. Thus, the arithmetic mean of a set of n numbers $x_1, x_2, \ldots, x_n$ is denoted by $\bar{x}$ and defined as

$$\bar{x} = \frac{x_1 + x_2 + x_3 + \cdots + x_n}{n} \qquad (1\text{-}6)$$

Deviation is the difference between each piece of test data and the arithmetic mean. The deviation of $x_1, x_2, \ldots,$ from their arithmetic mean $\bar{x}$ is denoted by $d_1, d_2, \ldots, d_n$ and is defined as

$$d_1 = x_1 - \bar{x}$$

$$d_2 = x_2 - \bar{x}$$

$$d_n = x_n - \bar{x}$$

The *algebraic sum of the deviations* of a set of numbers from their arithmetic mean is zero!

EXAMPLE 1-5

For the following data compute

(a) The arithmetic mean.

(b) The deviation of each value.

(c) The algebraic sum of the deviations.

$$x_1 = 50.1$$

$$x_2 = 49.7$$

$$x_3 = 49.6$$

$$x_4 = 50.2$$

Solution

(a) The arithmetic mean is computed as

$$\bar{x} = \frac{50.1 + 49.7 + 49.6 + 50.2}{4}$$

$$= \frac{199.6}{4} = 49.9$$

(b) The deviation of each value from the mean is computed as

$$d_1 = 50.1 - 49.9 = 0.2$$

$$d_2 = 49.7 - 49.9 = -0.2$$

$$d_3 = 49.6 - 49.9 = -0.3$$

$$d_4 = 50.2 - 49.9 = 0.3$$

(c) The algebraic sum of the deviations is

$$d_{tot} = 0.2 - 0.2 - 0.3 + 0.3 = 0$$

The degree to which numerical data spread about the average value is called variation or dispersion of the data. One measure of variation is the *average deviation*. The average deviation may be used as an expression of the precision of a measuring instrument. A low value for average deviation indicates a precise instrument. The average deviation D of a set of numbers is

$$D = \frac{|d_1| + |d_2| + \cdots + |d_n|}{n} \tag{1-7}$$

which states that the average deviation is the arithmetic sum of the *absolute* values of the individual deviations divided by the number of readings.[2]

EXAMPLE 1-6 Calculate the average deviation for the data from Example 1-5.

Solution The average deviation is computed as follows

$$D = \frac{|0.2| + |-0.2| + |-0.3| + |0.3|}{4}$$

$$= \frac{1.0}{4} = 0.25$$

The **standard deviation** S for a set of values is the degree to which the values vary about the average value. The standard deviation of a set of n numbers is

$$S = \sqrt{\frac{d_1^2 + d_2^2 + \cdots + d_n^2}{n}} \tag{1-8}$$

For small numbers of readings ($n < 30$), the denominator is frequently expressed as $n - 1$ to obtain a more accurate value for the standard deviation.

EXAMPLE 1-7 Compute the standard deviation for the data from Example 1-5.

Solution

$$S = \sqrt{\frac{(0.2)^2 + (0.2)^2 + (0.3)^2 + (0.3)^2}{4 - 1}}$$

$$= \sqrt{\frac{0.4 + 0.4 + 0.09 + 0.09}{3}}$$

$$= \sqrt{\frac{0.26}{3}} = 0.294 \tag{1-8}$$

1-8 LIMITING ERRORS

Most manufacturers of measuring instruments state that an instrument is accurate within a certain percentage of a full-scale reading. For example, the manufacturer of a certain voltmeter may specify the instrument to be accurate

[2]Note that if the algebraic sum of the deviations were used in Eq. 1.7, the numerator would always be zero.

within $\pm 2\%$ with full-scale deflection. This specification is called the *limiting error* and means that a full-scale reading is guaranteed to be within the limits of 2% of a perfectly accurate reading. However, with readings that are less than full scale, the limiting error will increase. Therefore, it is important to obtain measurements as close as possible to full scale.

EXAMPLE 1-8

A 300-V voltmeter is specified to be accurate within $\pm 2\%$ at full scale. Calculate the limiting error when the instrument is used to measure a 120-V source.

Solution

The magnitude of the limiting error is

$$0.02 \times 300\text{ V} = 6\text{ V}$$

Therefore, the limiting error at 120 V is

$$\frac{6}{120} \times 100\% = 5\%$$

EXAMPLE 1-9

A voltmeter and an ammeter are to be used to determine the power dissipated in a resistor. Both instruments are guaranteed to be accurate within $\pm 1\%$ at full-scale deflection. If the voltmeter reads 80 V on its 150-V range and the ammeter reads 70 mA on its 100-mA range, determine the limiting error for the power calculation.

Solution

The magnitude of the limiting error for the voltmeter is

$$0.01 \times 150\text{ V} = 1.5\text{ V}$$

The limiting error at 80 V is

$$\frac{1.5}{80} \times 100\% = 1.86\%$$

The magnitude of the limiting error for the ammeter is

$$0.01 \times 100\text{ mA} = 1\text{ mA}$$

The limiting error at 70 mA is

$$\frac{1}{70} \times 100\% = 1.43\%$$

The limiting error for the power calculation is the *sum* of the individual limiting errors involved; therefore,

$$\text{Limiting error} = 1.86\% + 1.43\% = 3.29\%$$

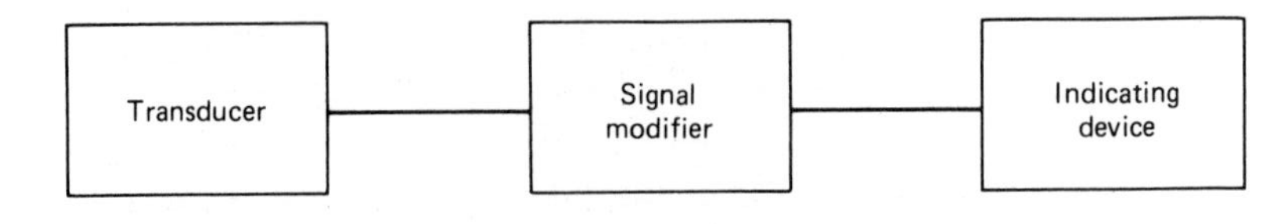

FIGURE 1-1 Element of an electronic instrument.

1-9 ELEMENTS OF ELECTRONIC INSTRUMENTS

In general, an electronic instrument is made up of the three elements shown in Fig. 1-1.

The **transducer** converts a nonelectrical signal into an electrical signal. Therefore, a transducer is required only if the quantity to be measured is nonelectrical (e.g., pressure).

The **signal modifier** is required to process the incoming electrical signal to make it suitable for application to the indicating device. The signal may need to be amplified until it is of sufficient amplitude to cause any appreciable change at the indicating device. Other types of signal modifiers might be voltage dividers, which are designed to reduce (attenuate) the amount of signal applied to the indicating device, or waveshaping circuits such as rectifiers, filters, or choppers.

The **indicating device** is generally a deflection-type meter for such general-purpose instruments as voltmeters, current meters, or choppers.

Electronic instruments may be used to measure current, voltage, resistance, temperature, sound level, pressure, or many other physical quantities. However, regardless of the units on the calibrated scale of the indicating meter, the pointer deflects up scale because of the flow of electrical current.

1-10 SELECTION, CARE, AND USE OF INSTRUMENTS

The finest instruments available may provide inaccurate results when mistreated or improperly used. Observance of several basic rules will generally ensure that instruments provide acceptable measurement results.

Most instruments are delicate, sensitive devices and should be treated with care. Before using an instrument, you should be thoroughly familiar with its operation. The best source of information is the operating and instructions manual, which is provided with any new instrument purchased. Electronics laboratories should have these manuals on file for easy access. If you are not already thoroughly familiar with an instrument's operation, specifications, functions, and limitations, read the manual before using the instrument.

You should select an instrument to provide the degree of accuracy required. Although a high degree of *accuracy* and good *resolution* are desirable, in general, the *cost* of the instrument is directly related to these properties.

Once an instrument has been selected for use, it should be visually inspected for any obvious physical problems such as loose knobs, damaged case, bent pointer, loose handle, and damaged test leads. If the instrument is powered by an internal battery, the condition of the battery should be checked prior to use. Many instruments have a "battery check" position for this purpose. When a battery must be replaced, make sure the proper replacement is used and that it is properly installed.

Before connecting the instrument into the circuit, make certain the *function* switch is set to the proper function and the *range-selector* switch to the proper range. If there is any question at all about the proper range, the instrument should be set to its *highest* range before connecting it into the circuit. Then it should be switched to lower ranges until an approximate midscale reading is obtained. The best range to use is the one that yields the *highest* scale deflection.

Many other considerations, including circuit loading, impedance matching, and frequency response, must be dealt with in order to obtain the most accurate results possible using test equipment. These and many other topics are discussed in subsequent chapters to enable the reader to become thoroughly familiar with the theory, use, and applications of basic electronic test instruments.

1-11 SUMMARY

The term *instrumentation* refers to instruments ranging from basic test instruments found on workbenches in many homes to complex scientific instruments used in research laboratories or automated systems used to control entire industrial plants.

All instruments serve certain common functions, although some are more suitable than others for particular applications. When selecting an instrument, you should consider several characteristics related to instrument quality. However, regardless of your choice, some error is inevitable. There are several sources and several categories of error.

1-12 GLOSSARY

Accuracy: The degree of exactness of a measurement when compared to the expected (most probable) value of the variable being measured.

Analog instrument: An instrument that produces a voltage or deflection in proportion to continuously variable physical quantites.

Arithmetic mean: The sum of a set of numbers divided by the total number of pieces of data in the set.

Deviation: The difference between any piece of data in a set of numbers and the arithmetic mean of the set of numbers.

Error: The deviation of a reading or set of readings from the expected value of the measured variable.

Expected value: The design value or the most probable value that calculations indicate you should expect to obtain.

Instrument: A device or mechanism used to determine the present value of a quantity under observation.

Measurement: The art (or process) of determining the amount, quantity, degree, or capacity by comparison (direct or indirect) with an accepted standard of the system of units employed.

Precision: A measure of the consistency or repeatability of measurements.

Standard: An instrument or device having a recognized permanent (stable) value that is used as a *reference* (or criterion).

Standard deviation: The degree to which the values of a set of numbers vary about the average value for the numbers.

Transducer: A device that converts one form of energy into another form.

1-13 REVIEW QUESTIONS

At the end of each chapter in this text there is a set of review questions related to the material in the chapter. The reader should answer these questions after a comprehensive study of the chapter to determine his or her comprehension of the material.

1. List the three basic functions of instruments.
2. What other name is used to identify deflection-type instruments?
3. What are the six commonly measured electrical quantities? List the quantities, their symbols, units, and unit abbreviations.
4. List three advantages that electronic instruments have over electrical instruments.
5. What are the four categories of standards?
6. Define the following:
   (a) Error.
   (b) Accuracy.
   (c) Precision.
   (d) Measurement.
   (e) Resolution.
   (f) Sensitivity.
7. Describe three major categories of error.
8. A person using an ohmmeter measured a 470-Ω resistor (actual value) as 47 Ω. What kind of error does this represent?
9. List three types of systematic errors and give an example of each.

10. What is the difference between the *accuracy* and the *precision* of the measurement?

11. Define the following terms:

(a) Average value.

(b) Arithmetic mean.

(c) Deviation.

(d) Standard deviation.

1-14 PROBLEMS

1-1 The current through a resistor is 1.5 A, but measurement yields a value of 1.46 A. Compute the absolute error and the percentage of error of the measurement.

1-2 The value of a resistor is 2 kΩ; however, measurement yields a value of 1.93 kΩ. Compute:

(a) The relative accuracy of the measurement.

(b) The percent accuracy of the measurement.

1-3 If the average of a set of voltage readings is 30.15 V, compute the precision of one of the readings that was equal to 29.9 V.

1-4 The output voltage of an amplifier was measured by six different students using the same oscilloscope with the following results.

(a) 20.20 V

(b) 19.90 V

(c) 20.05 V

(d) 20.10 V

(e) 19.85 V

(f) 20.00 V

Which is the most precise measurement?

1-5 Two resistors were selected from a supply of $2200 \pm 10\%\ \Omega$ resistors.

(a) Assume both resistors are $2200 \pm 0\%\ \Omega$ and compute the expected resistance of the parallel combination.

(b) Assume both resistors are $2200 + 10\%\ \Omega$ and compute the expected value of the parallel combination. What is the percent error when compared to the results of part a?

(c) Assume both resistors are $2200 - 10\%\ \Omega$ and compute the expected value of the parallel combination. What is the percent error when compared to the results of part a?

1-6 Three resistors, to be connected in series, were selected from a supply of $470{,}000 \pm 10\%\ \Omega$ resistors.

(a) If all three resistors are $470{,}000 \pm 0\%\ \Omega$, what is the expected value of the series combination?

(b) If all three resistors are $470{,}000 + 10\%\ \Omega$, what is the expected value of the combination? What is the percent error when compared to the results of part a?

(c) If all three resistors are $470{,}000 - 10\%\ \Omega$, what is the expected value of the combination? What is the percent error when compared to the results of part a?

1-7 A $160 \pm$ pF capacitor, an inductor of $160 \pm 10\%\ \mu$H, and a resistor of $1200 \pm 10\%\ \Omega$ are connected in series.

(a) If all three components are $\pm 0\%$, and $f_r = \frac{1}{2}\pi\sqrt{LC}$, compute the expected resonant frequency of the combination.

(b) If all three components are $+10\%$, compute the expected resonance frequency of the combination and the percent error when compared to the results of part a.

(c) If all three components are -10%, compute the expected resonance frequency and the percentage of error when compared to the results of part a.

1-8 Three resistors of $3300 \pm 10\%\ \Omega$ resistance are selected for use in a circuit. Two of the resistors are connected in parallel, and this parallel combination is then connected in series with the third resistor.

(a) Calculate the resistance of the combination if all three resistors are $3300 \pm 0\%\ \Omega$.

(b) Calculate the resistance of the combination and the combination's maximum percentage of error when compared to part a.

1-9 A $150 \pm 10\%\ \Omega$ resistor is connected to the terminals of a power supply operating at $200 \pm 0\ V_{dc}$. What range of current would flow if the resistor varies over the range of $\pm 10\%$ of its expected value? What is the range of error in the current?

1-10 The diameter of a copper conductor varies over its length as shown in the following table.

Reading No.	Diameter (mm)
1	2.21
2	2.18
3	2.20
4	2.21
5	2.17
6	2.19

(a) Calculate the precision of each measurement.
(b) Calculate the average precision.

1-11 A voltmeter is accurate to 98% of its full-scale reading.
(a) If a voltmeter reads 175 V on the 300-V range, what is the absolute error of the reading?
(b) What is the percentage of error of the reading in part a?

1-12 It is desired to determine a voltage to within one part in 300. What accuracy of the meter is required?

1-13 Eight resistors having a color-coded value of 5.6 kΩ were measured and found to have the following values. Determine the standard deviation of the batch.

Resistor No.	Value (kΩ)
1	5.75
2	5.60
3	5.65
4	5.50
5	5.70
6	5.55
7	5.80
8	5.55

1-14 A circuit was tuned for resonance by eight different students, and the following values for the resonant frequency of the circuit were recorded. Compute the arithmetic mean, the average deviation, and the standard deviation for the readings.

Reading No.	Resonant Frequency (kHz)
1	532
2	548
3	543
4	535
5	546
6	531
7	543
8	536

1-15 LABORATORY EXPERIMENTS

At the back of the text is a set of laboratory experiments. Experiments E1 and E2 apply the theory presented in Chapter 1. Their purpose is to provide students with "hands-on" experience, which is essential for a thorough understanding of the concepts involved.

Since Experiments E1 and E2 require no special equipment, the equipment that students will need should be found in any electronics laboratory. The contents of the laboratory report to be submitted by each student are listed at the end of each experimental procedure.

CHAPTER 2 Direct-Current Meters

2-1 Instructional Objectives

The purpose of this chapter is to familiarize the reader with the d'Arsonval meter movement, how it may be used in ammeters, voltmeters, and ohmmeters, some of its limitations, as well as some of its applications. After completing Chapter 2 you should be able to

1. Describe and compare constructions of the two types of suspension used in the d'Arsonval meter movement.
2. Explain the principle of operation of the d'Arsonval meter movement.
3. Describe the purpose of shunts across a meter movement.
4. Describe the purpose of multipliers in series with a meter movement.
5. Define the term *sensitivity*.
6. Analyze a circuit in terms of voltmeter loading or ammeter insertion errors.
7. Describe the construction and operation of a basic ohmmeter.
8. Perform required calculations for multipliers or shunts to obtain specific meter ranges of voltage and current.
9. Apply the concepts related to error studied in Chapter 1 to the circuits of Chapter 2.

2-2 INTRODUCTION

The history of the basic meter movement used in direct-current (dc) measurements can be traced to Hans Oersted's discovery in 1820 of the relationship between current and magnetism. Over the next half-century, various types of devices that made use of Oersted's discovery were developed. In 1881 Jacques d'Arsonval patented the *moving-coil galvanometer*. The same basic construction developed by d'Arsonval is used in meter movements today.

This basic moving-coil system, generally referred to as a d'Arsonval meter movement or a permanent magnet moving-coil (PMMC) meter movement, is

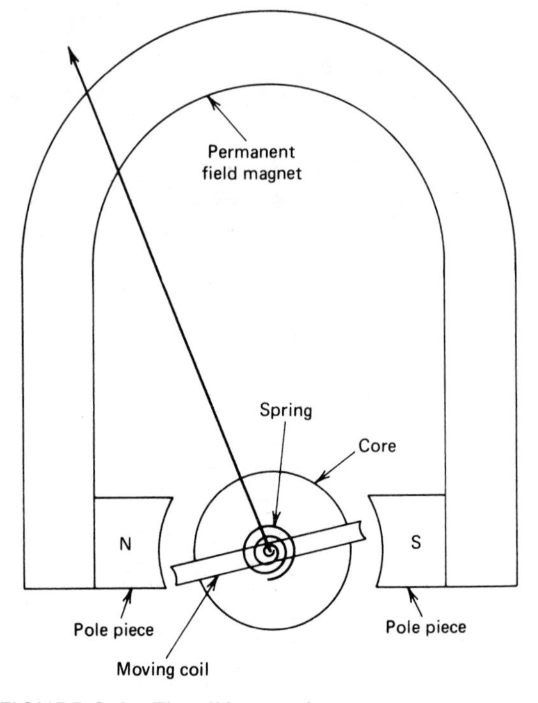

FIGURE 2-1 The d'Arsonval meter movement.

shown in Fig. 2-1. The moving-coil mechanism is generally set in a jewel and pivot suspension system to reduce friction. Another method of suspension is the "taut-band" suspension system which provides a more sensitive, but more expensive, meter movement. As a matter of comparison, a typical full-scale current for a jewel and pivot suspension system is 50 μA, whereas a full-scale current of 2 μA for a taut-band system is entirely practical.

2-3 THE D'ARSONVAL METER MOVEMENT

The **d'Arsonval meter movement** is in wide use even today. For this reason this chapter presents a detailed discussion of its construction and principle of operation. The typical commercial meter movement, shown in Fig. 2-2, operates on the basic principle of the dc motor. Figure 2-1 shows a horseshoe-shaped permanent magnet with soft iron pole pieces attached to it. Between the north–south pole pieces is a cylindrical-shaped soft iron core about which a coil of fine wire is wound. This fine wire is wound on a very light metal frame and mounted in a jewel setting so that it can rotate freely. A pointer attached to the moving coil deflects up scale as the moving coil rotates.

Current from a circuit in which measurements are being made with the meter passes through the windings of the moving coil. Current through the

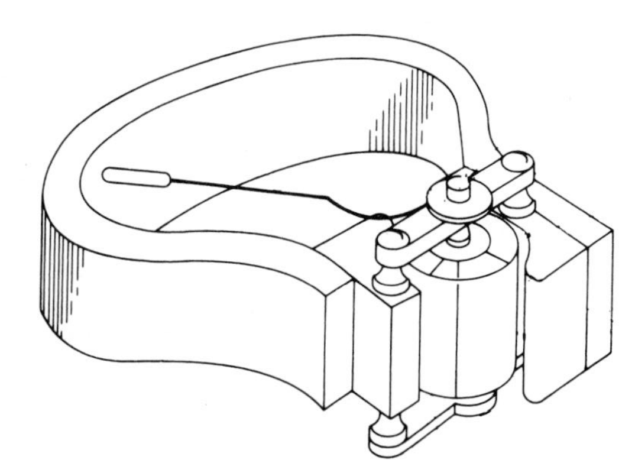

FIGURE 2-2 **Cutaway view of the d'Arsonval meter movement. (Courtesy Weston Instruments, a Division of Sangamo Weston, Inc.)**

coil causes it to behave as an electromagnet with its own north and south poles. The poles of the electromagnet interact with the poles of the permanent magnet, causing the coil to rotate. The pointer deflects up scale whenever current flows in the proper direction in the coil. For this reason, all dc meter movements show polarity markings.

It should be emphasized that the d'Arsonval meter movement is a *current* responding device. Regardless of the units (volts, ohms, etc.) for which the scale is calibrated, the moving coil responds to the amount of current through its windings.

2-4 D'ARSONVAL METER MOVEMENT USED IN A DC AMMETER

Since the windings of the moving coil shown in Fig. 2-2 are of very fine wire, the basic d'Arsonval meter movement has only limited usefulness without modification. One desirable modification is to increase the range of current that can be measured with the basic meter movement. This is done by placing a low resistance in *parallel* with the meter movement resistance, R_m. This low resistance is called a **shunt** (R_{sh}), and its function is to provide an alternate path for the total metered current I around the meter movement. The basic dc ammeter circuit is shown in Fig. 2-3. In most circuits I_{sh} is much greater than I_m, which flows in the movement itself. The resistance of the shunt is found by applying Ohm's law to Fig. 2-3

where

R_{sh} = resistance of the shunt
R_m = **internal resistance** of the meter movement (resistance of the moving coil)
I_{sh} = current through the shunt

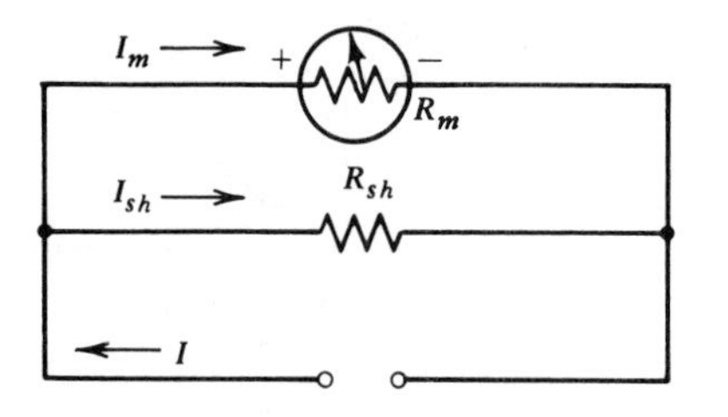

FIGURE 2-3 D'Arsonval meter movement used in an ammeter circuit.

I_m = full-scale deflection current of the meter movement
I = full-scale deflection current for the **ammeter**

The voltage drop across the meter movement is

$$V_m = I_m R_m$$

Since the shunt resistor is in parallel with the meter movement, the voltage drop across the shunt is equal to the voltage drop across the meter movement. That is,

$$V_{sh} = V_m$$

The current through the shunt is equal to the total current minus the current through the meter movement:

$$I_{sh} = I - I_m$$

Knowing the voltage across, and the current through, the shunt allows us to determine the shunt resistance as

$$R_{sh} = \frac{V_{sh}}{I_{sh}} = \frac{I_m R_m}{I_{sh}} = \frac{I_m}{I_{sh}} R_m = \frac{I_m}{I - I_m} \times R_m \ (\Omega) \tag{2-1}$$

EXAMPLE 2-1

Calculate the value of the shunt resistance required to convert a 1-mA meter movement, with a 100-Ω internal resistance, into a 0- to 10-mA ammeter.

Solution

$$V_m = I_m R_m = 1 \text{ mA} \times 100 \ \Omega = 0.1 \text{ V}$$

$$V_{sh} = V_m = 0.1 \text{ V}$$

$$I_{sh} = I - I_m = 10 \text{ mA} - 1 \text{ mA} = 9 \text{ mA}$$

$$R_{sh} = \frac{V_{sh}}{I_{sh}} = \frac{0.1 \text{ V}}{9 \text{ mA}} = 11.11 \ \Omega \tag{2-1}$$

The purpose of designing the shunt circuit is to allow us to measure a current I that is some number n times larger than I_m. The number n is called a multiplying factor and relates total current and meter current as

$$I = nI_m \qquad (2\text{-}2)$$

Substituting this for I in Eq. 2-1 yields

$$R_{sh} = \frac{R_m I_m}{nI_m - I_m}$$

$$= \frac{R_m}{n-1} \ (\Omega) \qquad (2\text{-}3)$$

Example 2-2 illustrates the use of Eqs. 2-2 and 2-3.

EXAMPLE 2-2

A 100-μA meter movement with an internal resistance of 800 Ω is used in a 0- to 100-mA ammeter. Find the value of the required shunt resistance.

Solution

The multiplication factor n is the ratio of 100 mA to 100 μA or

$$n = \frac{I}{I_m} = \frac{100 \text{ mA}}{100\ \mu\text{A}} = 1000 \qquad (2\text{-}2)$$

Therefore,

$$R_{sh} = \frac{R_m}{n-1} = \frac{800\ \Omega}{1000-1} = \frac{800}{999} \approx 0.80\ \Omega$$

2-5 THE AYRTON SHUNT

The shunt resistance discussed in the previous sections works well enough on a single-range ammeter. However, on a multiple-range ammeter, the **Ayrton shunt**, or the universal shunt, is frequently a more suitable design. One advantage of the Ayrton shunt is that it eliminates the possibility of the meter movement being in the circuit without any shunt resistance. Another advantage is that it may be used with a wide range of meter movements. The Ayrton shunt circuit is shown in Fig. 2-4.

The individual resistance values of the shunts are calculated by starting with the most sensitive range and working toward the least sensitive range. In Fig. 2-4, the most sensitive range is the 1-A range. The shunt resistance is $R_{sh} = R_a + R_b + R_c$. On this range the shunt resistance is equal to R_{sh} and can be computed by Eq. 2-3 where

$$R_{sh} = \frac{R_m}{n-1}$$

The equations needed to compute the value of each shunt, R_a, R_b, and R_c, can be developed by observing Fig. 2-5. Since the resistance $R_b + R_c$ is in parallel with $R_m + R_a$, the voltage across each parallel branch should be equal and can be written as

$$V_{R_b+R_c} = V_{R_a+R_m}$$

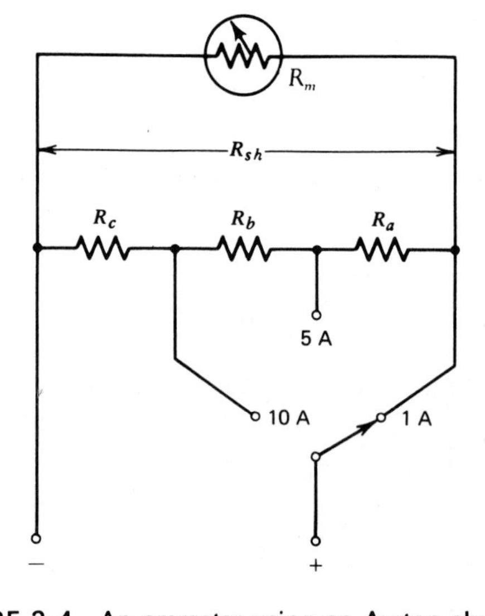

FIGURE 2-4 An ammeter using an Ayrton shunt.

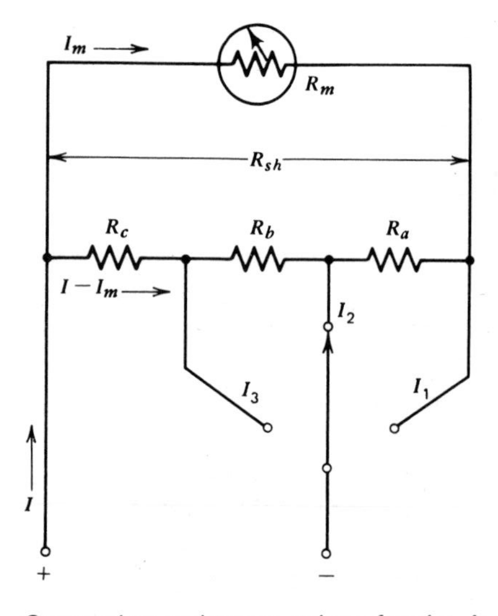

FIGURE 2-5 Computing resistance values for the Ayrton shunt.

In current and resistance terms we can write

$$(R_b + R_c)(I_2 - I_m) = I_m(R_a + R_m)$$

or

$$I_2(R_b + R_c) - I_m(R_b + R_c) = I_m[R_{sh} - (R_b + R_c) + R_m]$$

Multiplying through by I_m on the right yields

$$I_2(R_b + R_c) - I_m(R_b + R_c) = I_m R_{sh} - I_m(R_b + R_c) + I_m R_m$$

This can be rewritten as

$$R_b + R_c = \frac{I_m(R_{sh} + R_m)}{I_2} \ (\Omega) \qquad (2\text{-}4)$$

Having already found the total shunt resistance R_{sh}, we can now deterime R_a as

$$R_a = R_{sh} - (R_b + R_c) \ (\Omega) \qquad (2\text{-}5)$$

The current I is the maximum current for the range on which the ammeter is set. The resistor R_c can be determined from

$$R_c = \frac{I_m(R_{sh} + R_m)}{I_3} \qquad (2\text{-}6)$$

The only difference between Eq. 2-4 and Eq. 2-6 are the currents I_2 and I_3, which in each case is the maximum current for the range for which a shunt value is being computed.

The resistor R_b can now be computed as

$$R_b = (R_b + R_c) - R_c \ (\Omega) \qquad (2\text{-}7)$$

EXAMPLE 2-3 Compute the value of the shunt resistors for the circuit shown in Fig. 2-6.

Solution The total shunt resistance R_{sh} is found from

$$R_{sh} = \frac{R_m}{n-1} = \frac{1\ \text{k}\Omega}{100-1} = \frac{1\ \text{k}\Omega}{99} = 10.1\ \Omega \qquad (2\text{-}2)$$

This is the shunt for the 10 mA range. When the meter is set on the 100-mA range, the resistors R_b and R_c provide the shunt. The total shunt resistance is found by the equation

$$R_b + R_c = \frac{I_m(R_{sh} + R_m)}{I_2}$$

$$= \frac{(100\ \mu\text{A})(10.1\ \Omega + 1\ \text{k}\Omega)}{100\ \text{mA}} = 1.01\ \Omega \qquad (2\text{-}4)$$

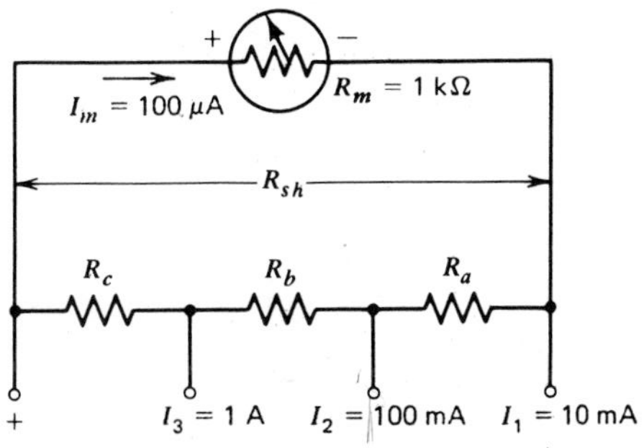

FIGURE 2-6 Ayrton shunt circuit.

The resistor R_c, which provides the shunt resistance on the 1-A range, can be found by the same equation; however, the current I will now be 1 A.

$$R_c = \frac{I_m(R_{sh} + R_m)}{I_3}$$

$$= \frac{(100\ \mu A)(10.1\ \Omega + 1\ k\Omega)}{1\ A} = 0.101\ \Omega \qquad (2\text{-}6)$$

The resistor R_b is found from Eq. 2-7 in which

$$R_b = (R_b + R_c) - R_c = 1.01\ \Omega - R_c = 1.01\ \Omega - 0.101\ \Omega = 0.909\ \Omega$$

The resistor R_a is found from

$$R_a = R_{sh} - (R_b + R_c)$$

$$= 10.1\ \Omega - (0.909\ \Omega + 0.101\ \Omega) = 9.09\ \Omega \qquad (2\text{-}5)$$

Check: $R_{sh} = R_a + R_b + R_c = 9.09\ \Omega - 0.909\ \Omega + 0.101\ \Omega = 10.1\ \Omega$

2-6 D'ARSONVAL METER MOVEMENT USED IN A DC VOLTMETER

The basic d'Arsonval meter movement can be converted to a dc **voltmeter** by connecting a **multiplier** R_s in *series* with the meter movement as shown in Fig. 2-7.

The purpose of the multiplier is to extend the voltage range of the meter and to limit current through the d'Arsonval meter movement to a maximum full-scale deflection current. To find the value of the multiplier resistor, we may first determine the **sensitivity**, S, of the meter movement. The sensitivity is found by taking the reciprocal of the full-scale deflection current, written as

$$\text{Sensitivity} = \frac{1}{I_{fs}}\ (\Omega/V) \qquad (2\text{-}8)$$

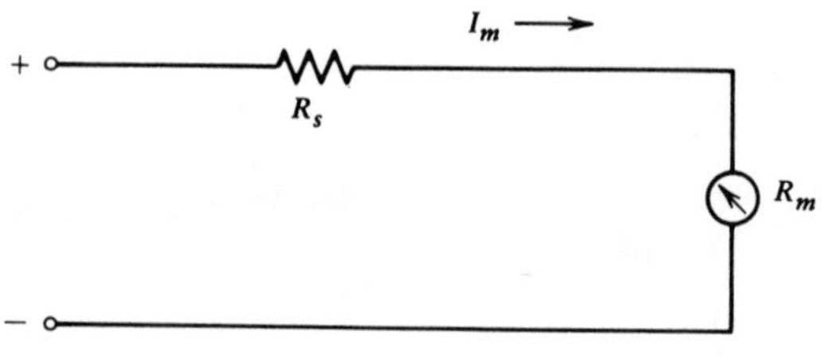

+
−
R_s
I_m
R_m

FIGURE 2-7 The d'Arsonval meter movement used in a dc voltmeter.

The units associated with sensitivity in Eq. 2-8 are ohms per volt, as may be seen from

$$\text{Sensitivity} = \frac{1}{\text{amperes}} = \frac{1}{\dfrac{\text{volt}}{\text{ohms}}} = \frac{\text{ohms}}{\text{volt}}$$

Voltage measurements are made by placing the voltmeter across the resistance of interest. This in effect places the total voltmeter resistance in parallel with the circuit resistance; therefore, it is desirable to make the voltmeter resistance much higher than the circuit resistance. Since different meter movements are used in various voltmeters and since the value of the multiplier is different for each range, total resistance is a difficult instrument rating to express. More meaningful information can be conveyed to the user via the sensitivity rating of the instrument. This rating, generally printed on the meter face, tells the resistance of the instrument for a *one*-volt range. To determine the total resistance that a voltmeter presents to a circuit, multiply the sensitivity by the range (see Example 2-5).

EXAMPLE 2-4

Calculate the sensitivity of 100-μA meter movement which is to be used as a dc voltmeter.

Solution

The sensitivity is computed as

$$S = \frac{1}{I_{fs}} = \frac{1}{100\ \mu\text{A}} = 10\ \frac{\text{k}\Omega}{\text{V}}$$

The units of sensitivity express the value of the multiplier resistance for the 1-V range. To calculate the value of the multiplier for voltage ranges greater than 1 V, simply multiply the sensitivity by the range and subtract the internal resistance of the meter movement, or

$$R_s = S \times \text{Range} - \text{Internal Resistance} \tag{2-9}$$

EXAMPLE 2-5

Calculate the value of the multiplier resistance on the 50-V range of a dc voltmeter that used a 500-μA meter movement with an internal resistance of 1 kΩ.

Solution

The sensitivity of the 500-μA meter movement in Fig. 2-8 is

$$S = \frac{1}{I_{fs}} = \frac{1}{500\ \mu\text{A}} = 2\ \frac{\text{k}\Omega}{\text{V}} \tag{2-8}$$

The value of the multiplier R_s is now calculated by multiplying the sensitivity by the range and subtracting the internal resistance of the meter movement.

$$R_s = S \times \text{Range} - R_m$$

$$= \frac{2\ \text{k}\Omega}{\text{V}} \times 50\ \text{V} - 1\ \text{k}\Omega = 100\ \text{k}\Omega - 1\ \text{k}\Omega = 99\ \text{k}\Omega \tag{2-9}$$

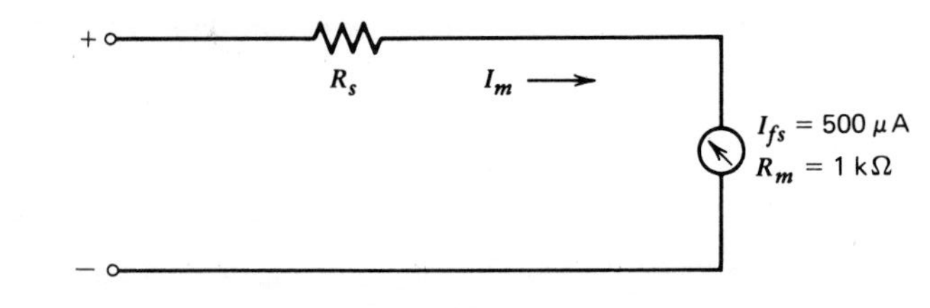

FIGURE 2-8 Basic dc voltmeter circuit.

By adding a rotary switch arrangement, we can use the same meter movement for ranges of dc voltage as shown in Example 2-6.

EXAMPLE 2-6 Calculate the value of the multiplier resistances for the multiple-range dc voltmeter circuit shown in Fig. 2-9.

Solution The sensitivity of the meter movement is computed as

$$S = \frac{1}{I_{fs}} = \frac{1}{50\ \mu\text{A}} = 20\ \frac{\text{k}\Omega}{\text{V}} \tag{2-8}$$

The value of the multiplier resistors can now be computed as follows.

(a) On the 3-V range,

$$R_{s1} = S \times \text{Range} - R_m$$
$$= \frac{20\ \text{k}\Omega}{\text{V}} \times 3\ \text{V} - 1\ \text{k}\Omega = 59\ \text{k}\Omega \tag{2-9}$$

(b) On the 10-V range,

$$R_{s2} = S \times \text{Range} - R_m$$
$$= \frac{20\ \text{k}\Omega}{\text{V}} \times 10\ \text{V} - 1\ \text{k}\Omega = 199\ \text{k}\Omega \tag{2-9}$$

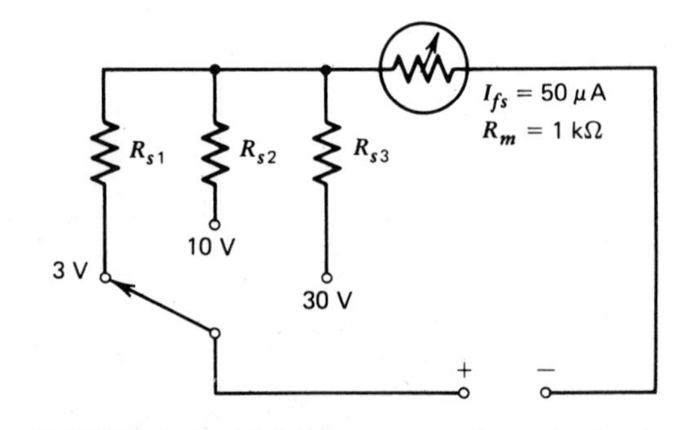

FIGURE 2-9 Multiple-range voltmeter circuit.

(c) On the 30-V range,

$$R_{s3} = S \times \text{Range} - R_m$$

$$= \frac{20\ \text{k}\Omega}{\text{V}} \times 30\ \text{V} - 1\ \text{k}\Omega = 599\ \text{k}\Omega \qquad (2\text{-}9)$$

A frequently used circuit for commercial multiple-range dc voltmeters is shown in Fig. 2-10. In this circuit the multiplier resistors are connected in series, and the range selector switches the appropriate amount of resistance into the circuit in series with the meter movement. The advantage of this circuit is that all multiplier resistors except the first (R_a) have standard resistance values and can be obtained commercially in precision tolerance.

EXAMPLE 2-7

Calculate the value of the multiplier resistors for the multiple-range dc voltmeter circuit shown in Fig. 2-10.

Solution

The sensitivity of the meter movement is computed as

$$S = \frac{1}{I_{fs}} = \frac{1}{50\ \mu\text{A}} = 20\ \text{k}\Omega/\text{V}$$

The value of the multiplier resistors can be computed as follows.

(a) On the 3-V range,

$$R_a = S \times \text{Range} - R_m = \frac{20\ \text{k}\Omega}{\text{V}} \times 3\ \text{V} - 1\ \text{k}\Omega$$

$$= 59\ \text{k}\Omega$$

(b) On the 10-V range,

$$R_b = S \times \text{Range} - (R_a + r_m) = \frac{20\ \text{k}\Omega}{\text{V}} \times 10\ \text{V} - 60\ \text{k}\Omega$$

$$= 140\ \text{k}\Omega$$

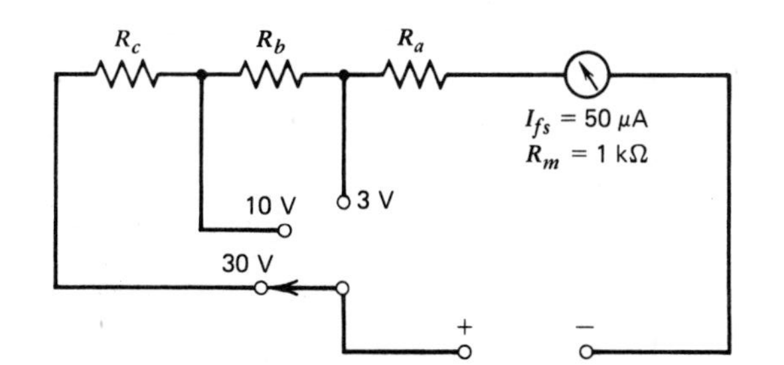

FIGURE 2-10 A commercial version of a multiple-range voltmeter.

(c) On the 30-V range,

$$R_c = S \times \text{Range} - (R_a + R_b + R_m) = \frac{20\ \text{k}\Omega}{\text{V}} \times 30\ \text{V} - 200\ \text{k}\Omega$$

$$= 400\ \text{k}\Omega$$

2-7 VOLTMETER LOADING EFFECTS

When a voltmeter is used to measure the voltage across a circuit component, the voltmeter circuit itself is in parallel with the circuit component. Since the parallel combination of two resistors is less than either resistor alone, the resistance seen by the source is less with the voltmeter connected than without. Therefore, the voltage across the component is less whenever *the voltmeter is connected*. The decrease in voltage may be negligible or it may be appreciable, depending on the *sensitivity* of the voltmeter being used. This effect is called *voltmeter loading*, and it is illustrated in the following examples. The resulting error is called a **loading error**.

EXAMPLE 2-8

Two different voltmeters are used to measure the voltage across resistor R_B in the circuit of Fig. 2-11. The meters are as follows.

$$\text{Meter } A\text{: } S = 1\ \text{k}\Omega/\text{V},\ R_m = 0.2\ \text{k}\Omega,\ \text{range} = 10\ \text{V}$$

$$\text{Meter } B\text{: } S = 20\ \text{k}\Omega/\text{V},\ R_m = 1.5\ \text{k}\Omega,\ \text{range} = 10\ \text{V}$$

Calculate

(a) Voltage across R_B without any meter connected across it.
(b) Voltage across R_B when meter A is used.
(c) Voltage across R_B when meter B is used.
(d) Error in voltmeter readings.

Solution

(a) The voltage across resistor R_B without either meter connected is found using the voltage divider equation:

$$V_{R_B} = E \frac{R_B}{R_A + R_B}$$

$$= 30\ \text{V} \times \frac{5\ \text{k}\Omega}{25\ \text{k}\Omega + 5\ \text{k}\Omega} = 5\ \text{V}$$

(b) Starting with meter A, the total resistance it presents to the circuit is

$$R_{T_A} = S \times \text{Range}$$

$$= \frac{1\ \text{k}\Omega}{\text{V}} \times 10\ \text{V} = 10\ \text{k}\Omega$$

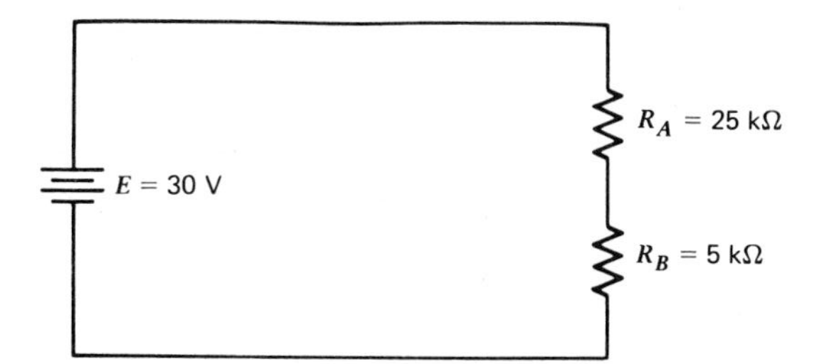

FIGURE 2-11 Circuit for Example 2-8 showing voltmeter loading.

The parallel combination of R_B and meter A is

$$R_{e1} = \frac{R_B \times R_{T_A}}{R_B + R_{T_A}}$$

$$= \frac{5\ \text{k}\Omega \times 10\ \text{k}\Omega}{5\ \text{k}\Omega + 10\ \text{k}\Omega} = 3.33\ \text{k}\Omega$$

Therefore, the voltage reading obtained with meter A, determined by the voltage divider equation, is

$$V_{R_B} = E \frac{R_{e1}}{R_{e1} + R_A}$$

$$= 30\ \text{V} \times \frac{3.33\ \text{k}\Omega}{3.33\ \text{k}\Omega + 25\ \text{k}\Omega}\ 3.53\ \text{V}$$

(c) The total resistance that meter B presents to the circuit is

$$R_{T_B} = S \times \text{Range}$$

$$= \frac{20\ \text{k}\Omega}{\text{V}} \times 10\ \text{V} = 200\ \text{k}\Omega$$

The parallel combination of R_B and meter B is

$$R_{e2} = \frac{R_B \times R_{T_B}}{R_B + R_{T_B}}$$

$$= \frac{5\ \text{k}\Omega \times 200\ \text{k}\Omega}{5\ \text{k}\Omega + 200\ \text{k}\Omega} = 4.88\ \text{k}\Omega$$

Therefore, the voltage reading obtained with meter B, determined by use of the voltage divider equation, is

$$V_{R_B} = E \frac{R_{e2}}{R_{e2} + R_A}$$

$$= 30\text{ V} \times \frac{4.88\text{ k}\Omega}{4.88\text{ k}\Omega + 25\text{ k}\Omega} = 4.9\text{ V}$$

(d)

$$\text{Voltmeter } A \text{ error} = \frac{5\text{ V} - 3.53\text{ V}}{5\text{ V}} \times 100\% = 29.4\%$$

$$\text{Voltmeter } B \text{ error} = \frac{5\text{ V} - 4.9\text{ V}}{5\text{ V}} \times 100\% = 2\%$$

Note in Example 2-8 that, although the reading obtained with meter B is much closer to the correct value, the voltmeter still introduced a 2% error due to loading of the circuit by the voltmeter. It should be apparent that, in electronic circuits in which high values of resistance are generally used, commercial volt–ohm–milliammeters (VOM) still introduce some circuit loading. Such instruments generally have a sensitivity of at least 20 kΩ/V. Instruments with a lower sensitivity rating generally prove unsatisfactory for most electronics work.

When a VOM is used to make voltage measurements, circuit loading due to the voltmeter is also minimized by using the highest range possible, as shown in Example 2-9.

EXAMPLE 2-9

Find the voltage reading and the percentage of error of each reading obtained with a voltmeter on

(a) Its 3-V range.

(b) Its 10-V range.

(c) Its 30-V range.

The instrument has a 20-kΩ/V sensitivity and is connected across R_B in Fig. 2-12.

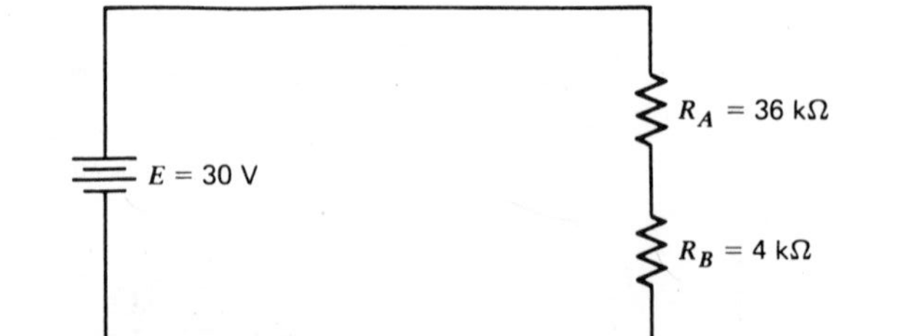

FIGURE 2-12 Circuit for Example 2-9 showing the effects of voltmeter loading on different ranges.

Solution

The voltage drop across R_B without the voltmeter connected is computed as

$$V_{R_B} = E\frac{R_B}{R_A + R_B}$$

$$= 30\ \text{V} \times \frac{4\ \text{k}\Omega}{36\ \text{k}\Omega + 4\ \text{k}\Omega} = 3\ \text{V}$$

(a) On the 3-V range,

$$R_T = S \times \text{Range}$$

$$= \frac{20\ \text{k}\Omega}{\text{V}} \times 3\ \text{V} = 60\ \text{k}\Omega$$

$$R_{\text{eq1}} = \frac{R_T R_B}{R_T + R_B} = \frac{60\ \text{k}\Omega \times 4\ \text{k}\Omega}{60\ \text{k}\Omega + 4\ \text{k}\Omega} = 3.75\ \text{k}\Omega$$

The voltmeter reading is

$$V_{R_B} = E\frac{R_{\text{eq1}}}{R_{\text{eq1}} + R_A}$$

$$= 30\ \text{V} \times \frac{3.75\ \text{k}\Omega}{3.75\ \text{k}\Omega + 36\ \text{k}\Omega} = 2.8\ \text{V}$$

The percentage of error on the 3-V range is

$$\text{Percent error} = \frac{3\ \text{V} - 2.8\ \text{V}}{3\ \text{V}} \times 100\% = 6.66\%$$

(b) On the 10-V range,

$$R_T = \frac{20\ \text{k}\Omega}{\text{V}} \times 10\ \text{V} = 200\ \text{k}\Omega$$

$$R_{\text{eq2}} = \frac{R_T R_B}{R_T + R_B} = \frac{200\ \text{k}\Omega \times 4\ \text{k}\Omega}{200\ \text{k}\Omega + 4\ \text{k}\Omega} = 3.92\ \text{k}\Omega$$

The voltmeter reading is

$$V_{R_B} = E\frac{R_{\text{eq2}}}{R_{\text{eq2}} + R_A}$$

$$= 30\ \text{V} \times \frac{3.92\ \text{k}\Omega}{3.92\ \text{k}\Omega + 36\ \text{k}\Omega} = 2.95\ \text{V}$$

The percentage of error on the 10-V range is

$$\text{Percent error} = \frac{3\text{ V} - 2.95\text{ V}}{3\text{ V}} \times 100\% = 1.66\%$$

(c) On the 30-V range,

$$R_T = \frac{20\text{ k}\Omega}{\text{V}} \times 30\text{ V} = 600\text{ k}\Omega$$

$$R_{eq3} = \frac{R_T R_B}{R_T + R_B} = \frac{600\text{ k}\Omega \times 4\text{ k}\Omega}{600\text{ k}\Omega + 4\text{ k}\Omega} = 3.97\text{ k}\Omega$$

The voltmeter reading is

$$V_{R_B} = E\frac{R_{eq3}}{R_{eq3} + R_A}$$

$$= 30\text{ V} \times \frac{3.97\text{ k}\Omega}{3.97\text{ k}\Omega + 36\text{ k}\Omega} = 2.98\text{ V}$$

The percentage of error on the 30-V range is

$$\text{Percent error} = \frac{3\text{ V} - 2.98\text{ V}}{3\text{ V}} \times 100\% = 0.66\%$$

We have learned the following from Example 2-9. The 30-V range introduces the least error because of loading. However, the voltage being measured causes only a 10% full-scale deflection, whereas on the 10-V range the applied voltage causes approximately one-third full-scale deflection with less than 2% error. The reading obtained on the 10-V range would be acceptable and less subject to gross error (Section 1-6). The percentage of error on the 10-V range is less than the average percentage of error for a mass-produced d'Arsonval meter movement.

We can experimentally determine whether the voltmeter is introducing appreciable error by changing to a higher range. If the voltmeter reading does *not* change, the meter is not loading the circuit appreciably. If loading is observed, select the range with the greatest deflection and yielding the most precise measurement (Section 1-6).

2-8 AMMETER INSERTION EFFECTS

A frequently overlooked source of error in measurements is the error caused by inserting an ammeter in a circuit to obtain a current reading. All ammeters contain some internal resistance, which may range from a low value for current meters capable of measuring in the ampere range to an appreciable value of 1 k**Ω** or greater for microammeters. Inserting an ammeter in a circuit always increases the resistance of the circuit and, therefore, always reduces the current in the circuit. The error caused by the meter depends on the relationship between the value of resistance in the original circuit and the value of resistance in the ammeter.

Consider the series circuit shown in Fig. 2-13 in which there is current through resistor R_1. The expected current, I_e, is the current *without* the ammeter in the circuit. Now, suppose we connect an ammeter in the circuit to measure the current as shown in Fig. 2-14. The amplitude of the current has now been reduced to I_m, as a result of the added meter resistance, R_m.

If we wish to obtain a relationship between I_e and I_m we can do so by using Thévenin's theorem. The circuit in Fig. 2-14 is in the form of a Thévenin equivalent circuit with a single-voltage source in series with a single resistor. With the output terminals X and Y shorted, the expected current flow is

$$I_e = \frac{E}{R_1} \qquad (2\text{-}10)$$

Placing the meter in series with R_1 causes the current to be reduced to a value equal to

$$I_m = \frac{E}{R_1 + R_m} \qquad (2\text{-}11)$$

Dividing Eq. 2-11 by Eq. 2-10 yields the following expression

$$\frac{I_m}{I_e} = \frac{R_1}{R_1 + R_m} \qquad (2\text{-}12)$$

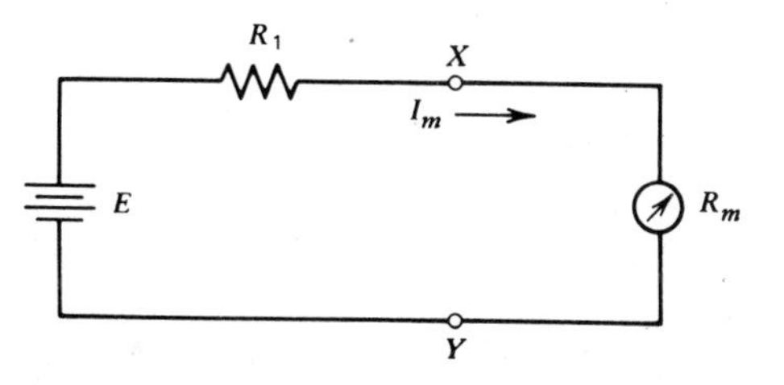

FIGURE 2-13 Expected current value in a series circuit.

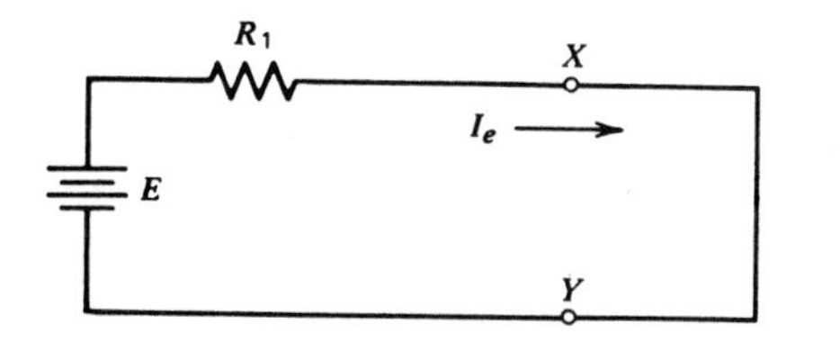

FIGURE 2-14 Series circuit with ammeter.

Equation 2-12 is quite useful in that it allows us to determine the error introduced into a circuit caused by ammeter insertion if we know the value of Thévenin's equivalent resistance and the resistance of the ammeter.

EXAMPLE 2-10

A current meter that has an internal resistance of 78 Ω is used to measure the current through resistor R_c in Fig. 2-15. Determine the percentage of error of the reading due to ammeter insertion.

Solution

The current meter will be connected into the circuit between points X and Y in the schematic in Fig. 2-16. When we look back into the circuit from terminals X and Y, we can express Thévenin's equivalent resistance as

$$R_{Th} = R_c + \frac{R_a R_b}{R_a + R_b}$$

$$= 1\ \text{k}\Omega + 0.5\ \text{k}\Omega = 1.5\ \text{k}\Omega$$

Therefore, the ratio of meter current to expected current is

$$\frac{I_m}{I_e} = \frac{R_{Th}}{R_{Th} + r_m} = \frac{1.5\ \text{k}\Omega}{1.5\ \text{k}\Omega + 78\ \Omega} = 0.95 \tag{2.12}$$

Solving for I_m yields

$$I_m = 0.95 I_e$$

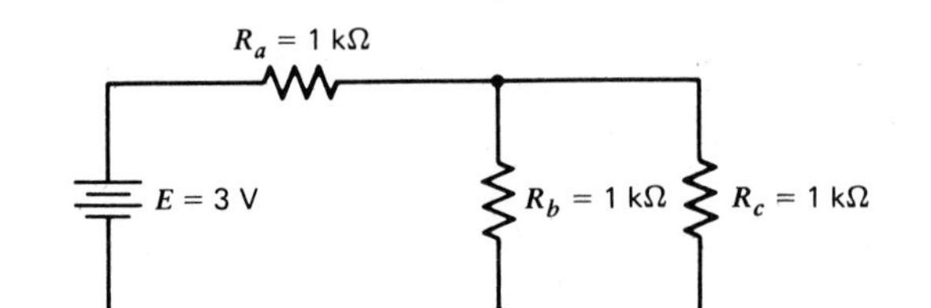

FIGURE 2-15 Series–parallel circuit for Example 2-10.

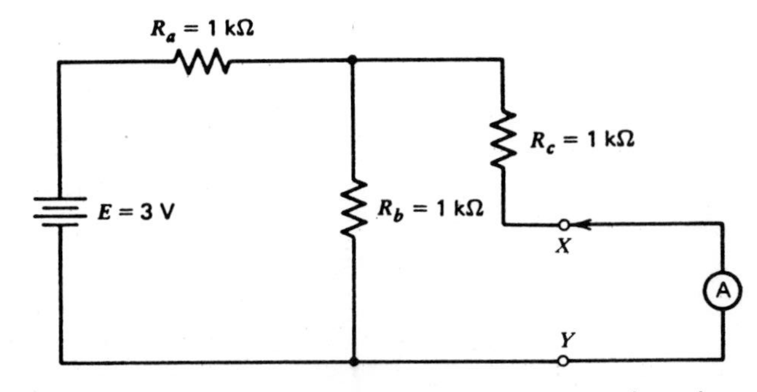

FIGURE 2-16 Circuit to demonstrate ammeter insertion.

The current through the meter is 95% of the expected current; therefore, the current meter has caused a 5% error as a result of its insertion. We can write an expression for the percentage of error attributable to ammeter insertion as

$$\text{Insertion error} = \left(1 - \frac{I_m}{I_e}\right) \times 100\% = 5.0\%$$

2-9 THE OHMMETER

The basic d'Arsonval meter movement may also be used in conjunction with a battery and a resistor to construct a simple **ohmmeter** circuit such as that shown in Fig. 2-17. If points X and Y are connected, we have a simple series circuit with current through the meter movement caused by the voltage source, E. The amplitude of the current is limited by the resistors R_z and R_m. Notice in Fig. 2-17 that resistor R_z consists of a fixed portion and a variable portion. The reason for this will be discussed toward the end of this section. Connecting points X and Y is equivalent to shorting the test probes together on an ohmmeter to "zero" the instrument before using it. This is normal operating procedure with an ohmmeter. After points X and Y are connected, the variable part of resistor R_z is adjusted to obtain exactly full-scale deflection on the meter movement.

The amplitude of the current through the meter movement can be determined by applying Ohm's law as

$$I_{fs} = \frac{E}{R_z + R_m} \tag{2-13}$$

To determine the value of the unknown resistor we connect the unknown, R_x, between points X and Y in Fig. 2-17, as shown in Fig. 2-18. The circuit current is now expressed as

$$I = \frac{E}{R_z + R_m + R_x}$$

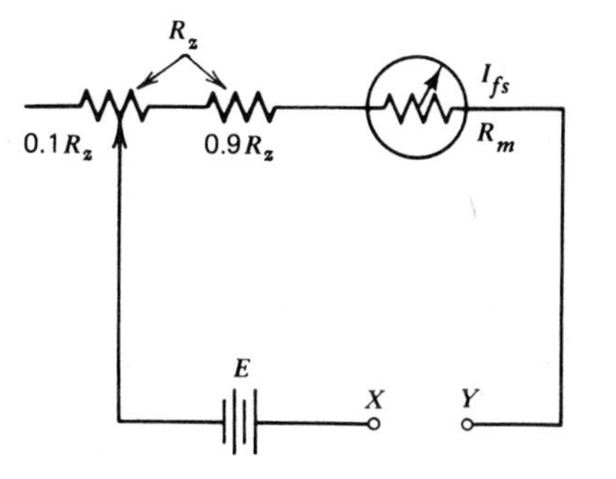

FIGURE 2-17 Basic ohmmeter circuit.

The current I is less than the full-scale current, I_{fs}, because of the additional resistance, R_x. The ratio of the current I to the full-scale deflection current I_{fs} is equal to the ratio of the circuit resistances and may be expressed as

$$\frac{I}{I_{fs}}=\frac{E/(R_z+R_m+R_x)}{E/(R_z+R_m)}=\frac{R_z+R_m}{R_z+R_m+R_x}$$

If we let P represent the ratio of the current I to the full-scale deflection current I_{fs}, we can say that

$$P=\frac{I}{I_{fs}}=\frac{R_z+R_m}{R_z+R_m+R_x} \qquad (2\text{-}14)$$

Equation 2-14 is very useful when marking off the scale on the meter face of the ohmmeter to indicate the value of a resistor being measured.

The following example illustrates the use of Eq. 2-14.

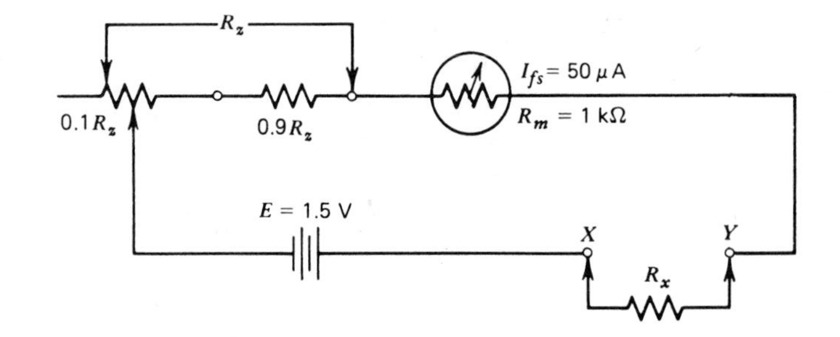

FIGURE 2-18 **Basic ohmmeter circuit with unknown resistor R_x connected between probes.**

EXAMPLE 2-11

A 1-mA full-scale deflection current meter movement is to be used in an ohmmeter circuit. The meter movement has an internal resistance, R_m, of 100 Ω, and a 3-V battery will be used in the circuit. Mark off the meter face for reading resistance.

Solution

The value of R_z, which will limit current to full-scale deflection current, is computed as

$$R_z=\frac{E}{I_{fs}}-R_m$$

$$R_z=\frac{3\text{ V}}{1\text{ mA}}-100\ \Omega=2.9\text{ k}\Omega$$

The value of R_x with 20% full-scale deflection is

$$R_x=\frac{R_z+R_m}{P}-(R_z+R_m)$$

$$=\frac{2.9\text{ k}\Omega+0.1\text{ k}\Omega}{0.2}-(2.9\text{ k}\Omega+0.1\text{ k}\Omega)$$

$$= \frac{3\ \text{k}\Omega}{0.2} - 3\ \text{k}\Omega = 12\ \text{k}\Omega$$

The value of R_x with 40% full-scale deflection is

$$R_x = \frac{R_z + R_m}{P} - (R_z + R_m)$$

$$= \frac{3\ \text{k}\Omega}{0.4} - 3\ \text{k}\Omega = 4.5\ \text{k}\Omega$$

The value of R_x with 50% full-scale deflection is

$$R_x = \frac{R_z + R_m}{P} - (R_z + R_m)$$

$$= \frac{3\ \text{k}\Omega}{0.5} - 3\ \text{k}\Omega = 3\ \text{k}\Omega$$

The value of R_x with 75% full-scale deflection is

$$R_x = \frac{R_z + R_m}{P} - (R_z + R_m)$$

$$= \frac{3\ \text{k}\Omega}{0.75} - 3\ \text{k}\Omega = 1\ \text{k}\Omega$$

The data are tabulated in Table 2-1. Using the data from Table 2-1, we can draw the ohmmeter scale shown in Fig. 2-19.

Two interesting and important facts may be seen from observing the ohmmeter scale in Fig. 2-19 and the data in Table 2-1. First, the ohmmeter scale is very nonlinear. This is due to the high internal resistance of an

TABLE 2-1
Scale of Ohmmeter in Example 2-11

P (%)	R_x (kΩ)	$R_z + R_m$ (kΩ)
20	12	3
40	4.5	3
50	3	3
75	1	3
100	0	3

ohmmeter. Second, at half-scale deflection, the value of R_x is equal to the value of the internal resistance of an ohmmeter. A variable resistor may be connected to the ohmmeter probes and then set to the value required for half-scale deflection of the pointer. The variable resistor may then be removed and measured. Its value should equal the internal resistance of the ohmmeter.

EXAMPLE 2-12

An ohmmeter uses a 1.5-V battery and a basic 50-μA movement. Calculate

(a) The value of R_z required.

(b) The value of R_x that would cause half-scale deflection in the circuit in Fig. 2-24.

Solution

(a) The proper value of R_z is computed as

$$R_z = \frac{E}{I_{fs}} - R_m$$

$$= \frac{1.5 \text{ V}}{50\ \mu\text{A}} - 1 \text{ k}\Omega = 29 \text{ k}\Omega$$

(b) With midscale deflection, R_x is equal to the internal resistance of the ohmmeter; therefore

$$R_x = R_z + R_m = 30 \text{ k}\Omega$$

In Example 2-11 the value of resistance that corresponded to 20% full-scale deflection was computed to be 12 kΩ. Suppose we connected a 97-kΩ resistor to our ohmmeter circuit. The resulting current flow would be

$$I = \frac{E}{R_x + R_z + R^m} = \frac{3 \text{ V}}{100 \text{ k}\Omega} = 30\ \mu\text{A}$$

This amount of current would cause the pointer to deflect 3% of full scale. It should be apparent that larger-value resistors would permit less deflection. Therefore, we can conclude that a 100-kΩ resistor would be about the maximum value of resistance that could be measured with any degree of accuracy with the particular ohmmeter circuit of Fig. 2-19.

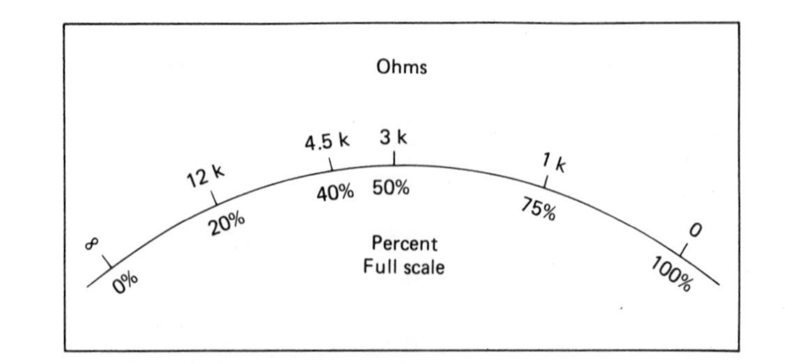

FIGURE 2-19 Ohmmeter scale showing nonlinear characteristics.

The variable portion of resistor R_z is frequently viewed as a means of compensating for battery aging. However, you should be aware of the consequences of this action. Consider the following calculations.

EXAMPLE 2-13

An ohmmeter is designed around a 1-mA meter movement and a 1.5-V cell. If the cell voltage decays to 1.3 V because of aging, calculate the resulting error at midrange on the ohmmeter scale.

Solution

The internal resistance of the ohmmeter is

$$R_{in} = \frac{E}{I} = \frac{1.5\ \text{V}}{1\ \text{mA}} = 1.5\ \text{k}\Omega$$

Therefore, the ohmmeter scale should be marked as 1.5 **kΩ** at midrange. An external resistance of 1.5 **kΩ** would cause the pointer to deflect to midscale. When the cell voltage decays to 1.3 V and the ohmmeter is adjusted for full-scale deflection by reducing R_z, the total internal resistance of the ohmmeter is now

$$R_{in} = \frac{E}{I} = \frac{1.3\ \text{V}}{1\ \text{mA}} = 1.3\ \text{k}\Omega$$

If a 1.3-**kΩ** resistor is now measured with the ohmmeter, we will expect less than midscale deflection. However, the pointer will deflect to midscale, which is marked as 1.5 **kΩ**. The aging of the cell has caused an incorrect reading. The percentage of error associated with the reading is

$$\text{Percent error} = \frac{1.5\ \text{k}\Omega - 1.3\ \text{k}\Omega}{1.5\ \text{k}\Omega} \times 100\% = 13.3\%$$

To prevent the ohmmeter from being zeroed if the battery has aged considerably, the variable portion of R_z is usually limited to a maximum of 10% of the total value of R_z (see Figs. 2-17 and 2-20).

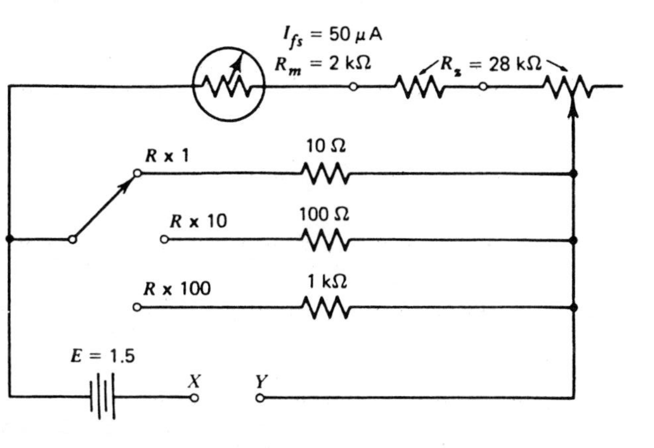

FIGURE 2-20 Multiple-range ohmmeter.

2-10 MULTIPLE-RANGE OHMMETERS

The ohmmeter circuit discussed in the previous section is not capable of measuring resistance over a wide range of values. Therefore, we need to extend our discussion of ohmmeters to include *multiple*-range ohmmeters.

One way to build a multiple-range ohmmeter is shown in Fig. 2-20. This instrument makes use of a basic 50-μA meter movement with an internal resistance of 2 k**Ω**. An additional resistance of 28 k**Ω** is provided by R_z, which includes a fixed resistance and the zeroing potentiometer. R_z is necessary to limit current through the meter movement to 50 μA when the test probes (not shown) connected to X and Y are shorted together. As may be seen, when the instrument is on the $R \times 1$ range, a 10-**Ω** resistor is in parallel with the meter movement. Therefore, the internal resistance of the ohmmeter on the $R \times 1$ range is 10 **Ω** in parallel with 30 k**Ω**, which is approximately 10 **Ω**. This means the pointer will deflect to midscale when a 10-**Ω** resistor is connected across X and Y.

When the instrument is set to the $R \times 10$ range, the total resistance of the ohmmeter is 100 **Ω** in parallel with 30 k**Ω**, which is now approximately 100 **Ω**. Therefore, the pointer deflects to midscale when a 100-**Ω** resistor is connected between the test probes. Midscale is marked as 10 **Ω**. Therefore, the value of the resistor is determined by multiplying the reading by the range multiplier of 10 producing a midscale value of 100 **Ω** ($R \times 10$).

When our ohmmeter is set on the $R \times 100$ range, the total resistance of the instrument is 1 k**Ω** in parallel with 30 k**Ω**, which is still approximately 1 k**Ω**. Therefore, the pointer deflects to midscale when we connect the test probes across a 1-k**Ω** resistor. This provides us a value for the midscale reading of 10 multiplied by 100, or 1 k**Ω** for our resistor.

EXAMPLE 2-14

(a) In Fig. 2-21 determine the current through the meter, I_m, when a 20-**Ω** resistor between terminals X and Y is measured on the $R \times 1$ range.
(b) Show that this same current flows through the meter movement when a 200-**Ω** resistor is measured on the $R \times 10$ range.
(c) Show that the same current flows when a 2 k**Ω** resistor is measured on the $R \times 100$ range.

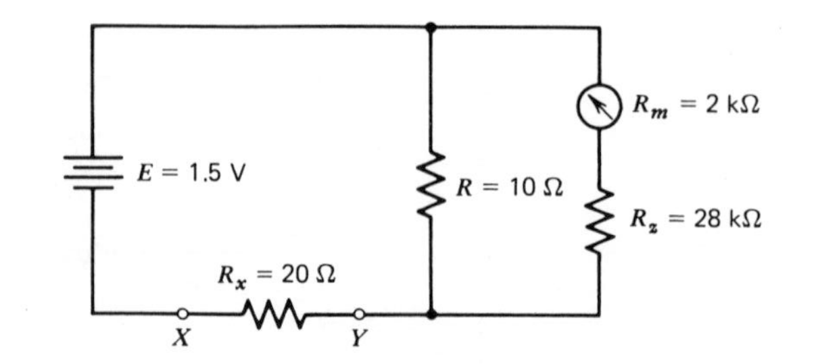

FIGURE 2-21 **Circuit for Example 2-14 with ohmmeter on $R \times 1$ range.**

Solution

(a) When the ohmmeter is set on the $R \times 1$ range, the circuit is as shown in Fig. 2-21. The voltage across the potential combination of resistance is computed as

$$V = 1.5\ \text{V} \times \frac{10\ \Omega}{10\ \Omega + 20\ \Omega} = 0.5\ \text{V}$$

The current through the meter movement is computed as

$$I_m = \frac{0.5\ \text{V}}{30\ \text{k}\Omega} = 16.6\ \mu\text{A}$$

(b) When the ohmmeter is set on the $R \times 10$ range, the circuit is as shown in Fig. 2-22. The voltage across the parallel combination of resistance is computed as

$$V = 1.5\ \text{V} \times \frac{100\ \Omega}{100\ \Omega + 200\ \Omega} = 0.5\ \text{V}$$

The current through the meter movement is computed as

$$I_m = \frac{0.5\ \text{V}}{30\ \text{k}\Omega} = 16.6\ \mu\text{A}$$

(c) When the ohmmeter is set on the $R \times 100$ range, the circuit is as shown in Fig. 2-23. The voltage across the parallel combination is computed as

$$V = 1.5 \times \frac{1\ \text{k}\Omega}{1\ \text{k}\Omega + 2\ \text{k}\Omega} = 0.5\ \text{V}$$

The current through the meter movement is computed as

$$I_m = \frac{0.5\ \text{V}}{30\ \text{k}\Omega} = 16.6\ \mu\text{A}$$

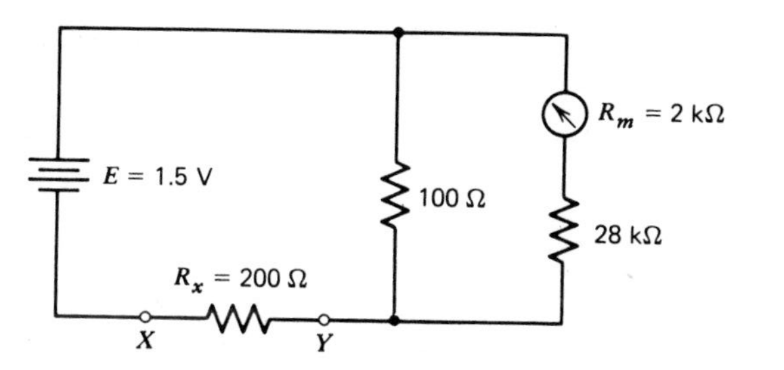

FIGURE 2-22 Ohmmeter on $R \times 10$ range.

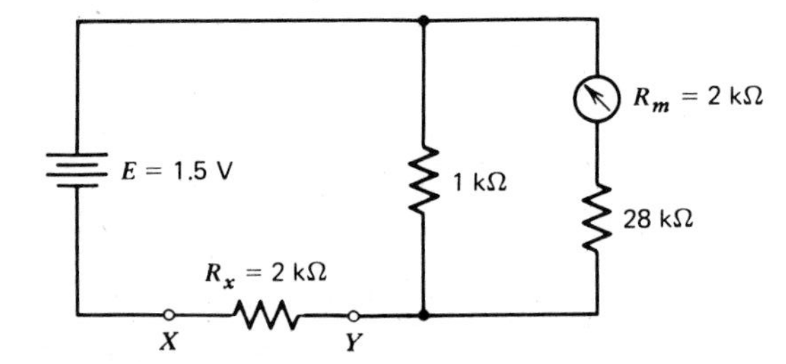

FIGURE 2-23 Ohmmeter on $R \times 100$ range.

As can be seen, the current through the meter movement is 16.6 μA in each situation in Example 2-14. This means the meter face is marked as 20 **Ω** at 33.2% of full-scale deflection.

When the ohmmeter is on the $R \times 1$ range, a reading of 20 **Ω** times the multiplier of 1 means the unknown resistor has a value of 20 **Ω**. When the ohmmeter is on the $R \times 10$ range, a reading of 20 **Ω** times the multiplier of 10 means the unknown resistor has a value of 200 **Ω**. Similarly, when the ohmmeter is on the $R \times 100$ range, a reading of 20 **Ω** times the multiplier of 100 means the unknown resistor has a value of 2 k**Ω**. The important thing to note is that a multiple-range ohmmeter may have a *single* scale for *all* ranges.

2-11 THE MULTIMETER

Thus far in Chapter 2 we have discussed the ammeter, the voltmeter, and the ohmmeter. All these instruments have one thing in common. Each uses the *same* basic current-sensitive d'Arsonval meter movement. Therefore, it might seem reasonable, given a proper switching arrangement, to combine the three circuits in a *single instrument*. The **multimeter** or volt–ohm–milliammeter (VOM) is such an instrument. It is a general-purpose test instrument that has the necessary circuitry to measure ac or dc voltage, direct current, or resistance.

A typical commercial VOM of laboratory quality is normally designed around a basic 50-μA meter movement. The Simpson, Model 260 (shown in Fig. 2-24), is a typical general-purpose VOM. The instrument uses a 50-μA meter movement and therefore has a sensitivity of 20 k**Ω**/V on the dc voltage ranges. It is capable of a wide range of measurements, as shown in Table 2-2,

TABLE 2-2
Simpson 260 Measurement Ranges

Direct current	0–50 μA, 0–1/10/100/500 mA, 0–10 A
Dc volts	0–250 mV, 0–2.6/10/50/250/1000/5000 V
Ac volts	0–2.5/100/50/250/1000/5000 V
Ohms	$R \times 1$, $R \times 100$, $R \times 10{,}000$

FIGURE 2-24 Typical laboratory-quality VOM (Simpson Model 260). (Courtesy Simpson Electric Company.)

which lists 22 ranges for measuring voltage, current, and resistance as well as additional ranges for measuring audio-frequency output voltage and sound level.

2-12 CALIBRATION OF DC INSTRUMENTS

Although the actual techniques for calibrating instruments using the d'Arsonval meter movement are covered in subsequent chapters, we introduce the topic here since we have just discussed dc instruments.

Calibration means to compare a given instrument against a *standard* instrument to determine its accuracy. A dc voltmeter may be calibrated by comparing it with one of the standards discussed in Section 1-5 or with a potentiometer as described in Section 4.3. The circuit shown in Fig. 2-25 may be used to calibrate a dc voltmeter, the test voltmeter reading, V, is compared to the voltage reading obtained with the standard instrument, M.

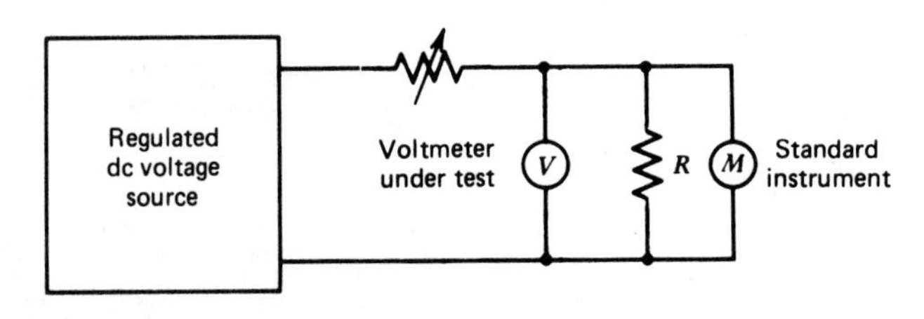

FIGURE 2-25 Calibration circuit for a dc voltmeter.

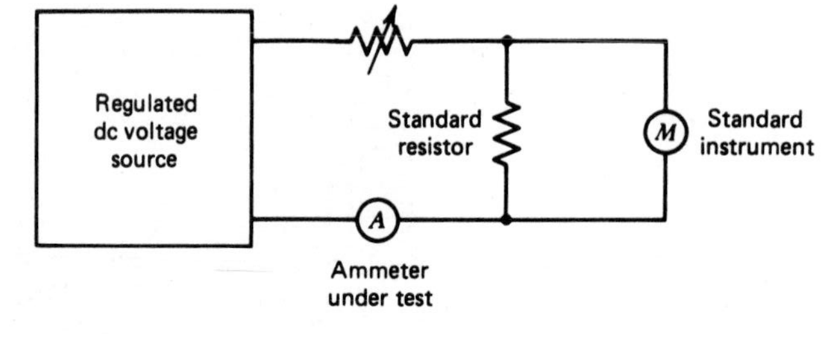

FIGURE 2-26 Calibration circuit for a dc ammeter.

A dc ammeter is usually calibrated by using a standard resistor R_s and either a standard voltmeter or a potentiometer M. The circuit shown in Fig. 2-26 may be used to calibrate an ammeter. The test ammeter reading, A, is compared to the calculated Ohm's law current from the voltage reading obtained across the known standard resistor using the standard voltmeter M.

The ohmmeter circuit designed around the d'Arsonval meter movement is usually considered to be an instrument of moderate accuracy. The accuracy of the instrument may be checked by measuring different values of standard resistance and noting the reading obtained. However, when precise resistance measurements are required, a comparison-type resistance measurement using a bridge is preferable (see Chapter 5 on bridges).

2-13 APPLICATIONS

The most fundamental applications of the instruments designed around the d'Arsonval meter movement are implied by the names of the instruments—voltmeter, ammeter, and ohmmeter. The purpose of this section is to point out some applications that may not be quite as obvious. These will show you the versatility of the instruments and help you to adapt them to your own needs.

2-13.1 Electrolytic Capacitor Leakage Tests

A current meter may be used to measure the leakage current of electrolytic capacitors. The leakage current depends on the voltage rating of the capacitor and its capacitance value. The test voltage applied to the capacitor should be near the dc-rated value for the capacitor. After the capacitor charges to the

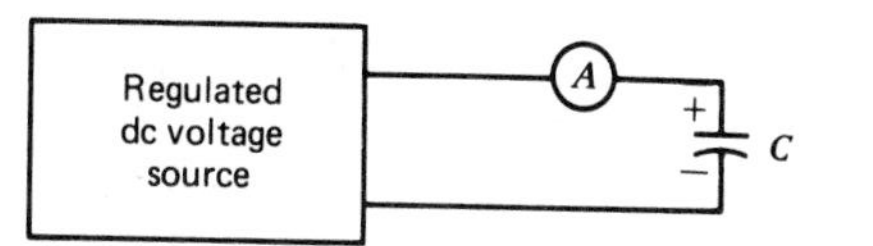

FIGURE 2-27 **Circuit for determining leakage current for an electrolytic capacitor.**

supply voltage, ideally the flow of current should stop; however, because of capacitor leakage, a small current continues to exist.

Because of the design of electrolytic capacitors, they tend to have a relatively high leakage current. As a rule of thumb, the *acceptable* leakage current for electrolytic capacitors when tested as in Fig. 2-27 is

1. Capactiors rated at 300 V or higher—0.5 mA.
2. Capacitors rated at 100 to 300 V—0.2 mA.
3. Capacitors rated at less than 100 V—0.1 mA.

2-13.2 Nonelectrolytic Capacitor Leakage Tests

A voltmeter may be used to check for leakage current across the plates of nonelectrolytic capacitors (paper, molded composition, mica, etc.). The leakage of a capacitor may be expressed in terms of its equivalent resistance. If we apply a dc voltage across a series circuit consisting of a capacitor suspected of being leaky, and a dc voltmeter, as shown in Fig. 2-28, the applied voltage will be divided across this voltage divider network according to the ratio of the resistance (after charging) in series with the input resistance of the voltage. Therefore, all the applied voltage will appear across the capacitor. If the capacitor is leaky, a voltage reading will be obtained on the voltmeter because of the flow of current. The equivalent resistance that the capacitor represents can be computed from

$$R = R_{in} \frac{E - V}{V} \ (\Omega) \tag{2-15}$$

where

R = capacitor's equivalent resistance, Ω
R_{in} = input resistance of the voltmeter, Ω
E = applied dc voltage, volts
V = voltmeter reading, volts

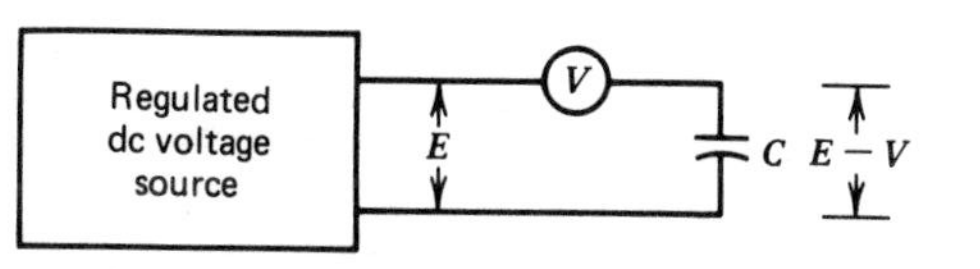

FIGURE 2-28 **Circuit for determining leakage current of a nonelectrolytic capacitor.**

The equivalent resistance R of a nonelectrolytic capacitor should be on the order of 100 MΩ or higher. Therefore, if an equivalent resistance value of less than, say 80 MΩ, is obtained, the capacitor should be suspect.

2-13.3 Using the Ohmmeter for Continuity Checks

An important application of the ohmmeter is to check *continuity* on such components as lamps or fuses when troubleshooting. An open filament on a lamp or a burned-out fuse, a switch contact, or a coil may appear acceptable upon visual inspections but may actually be faulty. A continuity check with an ohmmeter would indicate whether an "open" exists.

Continuity checks can also be made on test leads, oscilloscope probes, coaxial cables, multiconductor cables, ac cords, and many other devices. An ohmmeter check for continuity is made by setting the resistance switch to a suitable scale, and placing the test probes at two points between which continuity is being checked, as shown in Fig. 2-29. A full-scale reading on the ohmmeter indicates continuity.

2-13.4 Using the Ohmmeter to Check Semiconductor Diodes

The ohmmeter is frequently used to make quick checks on semiconductor diodes. The question of which (unmarked) terminal of a semiconductor diode is the anode and which is the cathode frequently comes up. The ohmmeter can be used to answer this question very easily. If the positive lead of the ohmmeter is connected to the anode (p material) of the diode and the common terminal of the ohmmeter is connected to the cathode (n material), the ohmmeter should indicate a relatively *low* value of *forward* (bias) resistance. If the ohmmeter leads are reversed, the ohmmeter should indicate a high value of *reverse* (bias) resistance. These measurements are shown in Fig. 2-30.

This test also distinguishes between a good and a defective diode. A good diode will have a high ratio of reverse to forward resistance, a defective diode a low ratio.

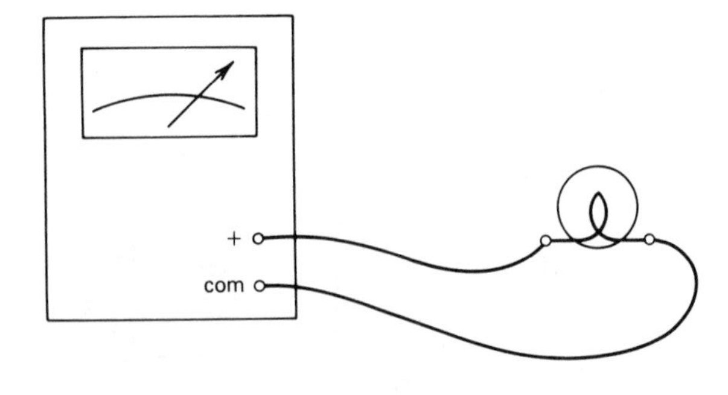

FIGURE 2-29 Continuity check on a lamp filament.

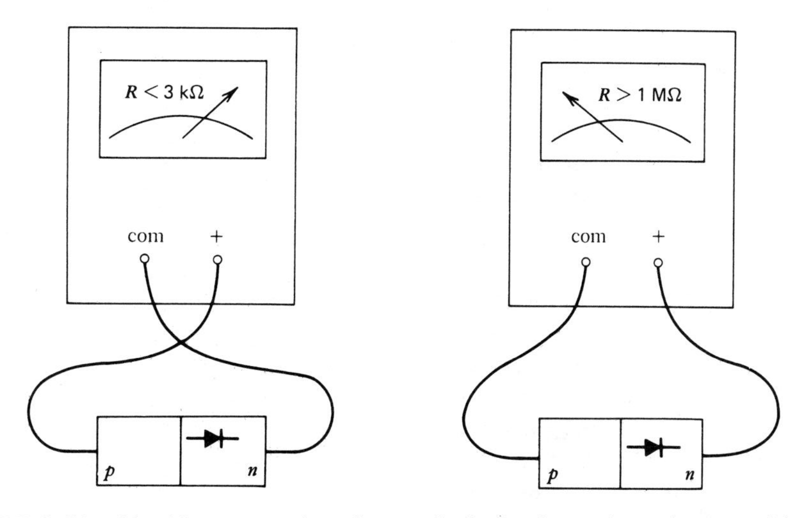

FIGURE 2-30 Checking a semiconductor diode for forward- and reverse-biased conditions.

2-14 SUMMARY

The basic d'Arsonval *meter movement* is a *current*-sensitive device capable of directly measuring only very small currents. Its usefulness as a measuring device is greatly increased with the proper external circuitry. Large currents can be measured by adding *shunts*. Voltages can be measured by adding *multipliers*. Resistance can be measured by adding a *battery* and a *resistance* network.

All ammeters and voltmeters introduce some error into any circuit under test because the meter *loads* the circuit—this is a common instrumentation problem. The effects of voltmeter loading may be reduced by using a voltmeter with a sensitivity rating of 20 kΩ/V or greater. Most laboratory-quality, commercial multimeters have sensitivity ratings of this value or higher.

2-15 GLOSSARY

Ammeter: An instrument for measuring current, using a basic movement and shunts.

Ayrton shunt: A shunt arrangement, also called a universal shunt, that prevents a meter movement from being used with no shunt.

d'Arsonval meter movement: A device consisting of a permanent horseshoe magnet and a movable electromagnetic coil that rotates about a magnetic core. A pointer is attached to the movable coil.

Internal resistance: The resistance within the meter movement caused mostly by the resistance of the fine wire used to wind the electromagnetic moving coil.

Loading error: The error (or disturbance to original conditions) caused by placing an ammeter in a circuit to obtain a measurement.

Multimeter: An instrument containing the proper circuitry in a single enclosure to obtain voltage, current, and resistance measurements. Usually designed around the d'Arsonval meter movement.

Multiplier: A resistor placed in series with a basic meter movement to extend the voltage range of a basic meter movement.

Ohmmeter: An instrument, designed around a basic meter movement, that is capable of measuring resistance.

Sensitivity: The reciprocal of the full-scale deflection current expressed in units of ohms per volt. A measure of the instrument indication (deflection) to a change in current.

Shunt: A resistor placed in parallel (shunt) with a basic meter movement to extend the current range of the basic meter movement.

Voltmeter: An instrument, using a basic meter movement and multipliers, that is capable of measuring voltage.

2-16 REVIEW QUESTIONS

The following review questions relate to the material in the chapter. Readers should answer these questions after study of the chapter to determine their comprehension of the material.

1. List two types of suspension system used with the d'Arsonval meter movement.
2. Describe briefly the principle of operation of d'Arsonval meter movements.
3. How can the basic d'Arsonval meter movement be used to measure high-amplitude currents?
4. How can the d'Arsonval meter movement be used to measure voltages?
5. How can the d'Arsonval meter movement be used to measure resistance?
6. What is meant by sensitivity? What are its units of measurement?
7. What effect, if any, does connecting a voltmeter across a resistor in a circuit have on the current through the resistor?
8. What effect, if any, does connecting an ammeter in series with a resistor in a circuit have on the current through the resistor?
9. What is the purpose of the zeroing resistor in an ohmmeter and does it always accomplish its intended purpose?

10. What is the significance of midscale deflection on any ohmmeter range?

11. How can a person check whether a voltmeter is introducing error through loading?

2-17 PROBLEMS

2-1 Calculate the voltage drop developed across a d'Arsonval meter movement having an internal resistance of 850 Ω and a full-scale deflection current of 100 μA.

2-2 Find the resistance of a multiplier required to convert a 200-μA meter movement into a 0- to 150-V dc voltmeter, if $R_m = 1$ kΩ.

2-3 Calculate the half-scale current of a meter movement that has a sensitivity of 20 kΩ/V.

2-4 Find the value of shunt resistance required to convert a 1-mA meter movement with an internal resistance of 105 Ω into a 0- to 150-mA meter current.

2-5 Which meter has a greater sensitivity: meter A having a range of 0 to 10 V and a multiplier resistor of 18 kΩ, or meter B with a range of 0 to 300 V and a 298-kΩ multiplier resistor? Both meter movements have an internal resistance of 2 kΩ.

2-6 Find the currents through meters A and B in the circuit of Fig. 2-31.

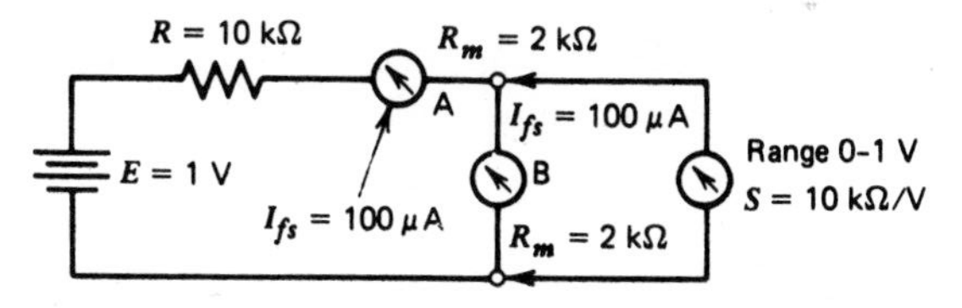

FIGURE 2-31 Circuit for Problem 2-6.

2-7 Calculate the value of the resistors R_1 through R_5 in the circuit of Fig. 2-32.

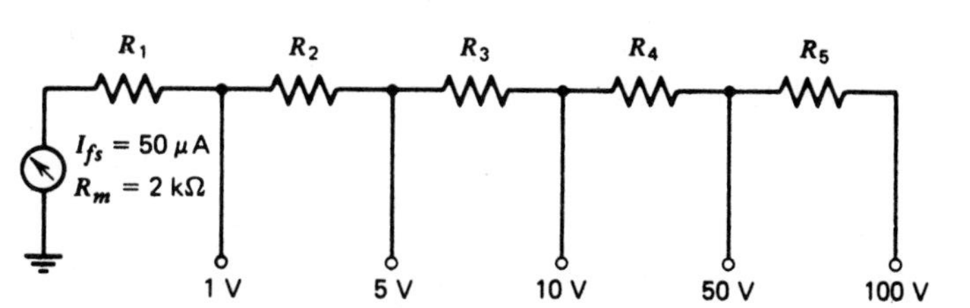

FIGURE 2-32 Circuit for Problem 2-7.

2-8 Calculate the value of the resistors R_1 through R_3 in the circuit of Fig. 2-33.

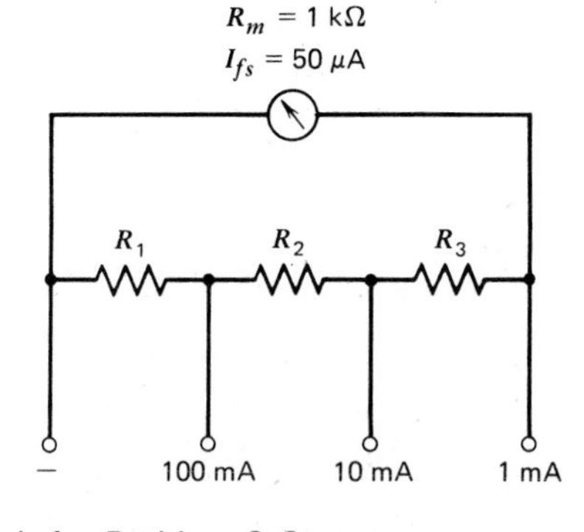

FIGURE 2-33 Circuit for Problem 2-8.

2-9 Calculate the value of the resistors R_1 through R_4 in the circuit of Fig. 2-34.

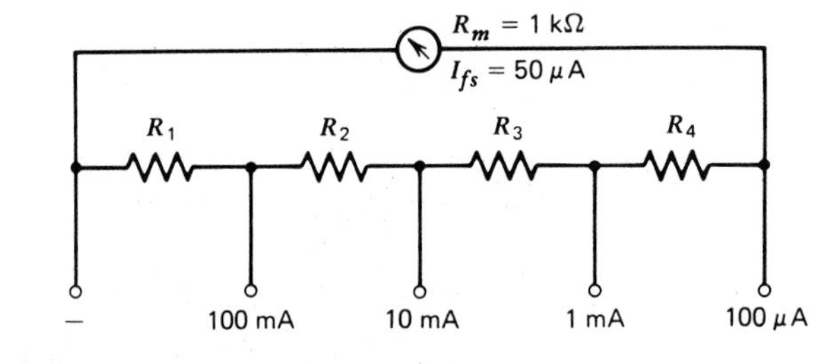

FIGURE 2-34 Circuit for Problem 2-9.

2-10 In the circuit of Fig. 2-35, what is the value of R_x if the meter reads half-scale?

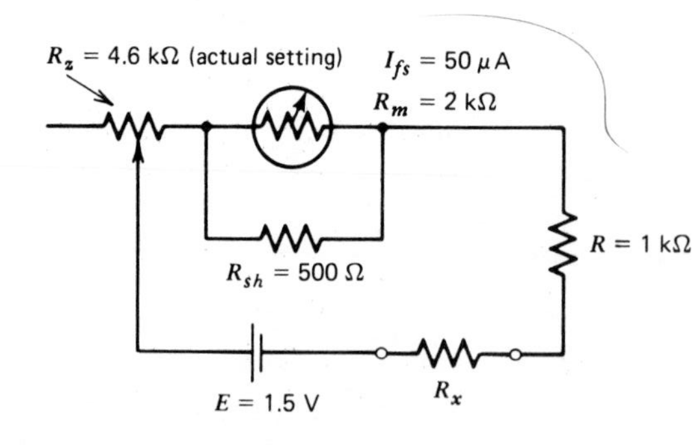

FIGURE 2-35 Circuit for Problem 2-10.

2-11 A voltage reading is to be taken across the 6-kΩ resistor in the circuit of Fig. 2-36. A voltmeter with a sensitivity of 10 kΩ/V is to be used. If the instrument has ranges of 1 V, 5 V, 10 V, and 100 V, what is the most sensitive range that may be used to obtain a reading having less than 3% error owing to voltmeter loading?

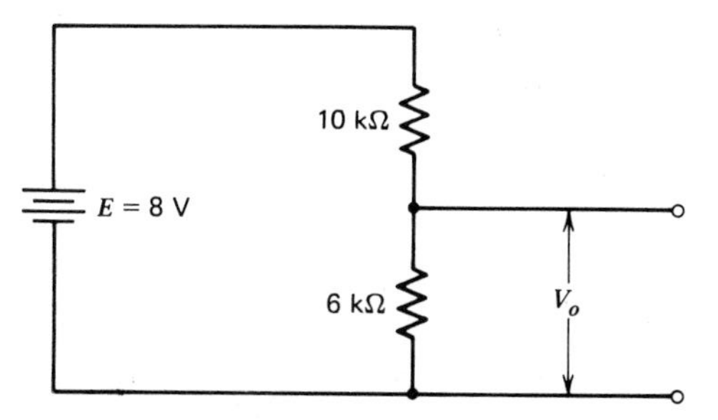

FIGURE 2-36 Circuit for Problem 2-11.

2-12 In the circuit of Fig. 2-37, voltmeter A having a sensitivity of 5 kΩ/V is connected to points X and Y and indicates 15 V on its 30-V range. Voltmeter B is then connected to points X and Y and indicates 16.13 V on its 50-V range. Find the sensitivity of voltmeter B.

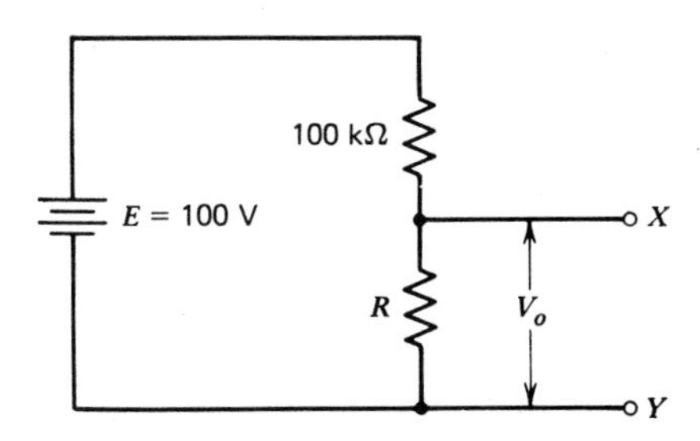

FIGURE 2-37 Circuit for Problem 2-12.

2-13 A voltage reading is to be taken across the 50-kΩ resistor in the circuit of Fig. 2-38. A voltmeter with a sensitivity of 20 kΩ/V and a guaranteed accuracy of ±2% at full scale is to be used on its 10-V range. What is the minimum voltmeter reading that could be expected?

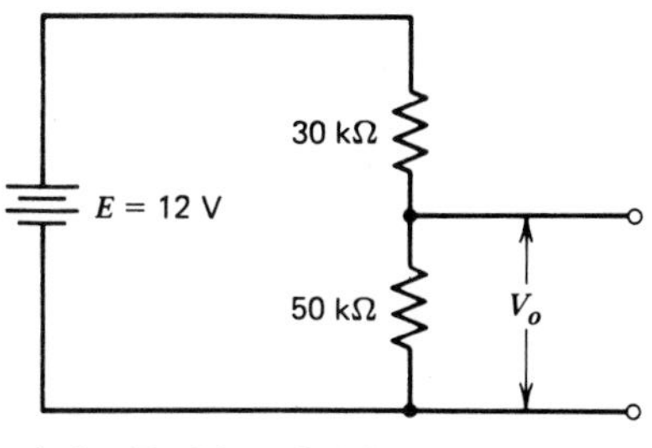

FIGURE 2-38 Circuit for Problem 2-13.

2-18 LABORATORY EXPERIMENTS

Experiments E3 and E4, which are found at the back of the text, deal specifically with the theory presented in Chapter 2. These experiments are intended to provide students with hands-on experience, which is essential for a thorough understanding of the concepts involved.

The experiments require no special equipment; therefore, the equipment should be found in almost any electronics laboratory. One comment might be in order—the resistance values for the shunts on the multimeter experiment are **standard** EIA commercial values for composition resistors. However, they may not be values that are ordinarily stocked in the laboratory. Because they are standard, they are easily purchased.

The contents of the laboratory report to be submitted by each student are listed at the end of each experimental procedure. The troubleshooting procedure is necessary only for problems requiring circuits or measurements.

CHAPTER 3

Alternating-Current Meters

3-1 Instructional Objectives

This chapter discusses the characteristics of the different types of meter movements used to measure alternating current (ac). The advantages, limitations, and applications of the various movements are emphasized. After completing the chapter you should be able to

1. List the four principal meter movements discussed in the chapter and an application for each.
2. Describe the purpose and operation of the diode in a half-wave rectifier circuit.
3. Trace the current path in a full-wave bridge rectifier.
4. Describe the purpose of the second diode in a three-lead instrument rectifier.
5. Describe the purpose of the shunt resistor, which is often used in rectifier circuits.
6. List five applications for the electrodynamometer movement.
7. List one disadvantage of the electrodynamometer movement for voltage measurements compared to the d'Arsonval meter movement.
8. List the typical frequency range of the iron-vane meter movement.
9. Calculate ac sensitivity and the value of multiplier resistors for half-wave and full-wave rectification.

3-2 INTRODUCTION

Several types of meter movements may be used to measure alternating current or voltage. The five principal meter movements used in ac instruments are listed in Table 3-1.

Although there are particular applications for which each type of meter movement in Table 3-1 is best suited, the d'Arsonval meter movement is by

TABLE 3-1
Application of Meter Movements

Meter Movement	Dc Use	Ac Use	Applications
Electrodynamometer	Yes	Yes	"Standards" meter, transfer instrument, wattmeter, frequency meter
Iron-vane	Yes	Yes	"Indicator" applications such as in automobiles
Electrostatic	Yes	Yes	Measurement of high voltage when very little current can be supplied by the circuit being measured
Thermocouple	Yes	Yes	Measurement of radio-frequency ac signals
d'Arsonval (PMMC)	Yes	Yes—with rectifiers	Most widely used meter movement for measuring direct current or voltage and resistance

far the most frequently used, even though it cannot directly measure alternating current or voltage. Therefore, we begin with a discussion of instruments for measuring alternating signals that use the d'Arsonval meter movement.

3-3 D'ARSONVAL METER MOVEMENT USED WITH HALF-WAVE RECTIFICATION

In Chapter 2 we discussed the measurement of direct current and voltage, as well as resistance measurements, using the d'Arsonval meter movement which is a dc-responding device. In this chapter we will discover that we can use the same d'Arsonval meter movement to measure alternating current and voltage.

In order to measure alternating current with the d'Arsonval meter movement, we must first **rectify** the alternating current by use of a **diode** rectifier to produce unidirectional current flow. Several types of rectifiers are selected, such as a copper oxide rectifier, a vacuum diode, or a semiconductor or "crystal" diode.

If we add a diode to the dc voltmeter circuit discussed in Chapter 2, as shown in Fig. 3-1, we will have a circuit that is capable of measuring ac voltage.

Recall from Chapter 2 that the sensitivity of a dc voltmeter is

$$S = \frac{1}{I_{fs}} = \frac{1}{1\text{ mA}} = 1\text{ k}\Omega/\text{V} \tag{2-8}$$

A multiplier of ten times this value means a 10-V dc input will cause exactly full-scale deflection when connected with the polarity indicated in

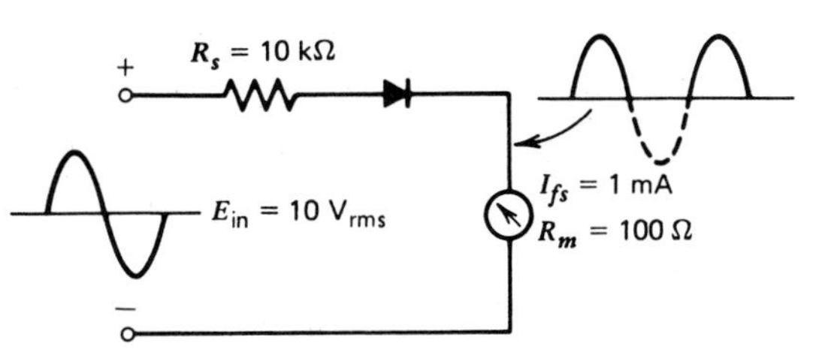

FIGURE 3-1 Dc voltmeter circuit modified to measure ac voltage.

Fig. 3-1. The forward-biased diode will have no effect on the operation of the circuit if we assume an ideal diode.

Now suppose we replace the 10-V dc input with a 10-V **rms** (**root-mean-square**) sine-wave input. The voltage across the meter movement is just the positive half-cycle of the sine wave because of the rectifying action of the diode. The peak value of the 10-V rms sine wave is

$$E_p = 10\ V_{rms} \times 1.414 = 14.14\ V_{peak} \qquad (3\text{-}1)$$

The dc meter movement will respond to the **average** value of the ac sine wave where the average, or dc value, is equal to 0.318 times the peak value, or

$$E_{ave} = E_{dc} = 0.318 \times E_p$$

This is sometimes written as

$$E_{ave} = \frac{E_p}{\pi} = 0.45 \times E_{rms}$$

The diode action produces an approximate half sine wave across the load resistor. The average value of this voltage is referred to as the dc voltage. This is the voltage to which a dc voltmeter connected across the load resistor would respond. For example, if the output voltage from a half-wave rectifier is 10 V, a dc voltmeter will provide an indication of approximately 4.5 V. Therefore, we can see that the pointer that deflected full scale when a 10-V dc signal was applied deflects to only 4.5 V when we apply a 10-V rms **sinusoidal** ac waveform. This means that the ac voltmeter is not as sensitive as the dc voltmeter. In fact, an ac voltmeter using half-wave rectification is only *approximately* 45% as sensitive as a dc voltmeter.

Actually, the circuit would probably be designed for full-scale deflection with a 10-V rms alternating current applied, which means the multiplier resistor would be only 45% of the value of the multiplier resistor for a 10-V dc voltmeter. Since we have seen that the equivalent dc voltage is equal to 45% of the rms value of the ac voltage, we can express this in the form of an equation for computing the value of the multiplier resistor,

$$R_s = \frac{E_{dc}}{I_{dc}} - R_m = \frac{0.45 E_{rms}}{I_{dc}} - R_m \qquad (3\text{-}2)$$

We can infer from Eq. 3-2, for a half-wave rectifier, that

$$S_{ac} = 0.45 S_{dc} \tag{3-3a}$$

EXAMPLE 3-1 Compute the value of the multiplier resistor for a 10-V rms ac range on the voltmeter shown in Fig. 3-2 using

(a) Equation 2-8.
(b) Equation 3-3a.
(c) Equation 3-2.

Solution We can approach the problem in several ways. Consider the following.

(a) We can first find the sensitivity of the meter movement.

$$S_{dc} = \frac{1}{I_{fs}} = \frac{1}{1\text{ mA}} = \frac{1\text{ k}\Omega}{\text{V}}$$

Multiplying the dc sensitivity by the dc range gives us the total resistance, from which we subtract the resistance of the meter movement as

$$R_s = S_{dc} \times \text{Range}_{dc} - R_m$$

$$= \frac{1\text{ k}\Omega}{\text{V}} \times \frac{0.45E_{rms}}{1} - R_m$$

$$= \frac{1\text{ k}\Omega}{\text{V}} \times \frac{4.5\text{ V}}{1} - 300\ \Omega = 4.2\text{ k}\Omega$$

(b) We may also choose to start by finding the ac sensitivity for a half-wave rectifier:

$$S_{ac} = 0.45 S_{dc} = 0.45 \times \frac{1}{I_{fs}} = \frac{450\ \Omega}{\text{V}} \tag{3-3a}$$

Then we can say

$$R_s = S_{ac} \times \text{Range}_{ac} - R_m$$

$$= \frac{450\ \Omega}{\text{V}} \times \frac{10\text{ V}}{1} - 300\ \Omega = 4.2\text{ k}\Omega$$

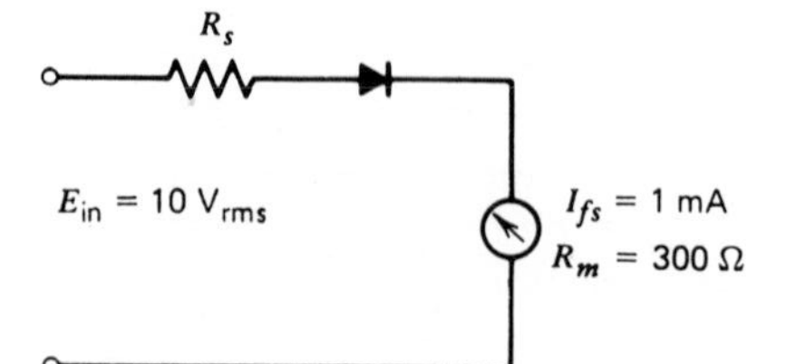

FIGURE 3-2 Ac voltmeter using half-wave rectification.

(c) If we have no interest in the sensitivity, we can use Eq. 3-2:

$$R_s = \frac{0.45E_{\text{rms}}}{I_{fs}} - R_m$$

$$= \frac{0.45 \times 10\ V_{rms}}{1\ \text{mA}} - 300\ \Omega$$

$$= \frac{4.5\ \text{V}}{1\ \text{mA}} - 300\ \Omega = 4.2\ \text{k}\Omega \qquad (3\text{-}2)$$

You should note in methods a and b of Example 3-1 that we must be consistent in working with ac or dc parameters. If, as in method a, you wish to work with dc sensitivity, you must work with dc voltage. Similarly, if you work with ac sensitivity, you must work with ac voltage.

Commercially produced ac voltmeters that use half-wave rectification also have an additional diode and a shunt as shown in Fig. 3-3. This double-diode arrangement in a single package is generally called an **instrument rectifier**. The additional diode D_2 is reverse-biased on the positive half-cycle and has virtually no effect on the behavior of the circuit. In the negative half-cycle, D_2 is forward-biased and provides an alternate path for reverse-biased leakage current that would normally flow though the meter movement and diode D_1. The purpose of the shunt resistor R_{sh} is to increase the current flow through D_1 during the positive half-cycle so that the diode is operating in a more linear portion of its characteristic curve.

Although this shunt resistor improves the linearity of the meter on its low-voltage ac ranges, it also further reduces the ac sensitivity.

EXAMPLE 3-2

In the half-wave rectifier shown in Fig. 3-4, diodes D_1 and D_2 have an average forward resistance of 50 Ω and are assumed to have an infinite resistance in the reverse direction. Calculate the following.

(a) The value of the multiplier R_s.

(b) The ac sensitivity.

(c) The equivalent dc sensitivity.

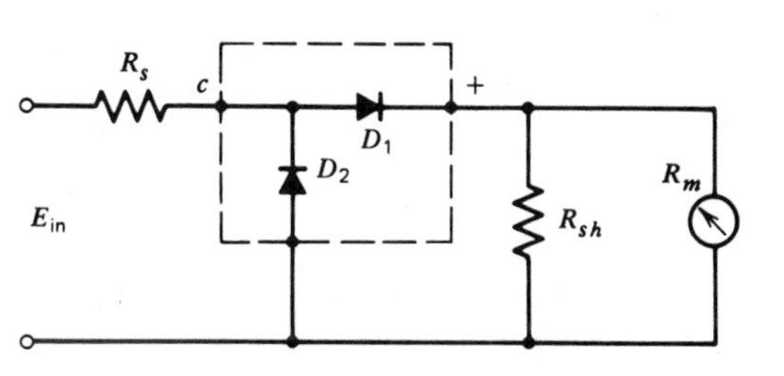

FIGURE 3-3 Half-wave rectification using an instrument rectifier and a shunt resistor for improved linearity.

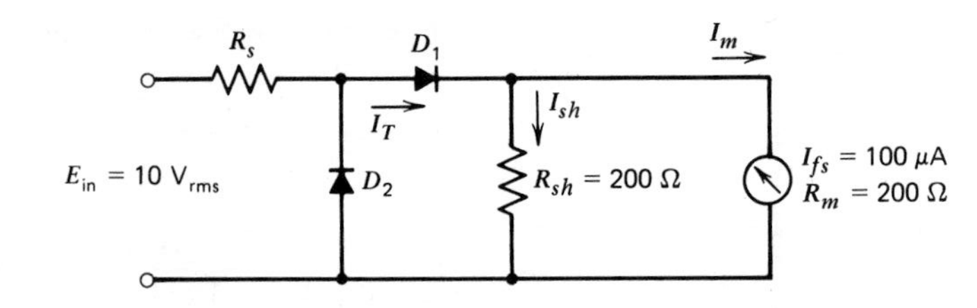

FIGURE 3-4 Half-wave rectifier with shunt resistor.

Solution

(a)

$$I_{sh} = \frac{E_m}{R_{sh}} = \frac{100\ \mu\text{A} \times 200\ \Omega}{200\ \Omega} = 100\ \mu\text{A}$$

$$I_T = I_{sh} + I_m = 100\ \mu\text{A} + 100\ \mu\text{A} = 200\ \mu\text{A}$$

$$E_{dc} = 0.45 \times E_{rms} = 0.45 \times 10\ \text{V} = 4.5\ \text{V}$$

The total resistance of the meter circuit is

$$R_T = \frac{E_{dc}}{I_T} = \frac{4.5\ \text{V}}{200\ \mu\text{A}} = 22.5\ \text{k}\Omega$$

The total resistance is made up of several separate resistances and is computed as

$$R_T = R_s + R_d + \frac{R_m R_{sh}}{R_m + R_{sh}}$$

Therefore, we can solve for R_s as

$$R_s = R_T - R_d - \frac{R_m R_{sh}}{R_m + R_{sh}}$$

$$= 22{,}500\ \Omega - 50\ \Omega - \frac{200\ \Omega \times 200\ \Omega}{200\ \Omega + 200\ \Omega}$$

$$= 22.35\ \text{k}\Omega$$

(b) The ac sensitivity is computed as

$$S_{ac} = \frac{R_T}{\text{Range}} = \frac{22{,}500\ \Omega}{10\ \text{V}} = 2250\ \Omega/\text{V}$$

(c) The dc sensitivity is computed as

$$S_{dc} = \frac{1}{I_T} = \frac{1}{200\ \mu\text{A}} = 5000\ \mu/\text{V}$$

or alternatively as

$$S_{dc} = \frac{S_{ac}}{0.45} = \frac{2250\ \Omega/\text{V}}{0.45} = 5000\ \Omega\text{V}$$

EXAMPLE 3-3

Using the E–I curve, you can determine the diode in the circuit in Fig. 3-5 to have 1-kΩ static resistance with full-scale deflection current of 100 μA through it. Compute the value of the multiplier resistor using the value R_d at full-scale deflection. Compute the diode resistance with 20-μA current and the value of input voltage that would cause 20 μA to flow.

Solution

The value of the multiplier resistor is found as

$$R_s = \frac{0.45E_{rms}}{I_{dc}} - (R_m + R_d) = 4.5\ \text{k}\Omega - 1.2\ \text{k}\Omega = 3.3\ \text{k}\Omega \tag{3-2}$$

The static resistance of the diode at 20 μA is

$$R_d = \frac{E_d}{I_d} = \frac{0.04\ \text{V}}{20\ \mu\text{A}} = 2\ \text{k}\Omega$$

The total resistance of the circuit is now

$$R_T = R_s + R_d + R_m$$
$$= 3.3\ \text{k}\Omega + 2\ \text{k}\Omega + 0.2\ \text{k}\Omega$$

The dc voltage that will cause 20 μA is

$$E_{dc} = I_{dc} \times R_T$$
$$= 20\ \mu\text{A} \times 5.5\ \text{k}\Omega = 0.11\ \text{V}$$

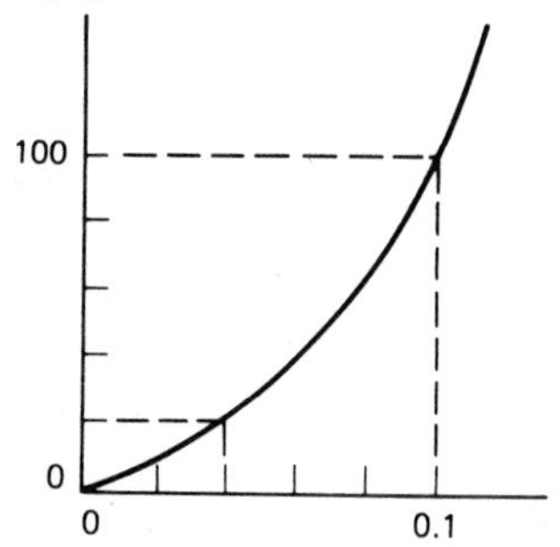

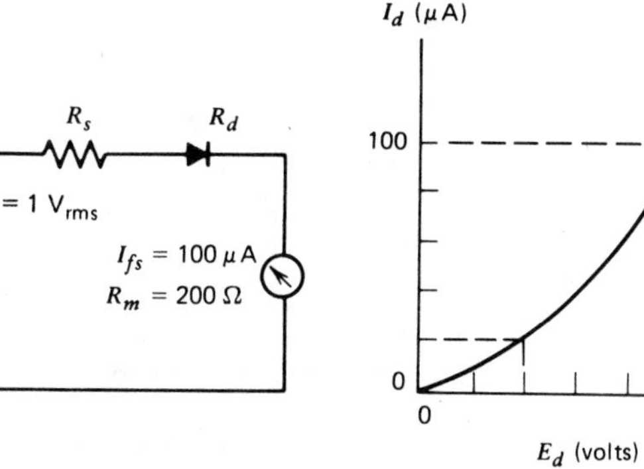

FIGURE 3-5 Circuit and E–I curve for Example 3-3.

The input voltage that will cause 20 μA is

$$E_{in} = \frac{E_{dc}}{0.45} = \frac{0.11}{0.45} = 0.23\ V_{rms}$$

If the diode resistance had not changed, the input voltage that would cause a 20-μA current to flow would be equal to 0.09 V. Therefore, an error of approximately 22% now exists.

3-4 D'ARSONVAL METER MOVEMENT USED WITH FULL-WAVE RECTIFICATION

Frequently, it is more desirable to use a full-wave rather than a half-wave rectifier in ac voltmeters because of the higher sensitivity rating. The most frequently used circuit for full-wave rectification is the bridge-type rectifier shown in Fig. 3-6.

During the positive half-cycle, current flows through diode D_2, through the meter movement from positive to negative, and through diode D_3. The polarities in circles on the transformer secondary are for the positive half-cycle. Since current flows through the meter movement on both half-cycles, we can expect the deflection of the pointer to be greater than with the half-wave rectifier, which allows current to flow only on every other half-cycle; if the deflection remains the same, the instrument using full-wave rectification will have a greater sensitivity.

Consider the circuit shown in Fig. 3-7. The peak value of the 10-V rms signal is computed as with the half-wave rectifier as

$$E_p = 1.414 \times E_{rms} = 14.14\ V_{peak}$$

The average, or dc, value of the pulsating sine wave is

$$E_{ave} = 0.636 E_p = 9\ V$$

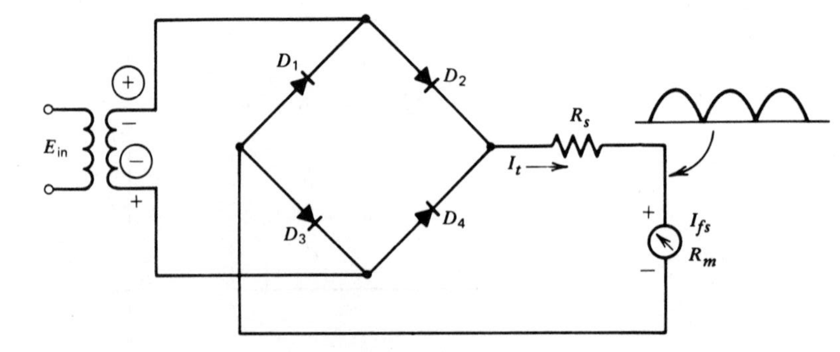

FIGURE 3-6 Full-wave bridge rectifier used in an ac voltmeter circuit.

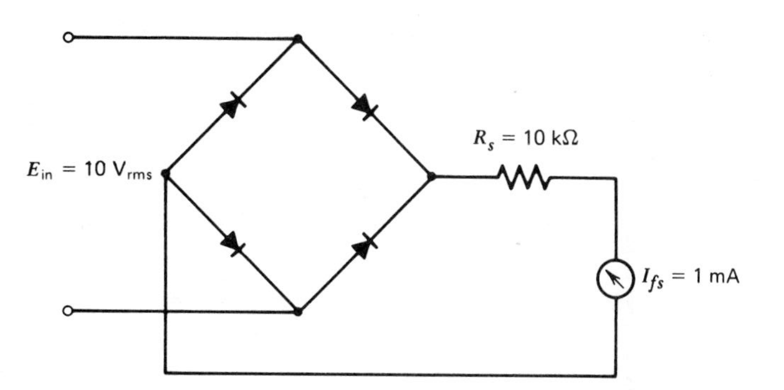

FIGURE 3-7 Ac voltmeter circuit using full-wave rectification.

Alternatively, this can be computed as

$$E_{ave} = 0.9 \times E_{rms} = 0.9 \times 10\text{ V} = 9\text{ V}$$

Therefore, we can see that the 10-V rms voltage is equivalent to 9 V_{dc}. When full-wave rectification is used, the pointer will deflect to 90% of full scale. This means an ac voltmeter using full-wave rectification has a sensitivity equal to 90% of the dc sensitivity, or it has twice the sensitivity of a circuit using half-wave rectification. As with the half-wave rectifier, the circuit would be designed for full-scale deflection, which means the value of the multiplier resistor would be only 90% of the value for a 10-V dc voltmeter. We may write this for a full-wave rectifier as

$$S_{ac} = 0.9S_{dc} \tag{3-3b}$$

EXAMPLE 3-4

Compute the value of the multiplier resistor for a 10-V rms ac range on the voltmeter in Fig. 3-8.

Solution

The dc sensitivity is

$$S_{dc} = \frac{1}{I_{fs}} = \frac{1}{1\text{ mA}} = \frac{1\text{ k}\Omega}{\text{V}} \tag{2-8}$$

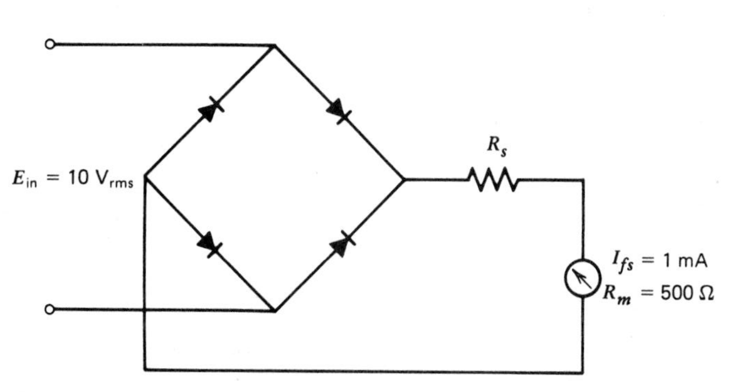

FIGURE 3-8 Ac voltmeter circuit using full-wave rectification.

The ac sensitivity is 90% of the dc sensitivity:

$$S_{ac} = 0.9 S_{dc} = 0.9 \times \frac{1\ \text{k}\Omega}{\text{V}} = \frac{900\ \Omega}{\text{V}} \tag{3-3b}$$

The multiplier resistor is therefore found to equal

$$R_s = S_{ac} \times \text{Range} - R_m$$
$$= \frac{900\ \Omega}{\text{V}} \times 10\ \text{V}_{rms} - 500\ \Omega = 8.5\ \text{k}\Omega$$

EXAMPLE 3-5

Each diode in the full-wave rectifier circuit shown in Fig. 3-9 has an average forward resistance of 50 Ω and is assumed to have an infinite resistance in the reverse direction. Calculate the following.

(a) The value of the multiplier R_s.

(b) The ac sensitivity.

(c) The equivalent dc sensitivity.

Solution

(a) We begin by computing the shunt current and the total current,

$$I_{sh} = \frac{E_m}{R_{sh}} = \frac{1\ \text{mA} \times 500\ \Omega}{500\ \Omega} = 1\ \text{mA}$$

and

$$I_T = I_{sh} + I_m = 1\ \text{mA} + 1\ \text{mA} = 2\ \text{mA}$$

The equivalent dc voltage is computed as

$$E_{dc} = 0.9 \times 10\ \text{V}_{rms} = 0.9 \times 10\ \text{V} = 9.0\ \text{V}$$

The total resistance of the meter circuit can now be computed as

$$R_T = \frac{E_{dc}}{I_T} = \frac{9.0\ \text{V}}{2\ \text{mA}} = 4.5\ \text{k}\Omega$$

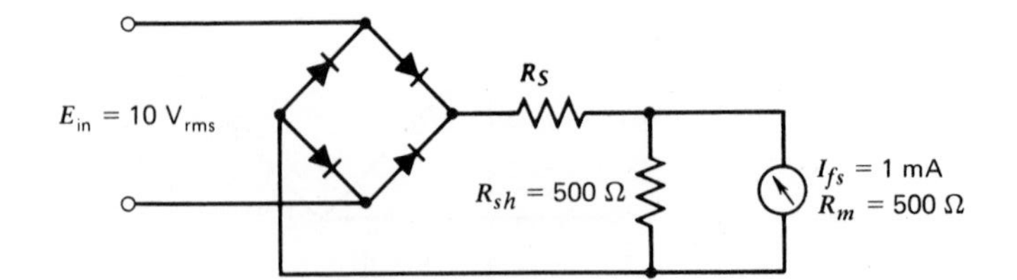

FIGURE 3-9 Ac voltmeter circuit using full-wave rectification and shunt.

and

$$R_s = R_T + 2R_d - \frac{R_m R_{sh}}{R_m + R_{sh}}$$

$$= 4500\ \Omega + 2 \times 50\ \Omega - \frac{500\ \Omega \times 500\ \Omega}{500\ \Omega + 500\ \Omega} = 4.15\ \text{k}\Omega$$

(b) The ac sensitivity is computed as

$$S_{\text{ac}} = \frac{R_T}{\text{Range}} = \frac{4500\ \Omega}{10\ \text{V}} = 450\ \Omega/\text{V}$$

(c) The dc sensitivity is computed as

$$S_{\text{dc}} = \frac{1}{I_T} = \frac{1}{2\ \text{mA}} = 500\ \Omega/\text{V}$$

or alternatively as

$$S_{\text{dc}} = \frac{S_{\text{ac}}}{0.9} = \frac{450\ \Omega\text{V}}{0.9} = 500\ \Omega/\text{V}$$

Take note that voltmeters using half-wave or full-wave rectification are suitable for measuring only sinusoidal ac voltages. In addition, the equations presented thus far are *not* valid for nonsinusoidal waveforms such as square, triangular, and sawtooth waves.

3-5 ELECTRODYNAMOMETER MOVEMENT

The **electrodynamometer movement** is the most fundamental meter movement in use today. Like the d'Arsonval movement previously discussed, the electrodynamometer is a current-sensitive device. That is, the pointer deflects up scale because of current flow through a moving coil. Even though this meter movement is the most fundamental in use, it is also the most versatile. Singe-coil movements may be used to measure direct or alternating current or voltage, or in a single-phase wattmeter or varmeter. Double-coil movements may be used in polyphase wattmeter or varmeter, and crossed-coil movements may be used as a power factor meter or as a frequency meter. Aside from all this, perhaps the most important applications for electrodynamometer movements are as voltmeter and ammeter **standards** and **transfer** instruments. Because of the inherent accuracy of the electrodynamometer movement, it lends itself well to use in **standards instruments**, those used for the calibration of other meters. The term **transfer instrument** is applied to an instrument that may be calibrated with a dc source and then used

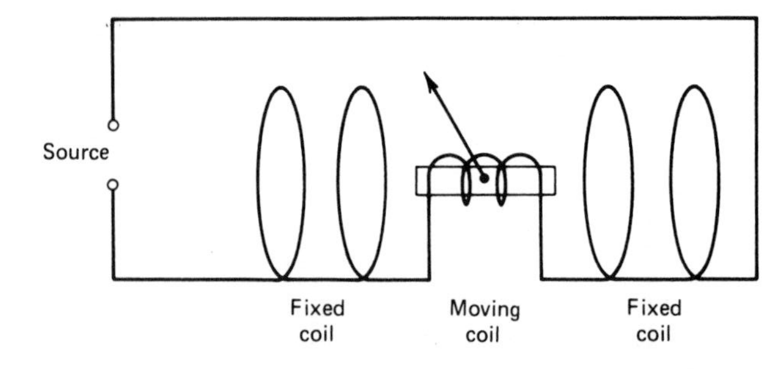

FIGURE 3-10 Electrodynamometer movement.

without modification to measure alternating current. This gives us a direct means of equating ac and dc measurements of voltage and current.[1]

The single-coil electrodynamometer movement consists of a fixed coil, divided into two equal halves, separated by a movable coil, as shown in Fig. 3-10. Both halves of the split fixed coil and the moving coil are connected in series, and current from the circuit being measured passes through all the coils causing a magnetic field around the fixed coils. The movable coil rotates in this magnetic field.

The basic electrodynamometer movement is capable of handling much more current than a d'Arsonval movement could handle without a shunt. A current flow of 100 mA is an approximate value for the maximum current without a shunt resistance. The increased current-handling capabilities are a direct result of the basic design of the meter movement. The magnetic coupling between the fixed coils and the moving coils is across an air gap that results in a weak magnetic field. For magnetic coupling to be sufficient, more current must flow through the coils, which means that a larger-diameter wire must be used. However, the larger-diameter wire has less resistance than a smaller-diameter wire. This causes the electrodynamometer movement to have a very *low* sensitivity rating of approximately 20 to 100 Ω/V.

EXAMPLE 3-6

An electrodynamometer movement that has a full-scale deflection current rating of 10 mA is to be used in a voltmeter circuit. Calculate the value of the multiplier for a 10-V range if R_m equals 50 **Ω**.

Solution

The sensitivity of the meter movement is

$$S = \frac{1}{I_{fs}} = \frac{1}{10\text{ mA}} = \frac{100\ \Omega}{\text{V}} \tag{2-8}$$

Therefore, the value of the multiplier resistor is

$$R_s = S \times \text{Range} - R_m$$

$$= \frac{100\ \Omega}{\text{V}} \times 10\text{ V} - 50\ \Omega = 950\ \Omega \tag{2-9}$$

[1]There is a frequency limitation to ac use, however. Most electrodynamometer movements are accurate over the frequency range from 0 to 125 Hz.

This resistor is placed in series with the meter movement in the same way as with the d'Arsonval meter movement.

When a shunt resistor is used with an electrodynamometer movement to expand current-measuring capabilities, the shunt resistor is normally placed in parallel with only the moving coil, as shown in Fig. 3-11. Since only the moving coil is shunted, the resistance of the moving coil would have to be known in order to compute the value of the shunt.

EXAMPLE 3-7

An electrodynamometer movement with a full-scale deflection current rating of 10 mA is to be used as a 1-A ammeter. If the resistance of the moving coil is 40 Ω, what is the value of the shunt?

Solution

The value of the shunt is computed in the same manner as discussed when using the d'Arsonval meter movement.

$$R_{sh} = \frac{R_m}{n-1}$$

$$= \frac{40\ \Omega}{100-1} = \frac{40\ \Omega}{99} = 0.404\ \Omega \qquad (2\text{-}3)$$

If the ammeter in Example 3-7 is connected to a 1-A dc source, the meter pointer should deflect to exactly full scale. The pointer should also deflect full scale if the 1-A dc source is replaced with a 1-A rms ac source.

Since the same current flows through the field coils and the moving coil, when the electrodynamometer movement is used as either an ammeter or a voltmeter, the pointer deflects as the square of the current. The result is a **square-law meter scale** such as is shown in Fig. 3-12.

Probably the most extensive application of the electrodynamometer movement is in wattmeters. The wattmeter may be used to measure either ac or dc power. The ac signals are not restricted to sinusoidal waveforms so that power developed by any ac waveform may be measured. When used as a wattmeter, the electrodynamometer is connected as shown in Fig. 3-13.

When used as a wattmeter, the fixed coils, called 'field' coils, are in series with the load and therefore conduct the same current as the load (plus a small current through the moving coil). The moving coil is connected as a

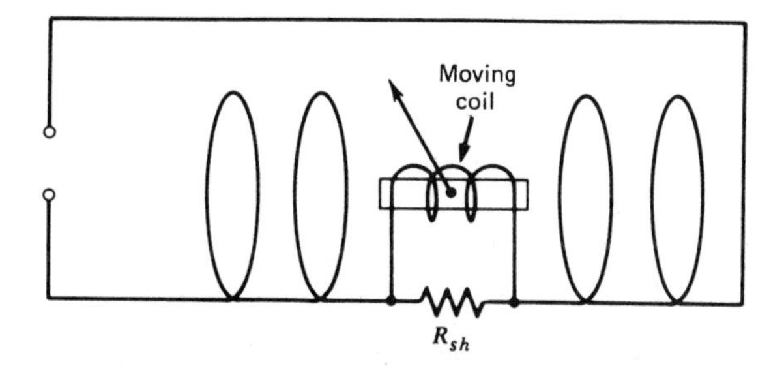

FIGURE 3-11 Electrodynamometer movement used as an ammeter.

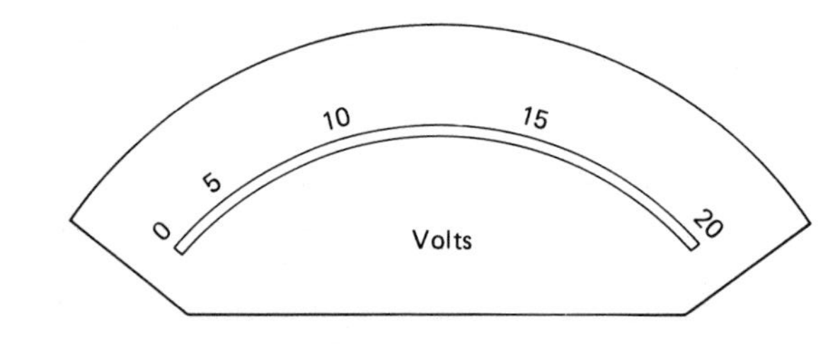

FIGURE 3-12 Square-law meter scale.

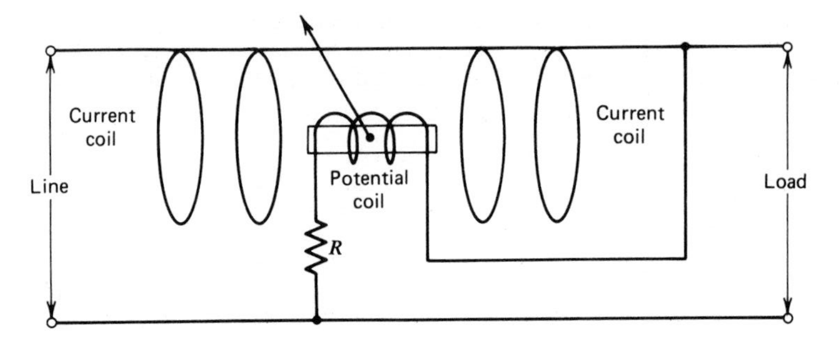

FIGURE 3-13 Electrodynamometer movement used as a wattmeter.

voltmeter across the load where the resistor R_s is the multiplier for the voltage-sensitive meter. The magnetic torque that causes the pointer to deflect up scale can be expressed in terms of the amount of deflection as

$$\theta_m = K_m EI \cos\theta \tag{3-4}$$

where

θ_m = angular deflection of the pointer
K_m = instrument constant, degrees per watt
E = rms value of source voltage
I = rms value of source current
$\cos\theta$ = power factor

EXAMPLE 3-8

A wattmeter that uses an electrodynamometer movement with $K_m = 8°/\text{W}$ is used to measure the power dissipated in an ac circuit. If the applied voltage of 100 V_{rms} produces a current of 0.5 A with a power factor of 0.8, how many degrees does the meter pointer deflect?

Solution

The angular deflection of the pointer may be calculated using

$$\theta_m = K_m EI \cos\theta$$

$$= \frac{8°}{\text{W}} \times \frac{110\ \text{V}}{1} \times \frac{0.05\ \text{A}}{1} \times \frac{0.8}{1} = 35.2° \tag{3-4}$$

Since volts times amperes equals watts, all units divide out except degrees, which are the correct units for angular deflection.

3-6 IRON-VANE METER MOVEMENT

The **iron-vane meter movement**, which consists of a fixed coil of many turns and two iron vanes placed inside the fixed coil, is widely used in industry for applications in which ruggedness is more important than a high degree of accuracy.

The current to be measured passes through the windings of the fixed coil, setting up a magnetic field that magnetizes the two iron vanes with the same polarity. This causes the iron vanes to repel one another. If one of the iron vanes is attached to the frame of a fixed coil, the other iron vane will then be repelled by an amount related to the square of the current. Therefore, the square-law meter scale shown in Fig. 3-12 is used with the basic iron-vane meter movement as well as with the electrodynamometer movement.

The basic iron-vane movement has a square-law response, but the fixed coil can be designed to provide a relatively linear response. The radial-vane design shown in Fig. 3-14 is just such a variation and does in fact have a nearly linear scale.

Although the iron-vane movement is responsive to direct current, the hysteresis, or magnetic lag, in the iron vanes causes appreciable error. Therefore, moving-vane instruments for measuring direct current are rarely used except for very inexpensive indicators, such as charge-discharge indicators on automobiles.

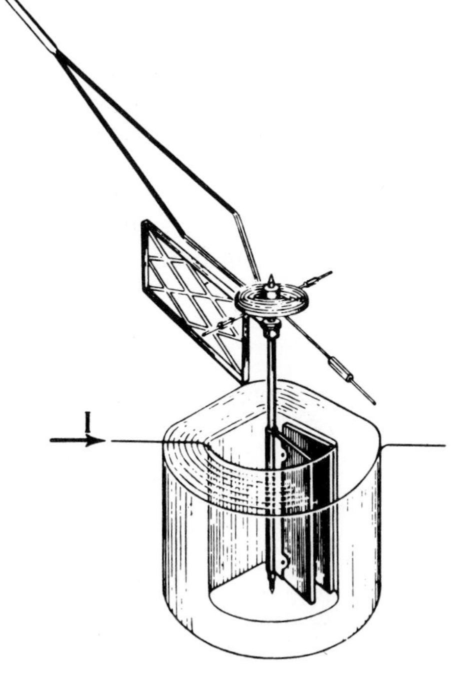

FIGURE 3-14 Radial-type iron-vane movement. (Courtesy Weston Instruments, a Division of Sangamo Weston, Inc.)

For ac applications, the magnetic lag presents no problems. Therefore, iron-vane meter movements are used extensively in industry for measuring alternating current when errors on the order of 5% to 10% are acceptable.

The basic current responding iron-vane meter movement can be used to measure voltage by adding a multiplier resistor as with the d'Arsonval movement. However, the iron-vane movement is very sensitive to frequency change and can be expected to provide accurate readings over a limited frequency range, approximately 25 to 125 Hz. When accurate measurements at higher frequencies are required, the thermocouple meter (see Section 3-7) is used. The iron-vane movement is sensitive to frequency primarily because the magnetization of the iron vane is nonlinear and because of losses incurred by eddy currents and hysteresis.

3-7 THERMOCOUPLE METER

A basic **thermocouple meter** is an instrument that consists of a heater element, usually made of fine wire, a thermocouple, and a d'Arsonval meter movement. This instrument can be used to measure both alternating current and direct current. The most attractive characteristic of the thermocouple meter is that it can be used to measure very high-frequency alternating currents. In fact, such instruments are very accurate well above 50 MHz. The schematic for a very basic thermocouple meter is shown in Fig. 3-15.

The instrument derives its name from the fact that its operation is based on the action of a thermocouple. A thermocouple, which consists of two dissimilar metals, develops a very small potential difference (0 to 10 mV) at the junction of the two metals. This potential difference, which is a function of the junction temperature, causes current to flow through the meter movement.

The thermocouple senses the temperature of the heater wire, which is a function of the current or the voltage being measured. Therefore, the thermocouple and the heater must be thermally coupled but electrically isolated.

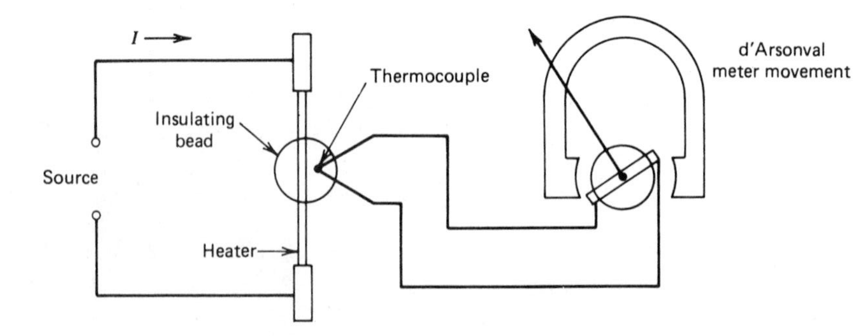

FIGURE 3-15 Schematic for a basic thermocouple meter.

Instruments for measuring over a wide range of currents (approximately 1 to 50 mA) are available. The following example illustrates the calculations involved in designing a basic three-range thermocouple voltmeter.

EXAMPLE 3-9

Design a basic three-range (5, 10, 25 V) thermocouple voltmeter around the following specifications.

- d'Arsonval meter movement:

$$I_{fs} = 50\ \mu A$$

$$R_m = 200\ \Omega$$

- Heater:

$$I_{max} = 5\ mA$$

$$R = 200\ \Omega$$

- Thermocouple: Thermocouple related specifications are shown in Figures 3-16 and 3-17.

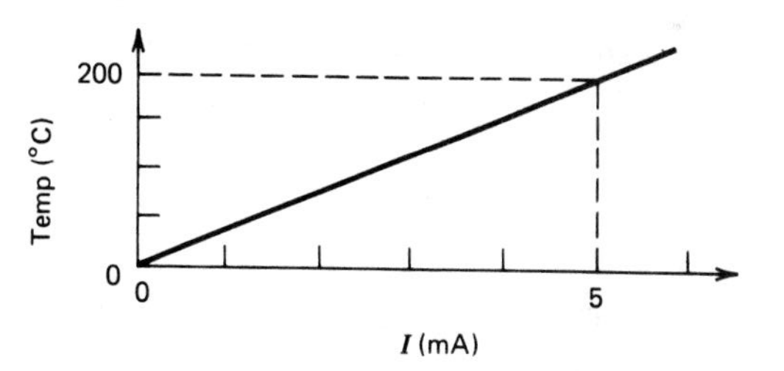

FIGURE 3-16 Heater current versus temperature graph.

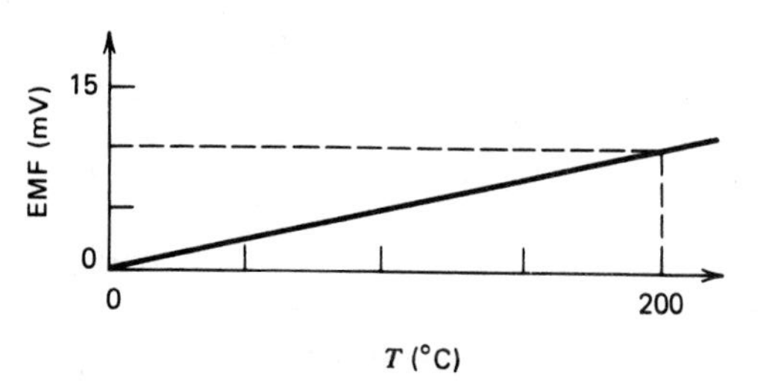

FIGURE 3-17 Graph of thermocouple junction temperature versus potential difference.

Solution

The value of the multiplier on each of the three ranges is calculated as follows.

(a) On the 5-V range,

$$R_s = \frac{E}{I_{\text{max}}} - R_n = \frac{5\text{ V}}{5\text{ mA}} - 200\ \Omega = 1\text{ k}\Omega - 200\ \Omega = 800\ \Omega$$

(b) On the 10-V range,

$$R_s = \frac{E}{I_{\text{max}}} - R_n = \frac{10\text{ v}}{5\text{ mA}} - 200\ \Omega = 1.8\text{ k}\Omega$$

(c) On the 25-V range,

$$R_s = \frac{E}{I_{\text{max}}} - R_n = \frac{25\text{ V}}{5\text{ mA}} - 200\ \Omega = 4.8\text{ k}\Omega$$

The graph of heater current versus temperature (Fig. 3-16) shows that the thermocouple temperature will be 200°C when the heater current is 5 mA. The graph of thermocouple junction temperature versus potential difference (Fig. 3-17) shows that a potential difference of 10 mV exists when the junction temperature is 200°C. A potential difference of 10 mV at the input terminals of the d'Arsonval meter movement causes full-scale deflection current flow. This is calculated as

$$I = \frac{E}{R_m} = \frac{10\text{ mV}}{200\ \Omega} = 50\ \mu\text{A}$$

The schematic diagram for the circuit is shown in Fig. 3-18.

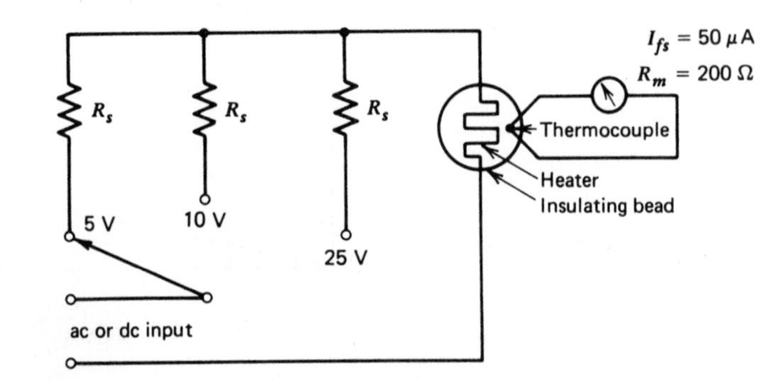

FIGURE 3-18 Multiple-range voltmeter using a thermocouple meter.

3-8 LOADING EFFECTS OF AC VOLTMETERS

As has already been discussed, the sensitivity of ac voltmeters, using either half-wave or full-wave rectification, is less than the sensitivity of dc voltmeters. Therefore, the loading effect of an ac voltmeter is greater than that of a dc voltmeter.

EXAMPLE 3-10

Determine the reading obtained with a dc voltmeter in the circuit in Fig. 3-19 when switch S is set to position A; then set the switch to position B and determine the reading obtained with a half-wave and a full-wave ac voltmeter. All the meters use a 100-μA full-scale deflection meter movement and are set on their 10-V dc or rms ranges.

Solution

The reading obtained with the dc voltmeter is computed as follows.

$$S_{\text{dc}} = \frac{1}{I_{fs}} = \frac{1}{100\ \mu\text{A}} = \frac{10\ \text{k}\Omega}{\text{V}} \tag{2-8}$$

$$R_s = S_{\text{dc}} \times \text{Range}$$

$$= \frac{10\ \text{k}\Omega}{\text{V}} \times \frac{10\ \text{V}}{1} = 100\ \text{k}\Omega \tag{2-9}$$

$$E = 20\ \text{V} \times \frac{100\ \text{k}\Omega \| 10\ \text{k}\Omega}{100\ \text{k}\Omega \| 10\ \text{k}\Omega + 10\ \text{k}\Omega}$$

$$= 20\ \text{V} \times \frac{9.09\ \text{k}\Omega}{9.09\ \text{k}\Omega + 10\ \text{k}\Omega} = 9.52\ \text{V}$$

The reading obtained with the ac voltmeter using half-wave rectification is computed as

$$S_{hw} = 0.45 S_{\text{dc}} = \frac{4.5\ \text{k}\Omega}{\text{V}} \tag{3-3a}$$

$$R_s = S_{hw} \times \text{Range} = 45\ \text{k}\Omega$$

$$E = 20\ \text{V} \times \frac{45\ \text{k}\Omega \| 10\ \text{k}\Omega}{45\ \text{k}\Omega \| 10\ \text{k}\Omega + 10\ \text{k}\Omega}$$

$$= 20\ \text{V} \times \frac{8.18\ \text{k}\Omega}{8.18\ \text{k}\Omega + 10\ \text{k}\Omega} = 9.0\ \text{V}$$

Finally, the reading obtained with the ac voltmeter using full-wave

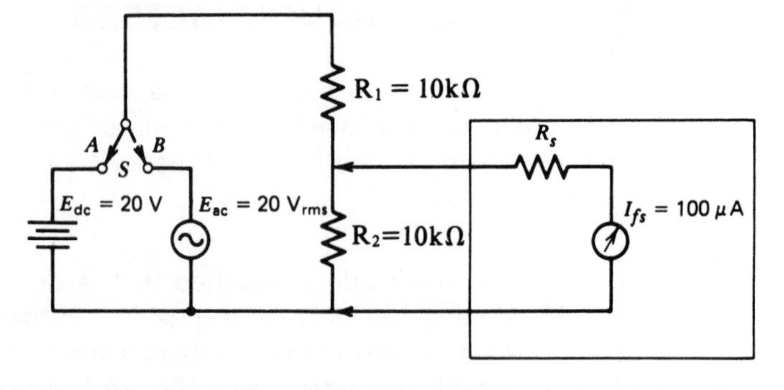

FIGURE 3-19 Circuit for comparing readings of ac and dc voltmeters.

rectification is computed as

$$S_{fw} = 0.90 S_{dc} = \frac{9.0\ \text{k}\Omega}{\text{V}}$$

$$R_s = S_{fw} \times \text{Range} = 90\ \text{k}\Omega$$

$$E = 20\ \text{V} \times \frac{90\ \text{k}\Omega \| 10\ \text{k}\Omega}{90\ \text{k}\Omega \| 10\ \text{k}\Omega + 10\ \text{k}\Omega}$$

$$= 20\ \text{V} \times \frac{9\ \text{k}\Omega}{9\ \text{k}\Omega + 10\ \text{k}\Omega} = 9.47\ \text{V}$$

As can be seen, the ac voltmeter using either half-wave or full-wave rectification has a greater loading effect than the dc voltmeter.

3-9 PEAK-TO-PEAK-READING AC VOLTMETERS

Frequently, it is desirable to measure nonsinusoidal waveforms. One way of taking this measurement is with peak-to-peak-reading ac voltmeters. The block diagram shown in Fig. 3-20 shows a basic peak-to-peak-reading ac voltmeter.

We have already discussed ac voltmeters using either half- or full-wave rectification. Therefore, our interest in this section is with the peak-to-peak detector. A circuit that is capable of detecting the peak-to-peak amplitude of ac signals, either sinusoidal or nonsinusoidal, is shown in Fig. 3-21.

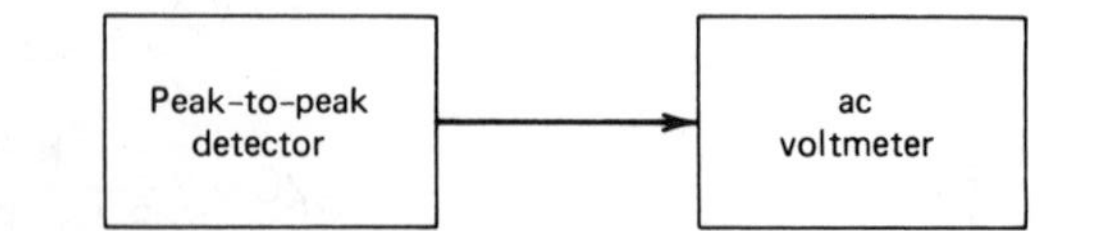

FIGURE 3-20 Block diagram for a peak-to-peak-reading ac voltmeter.

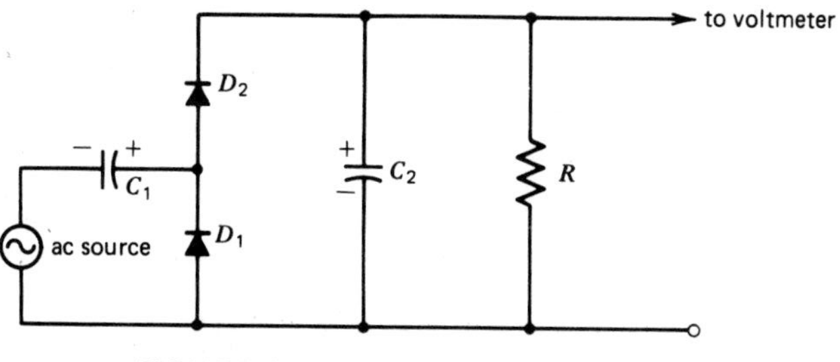

FIGURE 3-21 **Peak-to-peak detector.**

During the negative half-cycle of the input signal, diode D_1 is forward-biased and charges capacitor C_1 as shown in Fig. 3-21. During the positive half-cycle, diode D_1 is reverse-biased and D_2 is forward-biased. The positive going input signal and the voltage across C_1 are now of the same polarity. Capacitor C_2 charges to the sum of these voltages through D_2. The voltage across C_2 is now equal to the peak-to-peak value of the input signal. This voltage is now applied to an ordinary ac voltmeter. Peak-to-peak reading voltmeters are sometimes used to measure waveforms that either are nonsinusoidal or swing unevenly about a zero reference axis (e.g., 20 V positive and 5 V negative).

3-10 APPLICATIONS

Alternative-current voltmeters have many practical applications, both in the laboratory and around the home. One lab application is in transformer testing and in determining whether a waveform is sinusoidal. An ac voltmeter is also very useful around the home. Occasionally, appliances such as refrigerators or air conditioners fail to operate properly because of low ac line voltage. An ac voltmeter can be used to measure the line voltage during "peak" and "slack" demand periods. If the line voltage drops to less than about 100 V during the peak demand period, notify the power company.

To determine whether a waveform is sinusoidal requires a peak-to-peak-reading voltmeter and an rms-responding meter. The peak-to-peak value of the waveform may be obtained regardless of the type of waveform. If the waveform is sinusoidal, the reading obtained with the rms-responding meter will be equal to

$$E_{\mathrm{rms}} = \frac{E_{p-p}}{2} \times \frac{0.707}{1} \tag{3-5}$$

Several useful tests on transformers can be performed with an ac voltmeter. Tests to check phase relationships and polarity markings, to check the

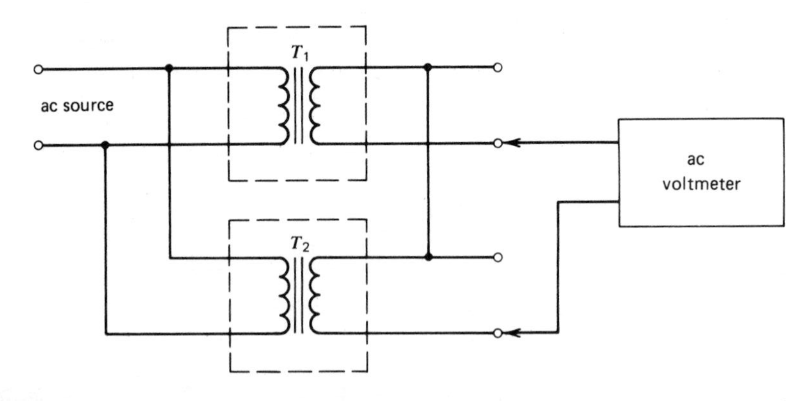

FIGURE 3-22 Test circuit for determining that two transformers are operating in phase.

impedance ratio between the primary and the secondary, to determine the regulating effect of a transformer, or to determine the Q of a tuned transformer are some of the tests that can be performed.

As an example, suppose that two transformers that are supposedly identical are to be operated in parallel and in phase. Many transformers are marked with dots or some similar system to indicate polarity. However, there is no standard system of marking transformers, nor will all transformers be marked in any manner. In such cases the test circuit shown in Fig. 3-22 may be used to check the phase relationship. The transformers are connected in phase if the ac voltmeter reads at, or near, zero. If the voltmeter reading is double the normal transformer secondary voltage for either transformer, the transformers are operating out of phase. The transformers can be made to operate in phase by reversing either the input or the output leads of one transformer. If the voltmeter still indicates a voltage, the transformer secondaries should be separated and the output voltage measured individually. If different secondary voltage readings are obtained, the transformers should not be connected as shown, since the transformers may be damaged.

3-11 SUMMARY

Several different types of meter movements are available for measuring alternating current or voltage. Each type has characteristics that make it most suitable for particular applications. For general purposes the d'Arsonval movement, with either a half-wave or a full-wave rectifier, is widely used.

Table 3-2 is a summary of the responses of the four current-responding meter movements to an ac sine wave or a dc voltage.

TABLE 3-2
Meter Movement Response to Ac or Dc Voltage

Meter Movement	Applied Voltage and Frequency	Reading Obtained
d'Arsonval	10 V_{rms}, 60 Hz	0 V
Iron vane	10 V_{rms}, 60 Hz	10 V
Electrodynamometer	10 V_{rms}, 60 Hz	10 V
Thermocouple	10 V_{rms}, 60 Hz	10 V
d'Arsonval with half-wave rectifier	10 V_{rms}, 60 Hz	4.5 V
d'Arsonval with full-wave rectifier	10 V_{rms}, 60 Hz	9.0 V
Iron vane	10 V_{dc}	10 V
Electrodynamometer	10 V_{dc}	10 V
Thermocouple	10 V_{dc}	10 V

3-12 GLOSSARY

Average: The value corresponding to the area under one-half cycle of a sinusoidal waveform divided by the distance of the curve along the horizontal axis.

Diode: An electronic device (usually a semiconductor *p-n* junction) that conducts current readily in only one direction.

Electrodynamometer movement: A basic but versatile meter movement consisting of a fixed coil divided into two equal halves, called field coils, and a moving coil between the field coils.

Electrostatic meter movement: An indicating mechanism resembling a variable capacitor and the only mechanism used for electrical indications that measures voltage directly rather than by the effect of current.

Instrument rectifier: A three-terminal molded package consisting of two diodes. One diode acts as a rectifier while the second diode provides a low-resistance path for leakage current of the rectifying diode.

Iron-vane meter movement: A meter movement in which the movable element in an iron vane is drawn into a magnetic field developed by the current being measured.

Rectify: To convert alternating current to a unidirectional current by removing or inverting the part of the waveform on one side of the zero-amplitude axis.

RMS v: Root-mean-square.

Sinusoidal: Having the form of a sine wave.

Square-law meter scale: The scale required for a meter movement, such as the iron-vane movement, for which the repelling force, and hence the pointer deflection, is proportional to the square of the current.

Standards instrument: An instrument used to calibrate other instruments.

Thermocouple meter: A meter that uses a thermocouple to sense the temperature of an element heated by a radio-frequency signal. The thermocouple emf is then applied to a d'Arsonval meter movement.

Transfer instrument: An instrument that is used to equate ac and dc measurements because it can be calibrated using direct current and then used to measure alternating current directly.

3-13 REVIEW QUESTIONS

The following questions should be answered after a thorough study of the chapter. The purpose of the questions is to check your comprehension of the material.

1. Which type of meter movement is most widely used in ac instruments for current and voltage measurements?
2. Which type of meter movement is most widely used in wattmeters?
3. How does the sensitivity of an ac voltmeter compare to the sensitivity of a dc voltmeter?
4. Define *transfer instrument*.
5. How does the sensitivity of an ac voltmeter using full-wave rectification compare with the sensitivity of one using half-wave rectification?
6. Which type of meter movement is best studied for use as a transfer instrument?
7. Show how the diodes are connected in an instrument rectifier and explain the purpose of each diode.
8. Which ac meter movement naturally has a square-law scale and why?
9. Compare the effects of circuit loading when using an ac voltmeter with half-wave rectification against those when using an ac voltmeter with full-wave rectification.

3-14 PROBLEMS

3-1 The current through a meter movement is 150 μA_{peak}. What is the dc value if the instrument uses half-wave rectification?

3-2 A d'Arsonval meter movement deflects to 0.8 mA. What is the peak value of the alternating current if the instrument uses full-wave rectification?

3-3 A d'Arsonval meter movement with a full-scale deflection current rating of 1 mA and an internal resistance of 500 Ω is to be used in a half-wave rectifier ac voltmeter. Calculate the ac and dc sensitivity and the value of the multiplier resistor for a 30-V rms range.

3-4 A d'Arsonval meter movement with a full-scale deflection current rating of 200 μA and an internal resistance of 500 Ω is to be used in an ac voltmeter using full-wave rectification. Calculate the value of the multiplier resistor for a 50-V peak-to-peak sine-wave range.

3-5 Calculate the ac and dc sensitivity and the value of the multiplier resistor required to limit current to the full-scale deflection current in the circuit shown in Fig. 3-23.

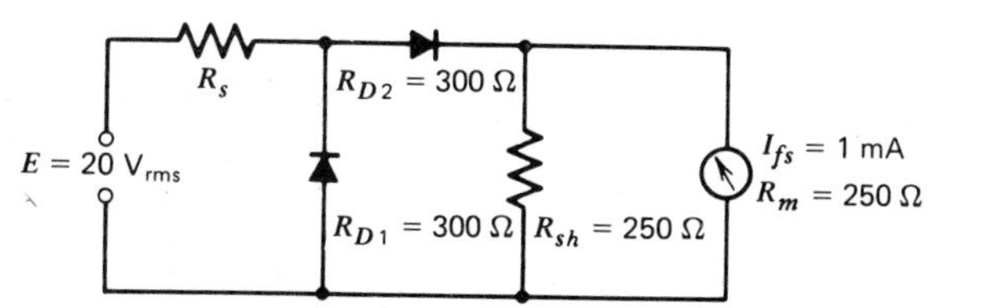

FIGURE 3-23 Circuit for Problem 3-5.

3-6 Calculate the ac and dc sensitivity and the value of the multiplier resistor required to limit current to the full-scale deflection current in the circuit shown in Fig. 3-24. All diodes have a forward resistance of 300 Ω and an infinite reverse resistance.

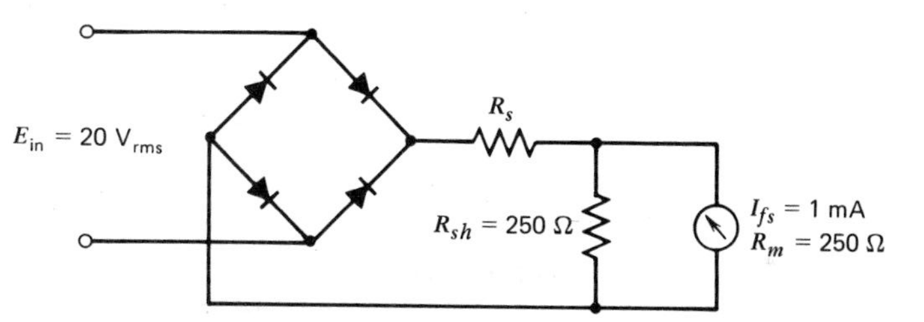

FIGURE 3-24 Circuit for Problem 3-6.

3-7 Figure 3-25 represents a meter face for an ac voltmeter with full-wave rectification. Compute the values of the peak-to-peak voltage and the

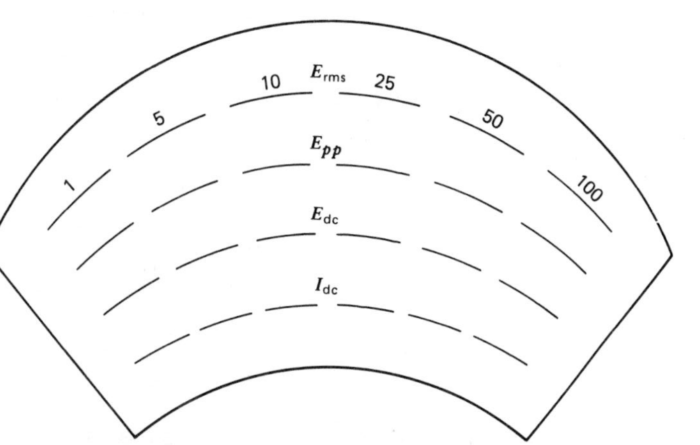

FIGURE 3-25 Meter scale for Problem 3-7.

dc voltage and current for the rms voltages shown if the dc sensitivity of the meter movement is 10 kΩ/V. Sketch the meter face and fill in the blanks.

3-8 Calculate the dc sensitivity and the value of the multiplier resistor required to limit current to the full-scale deflection current in the circuit shown in Fig. 3-26.

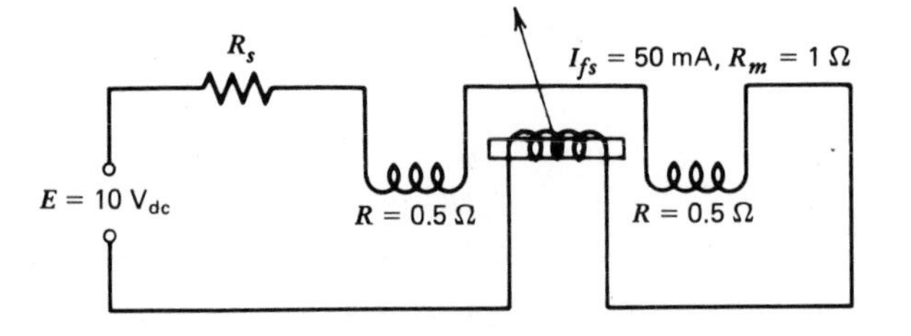

FIGURE 3-26 Circuit for Problem 3-8.

3-9 An rms ac voltmeter and a peak-to-peak-reading ac voltmeter are to be used to determine whether three ac signals are sinusoidal. Determine whether the signals are sinusoidal if the following readings are obtained

First signal	peak-to-peak reading	= 35.26 V
	rms reading	= 12.00 V
Second signal	peak-to-peak reading	= 11.31 V
	rms reading	= 4.00 V
Third signal	peak-to-peak reading	= 25.00 V
	rms reading	= 8.83 V

3-10 An ac voltmeter is to be used to measure the rms voltage across the 15-kΩ resistor in the circuit shown in Fig. 3-27. If the voltmeter uses half-wave rectification and a 100-μA d'Arsonval meter movement, if it is set on its 10-V range, and if R_m = 1.5 kΩ, what reading will be obtained?

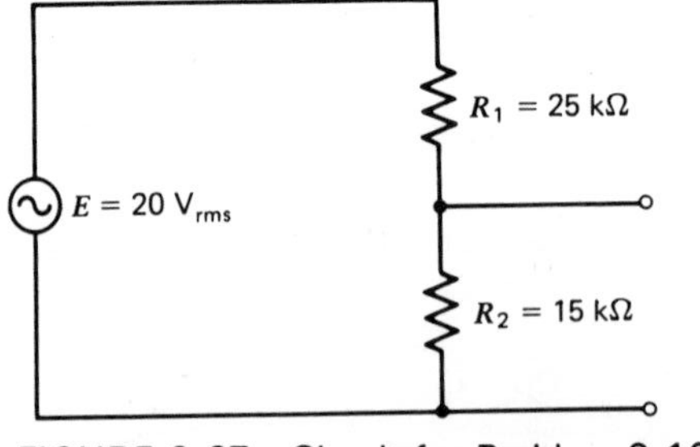

FIGURE 3-27 Circuit for Problem 3-10.

3-11 Two different ac voltmeters are used to measure the voltage across the 22-kΩ resistor in the circuit shown in Fig. 3-28. Meter A has ac

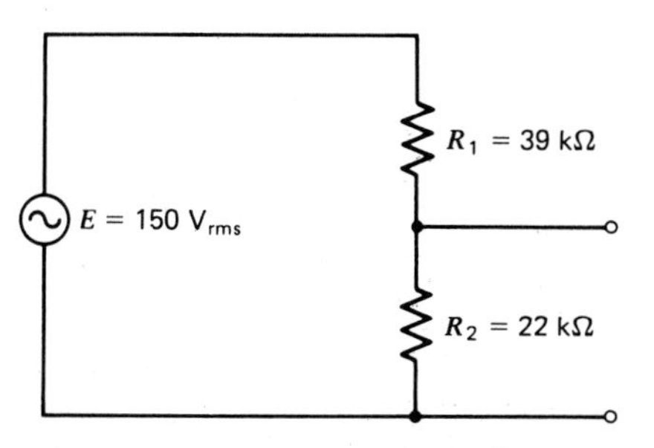

FIGURE 3-28 Circuit for Problem 3-11.

sensitivity of 10 kΩ/V, a guaranteed accuracy of 98% at full scale, and is set on its 200-V range. Meter B has an ac sensitivity of 4 kΩ/V, a guaranteed accuracy of 98.5% at full scale, and is set on its 100-V range. Which meter will provide a more accurate result?

3-12 The ac voltmeter described below is used to measure the voltage across the 68-kΩ resistor in the circuit shown in Fig. 3-29. What is the minimum voltage reading that should be observed? The ac voltmeter has

- Full-wave rectification.
- 100-μA meter movement.
- 150-V range.
- Limiting error of ±3% at full scale. (Refer to Chapter 1 for a discussion of limiting error.)

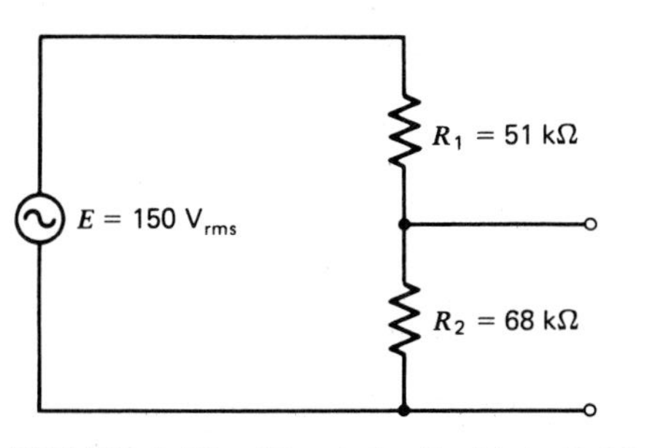

FIGURE 3-29 Circuit for Problem 3-12.

3-15 LABORATORY EXPERIMENTS

Laboratory experiments E5 and E6 apply the theory that has been presented in Chapter 3. The purpose of these experiments is to provide hands-on experience to reinforce the theory.

The equipment required to perform the experiments can be found in any well-equipped electronics laboratory. The second experiment calls for three specific types of meter movements. If these are not available, any meter movements that respond to alternating current can be used.

The contents of the laboratory report to be submitted by each student are listed at the end of each experimental procedure.

CHAPTER 4

Potentiometer Circuits and Reference Voltages

4-1 Instructional Objectives

In Chapter 4 we discuss the basic theory and applications of potentiometer circuits. Primary emphasis is on the principle of operation of potentiometer circuits and on the applications for such circuits in voltage measurements. After completing the chapter you should be able to

1. Describe the basic principle of operation of a potentiometer.
2. Describe why a potentiometer does not load a voltage source when used to measure the voltage source.
3. Describe the key operating condition when using a potentiometer.
4. Calculate the volt-box ratio when using a volt box.
5. Design a basic volt box, given the volt-box ratio and the desired output voltage.
6. Design a basic standard voltage source.
7. List two electronic components that make possible a simple standard voltage reference circuit.
8. Describe why the total resistance of a volt box should be very high.
9. Define the term *active circuit.*

4-2 INTRODUCTION

The loading effect of instruments always influences the ultimate value of a measured parameter. The effect is most important when making measurements in low-power circuitry or when high precision is necessary. Even when instruments with a high input impedance are used to measure voltage, a finite amount of current is drawn from the circuit under test, thereby loading the circuit and generating an internal "IR" drop.

One solution to the loading problem is to utilize a null or balanced circuit with which measurements are possible without drawing current from the circuit under test. The fundamental idea is to connect a second voltage source in such a way that a current is generated equal in magnitude, but opposite in direction, according to the superposition theorem, to the current that would otherwise be drawn from the circuit under test. Therefore, no current is actually drawn from the circuit under test. The two basic operational problems associated with such a system are

1. Establishing a precisely known voltage standard.
2. Establishing a detector for the zero-current condition.

The effects of the loading may be seen by considering a simple dry-cell battery whose terminal voltage is being measured. The circuit shown in Fig. 4-1 represents the situation. Here E is the electromotive force generated by the chemical action of the battery, R_s is the internal resistance of the battery, V_T is the terminal voltage of the battery, and R_L is the load connected to the battery. In this situation R_L is the resistance of the measuring voltmeter and I is the circuit current. Writing Kirchhoff's voltage law equation for the circuit, we have

$$E - IR_s - IR_L = 0 \tag{4-1}$$

or

$$IR_L = E - IR_s \tag{4-2}$$

But

$$IR_L = V_T$$

and therefore,

$$V_T = E - IR_s \tag{4-3}$$

The equation predicts that the terminal voltage as predicted by the voltmeter will be equal to the chemical electromotive force of the battery minus a voltage

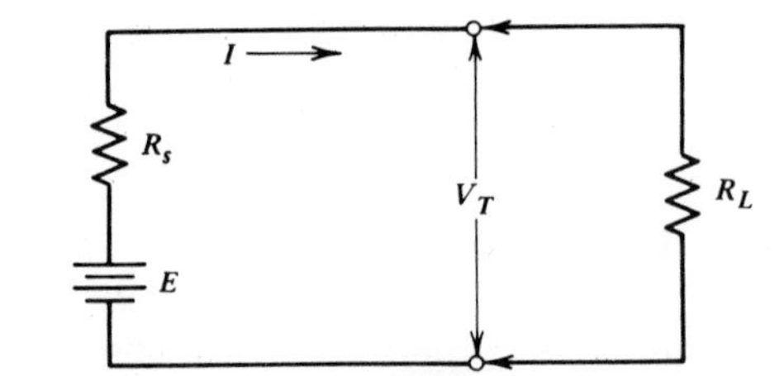

FIGURE 4-1 Terminal voltage of a dry cell being measured with a deflection-type instrument.

drop internal to the battery as given by IR_s. Solving for I, we get

$$I = \frac{E}{R_L + R_s} \tag{4-4}$$

In most cases, $R_L \gg R_s$ so that Eq. 4-4 can be approximated as

$$I \approx \frac{E}{R_L} \tag{4-5}$$

Equation 4-3 then becomes

$$V_T = E - \frac{ER_s}{R_L} \tag{4-6}$$

The internal voltage drop IR_s can be reduced by designing R_L as large as possible. However, with R_L large, error is still present because of the voltmeter loading effects.

4-3 BASIC POTENTIOMETER CIRCUITS

One solution to this measurement problem is to utilize an active (energy-supplied) circuit called a **potentiometer**. The key operating condition for the use of the potentiometer is *zero current* taken from the circuit being measured. A simple potentiometer circuit is illustrated in Fig. 4-2 which shows a relatively high-resistance wire connected in series with a voltage source E. A steady current in this wire produces a uniform voltage drop along its length. The voltage drop between the movable contacts X and Y is directly proportional to the length of the wire between points X and Y. The voltage drop across a unit length of the resistance wire depends on the value of the circuit current which can be changed by varying the rheostat R.

We can calibrate the potentiometer so that it becomes a direct-reading instrument. First, we adjust the current through the resistance wire so that the voltage drop becomes proportional to the length of the wire or numerically equal to a scale reading of the potentiometer. Next, we connect a standard cell and a galvanometer to terminals X and Y so that the voltage of the

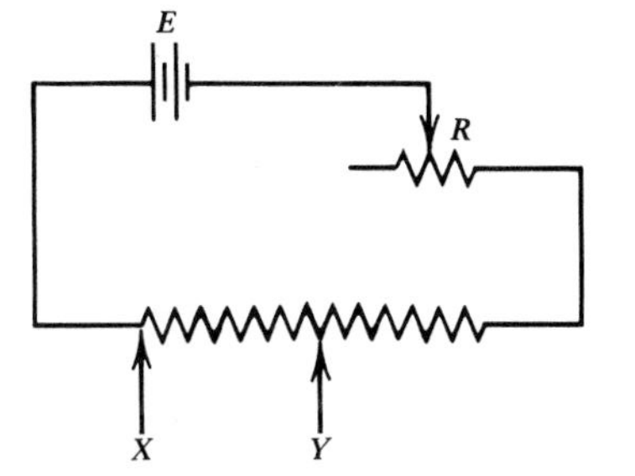

FIGURE 4-2 A basic potentiometer circuit.

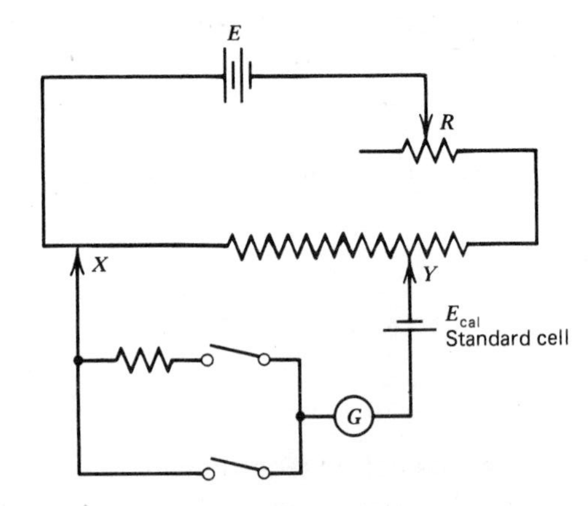

FIGURE 4-3 **Calibrating a potentiometer with a standard cell.**

standard cell opposes that of the working voltage, E, as shown in Fig. 4-3. The span from X to Y is adjusted to produce a voltage drop very nearly equal to the terminal voltage of the standard cell. The galvanometer G is then switched into the circuit in series with a protective resistor, and the rheostat is adjusted as needed to obtain a null indication on the galvanometer. A final adjustment is made by closing a second switch to increase sensitivity and readjusting for a null reading. This second switch is shown in Fig. 4-3. With the potentiometer calibrated, the standard cell is switched out of the measuring circuit. The potentiometer can now be used to measure the unknown emf or voltage drop without introducing any circuit loading.

The slide-wire potentiometer shown in Fig. 4-4 has a slide-wire with a total length of 200 cm and a resistance of 200 **Ω**. The voltage of the standard

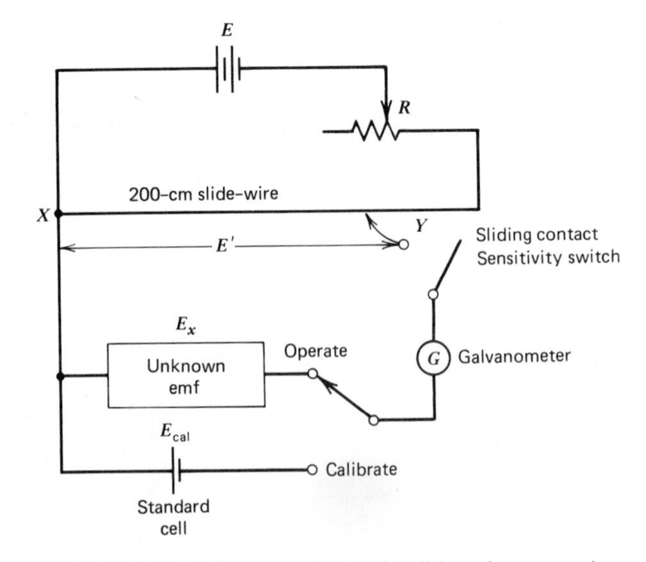

FIGURE 4-4 **Circuit diagram of a basic slide-wire potentiometer.**

cell equals 1.356 V and is used as a voltage reference. In the calibrate or standardize position, the sliding contact is set at the 135.6-cm mark on the slide-wire scale. The rheostat R is adjusted to provide a magnitude of slide-wire current that will cause no deflection of the galvanometer when the sensitivity switch contacts are closed. In this balance or null condition, the voltage drop across the 135.6-cm portion of the slide wire is exactly equal to 1.356 V, the calibrating voltage of the standard cell. Since this portion of the slide-wire represents a resistance of 135.5 Ω, the slide-wire current has been adjusted to 10 mA. The voltage at any point along the slide-wire is proportional to the length of the slide-wire and is obtained by converting length into a corresponding voltage. Once the potentiometer is calibrated, the working current should not be varied.

Once calibrated, the potentiometer can be used to measure any small dc voltage of approximately 2 V or less. When the potentiometer is set to its operate or EMF position, the sliding contact is moved along the resistance wire until the galvanometer shows no deflection as the sensitivity switch is closed. At this null condition, the unknown voltage V_x equals the voltage drop E across the $X-Y$ portion of the slide-wire. Therefore, the slide-wire scale reading can be converted directly to its corresponding voltage value.

EXAMPLE 4-1

The basic slide-wire potentiometer shown in Fig. 4-4 has a working battery of 3 V. The slide-wire has a resistance of 300 Ω and a length of 200 cm. A 200-cm scale placed alongside the slide-wire has 1-mm scale divisions, and interpolation can be made to one-half of a division. The instrument is standardized against a voltage reference source of 1.019 V, with the slider set at the 101.9-cm mark on the scale. Calculate the following.

(a) The working current.
(b) The resistance setting of the rheostat.
(c) The measurement range.
(d) The resolution expressed in mV.

Solution

(a) The total 200-cm length of the resistance wire represents 300 Ω. Therefore, the resistance of 101.9 cm is

$$R_{101.9\ \text{cm}} = \frac{101.9\ \text{cm} \times 300\ \Omega}{200\ \text{cm}} = 152.9\ \Omega$$

The working current is computed as

$$I = \frac{1.019\ \text{V}}{152.9\ \Omega} = 6.67\ \text{mA}$$

(b) The voltage across the slide-wire is computed as

$$V_{sw} = IR = 6.67\ \text{mA} \times 300\ \Omega = 2\ \text{V}$$

The resistance of the rheostat is therefore

$$R = \frac{E - V_{sw}}{I} = \frac{3\text{ V} - 2\text{ V}}{6.67\text{ mA}} = 150\ \Omega$$

(c) The measurement range is equal to the total voltage across the slide-wire and is equal to

$$V_{sw} = IR = 6.67\text{ mA} \times 300\ \Omega = 2\text{ V}$$

(d) The total number of scale divisions is computed as

$$\text{Scale divisions} = \frac{200\text{ cm} \times 10\text{ mm/1 cm}}{1/2} = 4000$$

Therefore, the resolution is

$$\text{Resolution} = \frac{2\text{ V}}{4000} = 0.5\text{ mV}$$

The slide-wire potentiometer just described is useful but somewhat impractical for industrial applications. Modern laboratory-type potentiometers use a calibrated circular slide-wire of one or more turns along with dial resistors. This reduces the size of the instrument significantly. Figure 4-5 shows a laboratory-type potentiometer which uses 15 precision resistors and a single-turn circular slide-wire. The total resistance of the slide-wire is 10 **Ω**, and the dial resistors have a total resistance of 150 **Ω**. The slide-wire has a scale with 200 divisions, and interpolation to one-fifth of a division is entirely feasible. The working current is maintained at 10 mA. Therefore, each step of the dial switch corresponds to a voltage step of 0.1 V, and each division of the slide-wire scale corresponds to 0.0005 V. By interpolation, readings can be estimated to approximately 0.0001 V.

EXAMPLE 4-2

The slide-wire potentiometer of Fig. 4-5 is equipped with a 10-turn slide-wire with a total resistance of 10 **Ω** and a 15-step dial switch with 10-**Ω** per step resistance. The circular slide-wire scale has 100 divisions, and interpolation can be made to one-fifth of a division. If the magnitude of the working voltage is 3 V, calculate the following.

(a) The measurement range.
(b) The resolution, expressed in μV.
(c) The working current.
(d) The resistance of the rheostat.

Solution

(a) To find the measurement range, we must first find the working current which is

$$I = \frac{0.1\text{ V/step}}{10\ \Omega\text{/step}} = 10\text{ mA}$$

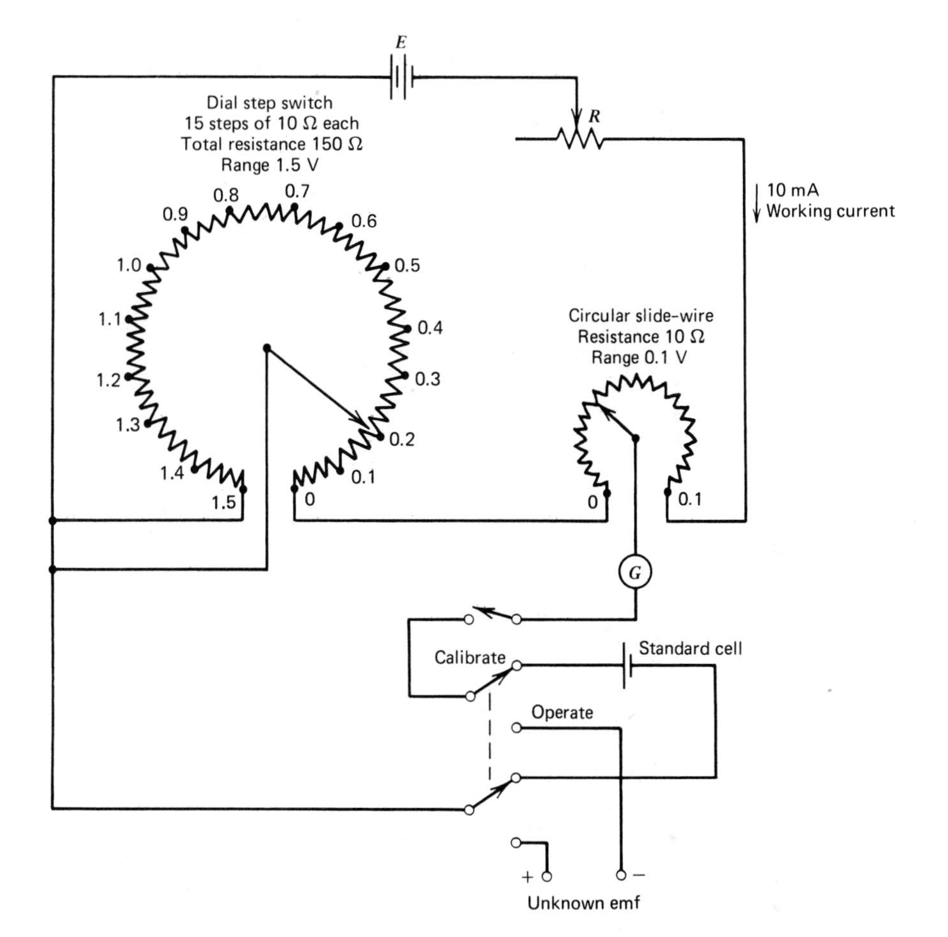

FIGURE 4-5 **Circuit diagram of a simple slide-wire potentiometer that uses dial resistors.**

and the total meter resistance which is

$$R_m = R_{\text{dial}} + R_{\text{slide-wire}} = \left(15 \text{ steps} \times \frac{10\ \Omega}{\text{step}}\right) + 10\ \Omega$$

$$= 160\ \Omega$$

The measurement range can now be computed as

$$\text{Measurement range} = 10 \text{ mA} \times 160\ \Omega = 1.6 \text{ V}$$

(b) The voltage drop across the slide-wire resistance of $10\ \Omega$ is

$$V = IR = 10 \text{ mA} \times 10\ \Omega = 0.1 \text{ V}$$

Therefore, each turn represents 0.1 V/10 Ω = 10 mV. Each scale division represents 1/100 × 10 mV = 100 μV. The resolution of the instrument is therefore

$$\text{Res} = \tfrac{1}{5} \times 100\ \mu\text{V} = 20\ \mu\text{V}$$

(c) The working current is

$$I = \frac{0.1\ \text{V}}{10\ \Omega} = 10\ \text{mA}$$

(d) The rheostat setting is

$$R = \frac{E - V_{\text{dial}} - V_{\text{slide-wire}}}{I}$$

$$= \frac{3\ \text{V} - 1.6\ \text{V}}{10\ \text{mA}} = 140\ \Omega$$

4-4 VOLTAGE REFERENCES

The development of solid-state devices such as constant-voltage (zener diode) and constant-current, or regulating, diodes, makes possible the design of standard voltage sources with a wide range of voltages. Thus, measurements have become more convenient. A **zener diode** is a diode to which the source voltage is applied in the reverse-biased mode. As the bias is increased in amplitude, a level is reached at which voltage remains essentially constant over a large range of bias current, A representative characteristic curve is shown in Fig. 4-6. A typical circuit for such a constant-voltage source is shown in Fig. 4-7.

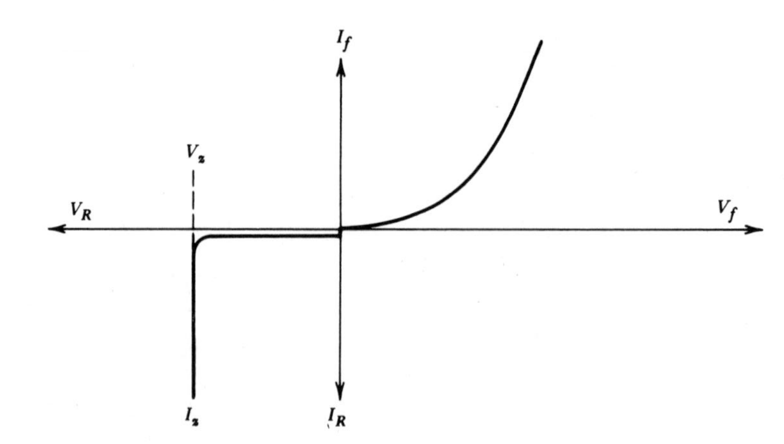

FIGURE 4-6 A representative characteristic curve for a zener diode.

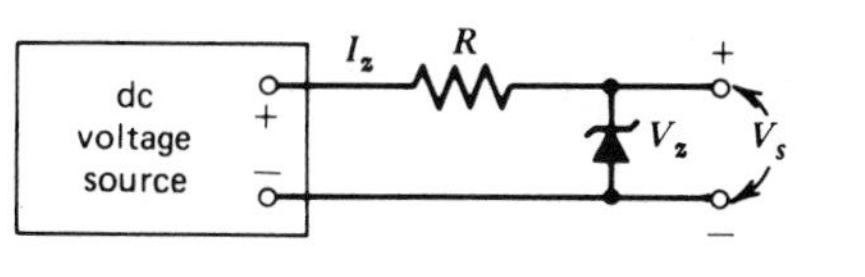

FIGURE 4-7 **Basic circuit for a constant-voltage source.**

In the design of such a circuit, the characteristic parameters of the zener diode must be available. The parameters that must be known are the specified zener current and the required zener voltage. Once the zener diode has been selected, the design of the circuit consists of determining the required resistance value of R to limit the zener current to a predetermined value, its suitable power rating, and the power rating of the zener diode.

The first step in the design is to write a voltage loop equation. According to Kirchhoff's voltage law, the equation for the circuit is

$$-V_{dc} + I_z R + V_z = 0 \tag{4-7}$$

from which we get

$$R = \frac{V_{dc} - V_z}{I_z} \tag{4-8}$$

where

R is the resistance
V_{dc} is the supply voltage
V_z is the zener operating voltage
I_z is the zener operating current

It is obvious from Eq. 4-8 that V_{dc} must be greater than V_z if R is to be a real value.

The power rating of the resistor can be calculated from the equation

$$P = \frac{2(V_{dc} - V_z)^2}{R} \tag{4-9}$$

The factor 2 is included to provide a safety factor.

EXAMPLE 4-3 Design a standard voltage source of 10 V when the supply voltage is 15 V.

Solution Choose a 10-V zener diode. A possible choice is the 1N 961. The power dissipation rating of the device is 400 mW, and it has a zener rating of 12.5 mA. Inserting these values into Eq. 4-8 gives

$$R = \frac{V_{dc} - V_z}{I_z}$$

$$= \frac{15\text{ V} - 10\text{ V}}{12.5 \times 10^3\text{ A}} = 400\ \Omega$$

The power rating of the resistor can be calculated using Eq. 4-9:

$$P(\text{watts}) = \frac{2(V_{dc} - V_z)^2}{R}$$

$$P = \frac{2(15\text{ V} - 10\text{ V})^2}{400\ \Omega} = 0.125\text{ W}$$

The selection of a $\frac{1}{8}$-W resistor would suffice.

An alternative design of a standard voltage source design is accomplished by maintaining a constant current through a fixed value as shown in Fig. 4-8. In this system the current is maintained by the constant-current device. **Constant-current diodes**, as well as other devices that will satisfy the requirements, are available. One such device is the **field-effect transistor.** The I_D–V_{DS} characteristics are such that the drain current is essentially constant over a relatively large change in drain-source voltage.

A convenient circuit connection of such an arrangement is to connect gate to source directly. In this case the bias is clamped at 0 V, eliminating the need for a high-quality bias supply. Such a practical circuit is shown in Fig. 4-9. The equation of interest in generating the voltage is

$$V_s = I_D R$$

The drain current, I_D, can be expressed in terms of the change in the drain-source voltage V_{DS} and in the drain resistance R_D.

$$I_D = \frac{\Delta V_{DS}}{\Delta R_D}$$

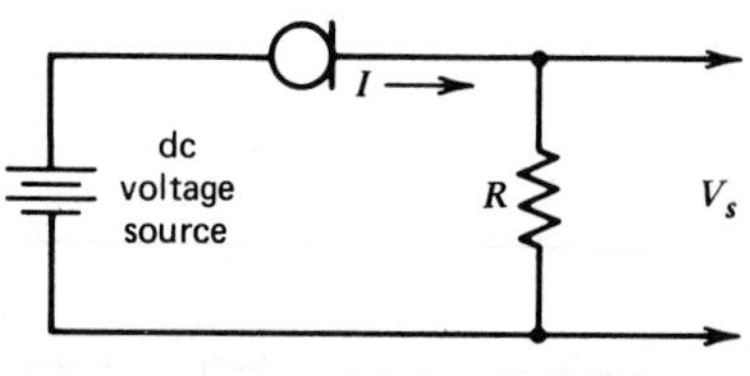

FIGURE 4-8 Standard voltage source maintaining a constant current.

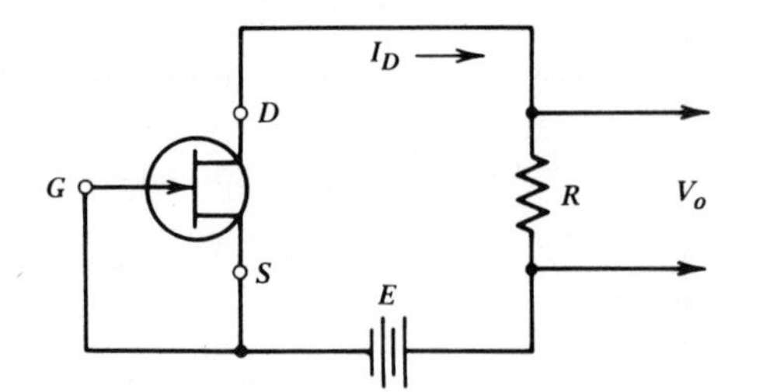

FIGURE 4-9 Standard voltage source using a field-effect transistor.

or

$$R_D = \frac{\Delta V_{DS}}{\Delta I_D} \tag{4-10}$$

The dynamic resistance R_D is very large for a field-effect transistor; therefore, the drain current I_D is essentially constant.

4-4.1 The Volt Box

The voltage range of a basic potentiometer is usually on the order of 1 to 3 V and sometimes even less. This feature limits its use to small-amplitude voltages. The range of a potentiometer can be increased very easily and conveniently by using a device called a **volt box**. The schematic of this device is shown in Fig. 4-10.

The circuit is rather self-explanatory in that a divider resistor is tapped at the appropriate location, so that only a small fraction of the total unknown voltage is applied to the potentiometer. Obviously, some current must be drawn from the unknown for the system to function. Therefore, $R_1 + R_2$ should be large enough to ensure that current drain is a minimum. Since the potentiometer itself draws no current, I_1 will equal I_2 in the balanced condition.

As seen in Fig. 4-10, V_2, which is the voltage applied to the potentiometer, is

$$V_2 = I_2 R_2 = I_1 R_2$$

but

$$I_1 = I_2 = I_T$$

and

$$V_2 = \frac{R_2}{R_1 + R_6} \times V_x \tag{4-11}$$

The volt box ratio is

$$\frac{V_x}{V_2} = \frac{R_1 + R_2}{R_2} \tag{4-12}$$

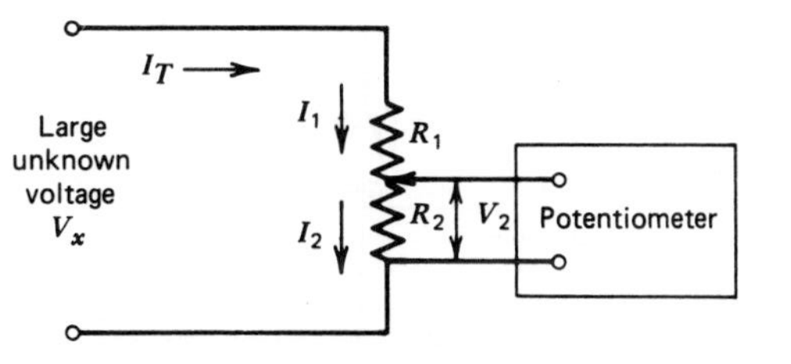

FIGURE 4-10 Basic volt-box circuit.

The divider resistor and the division ratio must be accurately known. Once known, the unknown voltage can be computed as

$$V_x = \frac{V_2(R_1 + R_2)}{R_2}$$

EXAMPLE 4-4 A volt box should be so designed that when an unknown voltage of 100 V is applied to the input terminals, an output voltage of 2 V is available at the volt box output terminals. An additional requirement is that $R_1 + R_2$ have a high value of 10 megohms. Determine the volt-box ratio and the relationship between R_1 and R_2.

Solution From Eq. 4-12 we see that

$$\frac{R_1 + R_2}{R_2} = \frac{V_x}{V_2} = \frac{100}{2}$$

or

$$\frac{R_1 + R_2}{R_2} = 50$$

Therefore,

$$R_1 + R_2 = 50R_2 \quad \text{or} \quad R_1 = 49R_2$$

4-5 APPLICATIONS

Two of the most frequent applications of potentiometers are for calibrating dc voltmeters and for measuring temperature, when used in conjunction with a thermocouple.

A dc voltmeter (or an electrodynamometer type of ac voltmeter) can be calibrated with the circuit shown in Fig. 4-11. If the voltmeter is receiving its initial calibration, the applied voltage is adjusted to exactly 1 V_{dc}, or a multiple of one volt, with the variable resistors. When the potentiometer

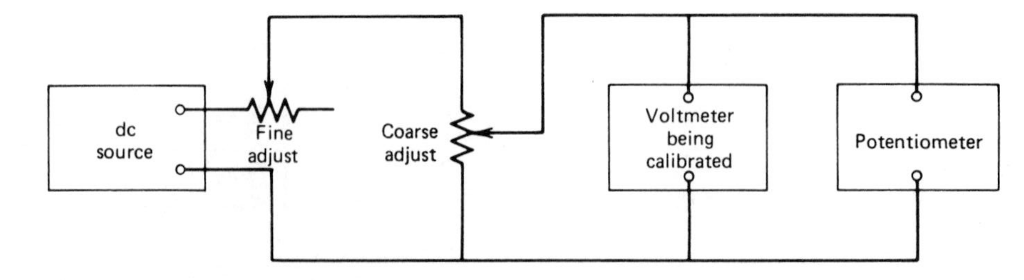

FIGURE 4-11 Circuit for calibrating a voltmeter with a potentiometer.

indicates the desired voltage, a major division mark is placed on the voltmeter scale. This procedure is repeated to obtain all the major division marks, and then intermediate marks are interpolated.

If a check of the calibration of a voltmeter is desired, the applied voltage is adjusted until the meter pointer rests at the first major division on the scale. The potentiometer is then adjusted for a null indication to provide a reading of the correct voltage. The correction factor for this point is the correct voltage value, determined with the potentiometer, minus the scale reading. The supply voltage is then adjusted for scale readings at each of the major divisions, and the above procedure with the potentiometer is repeated at each major division. After the readings have been taken at each of the major divisions with increasing voltage and then with decreasing voltage, a **calibration curve**, or correction curve, is plotted. Correction is plotted on the ordinate and scale reading along the abscissa as shown in Fig. 4-12.

The observed data points are connected by straight lines since nothing is known about the meter calibration between these points. Notice that the correction just described is the quantity that must be added to the observed value to obtain the true value.

A second very interesting application of potentiometers is for high-temperature measurement. In many high-temperature environments thermocouples are the most practical way to measure the temperature. Thermocouples, which are made by joining dissimilar metals, develop an electromotive force that is proportional to the temperature of the junction. This small potential difference can be measured very accurately with a potentiometer and then converted to a temperature reading. In many industrial applications in which high temperatures are involved, the measuring instrument used in conjunction with a thermocouple must be some distance from the thermocouple because of the heat. It is impractical to use an ordinary voltmeter with a d'Arsonval meter movement as the measuring instrument because the current required by the meter movement would cause an appreciable voltage drop across the resistance of the thermocouple junction as well as across the resistance of the interconnecting leads. Since, at null, a potentiometer draws no current from the circuit under test, it is ideal for such measurements.

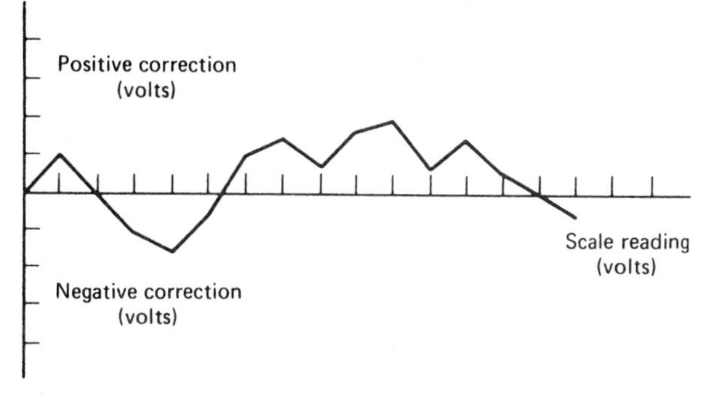

FIGURE 4-12 Typical calibration curve.

Thermocouple manufacturers provide graphs of thermocouple voltage versus temperature. By using these graphs you can easily convert the measured voltage to temperature.

EXAMPLE 4-5

What temperature is the thermocouple in Fig. 4-13 sensing if the current through the potentiometer galvanometer is 375 μA flowing from the thermocouple toward the potentiometer?

Solution

The current through the galvanometer is due, in part, to the thermocouple emf and, in part, to the voltage source within the potentiometer. The best approach to solving the problem is to find Thévenin's equivalent circuit to the right of points A–A'. Thévenin's equivalent voltage is computed as

$$V_{\text{Th}} = 6\text{ V} \times \frac{40\ \Omega}{1500\ \Omega}$$

Thévenin's equivalent resistance is computed as

$$R_{\text{Th}} = 40\ \Omega \| 1460\ \Omega = 38.93\ \Omega$$

Replacing the original circuit to the right of points A–A' with Thévenin's equivalent circuit, we obtain the total circuit shown in Fig. 4-14. From this circuit we can say

$$\begin{aligned} V_{\text{TC}} &= V_{\text{Th}} + (R_g + R_{\text{Th}})(I_g) \\ &= 160\text{ mV} + (200\ \Omega + 38.93\ \Omega) \times 375\ \mu\text{A} \\ &= 160\text{ mV} + 90\text{ mV} = 250\text{ mV} \end{aligned}$$

Figure 4-13 shows that, if the thermocouple voltage is 250 mV, the temperature being sensed by the thermocouple is 450°C.

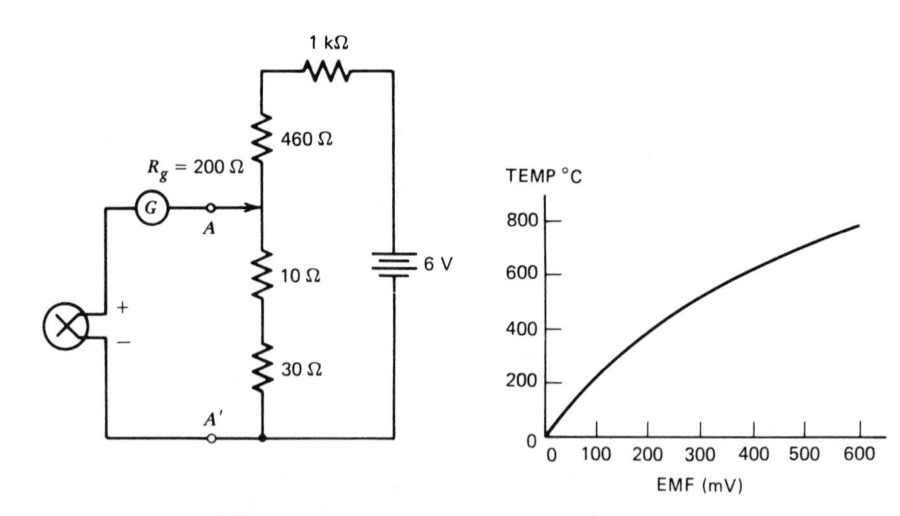

FIGURE 4-13 Circuit and graph for Example 4-3.

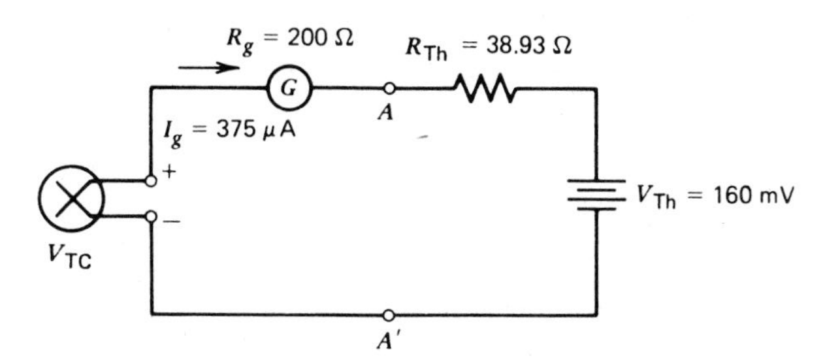

FIGURE 4-14 **Thévenin's equivalent circuit for the potentiometer in Fig. 4-13.**

4-6 SUMMARY

A potentiometer may be used in measurement applications that cannot tolerate circuit loading by the measuring instrument. Because the potentiometer draws no current from the circuit under test, circuit loading is eliminated. The primary applications for the potentiometer are for precise measurement of low-amplitude voltages and for calibration of voltmeters and ammeters.

A volt box is frequently used with the potentiometer to extend its range. This makes it possible to measure higher-amplitude voltage sources than can be measured with the potentiometer alone. Another way to increase the range of the potentiometer is to use a higher-amplitude referenced voltage source. Electronic components such as zener diodes, constant-current diodes, and field-effect transistors have made it possible to design accurate, higher-amplitude reference voltage sources.

4-7 GLOSSARY

Calibration curve: A graph obtained by plotting ammeter or voltmeter readings along the abscissa and positive or negative correction, determined with a potentiometer, along the ordinate.

Constant-current diode: A semiconductor device that within certain limits supplies a constant current regardless of external conditions. The device is very similar to a field-effect transistor with the gate connected to the source.

Field-effect transistor (FET): A voltage-operated semiconductor device, unlike bipolar transistors which are operated through current. There are two major categories of field-effect transistors: junction FETs and insulated-gate FETs.

Potentiometer: A precision instrument that operates on the null principle. Its primary application is for measuring unknown voltages.

Volt box: A precision voltage divider network, sometimes called a resistance ladder, used in conjunction with a potentiometer to extend its range.

Zener diode: A *p–n* junction device designed specifically to operate in the reverse breakdown region.

4-8 REVIEW QUESTIONS

The following questions should be answered after a thorough study of the chapter. The purpose of the questions is to determine the reader's comprehension of the material.

1. What is the principle of operation of a potentiometer circuit?
2. When is it desirable to use a potentiometer in a measurement application?
3. What is a volt box?
4. When should a volt box be used?
5. Why should the total resistance of a volt box be very high?
6. How is it possible to obtain a voltage measurement without drawing current from the circuit under test?
7. What problem, caused by a deflection-type instrument, is eliminated when making a voltage measurement with a potentiometer?
8. How can an internal voltage drop, IR_s, be made small?

4-9 PROBLEMS

4-1 The emf of a standard cell is measured with a potentiometer that gives a reading of 1.3562 V. When a 1-MΩ resistor is connected across the standard cell terminals, the potentiometer reading drops to 1.3560 V. Calculate the internal resistance of the standard cell.

4-2 A standard cell has an emf of 1.356 V and in internal resistance of 150 Ω. A dc voltmeter with a full-scale range of 3 V and an internal resistance of 30 kΩ is connected across the standard cell. Calculate the following.

(a) The voltmeter reading.
(b) The current drawn from the standard cell.
(c) The internal resistance of the voltmeter which would limit the standard cell current to 20 μA.

4-3 The potentiometer of Fig. 4-4 has a working battery with terminal voltage of 4.0 V. The 200-cm slide-wire has a resistance of 100 Ω, and the internal resistance of the galvanometer is 50 Ω. The standard cell has an emf of 1.356 V. The rheostat is adjusted so that the potentiometer is calibrated with the slider set at the 135.6-cm mark on the slide-wire. Calculate the following.

(a) The working current.
(b) The resistance of the rheostat setting.
(c) The current through the standard cell, if the connections to the standard cell are accidentally reversed.

4-4 The potentiometer of Problem 4-3 is standardized and then used to measure an unknown voltage. The slider is now set at the 76.3-cm mark on the slide-wire for a galvanometer null. Calculate the unknown voltage.

4-5 The potentiometer circuit of Fig. 4-5 has a dial step switch of 15 steps of 5 Ω each. The 11-turn slide-wire of total resistance 5.5 Ω is in series with the working battery and a rheostat. The maximum range of the instrument is 1.561 V. The galvanometer has an internal resistance of 50 Ω. The circular slide-wire has 100 divisions, and interpolation can be made to one-fifth of a division. Calculate the following

(a) The working current.

(b) The resistance setting of the rheostat.

(c) The resolution of the instrument.

4-6 A slide-wire potentiometer measures a voltage drop of 1 V dc across a resistor in a circuit. A dc voltmeter with a sensitivity of 20 kΩ/V now measures 0.5 V on its 2.5-V scale. Calculate the resistance of the resistor.

4-7 A volt box has a total resistance of 1 MΩ and its tapped at 5000 Ω. By what factor should the potentiometer reading be multiplied?

4-8 In a volt-box measurement the potentiometer voltage is 3 V. If R_2 is 10,000 Ω and the total resistance $R_1 + R_2$ is 100,000 Ω, what is the unknown voltage?

4-9 Design a volt box so that it has a total resistance of 5 MΩ and a ratio of 10:1.

4-10 The unknown voltage is 150 V and the potentiometer volt box is 3 V. What is the volt-box ratio?

4-11 Design a 7.5-V standard voltage source using a zener diode if the supply voltage is 20 V. Use a 1N702 diode.

4-12 Explain the operation of a potentiometer so that no current is drawn from the source.

4-13 To what resistance value should R_v be set for a null indication in Fig. 4-15?

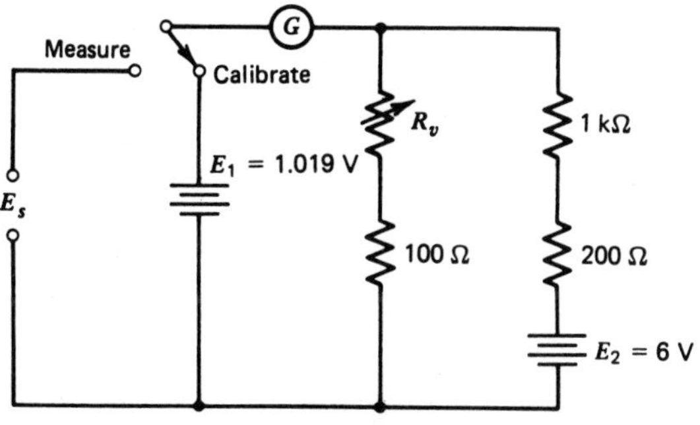

FIGURE 4-15 Circuit for Problem 4-13.

4-14 Determine the value of E for a null indication in the circuit in Fig. 4-16.

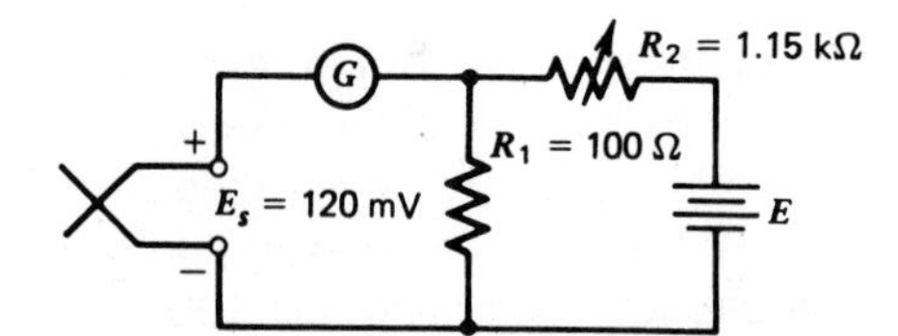

FIGURE 4-16 Circuit for Problem 4-14.

4-15 Determine the internal resistance of the galvanometer if 200 μA is flowing through it in the circuit of Fig. 4-17.

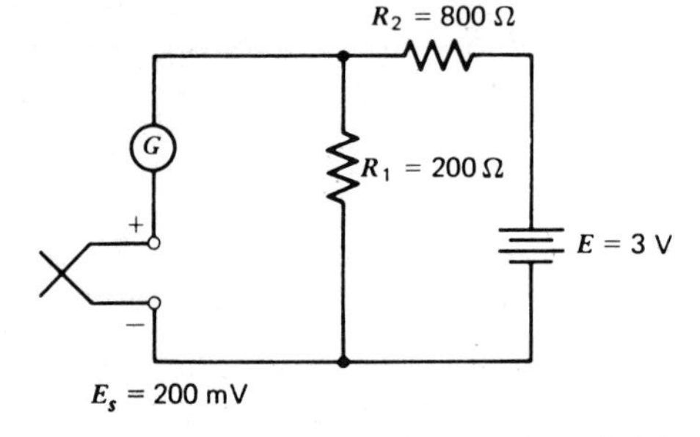

FIGURE 4-17 Circuit for Problem 4-15.

4-10 LABORATORY EXPERIMENTS

Experiments E7 and E8, which apply the theory presented in Chapter 4, are intended to provide students with hands-on experience, which is essential for a thorough understanding of the concepts involved.

Experiment E7 requires a basic potentiometer; however, if necessary, a satisfactory circuit can be constructed with a galvanometer. The contents of the laboratory report to be submitted by each student are listed at the end of each experimental procedure.

CHAPTER 5

Direct-Current Bridges

5-1 Instructional Objectives

Chapter 5 discusses the basic theory and applications of direct-current bridges. Primary attention is given to the principle of operation as well as to measurement and control applications and to the recent use of digital circuitry in bridges. After completing the chapter you should be able to

1. List and discuss the principal applications of Wheatstone bridges.
2. Describe the operation of the Wheatstone bridge.
3. List and discuss the principal applications of Kelvin bridges.
4. Describe the operation of the Kelvin bridge.
5. Solve for Thévenin's equivalent circuit for an unbalanced Wheatstone bridge.
6. Describe how a Wheatstone bridge may be used to control various physical parameters.
7. Define the term *null* as it applies to bridge measurements.
8. List an advantage of comparison-type measurements over deflection-type measurements.
9. Describe the difference between using minicomputers and microprocessors in test equipment.
10. Describe how microprocessors are being used in test equipment.

5-2 INTRODUCTION

Bridge circuits, which are instruments for making **comparison measurements**, are widely used to measure resistance, inductance, capacitance, and impedance.

Bridge circuits operate on a **null-indication** principle. This means the indication is *independent* of the calibration of the indicating device or any characteristics of it. For this reason, very high degrees of accuracy can be achieved using the bridges.

Bridge circuits are also frequently used in *control* circuits. When used in such applications, one arm of the bridge contains a resistive element that is sensitive to the physical parameter (temperature, pressure, etc.) being controlled.

This chapter discusses basic bridge circuits and their applications in measurement and control. It also introduces some very recent concepts regarding the use of digital principles in bridges.

5-3 THE WHEATSTONE BRIDGE

The **Wheatstone bridge** consists of two parallel resistance branches with each branch containing two series elements, usually resistors. A dc voltage source is connected across this resistance network to provide a source of current through the resistance network. A *null detector,* usually a **galvanometer**, is connected between the parallel branches to detect a condition of **balance**. This circuit, shown in Fig. 5-1, was first devised by S. H. Christie in 1833. However, it was little used until 1847 when Sir Charles Wheatstone, for whom the circuit is named, recognized its possibilities as a very accurate means of measuring resistance.

The Wheatstone bridge has been in use longer than almost any other electrical measuring instrument. It is still an accurate and reliable instrument and is heavily used in industry. Accuracy of 0.1% is quite common with the Wheatstone bridge as opposed to 3% to 5% error with the ordinary ohmmeter for resistance measurement.

In using the bridge to determine the value of an unknown resistor, say R_4, we vary one of the remaining resistors until the current through the null detector decreases to zero. The bridge is then in a balanced condition, which means the voltage across resistor R_3 is equal to the voltage drop across R_4. Therefore, we can say that

$$I_3R_3 = I_4R_4 \tag{5-1}$$

At balance the voltage drops across R_1 and R_2 must also be equal; therefore,

$$I_1R_1 = I_2R_2 \tag{5-2}$$

Since no current flows through the galvanometer G when the bridge is

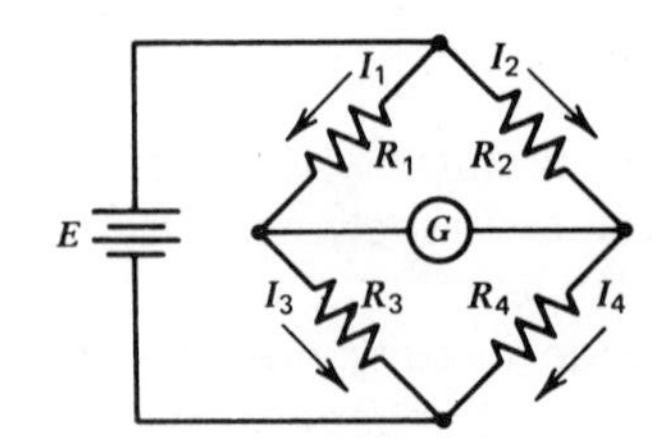

FIGURE 5-1 Wheatstone bridge circuit.

balanced, we can say that

$$I_1 = I_3$$

and

$$I_2 = I_4$$

Substituting I_1 for I_3 and I_2 for I_4 in Eq. 5-1 yields the following:

$$I_1 R_3 = I_2 R_4 \tag{5-3}$$

Now, if we divide Eq. 5-2 by Eq. 5-3, we obtain

$$\frac{R_1}{R_3} = \frac{R_2}{R_4}$$

This can be rewritten as

$$R_1 R_4 = R_2 R_3 \tag{5-4}$$

Equation 5-4 states the conditions for balance of a Wheatstone bridge and is useful for computing the value of an unknown resistor once balance has been achieved.

EXAMPLE 5-1

Determine the value of the unknown resistor, R_x, in the circuit of Fig. 5-2 assuming a null exists (current through the galvanometer is zero).

Solution

We see in Eq. 5-4 that the products of the resistance in opposite arms of the bridge are equal at balance. Therefore,

$$R_x R_1 = R_2 R_3$$

Solving for R_x yields

$$R_x = \frac{R_2 R_3}{R_1}$$

$$= \frac{15\ \text{k}\Omega \times 32\ \text{k}\Omega}{12\ \text{k}\Omega} = 40\ \text{k}\Omega$$

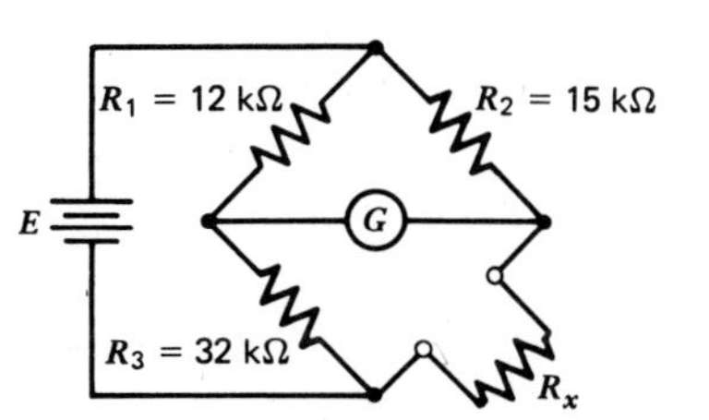

FIGURE 5-2 Circuit for Example 5-1.

5-4 SENSITIVITY OF THE WHEATSTONE BRIDGE

When the bridge is in an unbalanced condition, current flows through the galvanometer, causing a deflection of its pointer. The amount of deflection is a function of the sensitivity of the galvanometer. We might think of **sensitivity** as *deflection per unit current*. This means that a more sensitive galvanometer deflects a greater amount for the same current. Deflection may be expressed in linear or angular units of measure. Sensitivity S can be expressed in units of

$$S = \frac{\text{millimeters}}{\mu\text{A}} \text{ or } \frac{\text{degrees}}{\mu\text{A}} \text{ or } \frac{\text{radians}}{\mu\text{A}} \tag{5-5}$$

Therefore, it follows that total deflection D is

$$D = S \times I$$

where S is as defined and I is the current in microamperes (μA). We might naturally question how to determine the amount of deflection that will result from a particular degree of unbalance.

Although general circuit analysis techniques can be applied to this kind of problem, in our approach we will make frequent use of **Thévenin's theorem**. Since our interest is in finding the current through the galvanometer, we want to find Thévenin's equivalent circuit for the bridge as seen by the galvanometer. Thévenin's equivalent voltage is found by removing the galvanometer from the bridge circuit as shown in Fig. 5-3 and computing the "open-circuit" voltage between terminals a and b. Applying the voltage divider equation permits us to express the voltage at point a as

$$V_a = E \frac{R_3}{R_1 + R_3}$$

and the voltage at point b as

$$V_b = E \frac{R_4}{R_2 + R_4}$$

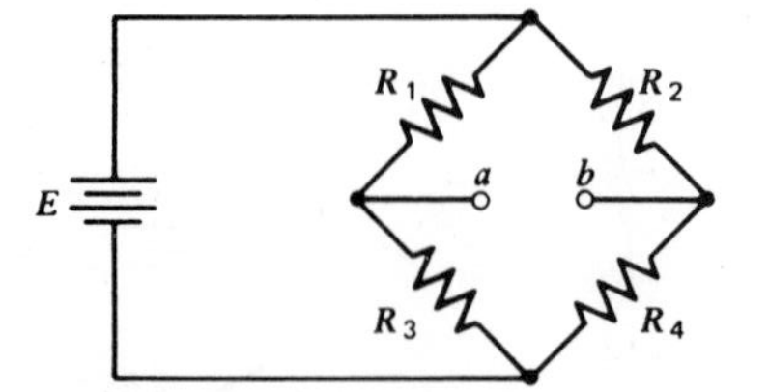

FIGURE 5-3 **Wheatstone bridge with the galvanometer removed to facilitate computation of Thévenin's equivalent voltage.**

The difference in V_a and V_b represents Thévenin's equivalent voltage. That is,

$$V_{Th} = V_a - V_b = E\frac{R_3}{R_1 + R_3} - E\frac{R_4}{R_2 + R_4}$$

$$= E\left(\frac{R_3}{R_1 + R_3} - \frac{R_4}{R_2 + R_4}\right) \tag{5-6}$$

Thévenin's equivalent resistance is found by replacing the voltage source with its internal resistance and computing the resistance seen looking back into the bridge at the terminals from which the galvanometer was removed. Since the internal resistance of the voltage source E is assumed to be very low, we will treat it as $0\,\Omega$ and redraw the bridge as shown in Fig. 5-4 to facilitate computation of the equivalent resistance. The equivalent resistance of the circuit in Fig. 5-4 is calculated as $R_1 \| R_3 + R_2 \| R_4$ or

$$R_{Th} = \frac{R_1 R_3}{R_1 + R_3} + \frac{R_2 R_4}{R_2 + R_4} \tag{5-7}$$

Thévenin's equivalent circuit for the bridge, as seen looking back into the bridge from terminals a and b in Fig. 5-3, is shown in Fig. 5-5. A galvanometer connected between the output terminals a and b of a Wheatstone bridge (Fig. 5-3) or its Thévenin's equivalent circuit (Fig. 5-5) will experience the same deflection. The magnitude of the current is limited by both Thévenin's equivalent resistance and any resistance connected between terminals a and b. The resistance between terminals a and b generally consists of only the resistance of the galvanometer R_g. The deflection current in the galvanometer is

$$I_g = \frac{V_{Th}}{R_{Th} + R_g} \tag{5-8}$$

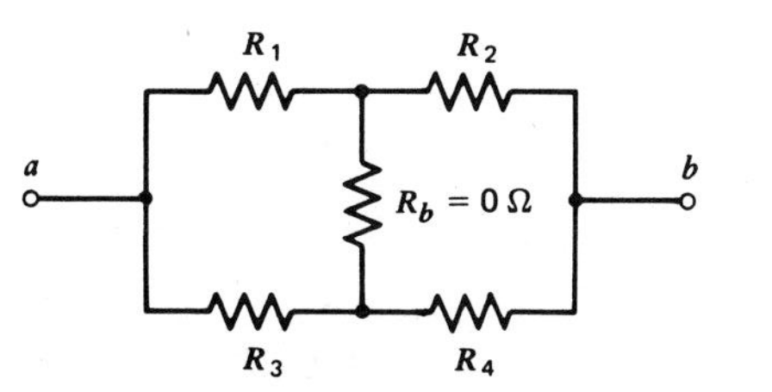

FIGURE 5-4 Circuit for finding Thévenin's equivalent resistance.

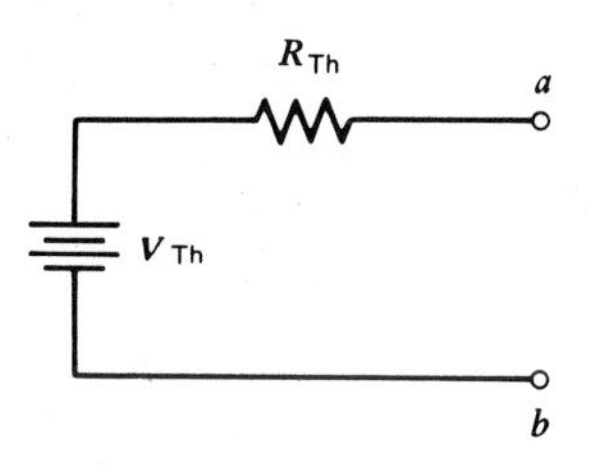

FIGURE 5-5 Thévenin's equivalent circuit for an unbalanced Wheatstone bridge.

EXAMPLE 5-2 Calculate the current through the galvanometer in the circuit of Fig. 5-6.

Solution The easiest way to solve for the current is to find Thévenin's equivalent circuit for the bridge as seen by the galvanometer. Thévenin's equivalent voltage is calculated as follows:

$$
\begin{aligned}
V_{Th} &= E\left(\frac{R_3}{R_3+R_1}-\frac{R_4}{R_4+R_2}\right) \\
&= 6\ \text{V} \times \left(\frac{3.5\ \text{k}\Omega}{3.5\ \text{k}\Omega + 1\ \text{k}\Omega}-\frac{7.5\ \text{k}\Omega}{7.5\ \text{k}\Omega + 1.6\ \text{k}\Omega}\right) \\
&= (6\ \text{V})(0.778-0.824) = 0.276\ \text{V}
\end{aligned}
$$

Thévenin's equivalent resistance is computed as

$$
\begin{aligned}
R_{Th} &= \frac{R_1 R_3}{R_1+R_3}+\frac{R_2 R_4}{R_2+R_4} \\
&= \frac{1\ \text{k}\Omega \times 3.5\ \text{k}\Omega}{1\ \text{k}\Omega + 3.5\ \text{k}\Omega}+\frac{1.6\ \text{k}\Omega \times 7.5\ \text{k}\Omega}{1.6\ \text{k}\Omega + 7.5\ \text{k}\Omega} = 2.097\ \text{k}\Omega
\end{aligned}
\tag{5-7}
$$

Thévenin's equivalent circuit can now be connected to the galvanometer as shown in Fig. 5-7. The current through the galvanometer is now calculated as

$$
\begin{aligned}
I_g &= \frac{V_{Th}}{R_{Th}+R_g} \\
&= \frac{0.276\ \text{V}}{2.097\ \text{k}\Omega + 200\ \Omega} = 120\ \mu\text{A}
\end{aligned}
$$

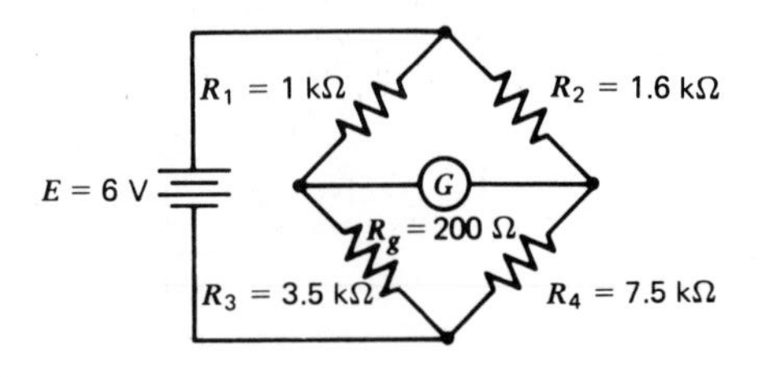

FIGURE 5-6 Unbalanced Wheatstone bridge.

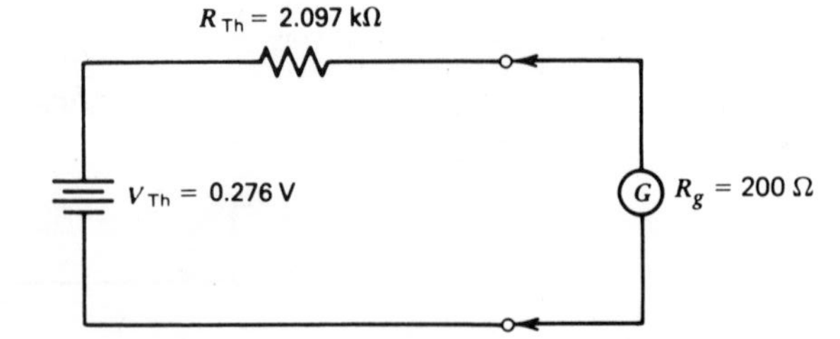

FIGURE 5-7 Thévenin's equivalent circuit for the unbalanced bridge of Fig. 5-6, connected to a galvanometer.

5-4.1 Slightly Unbalanced Wheatstone Bridge

If three of the four resistors in a bridge are equal in value to R and the fourth differs from R by 5% or less, we can develop an approximate but accurate expression for Thévenin's equivalent voltage and resistance.[1] Consider the circuit in Fig. 5-8. The voltage at point a is given as

$$V_a = E\frac{R}{R+R} = E\left(\frac{R}{2R}\right) = \frac{E}{2}$$

The voltage at point b is expressed as

$$V_b = E\frac{R+\Delta r}{R+R+\Delta r}$$

Thévenin's equivalent voltage is the difference in these voltages, or

$$V_{\text{Th}} = V_b - V_a = E\left(\frac{R+\Delta r}{R+R+\Delta r} - \frac{1}{2}\right) = E\left(\frac{\Delta r}{4R+2\Delta r}\right)$$

If Δr is 5% of R or less, then the Δr term in the denominator may be dropped without introducing appreciable error. If this is done, the expression for Thévenin's equivalent voltage simplifies to

$$V_{\text{Th}} \approx E\left(\frac{\Delta r}{4R}\right) \qquad (5\text{-}9)$$

Thévenin's equivalent resistance can be calculated by replacing the voltage source with its internal resistance (for all practical purposes a short circuit) and redrawing the circuit as seen from terminals a and b. The bridge, as it appears looking from the output terminals, is shown in Fig. 5-9. Thévenin's equivalent resistance is now calculated as

$$R_{\text{Th}} = \frac{R}{2} + \frac{(R)(R+\Delta r)}{R+R+\Delta r}$$

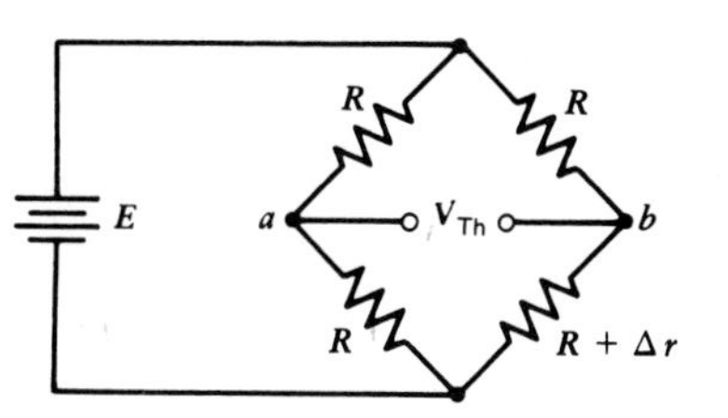

FIGURE 5-8 A Wheatstone bridge with three equal arms.

[1]This is an extremely practical circuit. It is commonly used in strain-gauge measurements. It is also used in transducers of control systems as an error detector. It is therefore important to develop equivalent voltage and resistance relations for it.

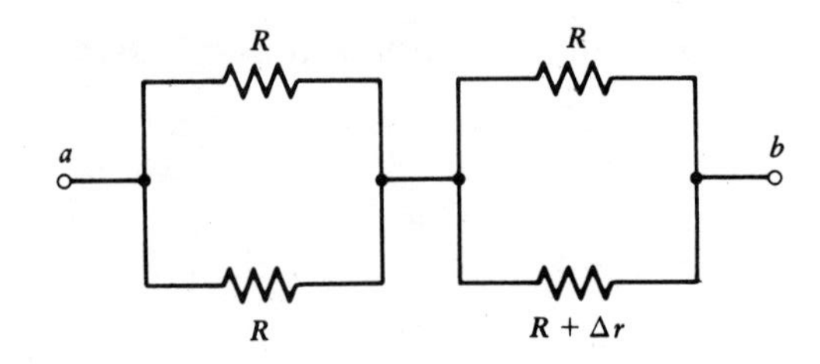

FIGURE 5-9 **Resistance of a Wheatstone bridge as seen from the output terminals *a* and *b*.**

Again, if Δr is small compared to R, the equation simplifies to

$$R_{\mathrm{Th}} \approx \frac{R}{2} + \frac{R}{2} \quad \text{or } R_{\mathrm{Th}} \approx R \tag{5-10}$$

Using these approximations, we have Thévenin's equivalent circuit as shown in Fig. 5-10. These approximations are about 98% accurate if $\Delta r \leq 0.05R$.

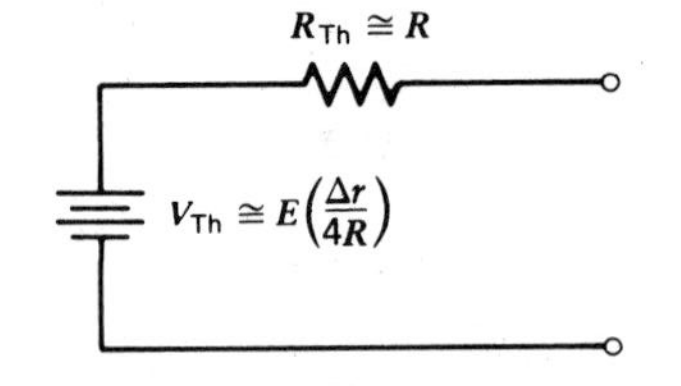

FIGURE 5-10 **An approximate Thévenin's equivalent circuit for a Wheatstone bridge containing three equal resistors and a fourth resistor differing by 5% or less.**

EXAMPLE 5-3

Use the approximation given in Eqs. 5-9 and 5-10 to calculate the current through the galvanometer in Fig. 5-11. The galvanometer resistance, R_g is 125 Ω and is a center-zero 200-0-200-μA movement.

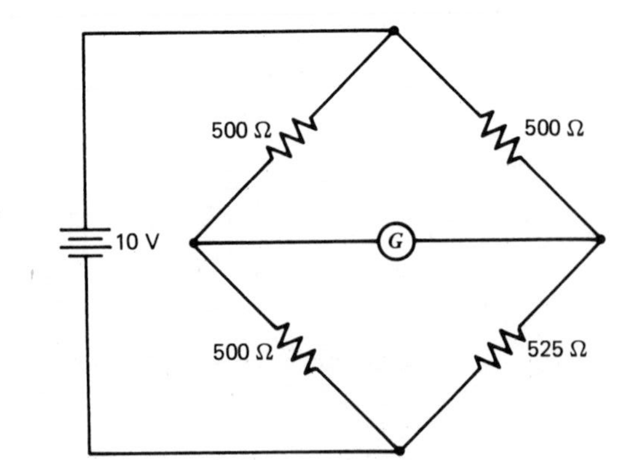

FIGURE 5-11 **Slightly unbalanced Wheatstone bridge.**

Solution

Thévenin's equivalent voltage is

$$V_{\text{Th}} = E\left(\frac{\Delta r}{4R}\right) = 10\text{ V} \times \frac{25\ \Omega}{2000\ \Omega} = 0.125\text{ V} \tag{5-9}$$

Thévenin's equivalent resistance is

$$R_{\text{Th}} \approx R \approx 500\ \Omega \tag{5-10}$$

The current through the galvanometer is

$$I_g = \frac{V_{\text{Th}}}{R_{\text{Th}} + R_g} + \frac{0.125\text{ V}}{500\ \Omega + 125\ \Omega} = 200\ \mu\text{A}$$

If the detector is a 200-0-200-μA galvanometer, we see that the pointer deflected full scale for a 5% change in resistance.

5-5 KELVIN BRIDGE

The **Kelvin bridge** (Fig. 5-12) is a modified version of the Wheatstone bridge. The purpose of the modification is to eliminate the effects of contact and lead resistance when measuring unknown low resistances. Resistors in the range of 1 Ω to approximately 1 $\mu\Omega$ may be measured with a high degree of accuracy using the Kelvin bridge. Since the Kelvin bridge uses a second set

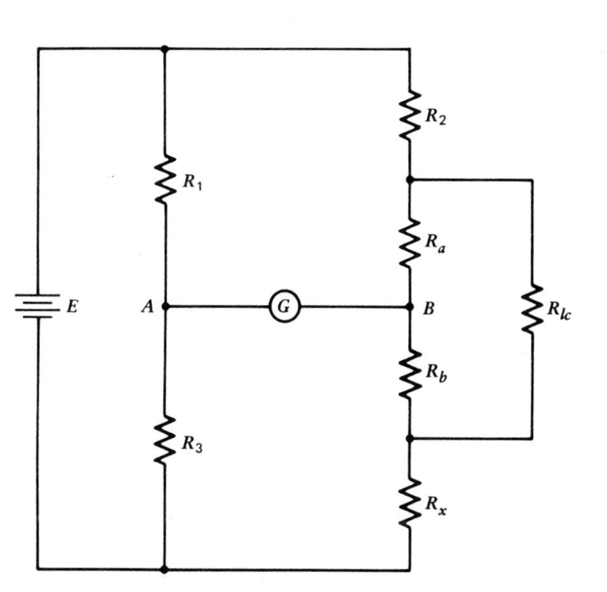

FIGURE 5-12 Basic Kelvin bridge showing a second set of ratio arms.

of ratio arms as shown in Fig. 5-12, it is sometimes referred to as the Kelvin double bridge.

The resistor R_{lc} shown in Fig. 5-12 represents the lead and contact resistance present in the Wheatstone bridge. The second set of ratio arms (R_a and R_b in Fig. 5-12) compensates for this relatively low lead–contact resistance. At balance the ratio of R_a to R_b must be equal to the ratio of R_1 to R_3. It can be shown (see Appendix A) that, when a null exists, the value for R_x is the same as that for the Wheatstone bridge, which is

$$R_x = \frac{R_2 R_3}{R_1}$$

This can be written as

$$\frac{R_x}{R_2} = \frac{R_3}{R_1}$$

Therefore, when a Kelvin bridge is balanced, we can say

$$\frac{R_x}{R_2} = \frac{R_3}{R_1} = \frac{R_b}{R_a} \tag{5-11}$$

EXAMPLE 5-4

If, in Fig. 5-12, the ratio of R_a to R_b is 1000, R_1 is 5 Ω, and $R_1 = 0.5R_2$, what is the value of R_x?

Solution

The resistance of R_x can be calculated using Eq. 5-11 as

$$\frac{R_x}{R_2} = \frac{R_a}{5\,\Omega} = \frac{1}{1000}$$

Since $R_1 = 0.5R_2$, the value of R_2 is calculated as

$$R_2 = \frac{R_1}{0.5} = \frac{5\,\Omega}{0.5} = 10\,\Omega$$

Now, we can calculate the value of R_x as

$$R_x = R_2\left(\frac{1}{1000}\right) = 10\,\Omega \times \frac{1}{1000} = 0.01\,\Omega$$

5-6 DIGITAL READOUT BRIDGES

The tremendous increase in the use of digital circuitry has had a marked effect on electronic test instruments. Early use of digital circuits in bridges was to provide digital readout. The actual measuring circuitry of the bridge remained the same. But operator error in observing the reading was eliminated by incorporating digital readout capabilities. The block diagram for a Wheatstone bridge with digital readout is shown in Fig. 5-13. Note that a logic circuit is used to provide a signal to R_3, sense the null, and provide a digital readout representing the value of R_x.

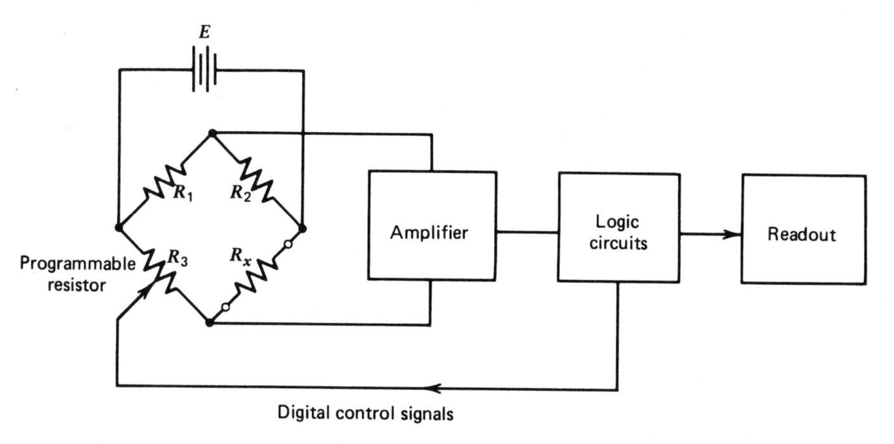

FIGURE 5-13 Block diagram for a Wheatstone bridge with a digital readout.

5-7 MICROPROCESSOR-CONTROLLED BRIDGES

Digital computers have been used in conjunction with test systems, bridges, process controllers, and in other applications for several years. In these applications computers are used to give instructions and perform operations on the measurement data. When the **microprocessor** was first developed, it was used in much the same way. However, real improvements in performance did not occur until microprocessors were truly integrated into the instrument. With this accomplished, not only do microprocessors give instructions about measurements, but they can also change the way in which measurements are made. This innovation has brought about a whole new class of instruments called *intelligent instruments*.

The complexity and cost of making analog measurements can be reduced using a microprocessor. This reduction of analog circuitry is important, even if additional digital circuitry must be added, because precision analog components are expensive. In addition, adjusting, testing, and troubleshooting analog circuits is time-consuming and costly. Often, digital circuits can replace analog circuits because various functions can be done either way.

The following are some of the ways in which microprocessors are reducing the cost and complexity of analog measurements.

- Replacing sequential control logic with stored control programs.
- Eliminating some auxiliary equipment by handling interfacing, programming, and other system functions.
- Giving wider latitude in the selection of measurement circuits, thereby making it possible to measure one parameter and calculate another parameter of interest.
- Reducing accuracy requirements by storing and applying correction factors.

Instruments in which microprocessors are an integral part can take the results of a measurement that is easiest to make in a circuit and then calculate and display the desired parameter, which may be much more difficult to measure directly. For example, conventional counters can measure the period of a low-frequency waveform. The frequency is then calculated by hand, or extensive circuitry is required to perform the required division. Such calculations are done very easily by the microprocessor.

Resistance and conductance are also reciprocals of each other. Some hybrid digital–analog bridges are designed to measure conductance by a current measurement. This measurement is then converted to a resistance value by rather elaborate circuitry. With a microprocessor-based instrument, a resistance value is easily obtained from the conductance measurement.

Many other similar examples could be presented. However, the important thing to remember is that the microprocessor is an integral part of the measuring instrument. As such, it produces an intelligent instrument that allows the choice of the easiest method of measurement and requires only one measurement circuit to obtain various results. Specifically, one quantity can be measured in terms of another, or several others, with completely different dimensions, and the desired results can be calculated with the microprocessor.

The General Radio, Model 1658 RLC Digibridge, shown in Fig- 5-14 is a microprocessor-based instrument. Such intelligent instruments represent a new era in impedance-measuring instruments. The following are some of the features of the instrument.

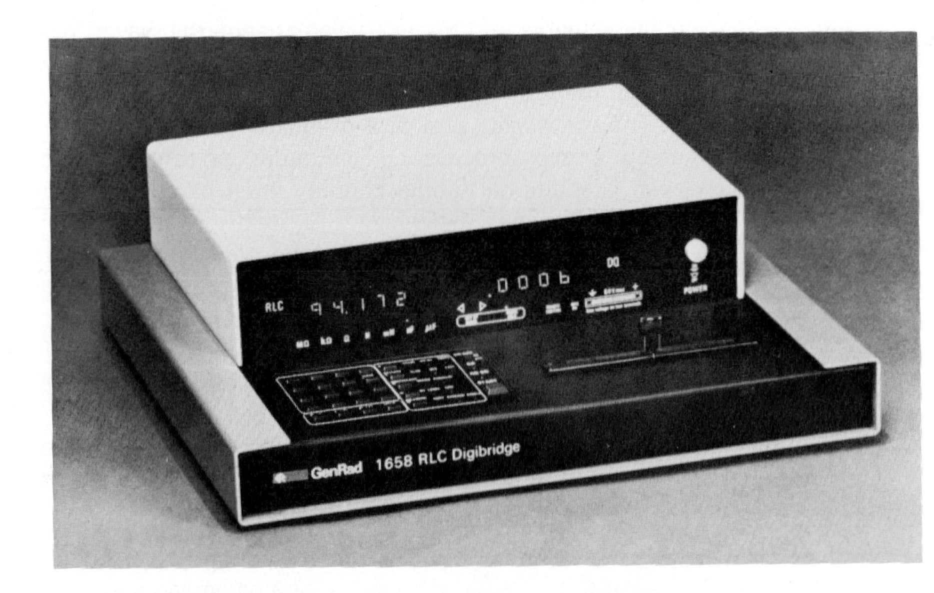

FIGURE 5-14 General Radio, Model 1658 RLC Digibridge. (Courtesy GenRad Inc., Concord, Mass.)

- Automatic measurement of resistance, *R*, inductance, *L*, capacitance, *C*, dissipation factor for capacitors, *D*, and storage factor for inductors, *Q*.
- 0.1% basic accuracy.
- Series or parallel measurement mode.
- Autoranging.
- No calibration ever required.
- Ten bins for component sorting and binning.
- Three test speeds.
- Three types of display-programmed bin limits, measured values, or bin number.

Most of these features are available as a result of the microprocessor. For example, the component sorting and binning feature is achieved by programming the microprocessor. When the instrument is used in this mode, bins are assigned a tolerance range. As a component is measured, a digital readout (labeled Bin No.) indicating the proper bin for the component is displayed on the keyboard control panel of Fig. 5-15. The theory of how the bridge circuit operates will be discussed in Chapter 13.

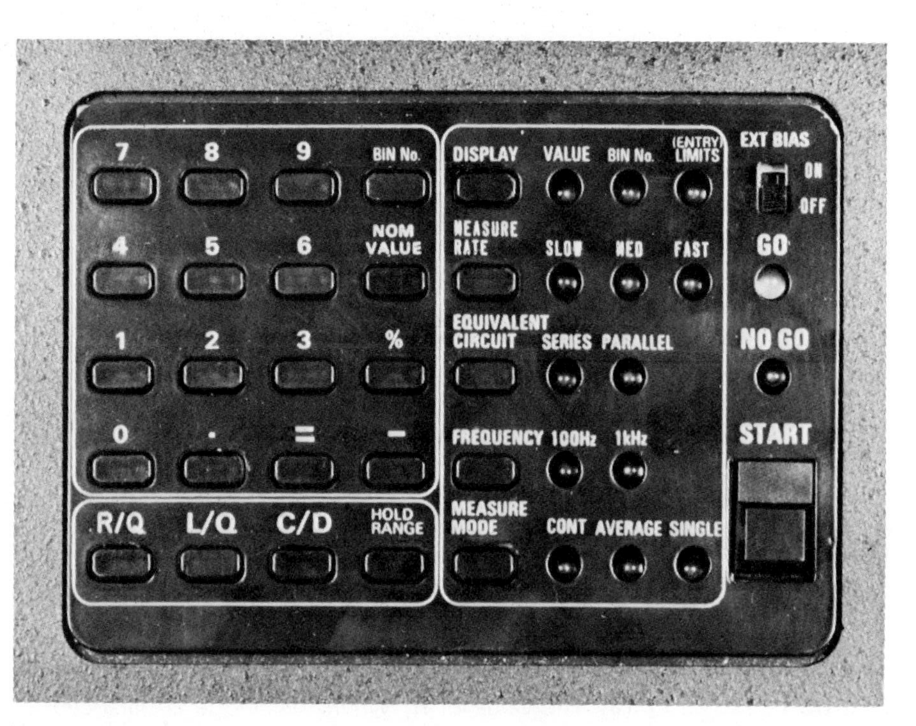

FIGURE 5-15 Control panel for Model 1658 Digibridge. (Courtesy GenRad Inc., Concord, Mass.)

5-8 BRIDGE-CONTROLLED CIRCUITS

We have seen that whenever a bridge is unbalanced, a potential difference exists at its output terminals. This potential difference causes current through a detector, such as a galvanometer, when the bridge is used as part of a measuring instrument. When a bridge is used as an error detector in a control circuit, the potential difference at the output of the bridge is called an error signal (see Fig. 5-16).

Passive circuit elements such as strain gauges, temperature-sensitive resistors (thermistors), or light-sensitive resistors (photoresistors) produce no output voltage. However, when they are used as one arm of a Wheatstone bridge, a change in their sensitive parameter (heat, light, pressure) produces a change in their resistance. This causes the bridge to be unbalanced, thereby producing an output voltage or an error signal.

Resistor R_v in Fig. 5-16 may be sensitive to one of many different physical parameters such as heat or light. If the particular parameter to which the resistor is sensitive is of such a magnitude that the ratio of R_2 to R_v equals the ratio of R_1 to R_3, then the error signal is zero. If the physical parameter changes, then R_v also changes. The bridge then becomes unbalanced and an error signal exists. In most control applications, the measured and controlled parameter is corrected, restoring R_v to the value that creates a null condition at the output of the bridge. Since R_v varies by only a small amount, Eqs. 5-9 and 5-10 generally apply. Because R_v is corrected rapidly back to the null, the amplitude of the error signal is normally quite low. Therefore, it is usually amplified before being used for control purposes.

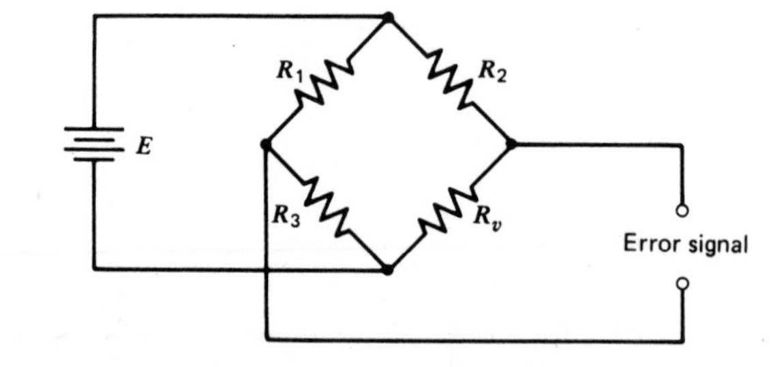

FIGURE 5-16 Wheatstone bridge error detector with R_v sensitive to some physical parameter.

EXAMPLE 5-5

Resistor R_v in Fig. 5-17*a* is temperature-sensitive, with the relation between its resistance and temperature as shown in Fig. 5-17*b*. Calculate

(a) At what temperature the bridge is balanced.
(b) The amplitude of the error signal at 60°C.

Solution

(a) The value of R_v when the bridge is balanced is calculated as

$$R_v = \frac{R_2 R_3}{R_1} = \frac{5\ \text{k}\Omega \times 5\ \text{k}\Omega}{5\ \text{k}\Omega} = 5\ \text{k}\Omega \qquad (5\text{-}4)$$

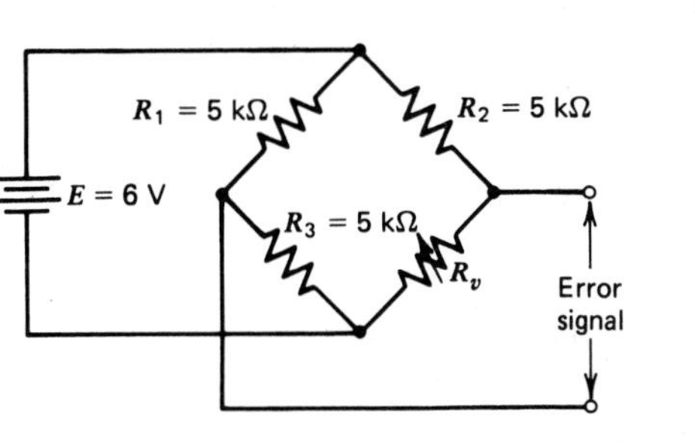

(a) Circuit

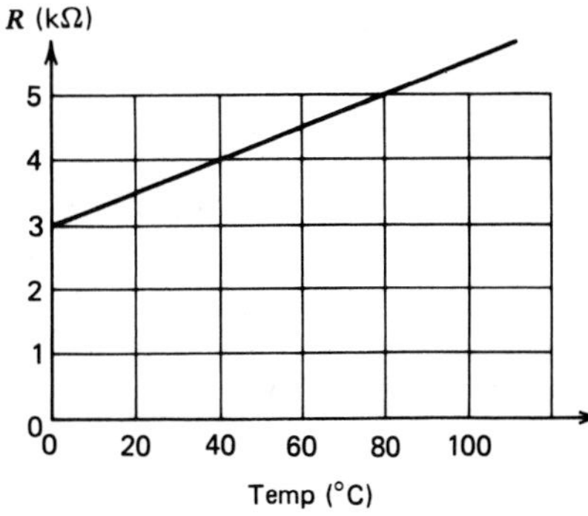

(b) Variation of R_v with temperature

FIGURE 5-17 Wheatstone bridge in which one arm (R_v) is temperature-sensitive.

The bridge is balanced when the temperature is 80°C. This is read directly from the graph of Fig. 5-17*b*.

(b) We can also determine, by reading directly from the graph, the resistance of R_v when its temperature is 60°C. This resistance value is 4.5 **kΩ**; therefore, the error signal, e_g, is

$$e_g = E\frac{R_3}{R_1 + R_3} - E\frac{R_v}{R_2 + R_v}$$

$$= 6\text{ V} \times \frac{5\text{ k}\Omega}{5\text{ k}\Omega + 5\text{ k}\Omega} - 6\text{ V} \times \frac{4.5\text{ k}\Omega}{5\text{ k}\Omega + 4.5\text{ k}\Omega} = 0.158\text{ V}$$

The error signal can also be determined using Eq. 5-9 as

$$e_g = V_{\text{Th}} = E\left(\frac{\Delta r}{4R}\right)$$

where Δr is 5 **kΩ** − 4.5 **kΩ** or 500 **Ω**; therefore,

$$e_g = 6\text{ V} \times \frac{500\ \Omega}{20\text{ k}\Omega} = 0.150\text{ V}$$

5-9 APPLICATIONS

There are many industrial applications for bridge circuits in the areas of measurement and control. A few of these applications are discussed here.

A Wheatstone bridge may be used to measure the dc resistance of various types of wire for the purpose of quality control either of the wire itself or of some assembly in which a quantity of wire is used. For example, the resistance of motor windings, transformers, solenoids, or relay coils may be measured.

Telephone companies and others use the Wheatstone bridge extensively to locate faults in cables. The fault may be two lines "shorted" together or a single line shorted to ground.

A portable **Murray loop** test method is one of the best known and simplest of loop tests and is used principally to locate ground faults in sheathed cables. Figure 5-18 shows a test setup. The defective conductor of length L_b is connected at its cable terminals to a healthy conductor of length L_a. The loop formed by these two conductors is connected to the test set as shown, and the bridge is balanced with the adjustable resistor R_2. The ratio of R_2 to R_1 is generally known as the ratio arms. At balance,

$$\frac{R_2}{R_1} = \frac{R_a + (R_b - R_x)}{R_x}$$

where R_a, R_b, and R_x are the resistances of L_a, L_b, and L_x, respectively.

$$R_2 R_x = R_1 R_a + R_1 R_b - R_1 R_x$$

$$R_2 R_x + R_1 R_x = R_1 R_a + R_1 R_b$$

$$R_x (R_1 + R_2) = R_1 (R_a + R_b)$$

$$R_x = \frac{R_1}{R_1 + R_2}(R_a + R_b)$$

But

$$R = \frac{p l}{A}$$

where

p is resistivity
l is length
A is cross-sectional area

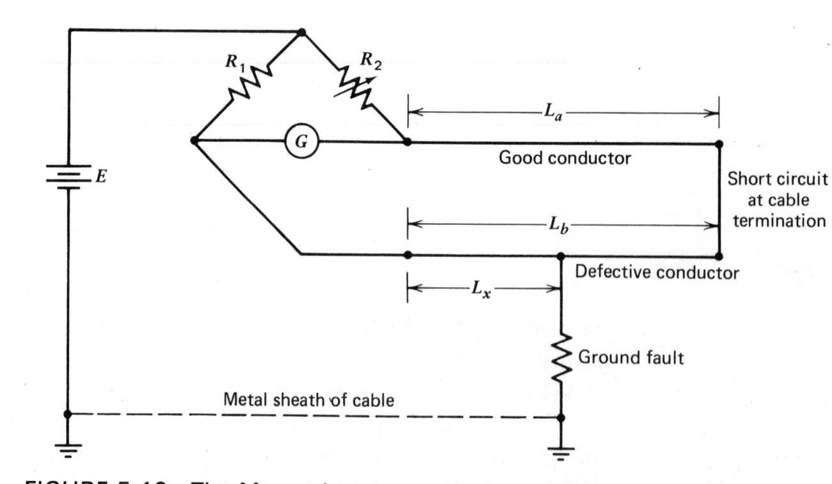

FIGURE 5-18 The Murray loop test to locate a ground fault (short circuit).

Therefore

$$\frac{P_x L_x}{A_x} = \frac{R_1}{R_1 + R_2} \left(\frac{P_a L_a}{A_a} + \frac{P_b L_b}{A_b} \right) \tag{5-12}$$

If both conductors consist of the same material and cross-sectional area, then

$$L_x = \frac{R_1}{R_1 + R_2}(L_a + L_b) \tag{5-13}$$

In a multicore cable the healthy conductor has the same length and same cross section as the faulty cable, so that $L_a = L_b$. Therefore,

$$L_x = \frac{R_1}{R_1 + R_2}(2_L) = \frac{2R_1 L}{R_1 + R_2} \tag{5-14}$$

The portable Wheatstone bridge can reasonably measure low-resistance ground faults. If, however, the fault resistance is high, the battery-operated test set is not adequate, and a high-voltage measurement must be made.

EXAMPLE 5-6

The Murray loop test set of Fig. 5-18 consists of two conductors of the same material and the same cross-sectional area. Both cables are connected 5280 feet from the test setup at the cable terminal. The bridge is balanced, when R_1 is 100 **Ω** and R_2 is 300 **Ω**. Find the distance from the ground fault to the test set.

Solution

$$L_x = \frac{2R_1 L}{R_1 + R_2} = \frac{2 \times 100\ \Omega \times 5280\ \text{ft}}{100\ \Omega + 300\ \Omega} = 240\ \text{ft}$$

The **Varley loop** test is one of the most accurate methods of locating ground faults and short circuits in a multiconductor cable. It is essentially a modification of the Murray loop test. This method uses a Wheatstone bridge, with two fixed ratio arms R_2 and R_1 and a rheostat R_3 in the third arm. Figure 5-19 shows a commonly used test method.

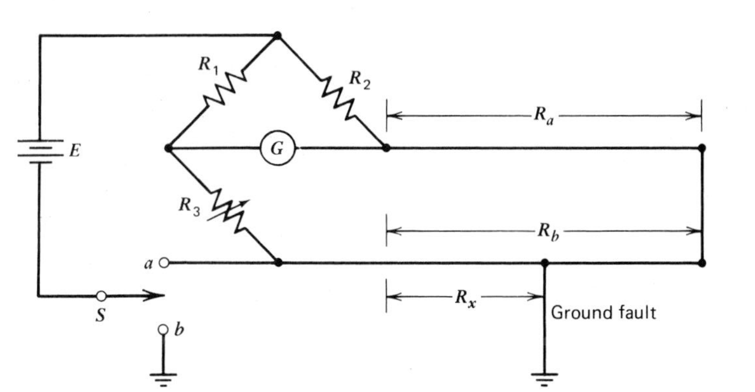

FIGURE 5-19 Wheatstone bridge connected for a Varley loop test.

Suppose a "short to ground" has occurred in the conductor represented by resistance R_b. A good conductor, in the multicore cable, is connected to the defective cable at the cable termination. The healthy conductor is represented by resistance R_a. To locate the fault, first set switch S to position a. Balance the bridge by adjusting R_3. When the bridge is balanced,

$$\frac{R_2}{R_1} = \frac{R_a + R_b}{R_3}$$

$$R_2 R_3 = R_1 R_a + R_1 R_b$$

$$R_1 (R_a + R_b) = R_2 R_3$$

and

$$R_a + R_b = \frac{R_2 R_3}{R_1} \tag{5-15}$$

Now set the switch to position b and balance the bridge again. The equation for balance is now

$$\frac{R_2}{R_1} = \frac{R_a + (R_b - R_x)}{R_x + R_3}$$

$$R_2 R_x + R_2 R_3 = R_1 R_a + R_a R_b - R_1 R_x$$

$$R_2 R_x + R_1 R_x = R_1 R_a + R_1 R_b - R_2 R_x$$

$$R_x (R_1 + R_2) = R_1 (R_a + R_b) - R_2 R_3$$

Solving for R_x yields

$$R_x = \frac{R_1 (R_a + R_b) - R_2 R_3}{R_1 + R_2}$$

The value of $R_a + R_b$ can now be obtained from Eq. 5-15.

EXAMPLE 5-7

The Varley loop test set of Fig. 5-20 consists of a defective conductor and a healthy conductor connected at the cable terminal located 10 miles from the test set. The cables have a resistance of 0.05 ohm per 1000 ft. When the switch is in position a and the circuit is balanced, the balancing resistance is

$$R_3 = 100\ \Omega$$

When the switch is in position b and the circuit is rebalanced, the balancing resistor becomes

$$R_3 = 99\ \Omega$$

Find the distance from the ground fault to the test set.

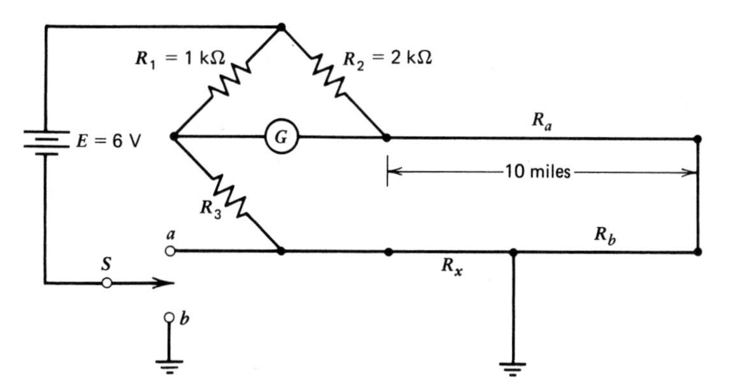

FIGURE 5-20 Varley loop tests to locate grounds or short circuits.

Solution

With the switch in position *a*,

$$R_a + R_b = \frac{R_2 R_3}{R_1} \times \frac{2000 \times 100}{1000} = 200\ \Omega$$

With the switch in position *b*,

$$R_x = \frac{R_1(R_a + R_b) - R_2 R_3}{R_1 + R_2} = \frac{1000 \times 200 - 2000 \times 99}{1000 + 2000}$$

$$= 0.67\ \Omega$$

A cable resistance of 0.05 **Ω** represents 1000 feet. Therefore, a cable resistance of 0.67 **Ω** represents

$$\frac{0.67 \times 1000}{0.05} = 13{,}333 \text{ ft}$$

The Wheatstone bridge is also widely used in control circuits such as those shown in Fig. 5-21.

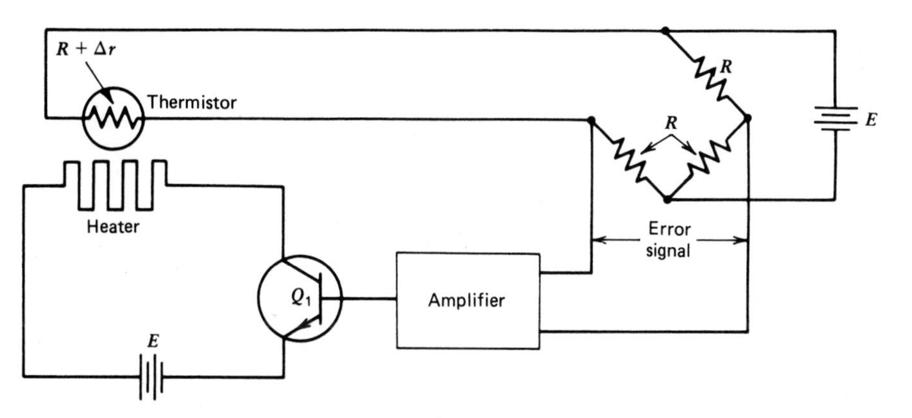

FIGURE 5-21 Basic bridge-controlled heater circuit.

The circuit is a basic heater circuit. At the desired temperature the resistance of the thermistor equals R. This causes the bridge to be balanced, and therefore, there is no error voltage. Since transistor Q_1 is biased off, no current flows through the heating element. If the temperature decreases, the thermistor resistance decreases, the bridge is unbalanced, and the error signal is amplified, which forward-biases transistor Q_1. This allows current to flow through the heating element until the thermistor resistance increases to a value of R, which balances the bridge and turns off the circuit. Such circuits are of use in many industrial applications in which the temperature must be maintained with close tolerance.

5-10 SUMMARY

The Wheatstone bridge is the most basic bridge circuit. It is widely used in measuring instruments and control circuits. Bridge circuits have a high degree of accuracy, limited only by the accuracy of the components used in the circuit. The Kelvin bridge is a modification of the Wheatstone bridge and is widely used to measure very low resistances.

Recent innovations in bridge-type instruments include digital readout and microprocessor-controlled bridges. Whenever the microprocessor becomes an integral part of the measuring circuitry, the instrument becomes highly sophisticated (intelligent).

The most frequently used analytical tool for analyzing an unbalanced Wheatstone bridge is Thévenin's theorem.

5-11 GLOSSARY

Balance: The condition of a bridge when no current flows through the detector (usually a galvanometer).

Comparison measurement: A measurement made with an instrument in which a standard against which an unknown is being compared is physically present within the instrument.

Galvanometer: A laboratory instrument using a d'Arsonval meter movement but with zero at center scale. Used to measure very small currents of either positive or negative polarity.

Kelvin bridge: A modification of the Wheatstone bridge. Contains an additional set of ratio arms to compensate for lead and contact resistors of 1 Ω or less.

Microprocessor: A "data processor" or "controller" contained on a single integrated-circuit chip.

Murray loop: A special Wheatstone bridge used to measure shorts between lines or to ground.

Null indication: A term used to indicate that no current is flowing through a galvanometer; hence, the pointer is resting at center scale zero, indicating the bridge is balanced.

Sensitivity: Deflection per unit of current.

Thévenin's theorem: An analytical tool used extensively to analyze an unbalanced bridge. The theorem states that a complex circuit may be replaced with a single equivalent voltage source and a single equivalent series resistance as seen looking back into the circuit from the output terminals with no load connected.

Varley loop: A special Wheatstone bridge configuration used to locate shorts between conductors or faults to ground in conductors.

Wheatstone bridge: A basic circuit configuration used in measuring instruments or control instruments. The bridge is balanced when the products of the resistors in opposite arms are equal.

5-12 REVIEW QUESTIONS

The following questions should be answered after a comprehensive study of the chapter. The purpose of the questions is to determine the reader's comprehension of the material.

1. How does the measuring accuracy of a Wheatstone bridge compare with that of an ordinary ohmmeter?
2. What are the criteria for balance of a Wheatstone bridge?
3. In what two types of circuits do Wheatstone bridges find most of their applications?
4. What are the criteria for balance of a Kelvin bridge?
5. What is the primary use of the Kelvin bridge?
6. How does the basic circuit for the Kelvin bridge differ from that for the Wheatstone bridge?
7. How does the use of microprocessors in bridge circuits differ from the use of minicomputers with bridges?
8. What are some ways in which microprocessors are reducing the cost and complexity of analog measurements?
9. What technique lends itself well to analyzing the unbalanced Wheatstone bridge?

5-13 PROBLEMS

5-1 Calculate the value of R_x in the circuit of Fig. 5-22 if $R_1 = 400\ \Omega$, $R_2 = 5\ k\Omega$, and $R_3 = 2\ k\Omega$.

5-2 Calculate the value of R_x in Fig. 5-22 if $R_1 = 10\ k\Omega$, $R_2 = 60\ k\Omega$, and $R_3 = 18.5\ k\Omega$.

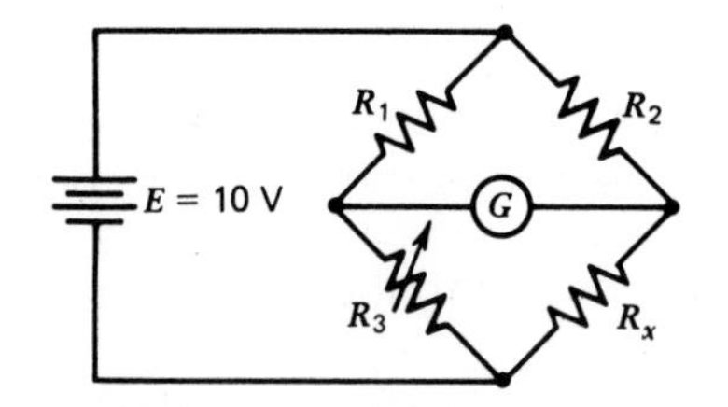

FIGURE 5-22 Circuit for Problem 5-1

5-3 Calculate the value of R_g in Fig. 5-22 if $R_1 = 5\ \text{k}\Omega$, $R_2 = 40\ \text{k}\Omega$, and $R_3 = 10\ \Omega$.

5-4 What resistance range must resistor R_3 of Fig. 5-23 have in order to measure unknown resistors in the range of 1 to 100 kΩ?

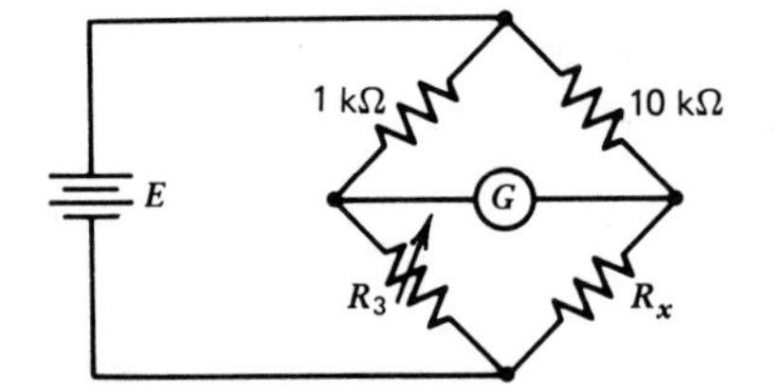

FIGURE 5-23 Circuit for Problem 5-4.

5-5 Calculate the value of R_x in the circuit of Fig. 5-24 if $R_a = 1200\ \Omega$, $R_a = 1600R_b$, $R_1 = 800R_b$, and $R_1 = 1.25R_2$.

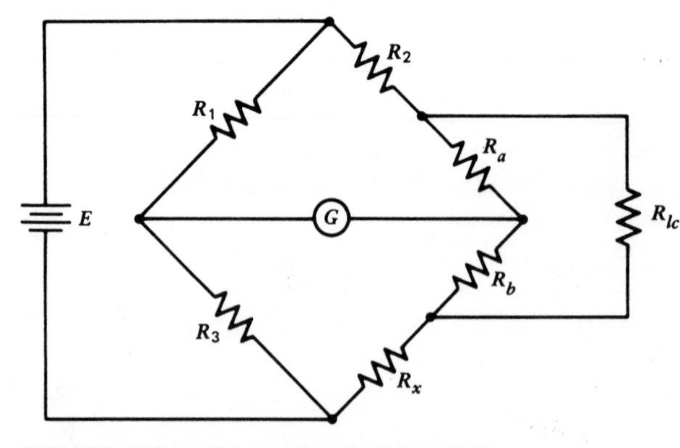

FIGURE 5-24 Circuit for Problem 5-5.

5-6 Calculate the current through the galvanometer in the circuit of Fig. 5-25.

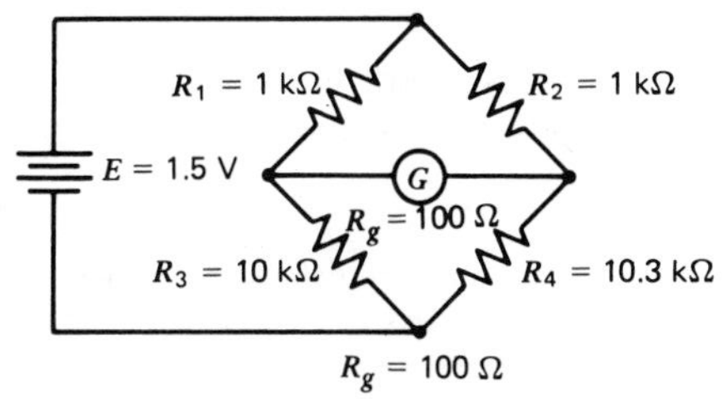

FIGURE 5-25 **Circuit for Problem 5-6.**

5-7 Three arms of a Wheatstone bridge contain resistors of known value that have a limiting error of ±0.2%. Calculate the limiting error of an unknown resistor when measured with this instrument.

5-8 Calculate the percentage of error in the value of the current through the galvanometer in Fig. 5-26 when the approximate Eqs. 5-9 and 5-10 are used to find Thévenin's equivalent circuit.

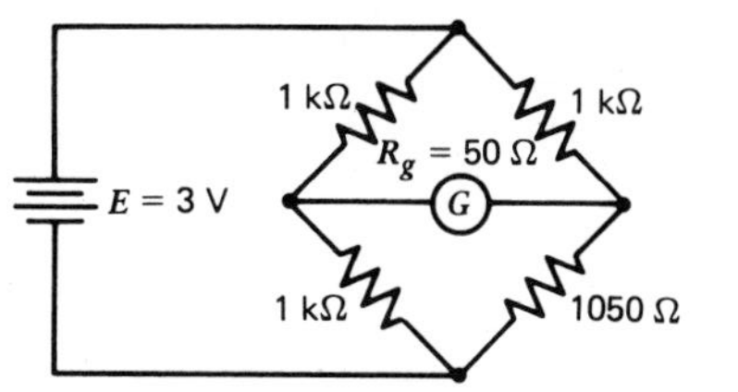

FIGURE 5-26 **Circuit for Problem 5-8.**

5-9 Calculate the value of R_x in the circuit of Fig. 5-27 if $V_{Th} = 24$ mV and $I_g = 13.6\ \mu A$.

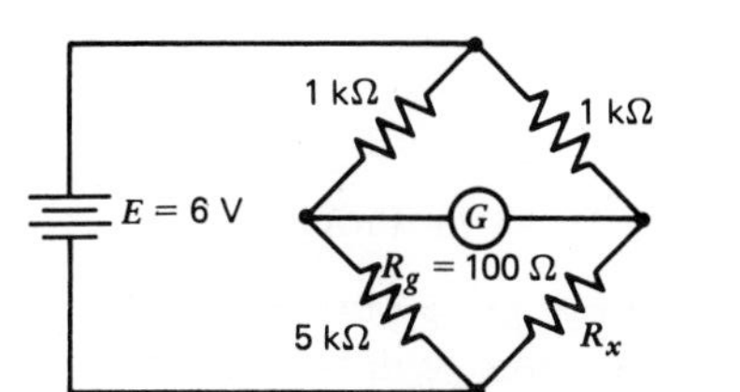

FIGURE 5-27 **Circuit for Problem 5-9.**

5-10 If the sensitivity of the galvanometer in the circuit of Fig. 5-28 is 10 mm/μA, determine its deflection.

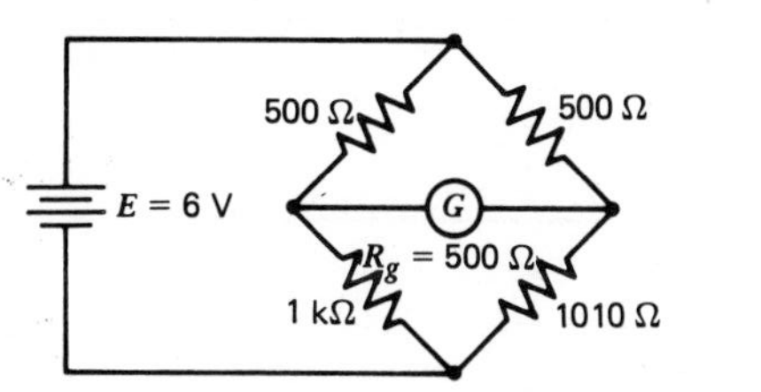

FIGURE 5-28 **Circuit for Problem 5-10.**

5-11 If the light beam that is directed on the photocell R in the circuit of Fig. 5-29 is interrupted, the resistance of the photocell increases from 10 to 40 kΩ. Calculate the current I_{in} when the beam is interrupted.

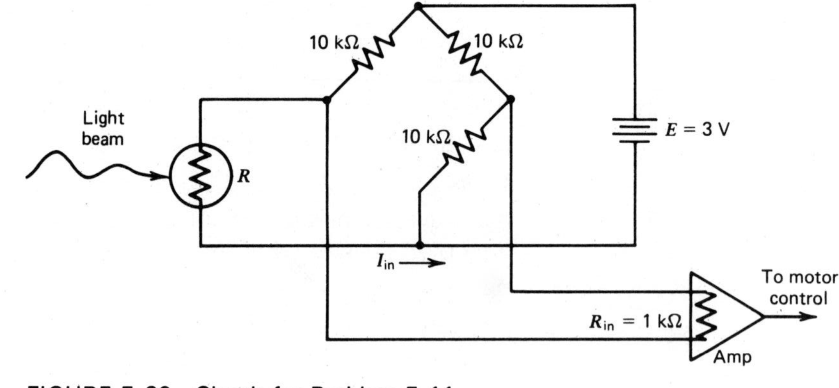

FIGURE 5-29 Circuit for Problem 5-11.

5-12 Calculate the current I_{in} in the circuit of Fig. 5-30 when the temperature is 50°C if $R = 500\ \Omega$ at 25°C and its resistance increases 0.7 Ω per degree.

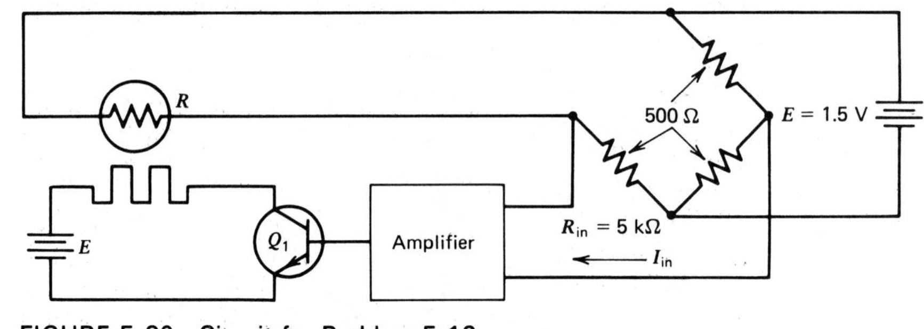

FIGURE 5-30 Circuit for Problem 5-12.

5-13 A Wheatstone bridge is connected for a Varley loop test as shown in Fig. 5-31. When the switch S is in position a, the bridge is balanced

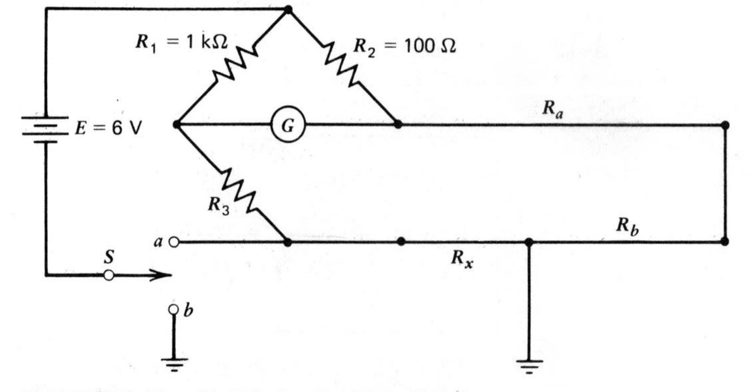

FIGURE 5-31 Circuit for Problem 5-13.

with $R_1 = 1000\ \Omega$, $R_2 = 100\ \Omega$, and $R_3 = 53\ \Omega$. When S is in position b, the bridge is balanced with $R_1 = 1000\ \Omega$, $R_2 = 100\ \Omega$, and $R_3 = 52.9\ \Omega$. If the resistance of the shorted wire is $0.015\ \Omega$/m, how many meters from the bridge has a short to ground occurred?

5-14 A Wheatstone bridge is connected for a Murray loop test as shown in Fig. 5-32 and balanced. Cable a is an aerial cable with a resistance of 0.1 ohm per 1000 ft. Cable b is an underground cable with a resistance of 0.005 ohm per 1000 ft. Neglecting temperature differences, calculate the distance L_x from the ground fault to the test set if $L_a = L_b$.

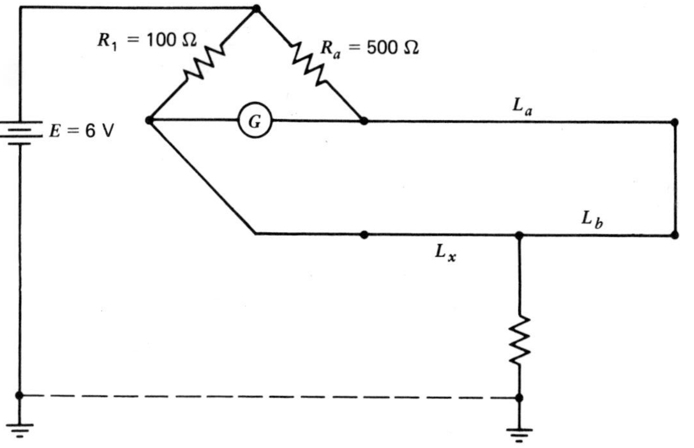

FIGURE 5-32 Circuit for Problem 5-14.

5-14 LABORATORY EXPERIMENTS

Experiments E9, E10, and E11, pertain to the theory presented in Chapter 5. The purpose of the experiments is to provide hands-on experience to reinforce the theory.

The experiments make use of components and equipment that are found in any well-equipped electronics laboratory. The contents of the laboratory report to be submitted by each student are included at the end of each experimental procedure.

CHAPTER 6

Alternating-Current Bridges

6-1 Instructional Objectives

This chapter examines various types of alternating-current bridges by considering those that have reactive components and a sinusoidal alternating-current voltage or current applied. After completing Chapter 6 you should be able to

1. Explain how a simple ac bridge circuit operates and derive an expression for the unknown parameters.
2. Explain the steps involved in balancing an ac bridge and derive the balance equations.
3. Identify each bridge by name.
4. Compute the values for the unknown impedance for the following ac bridges:
   (a) Similar-angle bridge.
   (b) Opposite-angle bridge.
   (c) Maxwell bridge.
   (d) Wien bridge.
   (e) Radio-frequency bridge.
   (f) Schering bridge.

6-2 INTRODUCTION

Alternating-current bridges are used to measure inductance and capacitances, and all ac bridge circuits are based on the Wheatstone bridge. The **Wheatstone bridge** is really a special case of a more general bridge. In the Wheatstone bridge, each of the four arms of the bridge contains a pure resistance. However, the general **bridge circuit** consists of four impedances, an ac voltage source, and a detector, as shown in Fig. 6-1. In the general bridge circuit, the impedances can be either pure resistances or complex

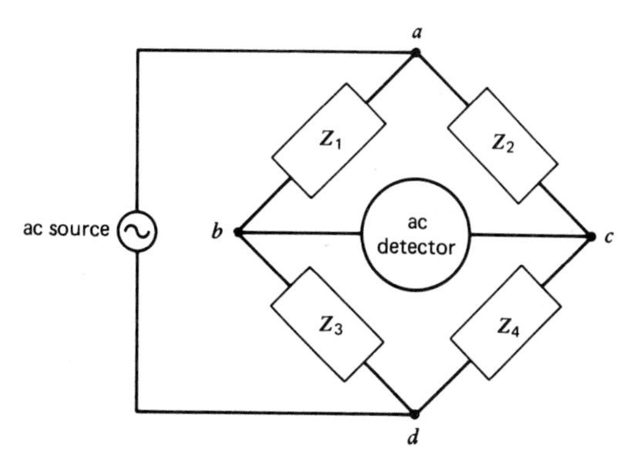

FIGURE 6-1 General ac bridge circuit.

impedances. The usefulness of ac bridge circuits is not restricted to the measurement of an unknown impedance. These circuits find other applications in many communication systems and complex electronic circuits. Alternating-current bridge circuits are commonly used for shifting phase, providing feedback paths for oscillators or amplifiers, filtering out undesired signals, and measuring the frequency of audio signals.

The operation of the bridge depends on the fact that when certain specific circuit conditions apply, the detector current becomes zero. This is known as the **null** or **balanced** condition. Since zero current means that there is no voltage difference across the detector, the bridge circuit may be redrawn as in Fig. 6-2. The dash line in the figure indicates that there is no potential difference and no current between points *b* and *c*. The voltages from point *a* to point *b* and from point *a* to point *c* must now be equal, which allows us to write

$$I_1 Z_1 = I_2 Z_2 \tag{6-1}$$

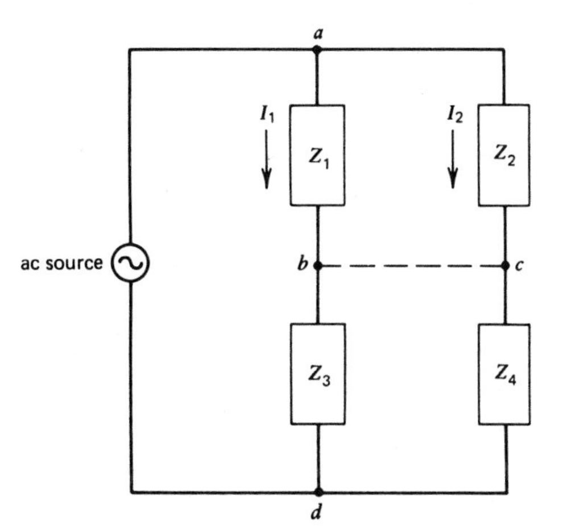

FIGURE 6-2 Equivalent of balanced (nulled) ac bridge circuit.

Similarly, the voltages from point *d* to point *b* and point *d* to point *c* must also be equal, leading to

$$I_1 Z_3 = I_2 Z_4 \tag{6-2}$$

Dividing Eq. 6-1 by Eq. 6-2 results in

$$\frac{Z_1}{Z_3} = \frac{Z_2}{Z_4} \tag{6-3}$$

which can also be written as

$$Z_1 Z_4 = Z_2 Z_3 \tag{6-4}$$

This equation is known as the general bridge equation and applies to any four-arm bridge circuit at balance, whether the branches are pure resistances or combinations of resistance, capacitance, and inductance. Notice that the ratios of impedances are not affected by the magnitude of the ac source voltage or the actual values of the branch currents. However, in general, the impedances are complex and therefore are functions of frequency. By careful choice of the various impedances, it is possible to remove from the balance equation the dependence on frequency, although most bridges are not independent of frequency. This will become clear after several types of ac bridges have been discussed.

Both the magnitude and phase angle of each of the four impedances must satisfy Eqs. 6-3 and 6-4 for there to be a null or balanced condition. Another way of saying this is that, if the bridge is to be balanced, both the real components and the imaginary (or *i*) components of the impedances must be balanced at the same time. When the bridge is not balanced, the equations do not hold, the circuit becomes more complicated, and conventional circuit techniques must be used to solve for the voltages and currents.

If the impedance is written in the form $\mathbf{Z} = Z\underline{/\theta}$ where Z represents the magnitude and θ the phase angle of the complex impedance, Eq. 6-4 can be written in the form

$$(Z_1\underline{/\theta_1})(Z_4\underline{/\theta_4}) = (Z_2\underline{/\theta_2})(Z_3\underline{/\theta_3}) \tag{6-5}$$

where

$$Z_1 Z_4\underline{/(\theta_1 + \theta_4)} = Z_2 Z_3\underline{/(\theta_2 + \theta_3)} \tag{6-6}$$

Therefore, two conditions must be met simultaneously when balancing an ac bridge. The first condition shows that the products of the magnitudes of the opposite arms must be equal:

$$Z_1 Z_4 = Z_2 Z_3 \tag{6-7}$$

The second requirement is that the sum of the phase angles of the opposite arms be equal:

$$\underline{/\theta_1} + \underline{/\theta_4} = \underline{/\theta_2} + \underline{/\theta_3} \tag{6-8}$$

EXAMPLE 6-1

The impedances of the ac bridge in Fig. 6-1 are given as follows:

$$Z_1 = 200\ \Omega\angle 30^\circ$$

$$Z_2 = 150\ \Omega\angle 0^\circ$$

$$Z_3 = 250\ \Omega\angle -40^\circ$$

$$Z_x = Z_4 = \text{unknown}$$

Determine the constants of the unknown arm.

Solution

The first condition for bridge balance requires that

$$Z_1 Z_x = Z_2 Z_3$$

or

$$Z_x = \frac{Z_2 Z_3}{Z_1} = \frac{150\ \Omega \times 250\ \Omega}{200\ \Omega} = 187.5\ \Omega$$

The second condition for balance requires that the sums of the phase angles of opposite arms be equal,

$$\theta_1 + \theta_x = \theta_2 + \theta_3$$

or

$$\theta_x = \theta_2 + \theta_3 - \theta_1$$

$$= 0^\circ + (-40^\circ) - 30^\circ = -70^\circ$$

Hence, the unknown impedance Z_x can be written as

$$Z_x = 187.5\ \Omega\angle -70^\circ = (64.13 - j176.19)\ \Omega$$

indicating that we are dealing with a capacitive element, possibly consisting of a series resistor and a capacitor.

EXAMPLE 6-2

Given the ac bridge of Fig. 6-3 in balance, find the components of the unknown arm Z_x.

Solution

$$\omega = 2\pi f = 2 \times \pi \times 1000\ \text{Hz} = 6283.19\ \text{rad/sec}$$

$$X_{L2} = \omega L_2 = 6283.19 \times 15.92 \times 10^{-3} = 100\ \Omega$$

$$X_{C3} = \frac{1}{\omega C_3} = \frac{1}{6283.19 \times 0.4 \times 10^{-6}}$$

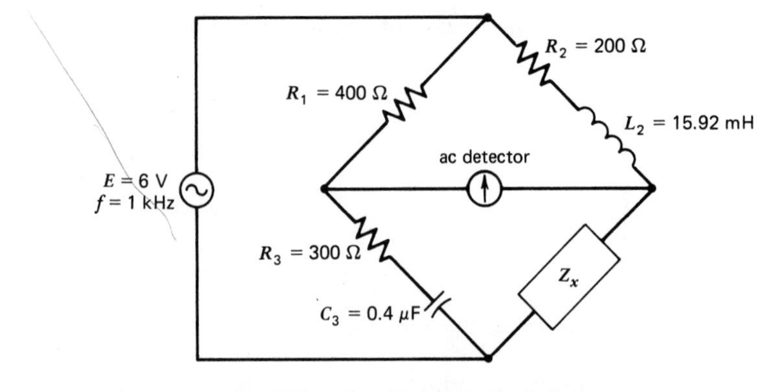

FIGURE 6-3 Ac bridge in balance.

The impedances of the bridge arms are

$$Z_1 = R_1 = 400\ \Omega\angle 0^\circ$$

$$Z_2 = R_2 + jX_{L2} = 200 + j100 = 223.6\angle 26.6^\circ$$

$$Z_3 = R_3 - jX_{C3} = 300 - j400 = 500\angle -53^\circ$$

$$Z_x = \frac{Z_2 Z_3}{Z_1} = \frac{(223.6\angle 26.6^\circ)(500\angle -53^\circ)}{400\angle 0^\circ} = 279.5\ \Omega\angle -26.4^\circ$$

$$= (250.35 - j124.28)\ \Omega$$

Therefore,

$$C = \frac{1}{\omega X_C} = \frac{1}{6283.19 \times 124.28} = 1.28\ \mu\text{F}$$

Thus, the equivalent-series resistance is 250.35 Ω, and the equivalent capacitance is 1.28 μF.

6-3 SIMILAR-ANGLE BRIDGE

A simple form of ac bridge is shown in Fig. 6-4. This is known as the similar-angle bridge and is used to measure the impedance of a capacitive circuit. This bridge is sometimes called the capacitance comparison bridge or the series resistance capacitance bridge. The impedance of the arms of this bridge can be written as

$$Z_1 = R_1 \qquad Z_2 = R_2$$

$$Z_3 = R_3 - jX_{C3} \qquad Z_4 = R_x - jX_{Cx}$$

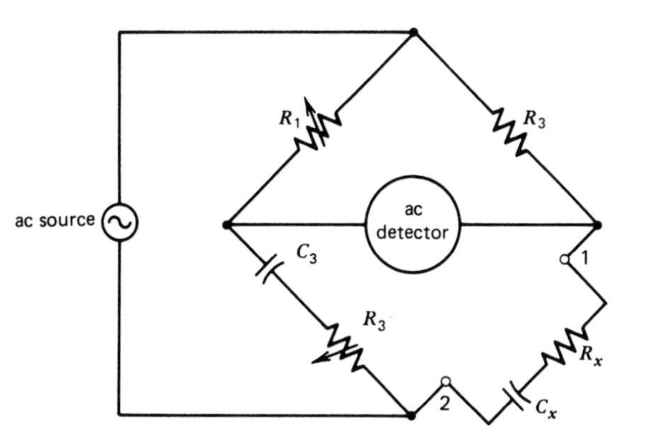

FIGURE 6-4 Similar-angle bridge.

Substituting these values in Eq. 6-4 gives the balance equation

$$R_1(R_x - jX_{Cx}) = (R_3 - jX_3)R_2$$

This equation can be simplified by multiplying through and then grouping the real and imaginary terms, yielding

$$R_1R_x - jR_1X_{Cx} = R_2R_3 - jR_2X_3$$

The only way in which the equation can be satisfied is for the real terms on each side of the equation to be equal and at the same time for the imaginary terms on each side to be equal. Thus, the two equations that must be satisfied are

$$R_1R_x = R_2R_3 \tag{6-9}$$

$$-jR_1X_{Cx} = -jR_2X_3 \tag{6-10}$$

From Eq. 6-10 we get

$$-jR_1\frac{1}{\omega C_x} = -jR_2\frac{1}{\omega C_3}$$

$$R_1C_3 = R_2C_x \tag{6-11}$$

Solving Eqs. 6-9 and 6-11 for the unknown quantities R_x and C_x leads to

$$R_x = \frac{R_2}{R_1}R_3 \tag{6-12}$$

$$C_x = \frac{R_1}{R_2}C_3 \tag{6-13}$$

In this case, the frequency dependence mentioned earlier has canceled out of the equations. Therefore, the similar-angle bridge is not dependent on either the magnitude or the frequency of the applied voltage.

In Fig. 6-4 note that the unknown impedance, Z_4, can be any impedance whose reactance is more capacitive than inductive. In other words, the unknown can be either an *RC* circuit or an *RLC* combination whose reactive component is negative. For this reason, the unknown resistance and capacitance obtained are referred to as the **equivalent-series resistance** and the **equivalent-series capacitance**. Similarly, assuming that the unknown impedance consists of a capacitor and resistor in parallel would result in finding the equivalent-parallel resistance and the equivalent-parallel capacitance.

EXAMPLE 6-3

A similar angle bridge is used to measure a capacitive impedance at a frequency of 2 kHz. The bridge constants at balance are

$$C_3 = 100\ \mu\text{F} \qquad R_1 = 10\ \text{k}\Omega$$

$$R_2 = 50\ \text{k}\Omega \qquad R_3 = 100\ \text{k}\Omega$$

Find the equivalent-series circuit of the unknown impedance.

Solution

Find R_x using Eq. 6-12.

$$R_x = \frac{R_2}{R_1} R_3 = \frac{(50 \times 10^3\ \Omega)(100 \times 10^3\ \Omega)}{10 \times 10^3\ \Omega} = 500\ \text{k}\Omega$$

Then find C_x using Eq. 6-13.

$$C_x = \frac{R_1}{R_2} C_3 = \frac{(10 \times 10^3\ \Omega)(100 \times 10^{-6}\ \text{F})}{50 \times 10^3\ \Omega} = 20\ \mu\text{F}$$

The equivalent-series circuit is shown in the illustration below.

500 kΩ 20 μF

6-4 MAXWELL BRIDGE

It is possible to determine an unknown inductance with capacitance standards. This is accomplished in a circuit known as a Maxwell bridge circuit and sometimes called a Maxwell–Wien bridge. Using capacitance as a standard has several advantages. Capacitance is influenced to a lesser degree by external fields and capacitors set up virtually no external field. Furthermore, capacitors are small and inexpensive. The Maxwell bridge is shown in Fig. 6-5.

The impedance of the arms of the bridge can be written as

$$Z_1 = \frac{1}{1/R_1 + j\omega C_1} \qquad Z_2 = R_2$$

$$Z_3 = R_3 \qquad Z_4 = R_x + jX_{Lx}$$

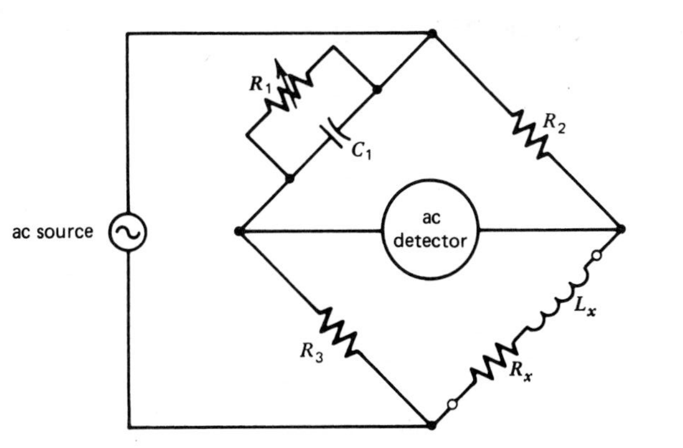

FIGURE 6-5 Maxwell bridge.

Substituting these values in Eq. 6-4 gives the balance equation

$$\frac{1}{1/R_1 + j\omega C_1}(R_x + jX_{Lx}) = R_2 R_3 \tag{6-14}$$

$$R_x + jX_{Lx} = \frac{R_2 R_3}{R_1} + j\omega R_2 R_3 C_1 \tag{6-15}$$

By setting both the real and imaginary parts equal to zero, we get

$$R_x = \frac{R_2 R_3}{R_1} \tag{6-16}$$

$$j\omega L_x = j\omega R_2 R_3 C_1$$

$$L_x = R_2 R_3 C_1 \tag{6-17}$$

EXAMPLE 6-4

A Maxwell bridge is used to measure an inductive impedance. The bridge constants at balance are

$$C_1 = 0.01\ \mu\text{F} \qquad R_1 = 470\ \text{k}\Omega$$

$$R_2 = 5.1\ \text{k}\Omega \qquad R_3 = 100\ \text{k}\Omega$$

Find the series-equivalent resistance and inductance.

Solution

Find R_x and L_x using Eqs. 6-16 and 6-17.

$$R_x = \frac{R_2 R_3}{R_1} = \frac{(5.1 \times 10^3\ \Omega)(100 \times 10^3\ \Omega)}{470 \times 10^3\ \Omega} = 1.09\ \text{k}\Omega$$

$$L_x = R_2 R_3 C_1 = (5.1 \times 10^3)(100 \times 10^3)(0.01 \times 10^{-6}) = 5.1\ \text{H}$$

6-5 OPPOSITE-ANGLE BRIDGE

For measurement of inductance, the similar-angle bridge could be used by replacing the standard capacitor with an inductance. However, since standard inductances are large and expensive to manufacture, inductive circuits are generally measured by using a form of the bridge circuit known as the opposite-angle bridge. This bridge, shown in Fig. 6-6, is sometimes known as a Hay bridge. This particular network is used for measuring the resistance and inductance of coils in which the resistance is a small fraction of the reactance X_L, that is, a coil having a high Q, meaning a Q greater than 10. The symbol Q designates the ratio of X_L to R for a coil. Otherwise the Maxwell bridge is used for measuring low Q coils ($Q < 10$). Solving the bridge equation results in the equivalent-series values of inductance and resistance. From Appendix B-1 we get

$$R_x = \frac{\omega^2 R_1 R_2 R_3 C_1^2}{1 + \omega^2 R_1^2 C_1^2} \tag{6-18}$$

$$L_x = \frac{R_2 R_3 C_1}{1 + \omega^2 R_1^2 C_1^2} \tag{6-19}$$

For the opposite-angle bridge, it can be seen that the balance conditions depend on the frequency at which the measurement is made.

EXAMPLE 6-5

Find the series-equivalent inductance and resistance of the network that causes an opposite-angle bridge to null with the following component values:

$$\omega = 3000 \text{ rad/s}, \quad R_2 = 10 \text{ k}\Omega, \quad R_1 = 2 \text{ k}\Omega,$$

$$R_3 = 1 \text{ k}\Omega, \quad C_1 = 1\ \mu\text{F}$$

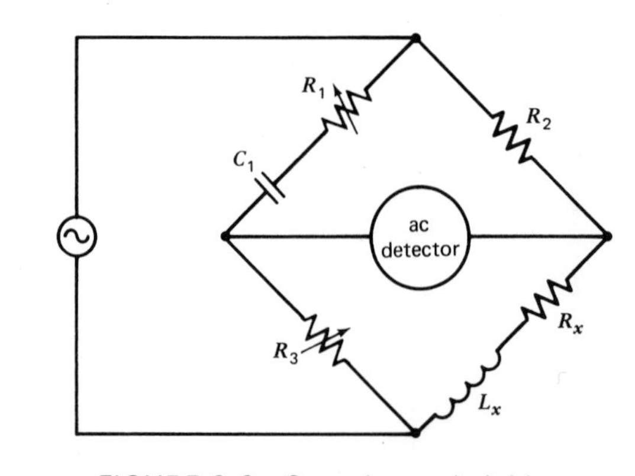

FIGURE 6-6 Opposite-angle bridge.

Solution

Find R_x and L_x, using Eqs. 6-18 and 6-19.

$$R_x = \frac{\omega^2 R_1 R_2 R_3 C_1^2}{1+\omega^2 R_1^2 C_1^2}$$

$$R_x = \frac{(3\times10^3)^2(2\times10^3)(10\times10^3)(1\times10^{-6})^2}{1+(3\times10^3)^2(2\times10^3)^2(1\times10^{-6})^2}$$

$$= \frac{180\times10^3}{1+36} = 4.86\ \mathbf{k\Omega}$$

$$L_x = \frac{R_2 R_3 C_1}{1+\omega^2 R_1^2 C_1^2}$$

$$= \frac{(10\times10^3)(1\times10^3)(1\times10^{-6})}{1+(3\times10^3)^2(2\times10^3)^2(1\times10^{-6})^2}$$

$$= \frac{10}{1+36} = 0.27 = 270\ \text{mH}$$

$$R_x = 4.86\ \mathbf{k\Omega}, \qquad L_x = 270\ \text{mH}$$

6-6 WIEN BRIDGE

A type of bridge called the Wien bridge is shown in Fig. 6-7. It is important because of its versatility, since it can measure either the equivalent-series components or the equivalent-parallel components of an impedance. This bridge is also used extensively as a feedback arrangement for a circuit called the Wien bridge oscillator. From Fig. 6-7 we see that the choice of whether the series or parallel components of an *RC* impedance are measured is made by selecting the terminal to which the unknown impedance is connected. If an unknown impedance is connected between points *b* and *d*, we find an

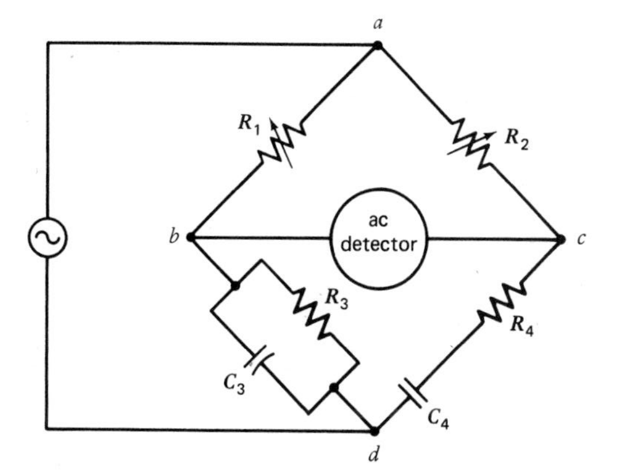

FIGURE 6-7 Wien bridge.

equivalent-parallel resistance and capacitance of the unknown impedance. If it is connected between points c and d, we find the equivalent-series resistance and capacitance. From Appendix B-2 we obtain the equivalent-parallel components of the unknown impedance.

$$R_3 = \frac{R_1}{R_2}\left(R_4 + \frac{1}{\omega^2 R_4 C_4^2}\right) \tag{6-20}$$

$$C_3 = \frac{R_2}{R_1}\left(\frac{1}{1+\omega^2 R_4^2 C_4^2}\right)C_4 \tag{6-21}$$

Appendix B-3 also gives the equivalent-series components of the unknown impedance.

$$R_4 = \frac{R_2}{R_1}\left(\frac{R_3}{1+\omega^2 R_3^2 C_3^2}\right) \tag{6-22}$$

$$C_4 = \frac{R_1}{R_2}\left(C_3 + \frac{1}{\omega^2 R_3^2 C_3^2}\right) \tag{6-23}$$

EXAMPLE 6-6

Find the equivalent-parallel resistance and capacitance that cause a Wien bridge to null with the following component values:

$$R_1 = 100\ \text{k}\Omega \qquad R_2 = 25\ \text{k}\Omega$$

$$R_4 = 3.1\ \text{k}\Omega \qquad C_4 = 5.2\ \mu\text{F}$$

$$f = 2.5\ \text{kHz}$$

Solution

Find R_3 and C_3 using Eqs. 6-20 and 6-21.

$$\omega = 2\pi f = 2\pi(2.5\times 10^3) = 15.71 \times \frac{10^3\ \text{rad}}{\text{sec}}$$

$$R_3 = \frac{R_1}{R_2}\left(R_4 + \frac{1}{\omega^2 R_4 C_4^2}\right)$$

$$= \frac{100\times 10^3}{25\times 10^3}\left(3.1\times 10^3 + \frac{1}{(15.71\times 10^3)^2(3.1\times 10^3)(5.2\times 10^{-6})^2}\right)$$

$$= 4\left(3.1\times 10^3 + \frac{1}{20.65}\right) = 12.4\ \text{k}\Omega$$

$$C_3 = \frac{R_2}{R_1}\left(\frac{1}{1+\omega^2 R_4^2 C_4^2}\right)C_4 = \frac{25\times 10^3}{100\times 10^3}$$

$$\cdot\left(\frac{1}{1+(15.71\times 10^3)^2(3.1\times 10^3)^2(5.2\times 10^{-6})^2}\right)(5.2\times 10^{-6})$$

$$= (1.3\times 10^{-6})\left(\frac{1}{1+64132.07}\right) = 20.3\ \text{pF}$$

6-7 RADIO-FREQUENCY BRIDGE

The radio-frequency bridge shown in Fig. 6-8 is often used in laboratories to measure the impedance of both capacitance and inductive circuits at higher frequencies. The measurement technique used with this bridge is known as the substitution technique. The bridge is first balanced with the Z_x terminals shorted. After the values of C_1 and C_4 are noted, the unknown impedance is inserted at the Z_x terminals, where $Z_x = R_x \pm jX_x$. Rebalancing the bridge gives new values of C_1 and C_4, which can be used to determine the unknown impedance. From Appendix B-3 we get

$$R_x = \frac{R_3}{C_2}(C_1' - C_1) \tag{6-24}$$

$$X_x = \frac{1}{\omega}\left(\frac{1}{C_4'} - \frac{1}{C_4}\right) \tag{6-25}$$

Notice that X_x can be either capacitive or inductive. If $C_4' > C_4$, and thus $1/C_4' < 1/C_4$, then X_x is negative, indicating a capacitive reactance. Therefore,

$$C_x = \frac{1}{\omega X_x} \tag{6-26}$$

However, if $C_4' < C_4$, and thus $1/C_4' > 1/C_4$, then X_x is positive and inductive and

$$L_x = \frac{X_x}{\omega} \tag{6-27}$$

Thus, once the magnitude and sign of X_x are known, the value of inductance or capacitance can be found.

Notice that the unknown impedance is represented by $R_x \pm jX_x$, which indicates a series-connected circuit. Thus, Eqs. 6-24 and 6-25 apply to the equivalent-series components of the unknown impedance.

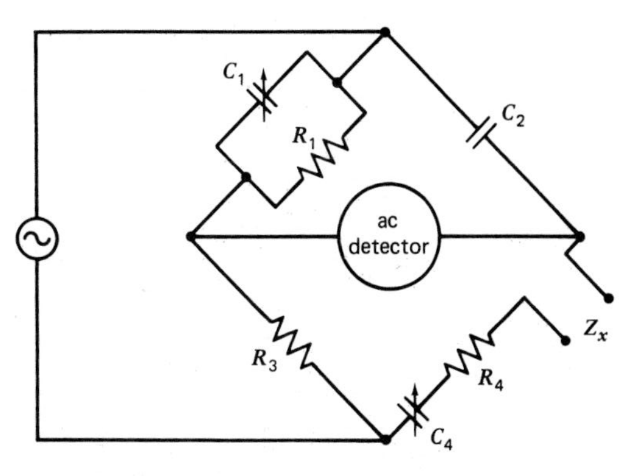

FIGURE 6-8 Radio-frequency bridge.

EXAMPLE 6-7

Find the equivalent-series elements for the unknown impedance of the network whose impedance measurements are made at null using a radio-frequency bridge.

$$f = 10\text{ MHz} \qquad R_1 = 120\text{ k}\Omega \qquad C_1 = 100\,\mu\text{F}$$
$$C_2 = 7.5\,\mu\text{F} \qquad R_3 = 7.5\text{ k}\Omega$$
$$C_4 = 120\,\mu\text{F} \qquad R_4 = 100\text{ k}\Omega$$

At rebalance:

$$C_1' = 110\,\mu\text{F} \qquad C_4' = 102.4\,\mu\text{F}$$

Solution

Find R_x and X_x from Eqs. 6-24 and 6-25; then find C_x or L_x.

$$\omega = 2\pi f = 2\pi(10\times 10^6) = 62.83 \times \frac{10^6\text{ rad}}{\text{sec}}$$

$$R_x = \frac{R_3}{C_3}(C_1' - C_1)$$
$$= \frac{7.5\times 10^3\,\Omega}{7.5\times 10^{-6}\text{ F}}(110\times 10^{-6}\text{ F} - 100\times 10^{-6}\text{ F}) = 10\text{ k}\Omega$$

$$X_x = \frac{1}{\omega}\left(\frac{1}{C_4'} - \frac{1}{C_4}\right)$$
$$= \frac{1}{62.83\times 10^6}\left(\frac{1}{102.4\times 10^{-6}} - \frac{1}{120\times 10^{-6}}\right)$$
$$= \frac{1}{62.83\times 10^6}(9765.63 - 8333.33)$$
$$= \frac{1}{62.83\times 10^6}(1432.3) = 22.8\times 10^{-6}$$

$$L_x = \frac{X_x}{\omega} = \frac{22.8\times 10^{-6}}{62.83\times 10^6} = 3.63\times 10^{-13}\text{ H}$$

6-8 SCHERING BRIDGE

The Schering bridge is one of the most important ac bridges. It is more advantageous than the similar-angle bridge. Although useful in measuring capacitance, the Schering bridge is particularly useful for measuring insulating properties, that is, for phase angles of very nearly 90°.

Figure 6-9 shows the basic circuit arrangement. Arm 1 contains a parallel combination of a resistor and a capacitor. The standard arm contains only a capacitor, C_3. This standard capacitor is usually a high-quality mica capacitor

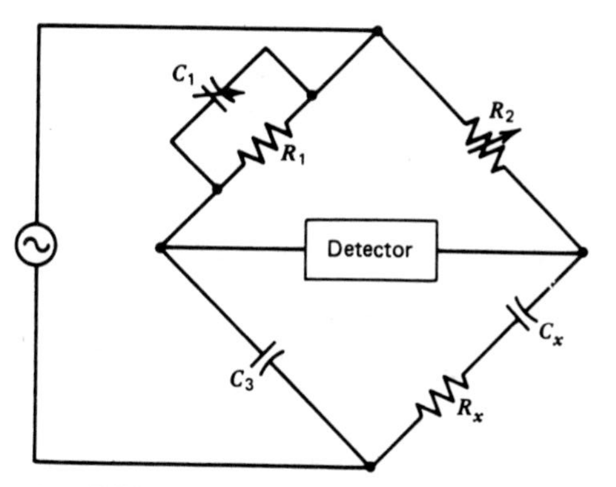

FIGURE 6-9 Schering bridge.

for general measurement work or an air capacitor for insulation measurements. A good-quality mica capacitor has very low losses (no resistance) and therefore a phase angle of approximately 90°. An air capacitor can have a very stable value and a very small electric field; the insulating material to be tested can easily be kept out of any strong fields.

The impedance of the arms of the Schering bridge can be written as

$$Z_1 = \frac{1}{1/R_1 + 1/-jX_{C1}} \qquad Z_2 = R_2$$

$$Z_3 = -jX_{C3} \qquad Z_4 = R_x - jX_x$$

Substituting these values in the general equation gives the balance equation

$$Z_4 = \frac{Z_2 Z_3}{Z_1} = \frac{R_2(-jX_{C3})}{\dfrac{1}{1/R_1 + 1/-jX_{C1}}}$$

$$= R_2(-jX_{C3})\left(\frac{1}{R_1} - \frac{1}{jX_{C1}}\right)$$

$$= R_2\left(\frac{-j}{\omega C_3}\right)\left(\frac{1}{R_1} + j\omega C_1\right)$$

and, expanding,

$$R_x - \frac{j}{\omega C_x} = \frac{R_2 C_1}{C_3} - \frac{jR_2}{\omega C_3 R_1} \tag{6-28}$$

Equating the real and imaginary terms, we find that

$$R_x = R_2 \frac{C_1}{C_3} \tag{6-29}$$

$$C_x = C_3 \frac{R_1}{R_2} \tag{6-30}$$

EXAMPLE 6-8

Find the equivalent-series elements for the unknown impedance of the Schering bridge network whose impedance measurements are to be made at null.

$$R_1 = 470\ \text{k}\Omega \qquad C_1 = 0.01\ \mu\text{F}$$

$$R_2 = 100\ \text{k}\Omega$$

$$C_3 = 0.1\ \mu\text{F}$$

Solution

Find R_x and C_x using Eqs. 6-29 and 6-30.

$$R_x = \frac{R_2 C_1}{C_3} = \frac{(100 \times 10^3)(0.01 \times 10^{-6})}{0.1 \times 10^{-6}} = 10\ \text{k}\Omega$$

$$C_x = \frac{C_3 R_1}{R_2} = \frac{(0.1 \times 10^{-6})(470 \times 10^3)}{100 \times 10^3} = 0.47 \times 10^{-6}\ \text{F}$$

$$= 0.47\ \mu\text{F}$$

6-9 SUMMARY

An ac bridge is a more general form of Wheatstone bridge, which handles only resistances and direct current. Different types of ac bridges differ in the types of impedances in the arms, which determine whether inductive or capacitive impedances, or both, can be measured, and whether the indicated value is series-equivalent or parallel-equivalent.

6-10 GLOSSARY

Balanced bridge: A Wheatstone bridge circuit which, when in a quiescent state, has an output voltage of zero.

Bridge circuit: An electrical network in which the value of an unknown component is obtained by balancing one circuit against another. It normally consists of four resistances connected in a diamond form with a current-detecting device connected between two opposite corners of the diamond.

Equivalent impedance: A single impedance equal in value to a combination of two or more impedances which it replaces. Any circuit, no matter how complicated, can be reduced to an equivalent impedance insofar as the voltage and current at the terminals of the circuit are concerned.

Equivalent-series capacitance: A capacitance value which, when placed in series with an equivalent-series resistance, causes the bridge to respond exactly the same as it does with the components actually in the bridge.

Equivalent-series resistance: A resistance value which, when placed in series with an equivalent-series capacitance, causes the bridge to respond exactly the same as it does with the components actually in the bridge.

Null: A balanced condition in a device or system whereby there is zero output.

Wheatstone bridge: A four-arm electric circuit, all arms of which are predominantly resistive. A divided electric circuit used to measure resistances.

6-11 REVIEW QUESTIONS

The following questions relate to the material in the chapter. These questions should be answered after studying the chapter to assess the reader's comprehension of the material.

1. What does bridge "null" or "balance" mean?
2. Why is an ac bridge a more general device than a Wheatstone bridge?
3. What happens in a bridge circuit to make it balance?
4. What two conditions must be satisfied to make an ac bridge balance?
5. Describe how the circuit of a similar-angle bridge differs from that of a Wheatstone bridge.
6. Do the balance conditions in a similar-angle bridge depend on frequency?
7. What determines whether a bridge measures equivalent-series or equivalent-parallel impedance components?
8. What function, not normally accomplished by the similar-angle bridge, is accomplished more readily by the opposite-angle bridge?
9. What is there about the opposite-angle bridge that makes it evident that balance is frequency-dependent?
10. What measuring application makes the Wien bridge particularly useful?
11. What are some other applications of the Wien bridge?

6-12 PROBLEMS

6-1 Using the balanced ac bridge of Fig. 6-10, find the constants of Z_x as R and C or L considered as a series circuit.

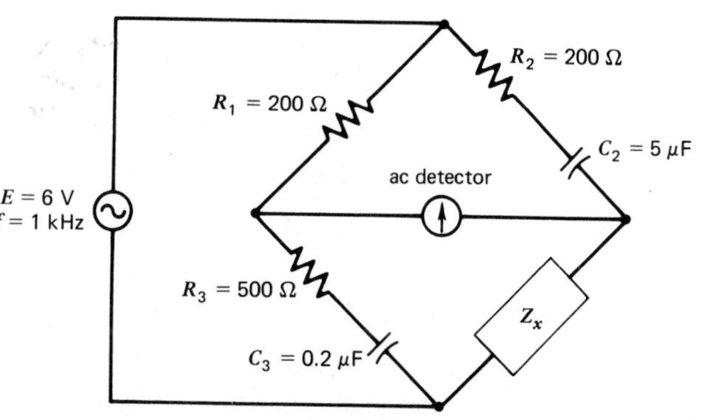

FIGURE 6-10 Circuit for Problem 6-1.

6-2 Using the balanced ac bridge of Fig. 6-10 with the following conditions—$Z_1 = 400\ \Omega\angle 0°$; $Z_2 = 300\ \Omega\angle -40°$; $Z_3 = 100\ \Omega\angle -20°$—find the constants of Z_x.

6-3 Given the similar-angle bridge of Fig. 6-11, find the equivalent-series values of R_x and C_x at balance.

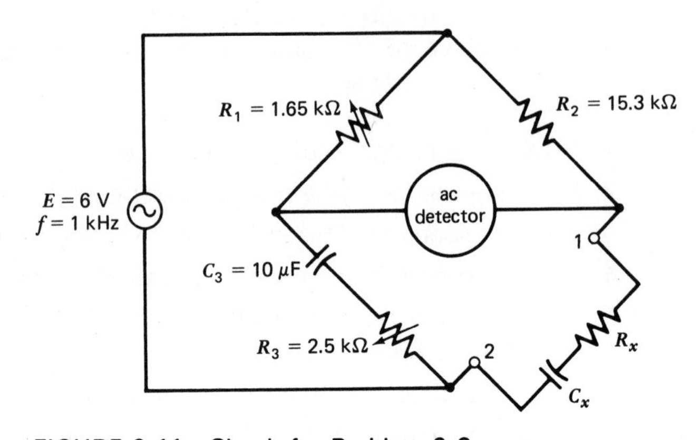

FIGURE 6-11 Circuit for Problem 6-3.

6-4 The similar-angle bridge of Fig. 6-11 is balanced with the following conditions: $Z_1 = 2000\ \Omega\angle 0°$; $Z_2 = 15{,}000\ \Omega\angle 0°$; $Z_3 = 1000\ \Omega\angle -50°$. Find the constants R_x and C_x.

6-5 Given the Maxwell bridge of Fig. 6-12, find the equivalent-series resistance and inductance of R_x and L_x, at balance.

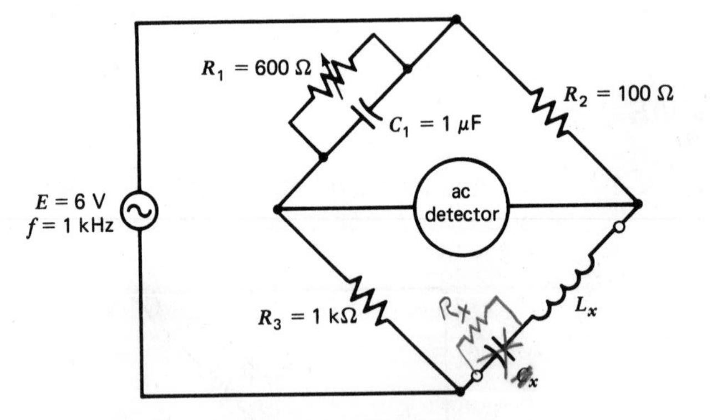

FIGURE 6-12 Circuit for Problem 6-5.

6-6 The Maxwell bridge of Fig. 6-12 is balanced with the following conditions: $Z_1 = 153.8\ \Omega\angle -75°$; $Z_2 = 100\ \Omega\angle 0°$; $Z_3 = 1000\ \Omega\angle 0°$. Find the constants R_x and L_x.

6-7 Given the opposite-angle bridge of Fig. 6-13, find the equivalent-series resistance and inductance (R_x and L_x) at balance.

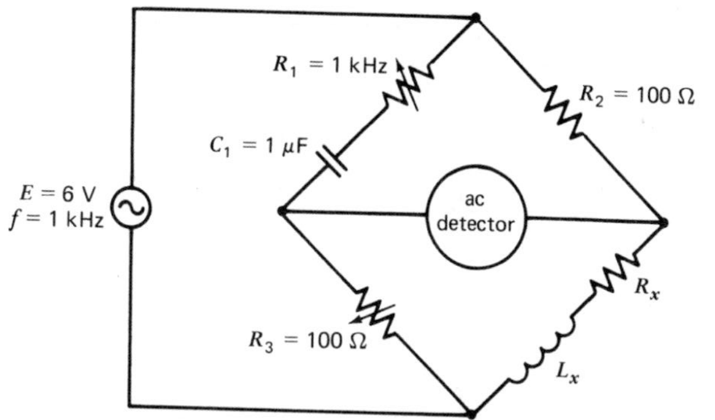

FIGURE 6-13 Circuit for Problem 6-7.

6-8 The opposite-angle bridge of Fig. 6-13 is balanced when $Z_1 = 1500\ \Omega\angle -50°$; $R_2 = 200\ \Omega\angle 0°$; $R_3 = 200\ \Omega\angle 0°$. Find the constants R_x and L_x.

6-9 Given the Wien bridge of Fig. 6-14, find the series-equivalent resistance and capacitance of R_4 and C_4 at balance, when Z_3 equals $7790\ \Omega\angle -72.8°$.

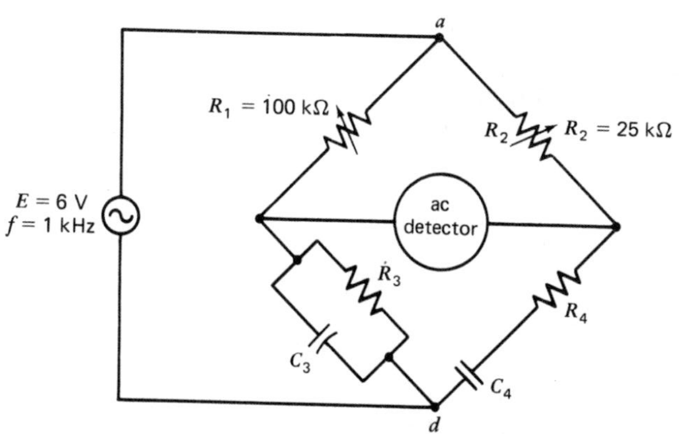

FIGURE 6-14 Circuit for Problem 6-9.

6-10 In the Wien bridge of Fig. 6-14, find the constants of the parallel arms of R_3 and C_3 for the following conditions:

$$Z_1 = 100{,}000\ \Omega\angle 0°$$
$$Z_2 = 25{,}000\ \Omega\angle 0°$$
$$Z_4 = 1050\ \Omega\angle -17.7°$$

6-11 Refer to the balanced radio-frequency bridge of Fig. 6-15 having the Z_x terminals short-circuited. When the unknown impedance is inserted across the Z_x terminals and the bridge is rebalanced, capacitor C_1' now equals 1.1 μF and C_4' equals 1.3 μF. Find the unknown equivalent-series constants of R_x and L_x or C_x.

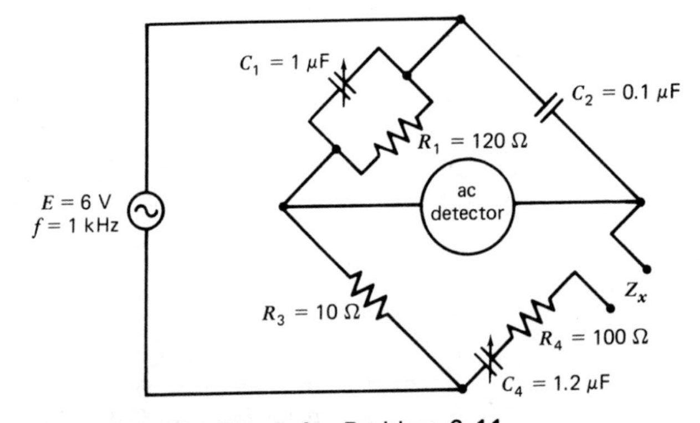

FIGURE 6-15 Circuit for Problem 6-11.

6-12 The radio-frequency bridge of Fig. 6-15 is balanced and Z_x is shorted with the following result conditions:

$$Z_1 = (120 - j159)\ \Omega$$
$$Z_2 = -j1592\ \Omega$$
$$Z_3 = 10\ \Omega$$
$$Z_4 = (100 - j132.6)\ \Omega$$

The unknown impedance is connected across the Z_x terminals and the bridge is rebalanced with the following conditions:

$$Z_1 = (120 - j144.7)\ \Omega$$
$$Z_2 = -j1592\ \Omega$$
$$Z_3 = 10\ \Omega$$
$$Z_4 = (100 - j144.7)\ \Omega$$

Find the equivalent-series elements for the unknown impedance.

6-13 The Schering bridge of Fig. 6-16 is operated at balance. Find the equivalent-series resistance and capacitance of R_x and C_x.

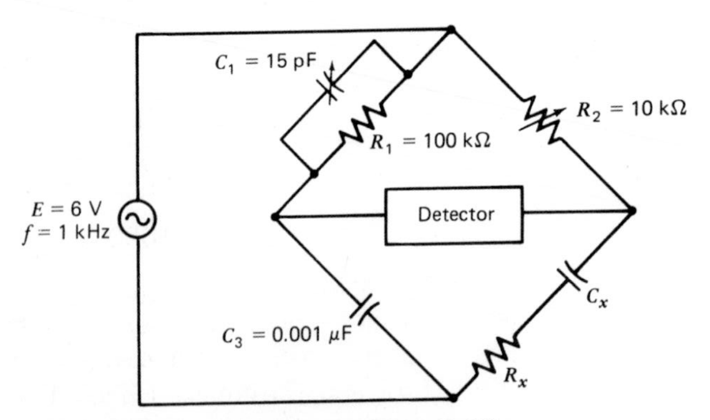

FIGURE 6-16 Circuit for Problem 6-13.

6-14 The balanced Schering bridge of Fig. 6-16 has the following conditions:

$$Z_1 = (2588.2 - j9659.3)\ \Omega$$

$$Z_2 = 10{,}000\ \Omega$$

$$Z_3 = -j159{,}000\ \Omega$$

Find the constants of R_x and C_x.

6-13 LABORATORY EXPERIMENTS

Experiments E12 and E13 are related to the theory presented in Chapter 6. The purpose of the experiments is to provide hands-on experience to reinforce the theory.

All the components and equipment required to perform the experiments, except the vector impedance meter, can be found in any well-equipped electronics laboratory. If your laboratory has no vector impedance meter, simply skip that measurement, for it is not vital to the success of the experiment.

The contents of the laboratory report to be submitted by each student are included at the end of each experimental procedure.

CHAPTER 7

Electronic Measuring Instruments

7-1 Instructional Objectives

This chapter will discuss some basic ideas behind electronic measuring instruments. We will see that electronic voltmeters have fewer loading errors and greater sensitivity than volt–ohm–milliammeters, and that they allow us to measure complex quantities. After completing Chapter 7, you should be able to

1. Define the principles of operation of the electronic voltmeter (EVM).
2. List the different types of EVMs.
3. Interpret readings from the EVM.
4. Analyze or design a basic electronic ohmmeter circuit.
5. Describe the basic principle of operation of the vector impedance meter and the vector voltmeter.

7-2 INTRODUCTION

The *volt–ohm–milliammeter*, or VOM, is a rugged, accurate instrument, but it suffers from certain disadvantages. The principal problem is that it lacks both sensitivity and high input resistance. A sensitivity of 20,000 Ω/V with a 0- to 0.5-V range has an input impedance of only (0.5 V) (20,000 Ω/V), or 10 kΩ.

The **electronic voltmeter** (EVM), on the other hand, can have an input resistance ranging from 10 to 100 MΩ, and the input resistance will remain constant over all ranges instead of being different on each range as in the VOM. The EVM presents less loading to circuits under test than the VOM.

The different types of electronic voltmeters go by a variety of names that tend to reflect the technology used in the bridge balance circuit. The original

EVMs used vacuum tubes, so they were called *vacuum-tube voltmeters* (VTVMs). With the introduction of the transistor and other semiconductor devices, vacuum tubes are no longer used in these instruments. Solid-state models with junction field-effect transistor (JFET) input stages are known as *transistor voltmeters* (TVM) or **field-effect transistor** voltmeters (FET VM). The basic circuit operates on the same principles, however, whatever the technology employed. EVMs fall into two categories, analog and digital, depending on the type of readout used. When a meter movement is used as a readout, the instrument is of the analog type, but if the readout used is one of the various numerical readouts available, the instrument is digital. Thus far, the analog type remains predominant since it is more economical. Furthermore, the analog meter is more suitable for special nonlinear scales such as logarithmic scales.

Although both analog and digital meters must perform the same function (measure voltage, current, and resistance), the principle of operation in each case is quite different. Digital meters will be discussed in a later chapter.

7-3 THE DIFFERENTIAL AMPLIFIER

In this section we examine a special type of amplifier called the **difference amplifier** or the **differential amplifier**. A block diagram of a difference amplifier is shown in Fig. 7-1*a*. It has two input signals (Fig. 7-1*b*) and one output signal (Fig. 7-1*d*). The output is an amplified version of the algebraic difference of the two input signals v_1 and v_2. If v_1 and v_2 are sine waves and in phase, then $v_1 - v_2$ is simply a new sine wave with a peak value equal to the difference in input peak values. The output signal is the difference voltage amplified as shown in Fig. 7-1*d*.

A FET version of the difference amplifier is shown in Fig. 7-2. Let us qualitatively consider a specific example. Suppose v_1 is a 1-V peak sine wave and v_2 is zero as shown in Fig. 7-3. When v_1 is alternating in a positive direction, the gate of the left FET is going positive so that drain current increases. The drain resistor voltage drop increases, producing a negative-going drain voltage from drain to ground, as shown. Recall the same action in a single-stage amplifier when a 180° phase shift occurs.

During the positive half of v_1, the current in the source resistor increases, causing a corresponding positive-going signal, as shown. The source signal is such that it *reverse-biases* the right FET. This reduces the current in the right FET, providing a lower voltage drop across the drain resistor, and we obtain a positive-going signal from the drain of the right FET to ground.

The signals at the drain of each FET are equal in magnitude but 180° out of phase. The output voltage from X to Y will be a sine wave with a peak value of twice the magnitude of either drain signal. If v_1 and v_2 are equal and in-phase sine waves, then $v_1 - v_2$ will be a zero-output signal.

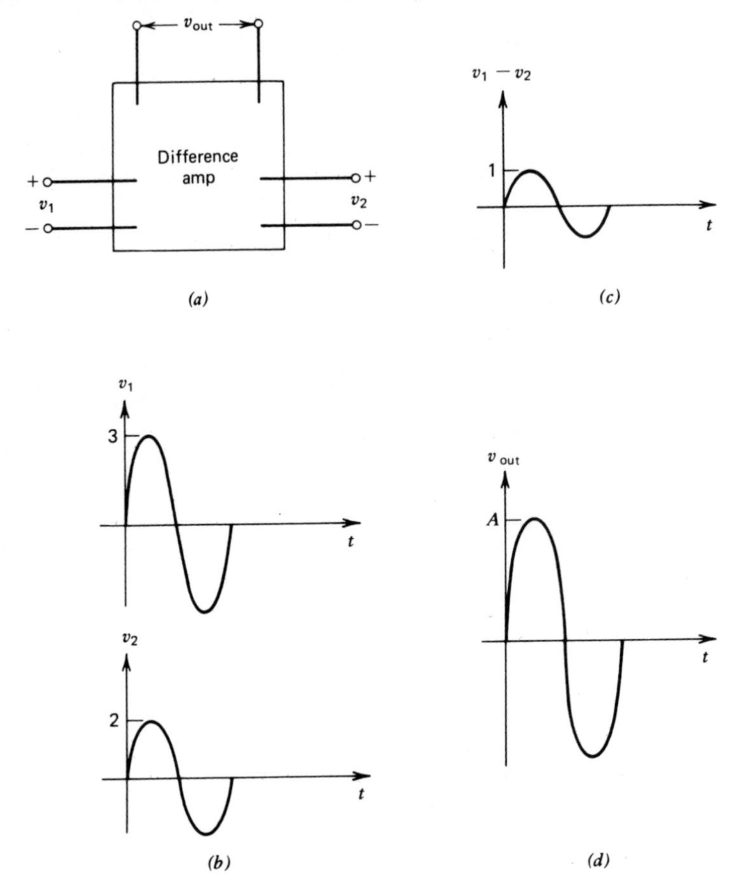

FIGURE 7-1 Difference amplifier. (*a*) Block diagram. (*b*) Input signals. (*c*) Difference of input signals. (*d*) Output of difference amplifier.

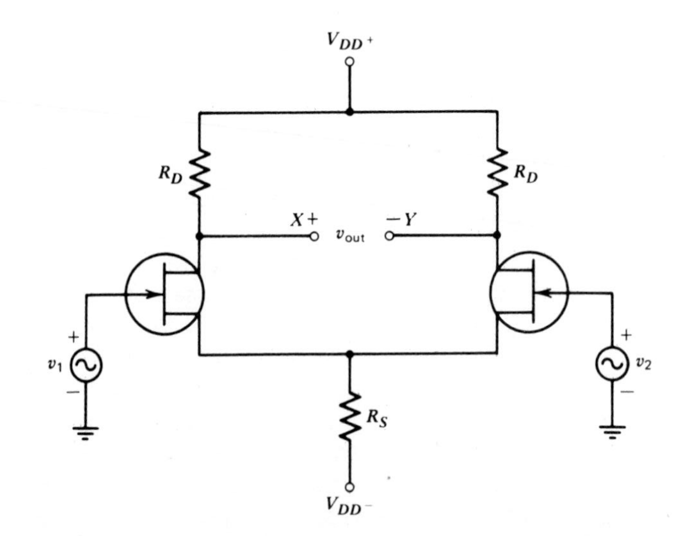

FIGURE 7-2 Difference amplifier of a field-effect transistor.

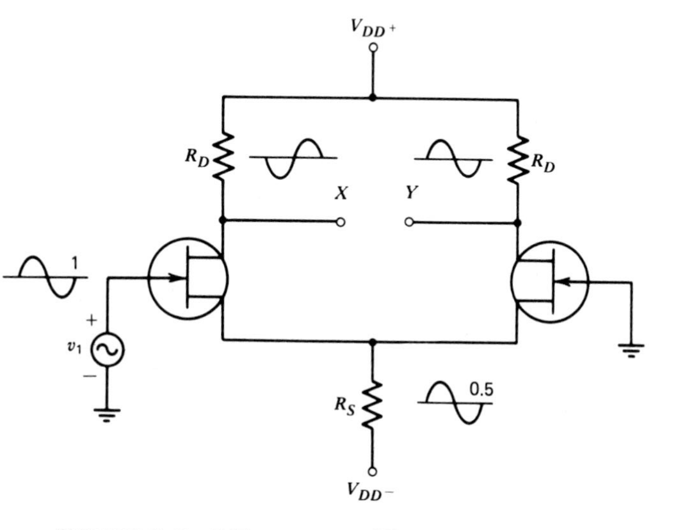

FIGURE 7-3 **Difference amplifier with one input.**

The difference amplifier can be represented as a bridge, as shown in Fig. 7-4. Each FET is represented by a variable dc resistance that is controlled by the input signals. When both input signals are equal, each FET resistance is equal, and we have a balanced bridge. If input signals are unequal, the bridge becomes unbalanced, and an output is produced. Using equivalent circuits for the FETs, we can show that

$$V_X = g_m\left(\frac{r_d R_D}{r_d + R_D}\right)v_1 \tag{7-1}$$

$$V_Y = g_m\left(\frac{r_d R_D}{r_d + R_D}\right)v_2 \tag{7-2}$$

$$v_{out} = V_x - V_y = g_m\left(\frac{r_d R_D}{r_d + R_D}\right)(v_1 - v_2) \tag{7-3}$$

where

r_d = ac drain resistance in ohms
g_m = transconductance in siemens

Notice that the coefficient of $v_1 - v_2$ is the voltage gain of a single common-source amplifier. Therefore, the output voltage is the algebraic difference of v_1 and v_2 multiplied by a constant.

Because of unmatched components, we can add an adjustment for near-perfect balance. A typical circuit uses a potentiometer, as shown in Fig. 7-5, which is adjusted until v_{out} is zero under the no-input signal conditions.

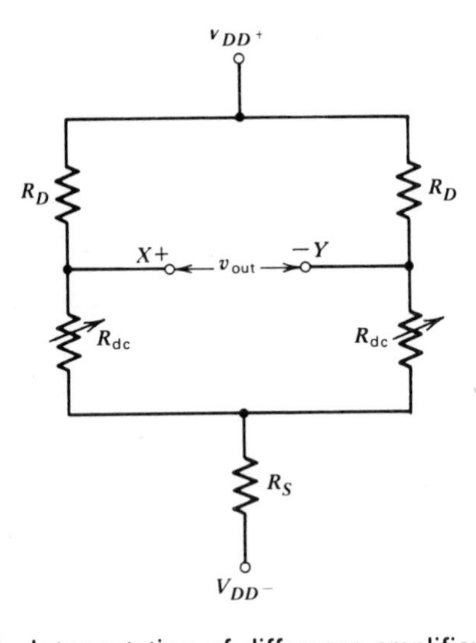

FIGURE 7-4 Interpretation of difference amplifier as a bridge.

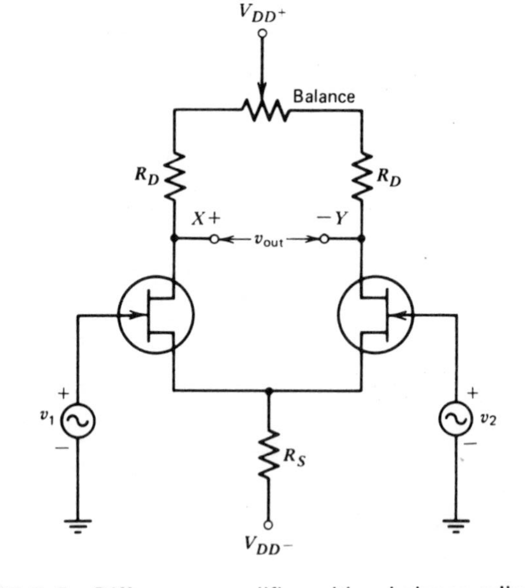

FIGURE 7-5 Difference amplifier with a balance adjustment.

7-4 THE DIFFERENTIAL-AMPLIFIER TYPE OF EVM

Field-effect transistors can be used to increase the input resistance of a dc voltmeter. This isolates the relatively low meter resistance from the circuit under test.

Figure 7-6 shows the schematic diagram of a difference amplifier using

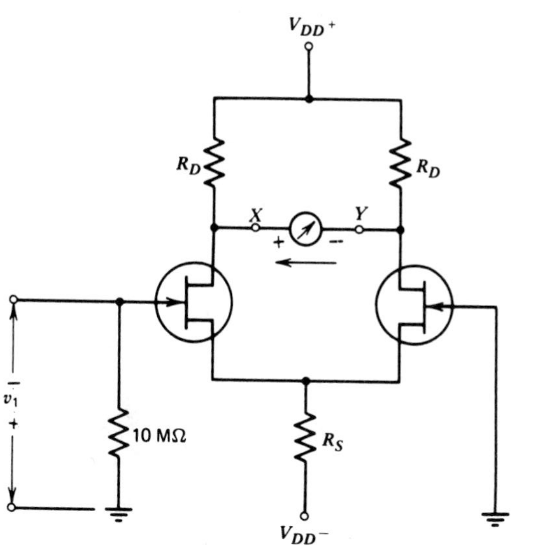

FIGURE 7-6 The difference-amplifier type of EVM.

field-effect transistors. This circuit also applies to a difference amplifier with ordinary bipolar junction transistors or BJTs. The circuit shown here consists of two FETs that should be reasonably matched for current gain to ensure thermal stability of the circuit. Therefore, an increase in source current in one FET is offset by a corresponding decrease in the source current of the other FET. The two FETs form the lower arms of the bridge circuit. Drain resistors R_D together form the upper arms. The meter movement is connected across the drain terminals of the FETs, representing two opposite corners of the bridge.

The circuit is balanced when identical FETs are used, so that for a zero input no current flows through the ammeter. If a negative dc voltage is applied to the gate of the left FET, a current will flow through the ammeter in the direction shown in Fig. 7-6. The size of this current depends on the magnitude of the input voltage. By properly designing the circuit, we can make the ammeter current directly proportional to the dc voltage across the input. Thus, the ammeter can be calibrated in volts to indicate the input voltage.

By using Thévenin's theorem, we find the relation between the ammeter current and the input dc voltage; the ammeter is considered the load. If we remove the ammeter, the circuit of Fig. 7-7*a* is seen. The output voltage is the voltage gain of a single FET times the difference of v_1 and v_2. Since v_2 is zero, the output voltage under open-circuit conditions is

$$v_{\text{out}} = g_m\left(\frac{r_d R_D}{r_d + R_D}\right)v_1 = g_m(r_d \| R_D)v_1 \tag{7-4}$$

To find the Thévenin resistance at terminals X and Y we first set v_1 and V_{DD} equal to zero. Under this condition both FETs have a resistance of r_d as

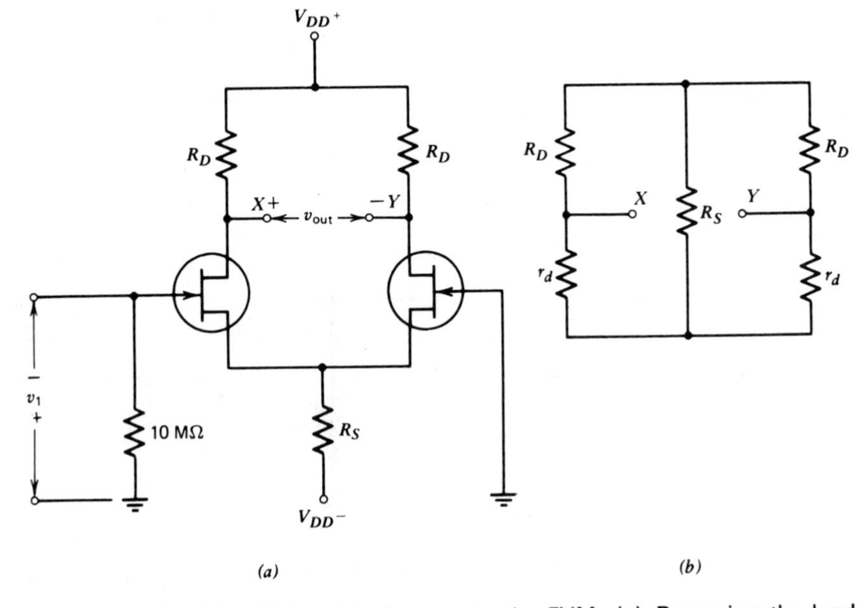

FIGURE 7-7 Applying Thévenin's theorem to the EVM. (*a*) Removing the load to V_{out}. (*b*) Setting all voltages equal to zero to find R_{Th}.

shown in Fig. 7-7*b*. From this balance circuit and assuming R_S is relatively large, we see that the resistance between X and Y terminals is

$$R_{Th} \approx 2r_d \| 2R_D = 2(r_d \| R_D) \tag{7-5}$$

The Thévenin equivalent circuit with the ammeter connected as a load is shown in Fig. 7-8. From the circuit the ammeter current is found as

$$i = \frac{V_{out}}{R_{Th} + R_m} = \frac{g_m (r_d \| R_D)}{2(r_d \| R_D) + R_m} V_1 \tag{7-6}$$

When $R_D \ll r_d$, Eq. 7-6 simplifies to

$$i = \frac{g_m R_D}{2R_D + R_m} V_1 \tag{7-7}$$

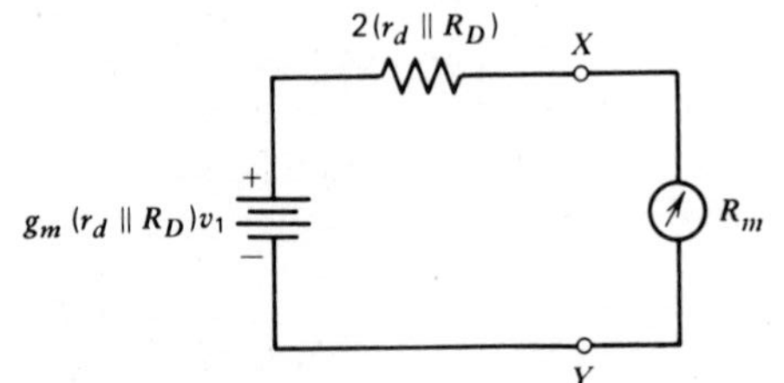

FIGURE 7-8 Equivalent circuit for the difference-amplifier type of EVM.

Equation 7-7 relates the ammeter circuit to the input dc voltage. It assumes that both FETs are identical. For nonidentical FETs, an approximate relation can be obtained by using the average value of g_m and r_d.

EXAMPLE 7-1

Given a difference-amplifier type of FET VM, find the ammeter current under the following conditions.

$$v_1 = 1\text{ V} \qquad R_D = 10\text{ k}\Omega$$

$$r_d = 100\text{ k}\Omega \qquad R_m = 50\ \Omega$$

$$g_m = 0.005\text{ siemens}$$

Solution

Find i using Eq. 7-6.

$$i = \frac{g_m(r_d \| R_D)}{2(r_d \| R_D) + R_m} v_1$$

$$= \frac{(0.005)[(100\times 10^3)\|(10\times 10^3)](1)}{2[(100\times 10^3)\|(10\times 10^3)] + 50}$$

$$= 2.50\text{ mA}$$

This final result indicates that the ammeter current is *directly proportional* to the input voltage. Hence, the ammeter may be marked off in uniform divisions. The result also tells us that if we wished to have a 0- to 1-V indicator, we would mark 1 V at full scale on an ammeter with a full-scale current of 2.5 mA. The remainder of the scale would be marked off in a linear manner.

The difference amplifier studied so far can be made practical by adding adjustments and switches, as shown in Fig. 7-9. Bridge balance, or zero meter current, is obtained by adjusting the **zero set** potentiometer, the purpose of which is to equalize both halves of the difference amplifier under zero-signal conditions.

Full-scale adjustment of **calibration** is effected by the potentiometer marked calibration, in series with meter movement internal resistance. This adjustment is necessary because g_m and r_d are different from FET to FET. From Eq. 7-6 it should be clear that for a fixed value of v_1, different values of current will flow if g_m and r_d change from FET to FET. Thus, if the ammeter is marked 1 V at full scale, we calibrate by measuring an input that is exactly 1 V. The calibration potentiometer is then adjusted to give a reading of exactly 1 V.

The range switch on the input allows several different full-scale ranges. The input resistance is 9 MΩ on any position. With the switch on the position shown, the voltmeter is on its most sensitive range and will read up to 1 V at full scale. For higher voltage, the range switch is moved to a lower attenuation point. For example, if the input voltage is 10 V, the range switch must be moved to the 10-V range. The voltage at the gate of the left FET is developed

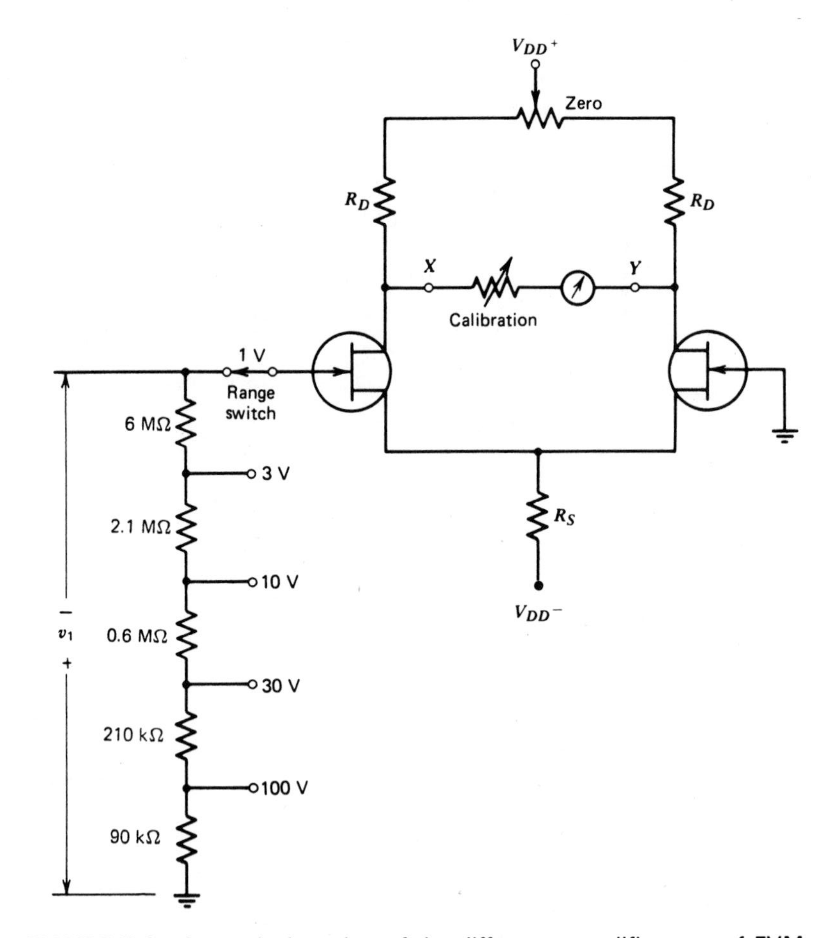

FIGURE 7-9 A practical version of the difference-amplifier type of EVM.

across 900 kΩ of the total resistance of 9 MΩ. The voltage at this gate is

$$V_G = \frac{(900 \times 10^3\ \Omega)(10\ \text{V})}{9 \times 10^6\ \Omega} = 1\ \text{V}$$

Thus, the meter deflects full scale with 10 V applied to the input, causing 1 V to be applied to the gate. With the range switch in the 100-V position, the gate voltage is developed across 90 kΩ of the total voltage resistance. The voltage at the gate is

$$V_G = \frac{(90 \times 10^3\ \Omega)(100\ \text{V})}{9 \times 10^6\ \Omega} = 1\ \text{V}$$

Again, the meter deflects full scale with 100 V applied to the input. The voltage divider provides 1 V at the gate of the FET.

7-5 THE SOURCE-FOLLOWER TYPE OF EVM

Another of the basic FET VM circuits is shown in Fig. 7-10. The two FETs form the upper arms of a bridge circuit, and source resistors together form the lower arms. When the input is zero, the bridge is balanced and there is no ammeter current. If v_1 is positive input voltage, current flows through the ammeter in the direction indicated.

By using Thévenin's theorem at terminals X and Y, we can find the relation between the ammeter current and the input voltage. The open-circuit voltage is found by first removing the ammeter, as shown in Fig. 7-11*a*.

Since the gate of the right FET is grounded, no signal is applied to its gate, and therefore no signal appears at terminal Y. The left FET, however, is a source follower. This is a **negative-feedback** amplifier with the feedback factor $\beta = -1$. Therefore, the gain can be expressed as

$$A' = \frac{A}{1 - A\beta} = \frac{A}{1 + A} = \frac{g_m(r_d \| R_S)}{1 + g_m(r_d \| R_S)} = \frac{g_m r_d R_S}{r_d + R_S + g_m r_d R_S} \qquad (7\text{-}8)$$

The output voltage from a source follower is

$$v_{\text{out}} = \frac{g_m r_d R_S}{r_d + R_S + g_m r_d R_S} v_1 = \frac{g_m \left(\dfrac{r_d}{g_m r_d + 1} \right) R_S}{\dfrac{r_d}{g_m r_d + 1} + R_S} v_1 \qquad (7\text{-}9)$$

Comparing this equation with Eq. 7-4 for the gain of the common-source circuit reveals that they are of the same form if $r_d/(g_m r_d + 1)$ replaces r_d in the common-source equation.

Thus, $r_d/(g_m r_d + 1)$ can be considered to be the output resistance of the FET when used in the source-follower circuit. The output equation is in

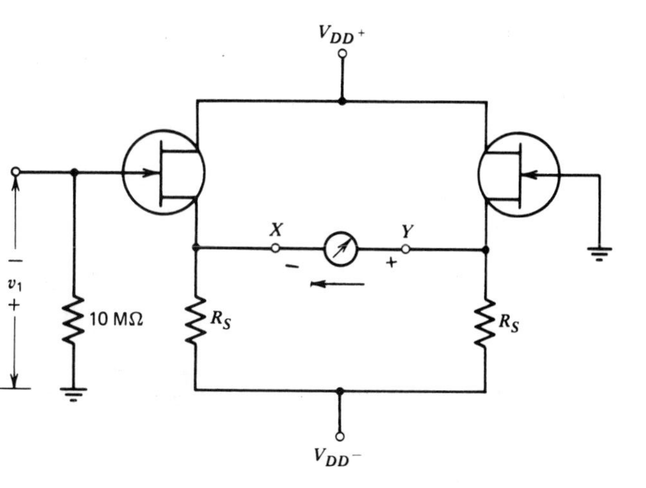

FIGURE 7-10 The source-follower EVM.

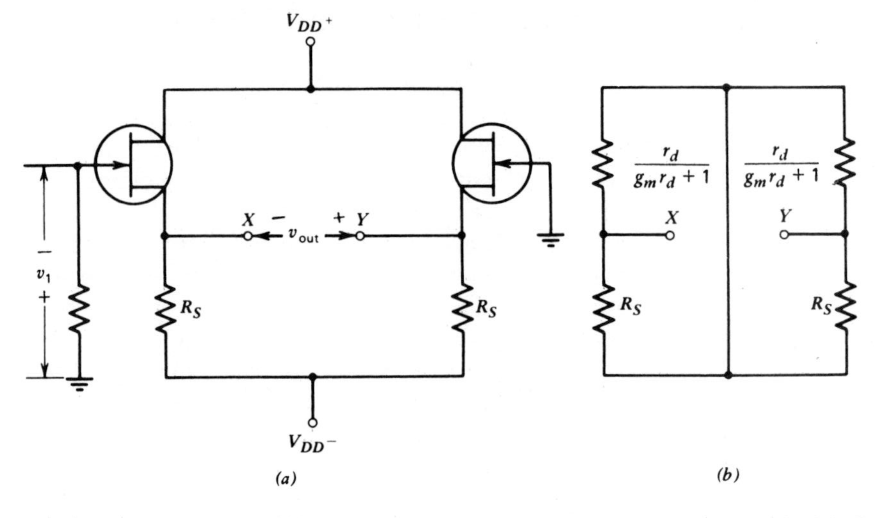

FIGURE 7-11 Applying Thévenin's theorem to the source-follower EVM. (*a*) Removing the ammeter to find V_{oc}. (*b*) Setting all voltages equal to zero to find R_{Th}.

the same simple form as that of the common-source amplifier, except that r_d as specified in the FET data sheet is replaced by $r'_d = r_d/(g_m r_d + 1)$, or

$$V_{out} = g_m \left(\frac{r'_d R_S}{r'_d + R_S} \right) V_1 \tag{7-10}$$

To find the Thévenin resistance looking back into the X and Y terminals, we see that the resistance looking into the source is $r_d/(g_m r_d + 1)$, as shown in Fig. 7-11*b*. From this circuit we find that the Thévenin resistance is

$$R_{Th} = R_S \left\| \frac{r_d}{g_m r_d + 1} + R_S \right\| \frac{r_d}{g_m r_d + 1} \tag{7-11}$$

$$= 2\left(R_S \left\| \frac{r_d}{g_m r_d + 1} \right. \right)$$

The circuit of Fig. 7-12 shows the Thévenin equivalent with the ammeter connected across the X and Y terminals. The current through the ammeter is

$$i = \frac{V_{out}}{R_{Th} + R_m}$$

$$= \frac{g_m r_d R_S}{r_d + R_S + g_m r_d R_S} \left[\frac{V_1}{2\left(R_S \left\| \frac{r_d}{g_m r_d + 1} \right. \right) + R_m} \right] \tag{7-12}$$

To obtain approximation, we observe that $g_m r_d R_S$ is much greater than r_d or R_S for most FETs. Furthermore, R_S is often much larger than $r_d/(g_m r_d + 1)$. Under these conditions, Eq. 7-12 can be simplified to

$$i \approx \frac{V_1}{2r_d/g_m r_d + R_m} = \frac{V_1}{2/g_m + R_m} \tag{7-13}$$

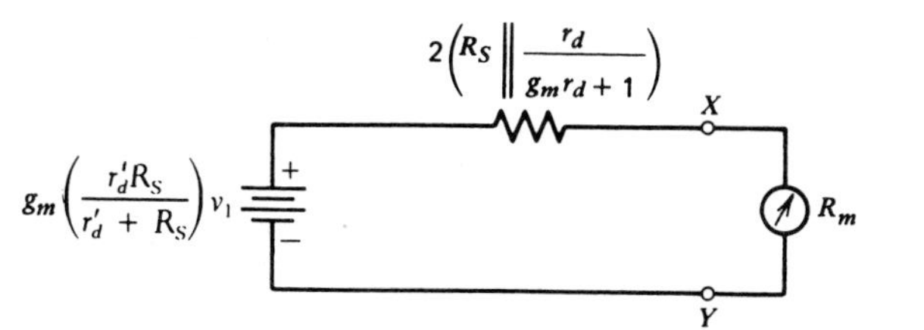

FIGURE 7-12 Equivalent circuit for source-follower EVM.

EXAMPLE 7-2

The EVM of Fig. 7-10 has the following values: $R_S = 50\ \text{k}\Omega$, $r_d = 100\ \text{k}\Omega$, and $g_m = 0.0057$ siemens. A 50-Ω ammeter meter movement is used. If the input voltage is exactly 1 V, find the approximate ammeter current.

Solution

$$i = \frac{V_1}{2/g_m + R_m} = \frac{1}{2/0.0057 + 50}$$
$$= 2.50\ \text{mA}$$

In this example we see that the ammeter current in an EVM is directly proportional to the input voltage. An input voltage of 1 V produces an ammeter current of 2.5 mA. If an ammeter with a 2.5-mA full-scale deflection could be found, it would be marked 1 V at full scale, 0.5 V at midscale, and so on, in a linear fashion to produce a 0- to 1-V indicator.

The circuit of Fig. 7-10 can be made practical by adding a zero adjust, a calibration adjust, and a voltage divider on the input, as shown in Fig. 7-13.

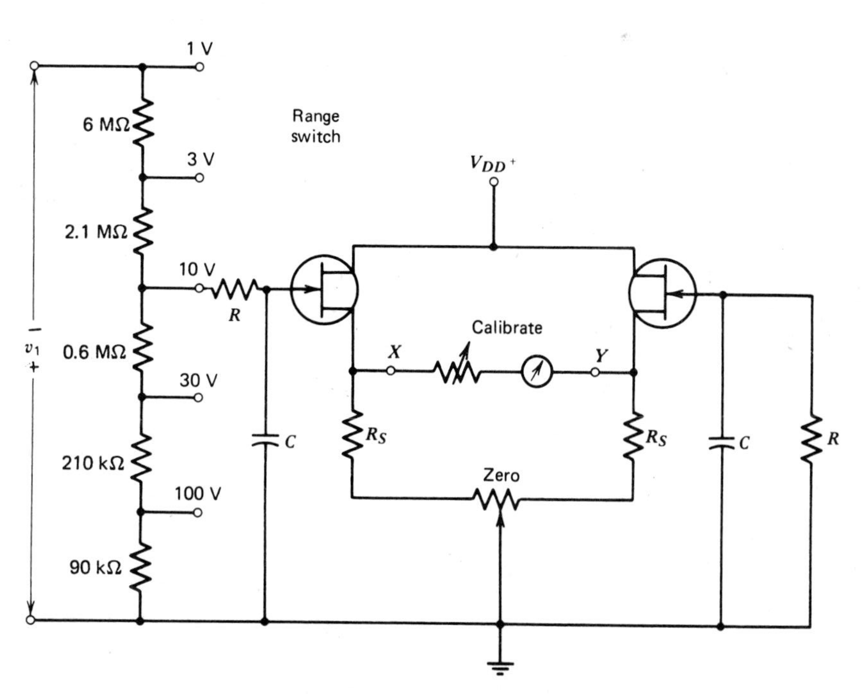

FIGURE 7-13 A practical version of the source-follower EVM.

The purpose of the adjustments and the voltage divider is the same as in our discussion in Section 7-4.

Notice that on any position of the range switch the input resistance seen from the measuring terminals is 9 MΩ. This is typical of most EVMs, although even higher input resistances are possible.

Many variations of the EVM shown in Fig. 7-13 appear in commercial instruments. Nevertheless, the circuit of Fig. 7-13 represents the basic idea behind the source-follower type of FET VM.

EXAMPLE 7-3

The EVM of Fig. 7-13 has the following values: $R_S = 50$ kΩ, $r_d = 100$ kΩ, and $g_m = 0.003$ siemens. A 50-Ω ammeter with a full-scale current of 1 mA is used. If the input voltage is exactly 1 V, what value of calibration resistance will produce full-scale current?

Solution

First, note that since $g_m r_d R_S$ is much larger than r_d or R_S and $r_d/(g_m r_d + 1)$ is much smaller than R_S, we can use the approximation given in Eq. 7-13. Therefore,

$$i = \frac{V_1}{2/g_m + R_m + R_{CAL}}$$

Rearranging and solving for R_{CAL}, we get

$$R_{CAL} = \frac{V_1}{i} - (2/g_m + R_m)$$

$$= 1/1 \times 10^{-3} - (2/0.003 + 50) \qquad (7\text{-}14)$$

$$= 283.33\ \Omega$$

Note that a 1-kΩ calibration resistor may be used in the circuit. Then as FETs are changed, the rheostat may be adjusted to take care of variations in g_m and r_d.

7-6 DC VOLTMETER WITH DIRECT-COUPLED AMPLIFIER

The dc electronic voltmeter usually consists of an ordinary dc meter movement preceded by a dc amplifier of one or more stages. When very high input resistance is required, it is convenient to use a FET for the input stage. The output of the FET can usually be directly coupled to the input of a BJT.

Direct-coupled amplifiers are one group that is commonly found in lower-priced dc voltmeters. Figure 7-14 shows a FET input direct-coupled dc amplifier. Bipolar transistor Q_2 along with resistors form a balanced bridge circuit. Field-effect transistor Q_1 serves as a source follower and is used to provide impedance transformation between the input and the base of Q_2.

The bias on Q_2 is such that $I_2 = I_3$ when V_{in} is zero. Under that condition, $V_x = V_y$, so that the the current through the dc movement is zero, $I_4 = 0$. The

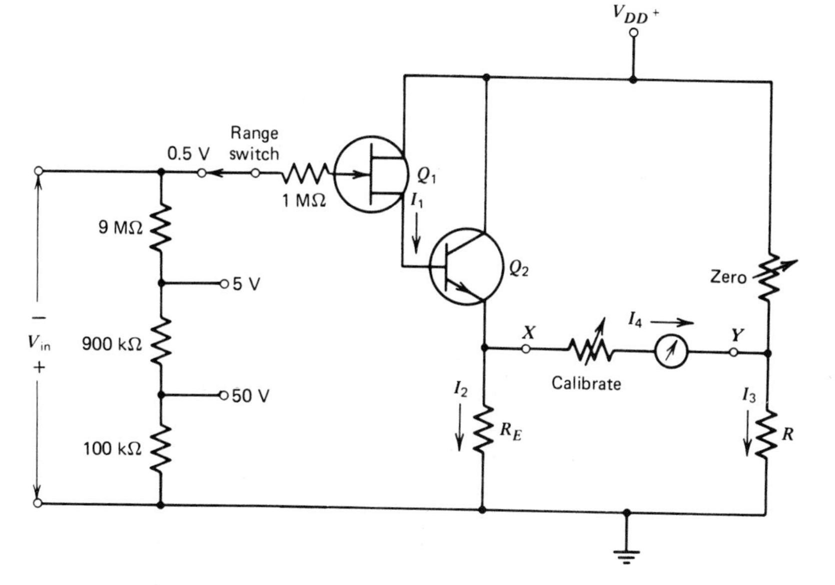

FIGURE 7-14 **Basic dc voltmeter circuit with FET input.**

bias on Q_2 is controlled by V_{in}, so that when an unknown input voltage V_{in} is applied, the bias on Q_2 increases, which causes voltage V_x to increase. Now, since V_x is greater than V_y, current I_4 is no longer zero. The magnitude of this meter current, hence the deflection of the meter, is *proportional* to V_{in}.

The value of V_{in} that causes maximum meter deflection is the basic range of the instrument. This is generally the lowest range on the Range switch in nonamplified models. Higher ranges can be obtained by using an input attenuator, and lower ranges can be obtained by a preamplifier.

The input attenuator in Fig. 7-14 is a calibrated front panel control in the form of a resistance voltage divider. The full-scale voltage appears across the divider so that the voltage at each tap is a progressively lower fraction of the full input voltage. (Analysis of the input attenuator was presented in Section 7-4.)

Bridge balance is obtained by adjusting the zero set potentiometer when V_{in} is zero. Full-scale calibration is obtained by adjusting the potentiometer marked calibration in series with the ammeter.

The principal advantage of this voltmeter is the high input resistance. The principal disadvantage is zero drift of the order of 1 mV/hr.

7-7 A SENSITIVE DC VOLTMETER

In order to overcome the drift limitation associated with direct-coupled dc amplifiers, we must resort to a different approach. One very popular method for voltmeters with a high-sensitivity rating is the use of the **chopped dc amplifier** as shown in Fig. 7-15.

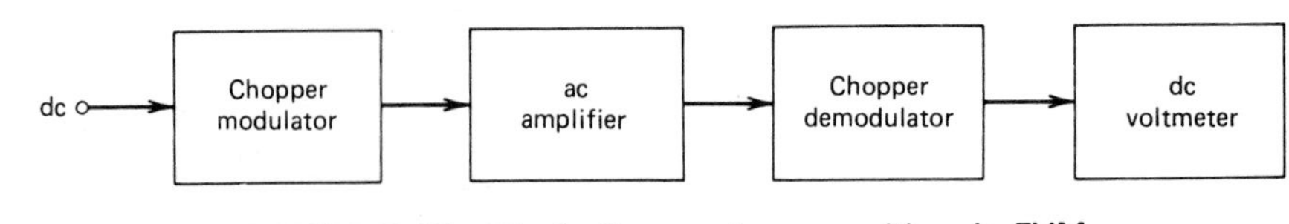

FIGURE 7-15 **Block diagram for a sensitive dc EVM.**

Notice that first the direct current is converted to an ac signal (modulation), which is then amplified in a standard ac amplifier and finally converted back to a dc voltage (**demodulation**) proportional to the original input signal. Many designers prefer ac amplifiers because dc amplifiers present more severe design problems (e.g., drift) than those of ac amplifiers. Choppers may be mechanical or electronic. The more interesting electronic choppers use **photocells** and photodiodes.

The circuit of Fig. 7-16 illustrates the operation of the electronic chopper amplifier. Photodiodes are used for modulation and demodulation. A photodiode changes from a high to a low resistance when illuminated by a light source such as a neon or incandescent lamp. The photodiode resistance increases greatly when it is not illuminated.

An oscillator causes two neon lamps to illuminate on alternate half-cycles. Each neon lamp illuminates one photodiode in the input and one in the output circuit. The two photodiodes in the input form a series shunt half-wave **modulator** or **chopper**. The action is like a switch that creates an ac voltage whose amplitude is proportional to the level of the input voltage and a frequency equal to the oscillator frequency.

A square wave is applied to the input of the amplifier which delivers an amplified square wave at its output. The two photodiodes in the amplifier

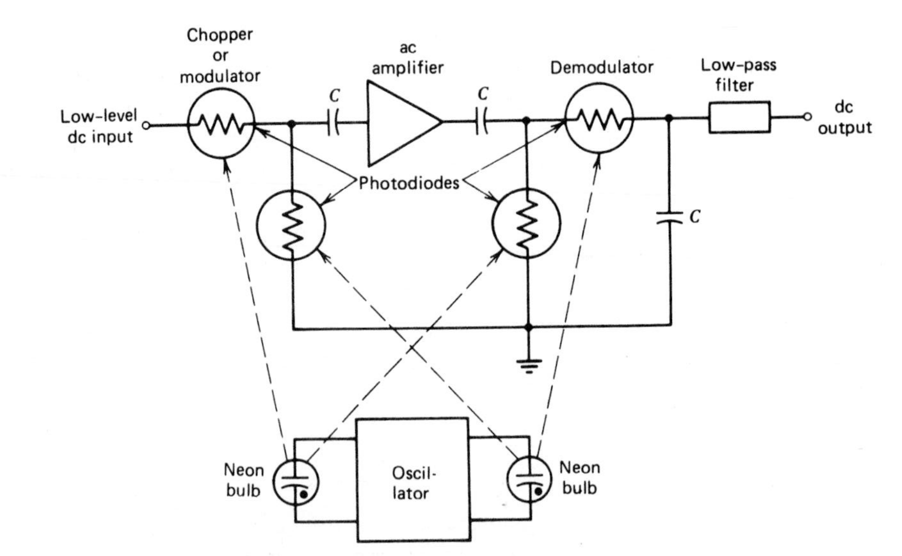

FIGURE 7-16 **Nonmechanical photoconductive chopper.**

output demodulate this signal which charges the capacitor to the peak of the output voltage. A low-pass filter removes any residual ac component, and the final voltage is applied to a meter movement.

The principal advantage of this meter system is the attainment of full-scale sensitivities of 1 μV or less and reduced zero drift. The principal disadvantages are the sacrifice of very high impedance and the presence of noise voltages, whether the chopping is done mechanically or by photoconductive means.

7-8 ALTERNATING-CURRENT VOLTMETER USING RECTIFIERS

The dc EVM discussed previously may be used to measure ac voltages by first detecting the alternating voltage, as shown in Fig. 7-17. In some situations, rectification takes place before amplification, in which case a simple diode circuit precedes the amplifier and meter as in Fig. 7-17*a*. This amplifier ideally requires zero-drift characteristics and unity voltage gain, a dc meter movement with adequate sensitivity.

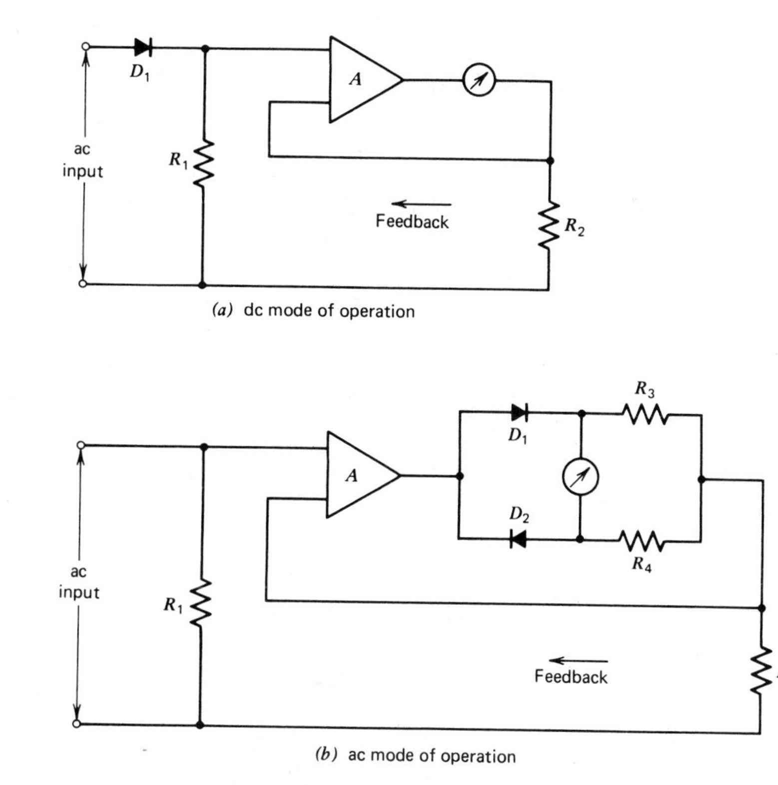

FIGURE 7-17 Basic ac voltmeter circuits. (*a*) The ac signal is first rectified, then amplified. (*b*) The ac signal is first amplified, then rectified.

In another method, rectification takes place after amplification as shown in Fig. 7-17*b*. This method generally uses a high open-loop gain and large negative feedback to overcome the nonlinearity of the rectifier diodes.

Alternating-current voltmeters that use half-wave or full-wave rectification are usually of the average-responding type, with the meter scale calibrated in terms of the rms value of a waveform instead of the average value. Thus, most meters are calibrated in terms of both rms and peak values. Since so many waveforms encountered in electronics are sinusoidal, these methods are satisfactory and much less expensive than a true rms-reading voltmeter. Nonsinusoidal waveforms, however, will cause this type of meter to read high or low, depending on the **form factor** of the waveform. The form factor is the ratio of the rms value to the average value of this waveform. It can be expressed as

$$k = \frac{V_{rms}}{V_{av}} \tag{7-15}$$

The main advantage of the ac voltmeter is that using negative feedback greatly reduces the response time. However, there is the disadvantage of reduced sensitivity, unless compensated by corresponding larger open-loop gains that are normally required.

7-9 TRUE RMS VOLTMETER

Complex waveforms are most accurately measured with a **true rms voltmeter**. This instrument indicates the rms value of any waveform (such as a sine wave, square wave, or sawtooth wave) by using an rms detector that responds directly to the heating value of the input signal.

To measure the rms value of an arbitrary waveform, we may feed an input signal to a heating element in close proximity to a **thermocouple**, as shown in Fig. 7-18.

Recall that a thermocouple is a junction of two dissimilar metals whose contact potential is a function of the temperature of the junction. The heater raises the temperature of the thermocouple and produces an output voltage that is proportional to the power delivered to the heater.

$$P = \frac{V^2_{rms}}{R_{heater}} \tag{7-16}$$

$$v_\theta = f(P) = f\left(\frac{V^2_{rms}}{R_{heater}}\right) = KV^2_{rms} \tag{7-17}$$

where K is the constant of proportionality.

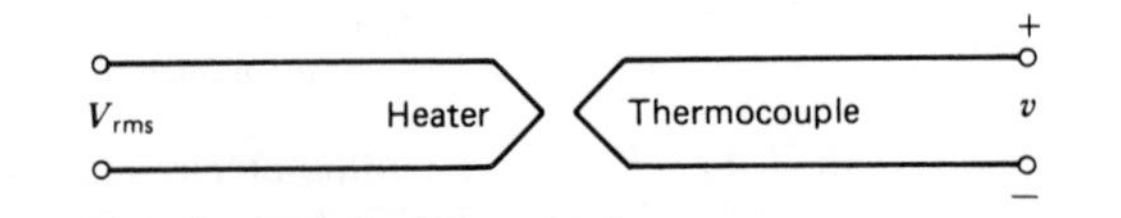

FIGURE 7-18 An rms detector using a thermocouple.

The value of K in Eq. 7-17 depends on the distance between the heater and the thermocouple and on the materials used in the heater and thermocouple. The difficulty with this method is that the thermocouple often displays a nonlinear characteristic. This problem is overcome in some instruments by placing two thermocouples in the same environment as shown in Fig. 7-19. The nonlinear characteristic of the input-measuring thermocouple is canceled by similar nonlinear effects of the balancing thermocouple in the feedback circuit. The two thermocouples form part of a balanced bridge applied to the input circuit of a dc amplifier. The ac input voltage is applied to the heater of the measuring thermocouple. An output voltage v_1 is produced that upsets the balance of the bridge. This unbalanced voltage is amplified by the dc amplifier and fed back to the heater of the balancing thermocouple. When the output of both thermocouples is equal, bridge balance will be reestablished. Thus, the dc feedback current is equal to the ac current in the input thermocouple. This dc current is therefore directly proportional to the rms value of the input voltage and is indicated on the dc voltmeter in the output circuit of the dc amplifier. To verify this we observe that

$$V_{out} = A(v_1 - v_2) \tag{7-18}$$

where A is the voltage gain of the dc amplifier. Rearranging Eq. 7-18, we get

$$v_1 - v_2 = \frac{V_{out}}{A} \approx 0 \tag{7-19}$$

when A is a very large number for a high-gain amplifier. Therefore

$$v_1 \approx v_2 \tag{7-20}$$

From Eq. 7-20 it is clear that

$$KV_{rms}^2 = KV_{out}^2$$

or

$$V_{rms} = V_{out} \tag{7-21}$$

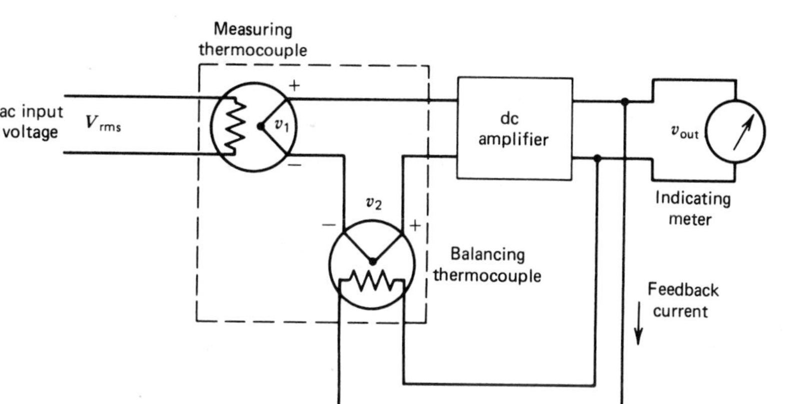

FIGURE 7-19 Block diagram of true rms-reading voltmeter. The measuring and balancing thermocouples are located in the same thermal environment.

The last result tells us that the voltage measured by the dc voltmeter is approximately equal to the rms value of the input signal. Hence, we have an rms voltmeter. The true rms value is measured independently of the waveform of the ac signal; thus, the waveform of the input signal is immaterial. If the ac input voltage is very small, an ac amplifier may be used to increase the signal level before applying it to the input thermocouple. Sensitivities in the millivolt region are possible with such an arrangement.

7-10 THE ELECTRONIC OHMMETER

A basic electronic ohmmeter is shown in Fig. 7-20. As can be seen, this circuit uses an operational amplifier (op-amp) configured as a noninverting amplifier with a gain of one. The very high input impedance seen looking into the noninverting input of the op-amp effectively isolates the meter movement from the resistive circuitry of the ohmmeter. The terminals to which a resistor of unknown value should be connected are identified as x and x'. The circuit is best analyzed by use of Thévenin's theorem to find an equivalent circuit for the resistive network as seen at points A and B. We must either disconnect the op-amp or treat it as an infinite impedance, which we can justifiably do since it does represent a very high impedance. Thévenin's equivalent voltage is computed as

$$V_{\mathrm{Th}} = V \frac{R_2}{R_1 + R_2} \tag{7-22}$$

and Thévenin's equivalent resistance is computed as

$$R_{\mathrm{Th}} = \frac{R_1 R_2}{R_2 + R_2} \tag{7-23}$$

The resistive circuitry is now replaced with Thévenin's equivalent circuit as shown in Fig. 7-21.

By observation we can see that if points x and x' are shorted together, which represents a measurement of zero ohms, the input voltage to the op-amp is 0 V. Therefore, the meter movement should show an indication of

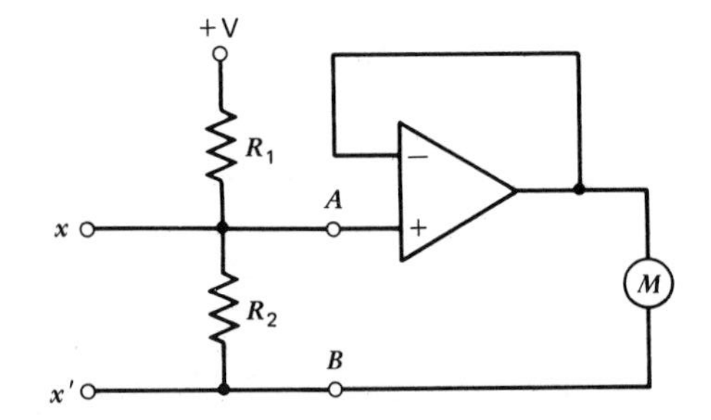

FIGURE 7-20 Basic electronic ohmmeter circuit.

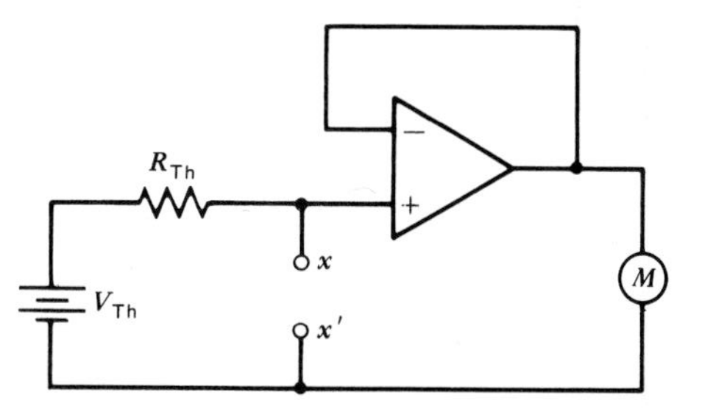

FIGURE 7-21 **Circuit of Fig. 7-20 with input Thévenized.**

0 Ω. In addition, if the test points are open indicating an infinite resistance, the meter movement should show full-scale deflection; if there is an unknown resistance, R_x, that is equal in value to R_{Th}, the meter movement should show exactly half-scale deflection.

The best way to show these relationships quantitatively is to start with a meter movement with known parameters and work back to the input as shown in the following example.

EXAMPLE 7-4

Assume a meter movement with 50-μA full-scale deflection current and 2-kΩ internal resisance and solve for R_1, R_2, and V for the circuit in Fig. 7-22.

Solution

Given the full-scale delfection current and internal resistance of the meter movement, we can solve for the output voltage of the op-amp that will cause the meter movement to deflect full scale.

$$V_o = I_{fs} R_m \tag{7-24}$$

$$= (50\ \mu\text{A})(2\ \text{k}\Omega) = 100\ \text{mV}$$

Since the op-amp is configured to have a gain of 1, its input voltage will equal its output voltage. The meter movement should deflect full scale when R_x equals an infinite resistance. When R_x equals infinite resistance, the input

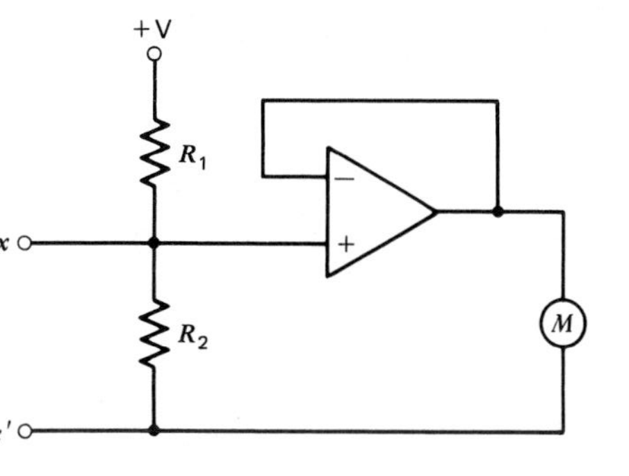

FIGURE 7-22 **Circuit for Example 7-4.**

voltage to the op-amp is Thévenin's equivalent voltage. Therefore, we can say

$$V_{\text{Th}} = V_o = 100 \text{ mV}$$

At midscale on the meter movement

$$I_m = 0.5 I_{fs} = 25\ \mu\text{A}$$

or

$$V_o = (25\ \mu\text{A})(2\ \text{k}\Omega) = 50 \text{ mV}$$

Since $V_o = \frac{1}{2} V_{\text{Th}}$ we can see that $R_x = R_{\text{Th}}$. Although R_{Th} can be any parallel combination of R_1 and R_2, we will make them equal in value. Their values will determine the midscale marking on the ohmmeter scale. Suppose we want the midscale reading to be 100 **Ω**, which means the pointer should deflect to midscale when $R_x = 100\ \Omega$. R_{Th} also equals 100 **Ω**, and both R_1 and R_2 equal 200 **Ω**. If $V_{\text{Th}} = 100$ mV, then we can solve for V by use of the voltage divider equation

$$V_{\text{Th}} = V \frac{R_2}{R_1 + R_2}$$

Solving for V yields

$$V = V_{\text{Th}} \frac{R_1 + R_2}{R_2}$$

or

$$= (100 \text{ mV})\left(\frac{200\ \Omega}{100\ \Omega}\right) = 200 \text{ mV}$$

We can construct a multiple-range electronic ohmmeter by incorporating a switching arrangement such as the one shown in Fig. 7-23. If both R_1 and R_2 equal 20 **Ω**, then

$$R_{\text{Th}} = R_1 \| R_2 = 10\ \Omega$$

An unknown resistor of 10 **Ω** will cause half-scale deflection. Therefore, midscale is marked as 10 **Ω**.

If R_1' and R_2' equal 200 **Ω**, then

$$R_{\text{Th}} = R_1' \| R_2' = 100\ \Omega$$

An unknown resistor of 100 **Ω** will cause half-scale deflection on the $R \times 10$ range, which means the unknown resistor equals a reading of 10 times 10 or 100 **Ω**.

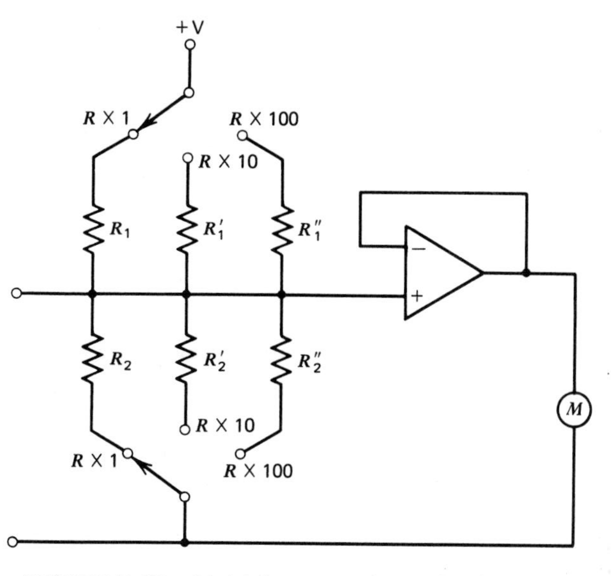

FIGURE 7-23 Multiple-range electronic ohmmeter.

If both R_1'' and R_2'' equal 2 kΩ, then

$$R_{Th} = R_1'' \| R_2'' = 1\ \text{k}\Omega$$

An unknown resistor of 1 kΩ will cause half-scale deflection on the $R \times 100$ range, which now means the unknown resistor equals a reading of 10×100 or 1 kΩ.

7-11 VECTOR IMPEDANCE METER

The value of an impedance is expressed in terms of magnitude Z and phase angle θ. The impedance may be due to a single component or a combination of components. At higher ratio frequencies (i.e., beyond 10 MHz), accurate calculations become very difficult because sources of inductance and capacitance, such as stray inductance and capacitance, become difficult to account for. Often it is more practical, and accurate, to measure the impedance of interest rather than to attempt the calculations, when working in the radio-frequency range or higher.

The vector impedance meter is an instrument designed to measure both the magnitude and the phase of an impedance and to display both values simultaneously. The instrument shown in Fig. 7-24 permits simultaneous measurements of the magnitude and the phase angle of impedance over the frequency range of 400 kHz to 110 MHz. The impedance to be measured is simply connected to the input terminals of the instrument; the instrument output is set to the desired frequency with front panel controls; and the

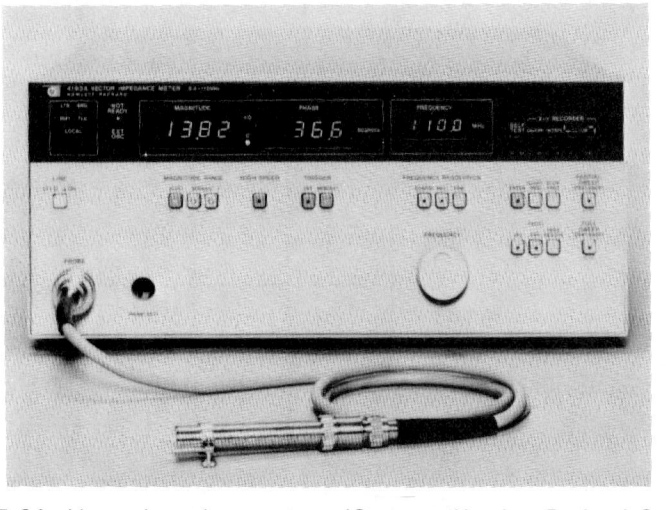

FIGURE 7-24 Vector impedance meter. (Courtesy Hewlett-Packard Company.)

magnitude and phase angle of the unknown are read directly from the panel meters on the front panel of the instrument.

There are two modes of operation by which vector impedance meters can be used to measure the magnitude of the impedance. These are

1. To pass a constant current of known value through an unknown impedance, measure the voltage drop across the impedance, and display the ratio of voltage-to-current as the magnitude of the impedance.
2. To apply a constant voltage to the unknown impedance, measure the resulting current, and display the ratio of current-to-voltage as the magnitude of the admittance, $1/Z$.

Phase measurements are made simultaneously with the measurement of the magnitude of the impedance. The phase angle is measured by applying both the voltage and the current signals to separate Schmitt trigger circuits and by adjusting the Schmitt trigger circuits to produce a positive spike each time the voltage or current sine wave passes through zero amplitude. These positive spikes are applied to a circuit called a *binary phase detector* which consists of a bistable multivibrator, a differential amplifier, and an integrating capacitor. The time interval between the zero crossings for current and voltage is directly related to the integrating capacitor voltage. This voltage is applied to the phase angle meter, thus providing an indication of the angular displacement between voltage and current.

7-12 VECTOR VOLTMETER

Vector voltmeters are used to measure voltages and provide measurement data regarding both magnitude and phase. Phase displacement may be with respect to a reference angle, which is usually 0°, or with respect to another point in a circuit. A phase displacement is measured by measuring the signals at the two points of interest simultaneously with probes A and B; the phase displacement between these points is displayed on the phase meter.

Vector voltmeters are useful in a wide variety of measurement applications. Some of these measurements are of

- Insertion losses.
- Complex impedance of mixers.
- *S*-parameters of transistors.
- Radio-frequency distortion.
- Amplitude modulation index.

The vector voltmeter converts two input radio-frequency signals of the same frequency to two intermediate-frequency signals. These intermediate-frequency signals have the same waveform, amplitude, and phase relationships as the input rf signals. Therefore, the fundamental components of the intermediate-frequency signals have the same amplitude and phase relationships as the fundamental components of the radio-frequency signals but are much easier to measure. These fundamental components are filtered from the intermediate-frequency signals and applied to the measuring circuitry of a voltmeter and phase meter.

7-13 SUMMARY

Simple dc field-effect transistor voltmeters can be made by using either a difference-amplifier or a source-follower type of EVM. A zero adjustment is required to equalize both halves of the EVM. Further calibration adjustment is necessary to offset the variation in FET parameters.

A simple ac EVM is made by rectifying the ac signal before applying it to the EVM. This signal may be amplified before or after rectification. If only sinusoidal signals are measured, the meter scale may be marked in rms values. Most ac voltmeters with an rms scale assume that the input signal is sinusoidal.

Thermocouples may be used to measure the rms value of any signal. By means of an amplifier and a feedback circuit, the rms value can be measured on a linear scale.

The following should be considered when choosing an analog voltmeter.

1. For dc measurements, select the meter with the widest capability to meet the circuits requirements.
2. For ac measurements involving sine waves with less than 10% distortion, the average-responding voltmeter provides the best accuracy and the most sensitivity per dollar investment.
3. For high-frequency measurements greater than 10 MHz, an ac voltmeter using a shunt-connected diode for peak reading is the most economical choice.

4. For measurements for which it is important to find the effective power of waveforms that depart from the true sinusoidal form, the rms-responding voltmeter is the appropriate choice.

Other very useful electronic measuring instruments include electronic ohmmeters, vector impedance meters, and vector voltmeters. Rather than being constructed as a separate measuring instrument, electronic ohmmeter circuits are normally included in instruments that are also capable of measuring voltage and current. Such instruments are usually called electronic multimeters.

Vector impedance meters can be used to measure complex impedances and display the value in terms of both magnitude and phase angle.

Vector voltmeters are used to make measurements similar to those made with vector impedance meters, except in terms of voltage. Voltage measurements are expressed in terms of both magnitude and direction.

7-14 GLOSSARY

Calibration adjustment: A rheostat in series with the ammeter of an EVM used to adjust the full-scale reading of the EVM.

Chopped dc amplifier: An amplifier that first converts the dc input to an ac signal before amplification. After amplification, the ac signal is demodulated to recover an amplified version of the dc input.

Chopper or modulator: A device used to interrupt a direct current or low-frequency alternating current to permit amplification of the signal by an ac amplifier.

Demodulation: Conversion of an ac signal to direct current with polarity reversals of the direct current for each 180° phase shift of the alternating current.

Difference amplifier: An amplifier with two inputs. The output is an amplified version of the algebraic difference of the two input signals.

Electronic voltmeter: A voltmeter which has an amplifying circuit, a very large input resistance, and a high sensitivity rating.

FET parameters: These are the g_m and r_d of an FET.

Field-effect transistor (FET): A semiconductor amplifying device in which the flow of charged particles through a bar of semiconductor material is controlled by the electric field of a reverse-biased junction (JFET), or an electrode insulated from the bar (insulated-gate field-effect transistor or metal oxide semiconductor field-effect transistor).

Form factor: The ratio of rms to average value of a waveform.

Modulator: See Chopper.

Negative feedback: The part of the output signal of an amplifier fed back to the input (180° out of phase) in order to stabilize the voltage gain.

Photocell: A transducer sensitive to light.

Thermocouple: A device made out of two dissimilar metals. A contact potential is developed across the junction of two metals. This potential is a function of the temperature of the junction.

True rms voltmeter: A detector that responds to the rms value (heat value) of the signal being detected.

Zero set: In an EVM the zero set is used to balance the difference amplifier or source-coupled amplifier.

7-15 REVIEW QUESTIONS

After studying the material in this chapter, try answering the questions given below. These questions will test your knowledge of the subject.

1. How does the FET VM differ from the VOM?
2. What is a difference amplifier?
3. What are the two basic FET VM circuits used for measuring dc voltages?
4. Why is it possible to mark the ammeter scale linearly in the basic EVM circuits discussed in this chapter?
5. What is the function of the zero adjust in a difference amplifier or a source-follower type of EVM?
6. What is the function of the calibration adjust in a difference amplifier or a source-follower type of EVM?
7. Why is the voltage divider used on the input of an EVM?
8. How does a true rms voltmeter measure the rms value of an ac waveform?
9. Why is negative feedback often used in voltmeters?
10. What approach is generally used in the construction of a very sensitive dc voltmeter?
11. What is a thermocouple?
12. What is the best technique for analyzing an electronic ohmmeter circuit?

7-16 PROBLEMS

7-1 Given the EVM of Fig. 7-6, assume the following values: $R_D = 15\ \text{k}\Omega$, $r_d = 100\ \text{k}\Omega$, and $g_m = 0.003$ siemens. If the meter has a resistance of $1800\ \Omega$ and a full-scale current of 5 mA, what value of v_1 produces full-scale current?

7-2 If the EVM of Fig. 7-6 is to have full-scale current for v_1 equal to 5 V, what size resistance must be added in series with the ammeter?

7-3 In the circuit of Fig. 7-9, if a 300-V range is to be added to the voltage divider, show the new voltage divider with appropriate resistance. The total resistance of the divider is to be 10 MΩ.

7-4 Given the EVM of Fig. 7-10, find the relation between ammeter current and input voltage if $r_d = 100$ kΩ, $g_m = 0.003$ siemens, $R_S = 15$ kΩ, and $R_m = 1800$ Ω.

7-5 In Fig. 7-10, $r_d = 10$ kΩ, $g_m = 0.003$ siemens, $R_S = 15$ kΩ, and $R_m = 1800$ Ω. How much current flows for a 1-V input? What size resistance must be added in series with the ammeter in order to have 0.1 mA of current for a 1-V input?

7-6 A peak-to-peak ac detector is connected to the input of the EVM shown in Fig. 7-10, with the parameter values given in Problem 7-4. An ammeter of 1800 Ω to 0.1 mA is used. The ac detector is an ideal peak-to-peak detector. If the input signal to the ac detector is a 1-V rms sine wave, what value of calibration resistor produces full-scale current?

7-7 In Problem 7-6 an ammeter of 1000 Ω to 50 μA is used. If the input signal to the ac detector is a 0.5-V rms sine wave, what value of calibration resistor produces full-scale current?

7-8 Given the EVM of Fig. 7-13, $R_S = 40$ kΩ, $r_d = 200$ kΩ, and $g_m = 0.004$ siemens. An ammeter of 1800 Ω − 0.1 mA is used. If the input signal is a 1-V rms sine wave, what value of calibration resistor produces full-scale deflection?

7-9 Repeat Problem 7-8 using a 1000-Ω to 50-mA ammeter.

7-17 LABORATORY EXPERIMENTS

Experiments E14 and E15 make use of the theory that has been presented in Chapter 7. The purpose of the experiments is to provide hands-on experience to reinforce the theory.

Both experiments can be performed with standard electronic components found in any electronics laboratory. The contents of the laboratory report to be submitted by each student are listed at the end of each experimental procedure.

CHAPTER 8

Oscilloscopes

8-1 Instructional Objectives

The purpose of this chapter is to familiarize the reader with the cathode-ray oscilloscope. The chapter discusses the theory of operation of this very versatile instrument, describes several features incorporated into special-purpose oscilloscopes, and discusses several applications of oscilloscopes. After completing the chapter you should be able to

1. List the major subsystems of an oscilloscope.
2. Draw a pictorial representation of a general-purpose cathode-ray tube (CRT) and label the components by name.
3. Define the following terms: deflection sensitivity, fluorescent, phosphorescence, and graticule.
4. Draw a block diagram of a basic cathode-ray oscilloscope (CRO) and label each block.
5. Make calculations involving bandwidth and rise time.
6. Describe the function of the following oscilloscope systems: horizontal amplifier, vertical amplifier, sweep generator, trigger circuit, and attenuator network.
7. Make calculations of voltage across a capacitor in an *RC* circuit during charge or discharge.
8. Calculate the value of attenuator resistors or of the attenuation factor.
9. Make calculations involving the resistance and capacitance of a high-impedance probe.
10. Describe the basic principle of the operation of a storage oscilloscope and of a sampling oscilloscope.
11. Determine the frequency or amplitude of a signal displayed on the CRT screen.
12. Determine the unknown frequency by means of a Lissajous pattern.
13. Compute phase angle from a Lissajous pattern.
14. Describe how to check the compensation of a probe with a square-wave signal.

8-2 INTRODUCTION

The cathode-ray oscilloscope, generally referred to as the oscilloscope or simply "scope," is probably the most versatile electrical measuring instrument available. Some of the electrical parameters that can be observed with the oscilloscope are ac or dc voltage, indirect measurement of ac or dc current, time, phase relationships, frequency, and a wide range of waveform evaluations such as rise time, fall time, ringing, and overshoot. Many nonelectrical physical quantities such as pressure, strain, temperature, and acceleration can be measured by using a transducer to convert the physical parameter to an equivalent voltage. The usefulness of the oscilloscope is limited only by the user's ability and ingenuity. The oscilloscope consists of the following major subsystems.

- Cathode-ray tube, or CRT.
- Vertical amplifier.
- Horizontal amplifier.
- Sweep generator.
- Trigger circuit.
- Associated power supplies.

The heart of the instrument is the cathode-ray tube. The remaining subsystems are necessary for signal conditioning, so that a visual representation of the input signal will be displayed on the face of the CRT.

Because of the importance of the oscilloscope as a measuring instrument and because of the extensive range of applications of the instrument, the remainder of the chapter will be subdivided to deal with each major subdivision of the instrument in detail as well as discussing several applications for this versatile instrument.

8-3 THE CATHODE-RAY TUBE

The cathode-ray tube used in an oscilloscope is very similar to the picture tube in a television set. A cross-sectional representation of a CRT, showing its major components, is given in Fig. 8-1. The major components of a general-purpose CRT are

- Evacuated glass envelope.
- Electron gun assembly.
- Deflection plate assembly.
- Accelerating anodes.
- Phosphor-coated screen.

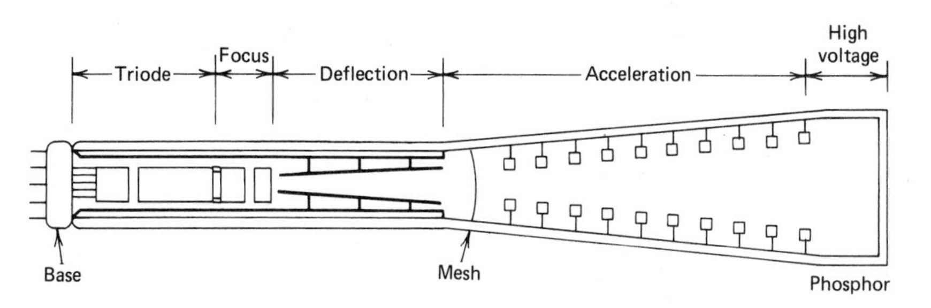

FIGURE 8-1 Cathode-ray tube with major components identified.

Cathode-ray tubes are manufactured in various sizes ranging from screen diameters of approximately 1 to 25 in. or larger. Most laboratory-quality oscilloscopes use a CRT that has a circular screen approximately 5 in. in diameter. All electrical connections except the high-voltage connection are made through the base of the CRT. The glass envelope is evacuated to a fairly high vacuum to permit the electron beam to traverse the tube easily.

The electron gun assembly consists of the triode section and the focus section which are shown in Fig. 8-1. The purpose of the electron gun assembly is to provide a source of electrons, converge and focus them into a well-defined beam, and accelerate them toward the **fluorescent** screen. The electrons that make up the beam are given off by **thermionic emission** from the heated cathode. The cathode is surrounded by a cylindrical cap that is at a negative potential. This cap, which has a small hole located along the longitudinal axis of the CRT, as shown in Fig. 8-2, acts as the control grid. Because the control grid is at a negative potential, electrons are repelled away from the cylinder walls and, therefore, stream through the hole where they move into the electric fields of the focusing anodes.

The focus lens consists of the first anode, focus ring, and astigmatism aperture, or second anode. The purpose of this section is to converge and **collimate** the beam to obtain the minimum-size and best-defined spot on the **phosphor** screen of the CRT.

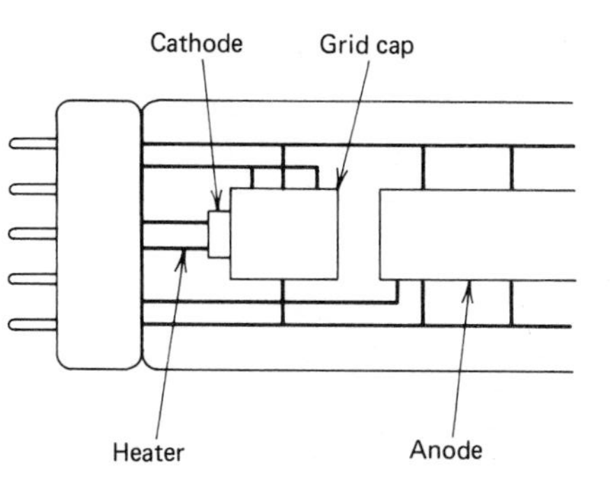

FIGURE 8-2 Triode section of the CRT.

The section of the CRT just beyond the electron gun assembly is the deflection system. Waveforms can be displayed on the CRT's phosphor screen only if there is some means of deflecting the electron beam both horizontally and vertically. This is the function of the deflection plates.

The deflection experienced by the electron beam in the CRT of an oscilloscope is called *electrostatic* deflection, which means that the electron beam is deflected by a force exerted on each electron by the electric field. Another method of deflection is magnetic as in the CRT of a television set. Magnetic deflection is more practical if the beam must be deflected a considerable distance as it is in a television set. Electrostatic deflection offers the advantages of higher-frequency operation as well as space saving inasmuch as the deflection plates are mounted inside the CRT. During the period of acceleration in the electron gun assembly, the electrons have gained kinetic energy as they gain velocity. The energy gain is a simple relationship involving only the second anode voltage of the focus lens and the electron charge and is given as

$$E_k = V_2 Q \tag{8-1}$$

Equating the expression for mechanical kinetic energy to the energy gain expression (Eq. 8-1) gives us

$$\tfrac{1}{2}mv^2 = V_2 Q \tag{8-2}$$

where

m = the mass of an electron
v = electron velocity
V_2 = accelerating voltage through the electron gun assembly, which equals the second-anode voltage of the focus lens

In CRTs that use electrostatic deflection, two sets of deflection plates are positioned at right angles to one another. These plates are usually positioned just forward of the second anode, with the vertical deflection plates first and the plates for horizontal deflection nearer the phosphor screen. The deflection plates may be parallel, angled, or single-bend plates as shown in Fig. 8-3. The angled and single-bend plates increase the beam scan by deflecting the electrons through a greater angle, thus making possible a larger-size screen or one somewhat shorter in length. The maximum frequency response of a single-plate CRT is limited by the time required for an electron to travel the

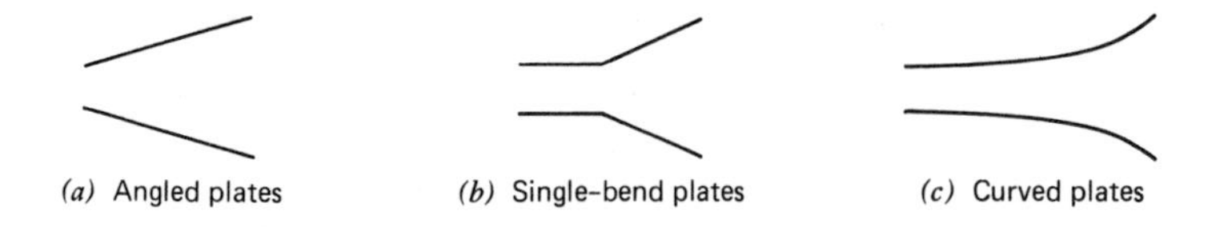

FIGURE 8-3 Deflection plate configurations.

length of the vertical deflection plates. At high frequencies a beam electron may be between the vertical deflection plates for more than one cycle of the signal applied to the deflection plates, which would cancel, or at least reduce, the net deflection of the beam electron. Transit time can be reduced either by reducing the length of the deflection plate or by increasing the electron velocity. However, doing either of these to reduce transit time causes degradation of other CRT parameters. The problem with transit time can be overcome by dividing the deflection plates into a number of smaller plates. Each of these plate segments is connected by *LC* delay elements. These elements effectively form a transmission line that matches the propagation time of the signal to the transit time of the beam electrons during the period they are between segmented deflection plates. This increases beam deflection at higher frequencies because, as an electron passes between the deflection plates, it experiences a continuous deflection. Figure 8-4 shows a segmented deflection plate system.

The deflection plates can be described by two geometric parameters: the *length* of the plates L, and the *plate separation*, d. The defecting action of the plates is *dependent* on the intenstiy of the electric field between the plates, which is expressed as

$$E_d = \frac{V_d}{d} \tag{8-3}$$

where

E_d = the intensity of the deflecting electric field, in volts per meter
V_d = the magnitude of the deflecting voltage
d = the plate separation, in meters

A lateral force expressed as

$$F = E_d Q \tag{8-4}$$

will be exerted on the electrons, deviating the beam from a straight-line trajectory. Expressed in terms of the deflecting voltage, the deflecting force is

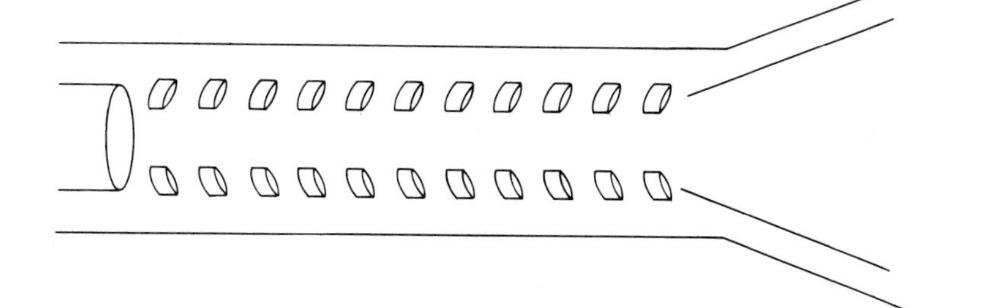

FIGURE 8-4 Segmented deflection plate system.

given as

$$F_d = \frac{V_d Q}{d} \tag{8-5}$$

which indicates that the deflecting force is directly proportional to the deflecting voltage. Equating F_d in Eq. 8-5 to the force in Newton's second law provides us with an expression for lateral acceleration, which is at right angles to the CRT axis:

$$a_y = \frac{V_d Q}{dm} \tag{8-6}$$

The lateral velocity of the deflected electron can be obtained from the kinematic equation

$$v_2^2 = 2a_y h + v_1^2 \tag{8-7}$$

However, since electrons enter the region between the deflection plates with zero lateral speed, Eq. 8-7 becomes

$$v_2^2 = 2a_y h \tag{8-8}$$

where

h = lateral distance traveled
a_y = lateral acceleration
v_2 = lateral speed at a distance, h

The deflection distance, expressed as a function of the time that the electron spends between the deflection plates, is given by the kinematic equation

$$h = \tfrac{1}{2}at^2 \tag{8-9}$$

If we substitute Eq. 8-6 into Eq. 8-9, we obtain the expression

$$h = \frac{V_d Q t^2}{2dm} \tag{8-10}$$

The time, t, required for the electrons to pass between the plates is given as

$$t = \frac{L}{v} \tag{8-11}$$

where

L = length of the deflection plates
v = electron velocity when electrons exit electron gun assembly

Combining Eqs. 8-10, 8-11, and 8-2, we obtain the expression

$$h = \frac{L^2 V_d}{4V_2 d} \tag{8-12}$$

This expression relates the lateral deflection, h, to L and d, which are part of the fixed geometry of the CRT, and to the external variables, V_d, the horizontal deflection voltage, and V_2, the accelerating voltage through the electron gun assembly. We can see from Eq. 8-12 that the horizontal deflection is directly proportional to the deflecting voltage and inversely proportional to the accelerating anode voltage.

Equation 8-12 expresses the amount of deflection only at the deflection plates. To determine the deflection of the beam on the face of the CRT, we must consider the distance from the deflection plates to the CRT screen. An approximate analysis is based on the geometry of the CRT shown in Fig. 8-5. If the deflection angle θ is expressed in radian measure, the value of θ is approximately equal to

$$\theta = \frac{h}{L/h} = \frac{2h}{L} \tag{8-13}$$

In addition, as can be seen in Fig. 8-5, we can express θ as

$$\theta = \frac{y}{R} \tag{8-14}$$

Equating the angle θ in Eqs. 8-13 and 8-14 yields

$$\frac{y}{R} = \frac{2h}{L} \tag{8-15}$$

Solving Eq. 8-15 for y gives us the expression

$$y = \frac{2hR}{L} \tag{8-16}$$

Substituting Eq. 8-12 into Eq. 8-16, we obtain the expression

$$y = \frac{RLV_d}{2V_2 d} \tag{8-17}$$

Equation 8-17 is very important in helping us understand the deflection of

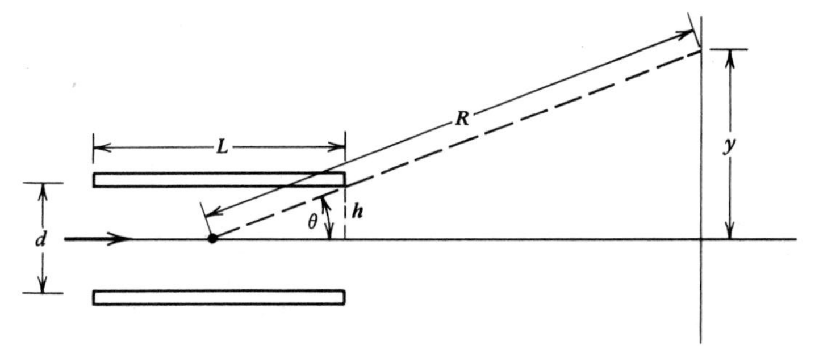

FIGURE 8-5 Deflection of the electron beam as a function of deflection voltage, accelerating anode voltage, and CRT geometry.

the electron beam on the CRT screen. The equation contains three fixed parameters which are related to the geometry of the CRT: R, L, and d. The accelerating anode voltage that appears in the equation is usually fixed, leaving only one externally controlled parameter, the deflection voltage, V_d. The beam deflection is a *linear function* of the deflection voltage. That is, doubling the deflection voltage doubles the deflection, or reducing the deflection voltage by half reduces the beam deflection by half.

An important parameter related to the operation of the CRT can be obtained from Eq. 8-17 by rearranging the equation for a ratio

$$\frac{V_d}{y} = \frac{2V_2 d}{RL} \tag{8-18}$$

which defines the **deflection sensitivity** as the voltage required per unit deflection. A typical deflection sensitivity can be obtained by considering Example 8-1.

EXAMPLE 8-1

Assume the distance R from the deflection plates to the screen of a CRT is 15 cm, the length of the deflection plates, L, is 2 cm, and the separation between the plates, d, is 1 cm. If the second anode voltage, V_2, is 500 V, what is the deflection sensitivity?

Solution

Equation 8-18 can be used to solve for the deflection sensitivity, which is the ratio V_d/y.

$$\frac{V_d}{y} = \frac{2(500\text{ V})(1\text{ cm})}{(15\text{ cm})(2\text{ cm})} = \frac{33.2\text{ V}}{\text{cm}}$$

The result of Example 8-1 is a typical value for the deflection sensitivity of a CRT and is an indication that an appreciable voltage is required to deflect the electron beam.

The next section of the CRT is the *postdeflection area*. After electrons pass beyond the deflection plates, they may or may not experience additional acceleration. This depends primarily on the maximum frequencies to be applied to the CRT. In general, if the maximum frequency to be displayed on the CRT is less than 10 MHz, no postdeflection acceleration is used. If higher-frequency signals are to be displayed, postdeflection acceleration is generally necessary to increase the brightness of the trace which otherwise may be quite dim. However, using postdeflection acceleration generally requires that the length of the CRT be extended to achieve the required beam deflection as a result of the increased velocity.

An alternative to increased CRT length is to add a dome-shaped mesh to the CRT just beyond the deflection plates, as shown in Fig. 8-6. However, the mesh reduces trace brightness and increases the size of spots unless the postdeflection accelerating voltage is increased significantly. Most modern oscilloscopes use a mesh CRT and a postdeflection accelerating voltage of about 20 kV. When the electron beam strikes the phosphor-coated face of the

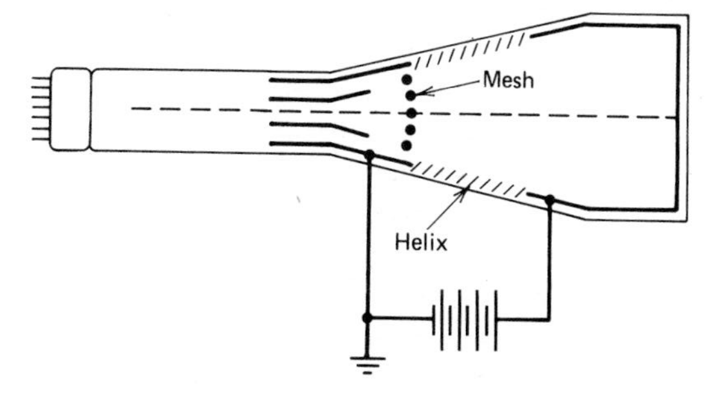

FIGURE 8-6 CRT with dome-shaped mesh to increase deflection.

CRT, a spot of light is produced because the phosphor absorbs kinetic energy from the electrons that strike it and then gives up the energy in the form of light. This property of emitting light when stimulated by electron bombardment is called *fluorescence*: therefore, we can say that phosphor is a fluorescent material.

Phosphor possesses a second desirable characteristic called **phosphorescence**, which means that the phosphor continues to emit light for a period of time after the source of excitation is removed. The length of time phosphorescence continues is a measure of the *persistence* of the fluorescent material. Persistence is usually classified as short, lasting for microseconds; medium, lasting for milliseconds; or long, lasting for seconds. Table 8-1 lists several different phosphors and some of their characteristics and applications.

TABLE 8-1
Phosphor Data

Phosphor Trace	Trace Color	Persistence	Application
P1	Yellow-green	Medium	General-purpose CRO
P2	Blue	Medium	Observation of low- and medium-speed signals
P4	White	Medium–short	Television picture tube
P7	Blue	Long	Observation of low- and medium-speed signals
P11	Blue	Medium–short	Photographic applications
P31	Green	Medium–short	Observations of low- and medium-speed signals. Most frequently used in general-purpose CRO

8-4 THE GRATICULE

The **graticule** is a grid of lines that serves as a scale when making time and amplitude measurements with an oscilloscope. The graticule may be etched or silk-screened on a plastic CRT faceplate, or it may be chemically deposited on the face of the CRT along with the phosphor. If the graticule is put on the plastic faceplate, it should be positioned on the inside surface, that is, the surface that will be in contact with the face of the CRT. This places the display produced by the electron beam and the graticule on the same plane, thereby eliminating measurement inaccuracies called *parallax errors*. Parallax

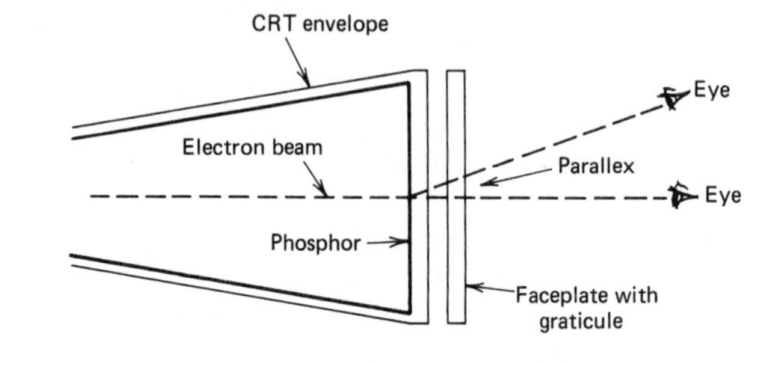

FIGURE 8-7 Illustration of parallax error.

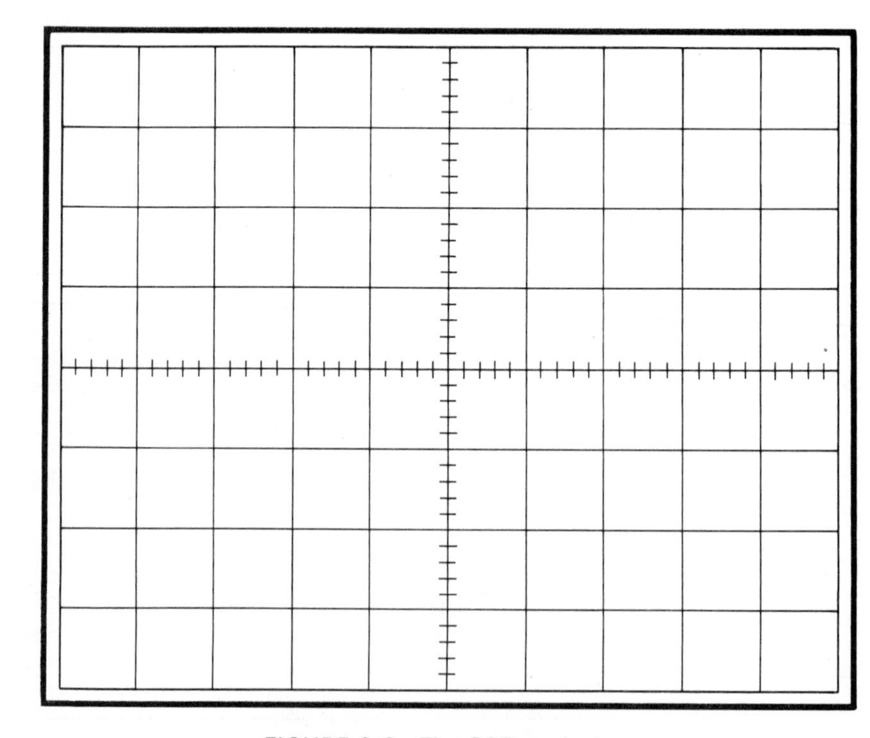

FIGURE 8-8 The CRT graticule.

errors occur when the trace and the graticule are in different planes and the observer's eye is shifted from the direct line of sight, as shown in Fig. 8-7.

Although different-size CRTs may be used, graticules are usually laid out in an 8×10 pattern as shown in Fig. 8-8. As can be seen, there are eight major vertical divisions and ten major horizontal divisions. The labeling on the front panel controls of an oscilloscope always refers to major divisions. Major divisions are marked off in inches or centimeters, with centimeters being much more widely used. The tick marks on the center graticule lines represent minor divisions or subdivisions. In addition to the standard graticule shown in Fig. 8-8, some graticules include marking for rise time measurements. A wide selection of graticules for specialized applications are available from major oscilloscope manufacturers.

8-5 BASIC OSCILLOSCOPE CONTROLS THAT DIRECTLY AFFECT THE BEAM

In normal operation, as part of a cathode-ray oscilloscope, a number of operating adjustments are required to control the beam characteristics. An *intensity control*, as shown in Fig. 8-9, is always connected to the control

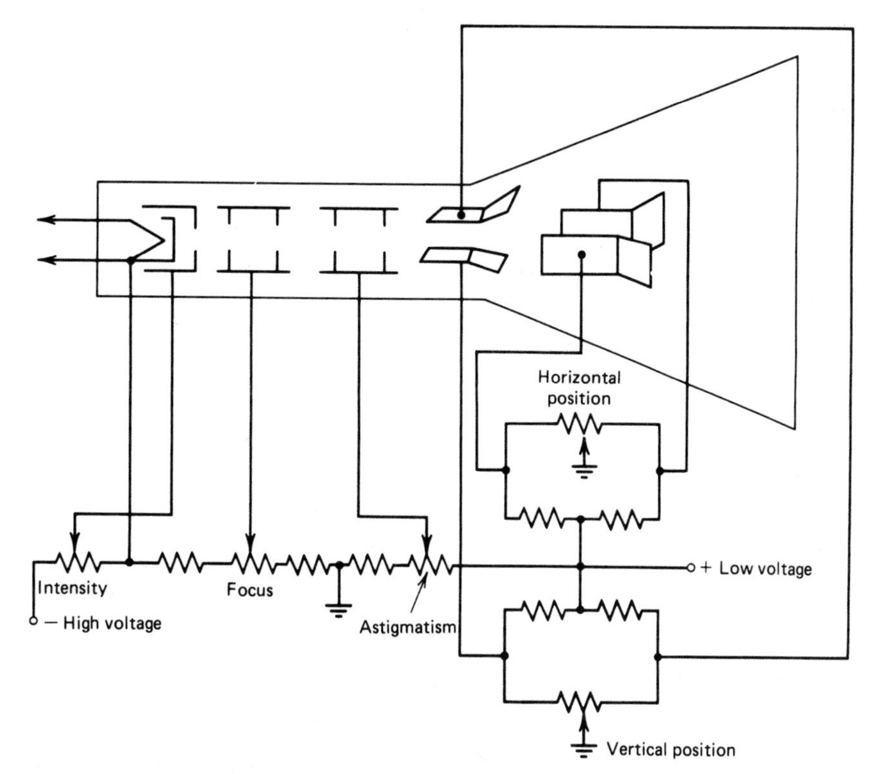

FIGURE 8-9 Basic circuit of controls that determine position, intensity, and focus of the electron beam on a CRT screen.

grid. The potential of the control grid is negative with respect to the cathode, as can be seen in Fig. 8-9. Therefore, as the intensity control is adjusted, the number of electrons that pass through the small hole in the control grid is affected, which in turn affects the brightness of the spot on the screen.

The *focus control*, also shown in Fig. 8-9, is connected to the focusing anode. The focusing anode and accelerating anode form an electrostatic lens to collimate the electrons into a well-defined beam.

Generally, a beam that is in sharp focus at the center of the screen will be out of focus near the edge of the screen because the lengths of the electron paths are different when the beam is deflected. Adjusting the *astigmatism control*, shown in Fig. 8-9, gives a sharp focus over the entire screen.

The beam can be positioned anywhere on the screen by adjusting the controls marked *horizontal position* and *vertical position*. When the horizontal and vertical position controls, shown in Fig. 8-9, are set to their midpoint position, the deflection voltages divide equally across both halves of the potentiometers. There is therefore no deflection of the beam; it simply travels along the axis of the CRT and strikes the center of the screen. Adjusting the horizontal or vertical position control deflects the beam to any desired position on the screen.

8-6 THE BASIC OSCILLOSCOPE

The CRT and the associated controls for accelerating, deflecting, and focusing the electron beam, which have been the topics of interest to this point in the chapter, permit us to obtain a lighted spot on the screen. For the CRT to be of practical use as part of a measuring instrument, we must connect to it additional electronic circuitry that can quickly deflect and control the electron beam. The purpose of the electronic circuits is to make the beam trace on the CRT screen a reproduction of the signal we apply to the input terminals of the oscilloscope. A block diagram of a basic oscilloscope is shown in Fig. 8-10.

In inexpensive, general-purpose oscilloscopes, the left horizontal deflection plate (looking toward the screen) and the lower vertical deflection plate are sometimes connected to ground. The beam is deflected upward and to the right by signals applied to the upper vertical deflection plate or to the right horizontal deflection plate. A signal to be displayed on the CRT screen is applied to the vertical input terminal where it is fed into the vertical amplifier. The signal is amplified and applied to the vertical deflection plate, which causes the beam to be deflected in the vertical plane. As can be seen in Fig. 8-10, the output of the vertical amplifier is connected to the *internal sync* position of switch S_1. With the switch set to internal sync, as it is for normal operation of the oscilloscope, the output of the vertical amplifier is applied to the **sweep generator**. This signal triggers the sweep generator, except in low-cost oscilloscopes with a free-running sweep generator. These oscilloscopes are of little practical use in modern electronics work and will therefore not be discussed. The purpose of the sweep generator is to develop

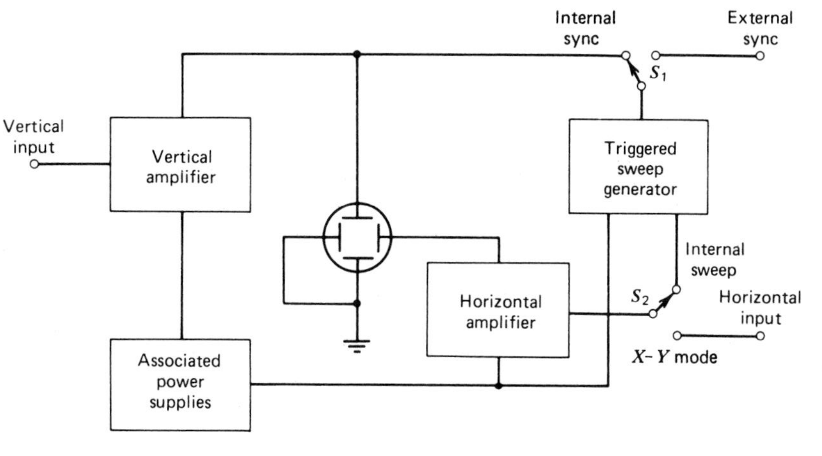

FIGURE 8-10 **Block diagram of a basic cathode-ray oscilloscope.**

a voltage at the horizontal deflection plate that increases linearly with time. This linearly increasing voltage, called a *ramp* voltage or a *sawtooth* waveform, causes the beam to be deflected equal distances horizontally per unit of time. The sweep generator will be discussed later in more detail.

The **horizontal amplifier** serves to amplify the signal at its input prior to the signal being applied to the horizontal deflection plates. The input signal to the horizontal amplifier depends on the position to which switch S_2 is set. In normal operation of the oscilloscope, the switch is set to *internal sweep*. When the instrument is used in the $X-Y$ mode, for phase-shift measurements or to determine the frequency of a signal, the signal that is applied to the *horizontal input* terminal is amplified by the horizontal amplifier.

8-7 BEAM DEFLECTION

Whether an oscilloscope is being used to display a reproduction of an input signal on the CRT screen, to determine an unknown frequency, or to measure the phase shift between two waveforms, the position of the beam on the screen is determined by the amplitude of the deflecting voltages on both the horizontal and vertical deflection plates. In this section we consider beam deflection with the instrument set to display a reproduction of an input signal, which is sometimes referred to as the amplitude versus time or $Y-t$ mode.

Before considering the effect of applying a signal to the vertical input terminal, we adjust the horizontal deflection voltage to position the beam to the left side of the screen, as we view it, and at the center of the screen vertically, which is referred to as the horizontal axis or the X-axis. If we now apply an input signal, for example, the sine wave shown in Fig. 8-11, we can trace the movement of the beam versus time and observe the resulting pattern on the CRT screen. Figure 8-11 shows the result of simultaneous horizontal

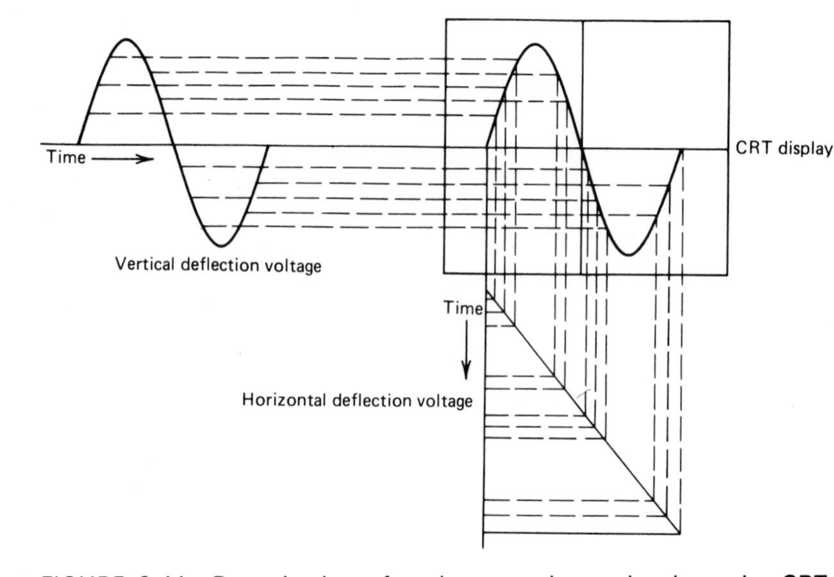

FIGURE 8-11 Reproduction of a sine-wave input signal on the CRT screen by deflecting the beam vertically with a sine wave and horizontally with a ramp voltage.

and vertical deflection of the electron beam when a sine wave is applied to the vertical deflection plates and a ramp voltage having the same period and phase as the sine wave is applied to the horizontal deflection plates.

8-8 OSCILLOSCOPE AMPLIFIERS

The purpose of an oscilloscope is to provide a faithful reproduction of signals applied to its input terminals. For the oscilloscope to be able to fulfill this purpose, considerable attention must be given to the design of the amplifiers in the oscilloscope. Amplifiers used in oscilloscopes may be categorized in several ways; however, probably the most clear-cut category is whether the amplifiers are *ac-coupled* or *dc-coupled*. Some low-cost oscilloscopes use ac-coupled amplifiers but, because of the advantages of dc-coupled amplifiers, they are commonly used in laboratory-quality oscilloscopes. Even though dc-coupled amplifiers are considerably more expensive, they offer the important advantge of responding to a dc voltage. This makes it possible to measure a pure dc voltage, a pure ac signal, or an ac signal riding on a dc voltage. Another advantage of the dc-coupled amplifier is elimination of the problem of low-frequency phase shift and the accompanying waveform distortion when observing a low-frequency pulse train with an oscilloscope.

Other less well-defined ways of categorizing oscilloscope amplifiers are according to bandwidth—either as a *narrowband* or *broadband* amplifier. These categories are not sharply defined, nor are they universally accepted.

Nonetheless, there is general agreement that if the frequency response does not extend up to the television color subcarrier frequency of 3.58 MHz, the amplifier is in the narrowband category. Conversely, if the frequency response curve of an amplifier is flat beyond 3.58 MHz, the amplifier is categorized as a broadband amplifier. Most general-purpose laboratory-quality oscilloscopes respond fully to frequencies in excess of 5 MHz. Therefore, the responsible amplifier, which is the vertical amplifier, is broadband.

8-9 VERTICAL AMPLIFIER

The vertical amplifier is the principal factor in determining the **sensitivity** and **bandwidth** of an oscilloscope. In general, greater sensitivity, which is expressed in terms of volts per centimeter of vertical deflection at the midband frequency, is obtained at the expense of bandwidth, since the product of gain times bandwidth is a constant for a given amplifier. As a rule of thumb, an amplifier at some specific cost may be obtained with a certain gain–bandwidth product, which is the product of the voltage gain of the amplifier and its bandwidth. Voltage gain may be sacrificed in favor of greater bandwidth, or vice versa, without significantly affecting the cost of the amplifier. However, if the gain–bandwidth product increases, the cost of the amplifier will increase.

The gain of the vertical amplifier determines the smallest signal that the oscilloscope can satisfactorily reproduce on the CRT screen. The sensitivity of an oscilloscope is directly proportional to gain of the vertical amplifier; that is, as gain increases sensitivity increases, which allows us to observe smaller-amplitude signals.

The vertical sensitivity is a measure of how much the electron beam will be deflected for a specified input signal. The CRT screen is covered with a plastic grid pattern called a graticule, as discussed in Section 8-4. The spacing between the grid lines is generally 1 cm. However, vertical sensitivity is generally expressed in volts per division. On the front panel of the oscilloscope shown in Fig. 8-12, one can see a knob attached to a rotary switch labeled Volts/Div. The rotary switch is electrically connected to the input attenuator network, which will be discussed in subsequent paragraphs. The setting of the rotary switch indicates what amplitude signal is required to deflect the beam vertically one division.

The vertical sensitivity of an oscilloscope is the smallest deflection factor that can be selected with the rotary switch. As an example, if the most sensitive position on the volts/division rotary switch is 5 mV/division, then the vertical sensitivity of the oscilloscope is 5 mV/division.

The *bandwidth* of an oscilloscope determines the range of frequencies that can be accurately reproduced on the CRT screen. The greater the bandwidth, the wider the range of frequencies that can be observed with the instrument. Ideally, the gain of the broadband amplifier should be constant from direct current to near the upper limit of the range of frequencies that can be

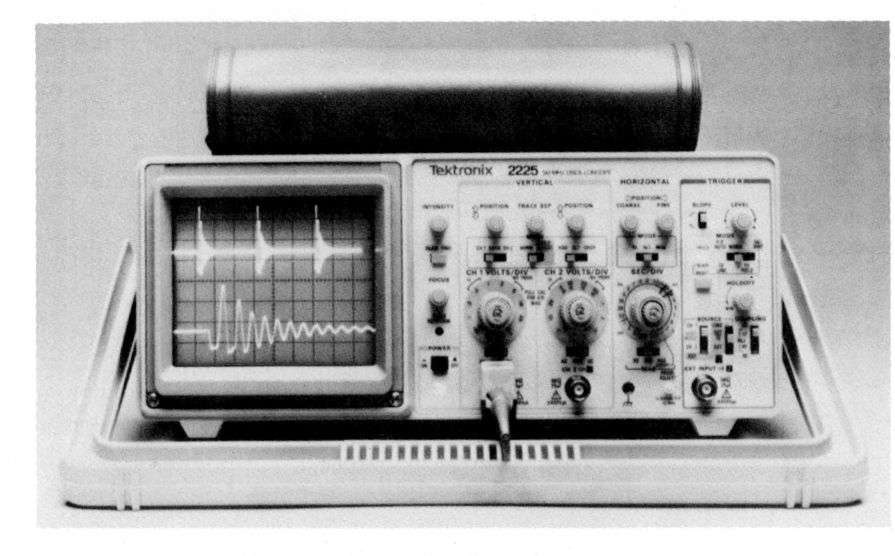

FIGURE 8-12 General-purpose triggered-sweep oscilloscope. (Copyright © 1989 Tektronix, Inc. All rights reserved. Reproduced by permission.)

observed with the oscilloscope. A typical curve for a broadband amplifier of gain versus frequency, called a frequency-response curve, is shown in Fig. 8-13.

The bandwidth of an oscilloscope is the range of frequencies over which the gain of the vertical amplifier is within 3 dB of the midband frequency gain. The upper limit of the bandwidth is the frequency, f_2, at which the gain has decreased by 3 dB. The bandwidth can be increased by using feedback in designing the amplifier. However, this reduces the gain, which is why the gain–bandwidth product for an amplifier is constant.

One of the specifications generally included on the specifications sheet for the vertical amplifier of a broadband oscilloscope is the *rise time*. The rise time of a pulse is defined as the time required for the edge to rise from 10% to 90% of its maximum amplitude. When an oscilloscope is used to observe

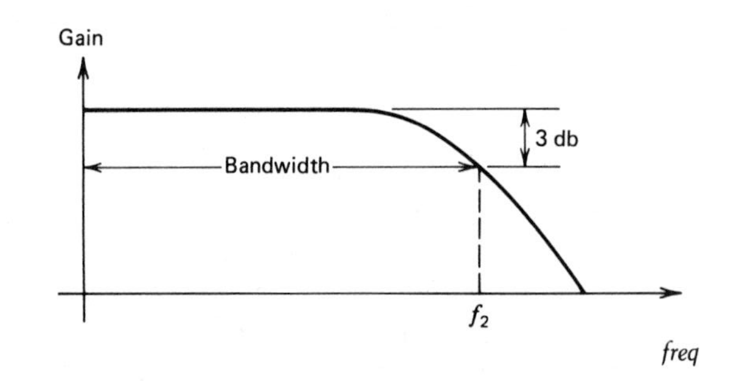

FIGURE 8-13 Typical frequency response curve for a broadband amplifier.

a pulse or a square wave, the rise time of the instrument must be faster than the rise time of the pulse or square wave. Otherwise the observed signal will not be accurately reproduced. Although the exact relationship between bandwidth and rise time varies slightly with amplifier design, an approximate relationship is given as

$$t_r \times \mathrm{BW} = 0.35 \qquad (8\text{-}19)$$

where

t_r = rise time measured in seconds
BW = bandwidth in hertz

EXAMPLE 8-2

If the bandwidth of an oscilloscope is given as direct current to 10 MHz, what is the fastest rise time a sine wave can have to be accurately reproduced by the instrument?

Solution

Using Eq. 8-19, we can compute the maximum rise time as

$$t_r = \frac{0.35}{10 \times 10^6\ \mathrm{Hz}} = 35\ \mathrm{ns}$$

The vertical amplifier of a laboratory-quality oscilloscope frequently consists of the two major circuit blocks shown in Fig. 8-14: a preamplifier and the main vertical amplifier or mainframe amplifier. The preamplifier is sometimes a separate, interchangeable plug-in unit that can quickly and easily be plugged into the mainframe of the oscilloscope. If different plug-in units are available for specific measurement applications, the measurement capabilities of an oscilloscope can be broadened considerably at a reasonable cost.

The active element in the first stage of the preamplifier is generally a field-effect transistor (FET) which provides a high input impedance for the oscilloscope. The final stage of the mainframe amplifier is generally a push–pull amplifier which provides voltages of equal amplitude but out of phase by 180° to achieve balanced deflection.

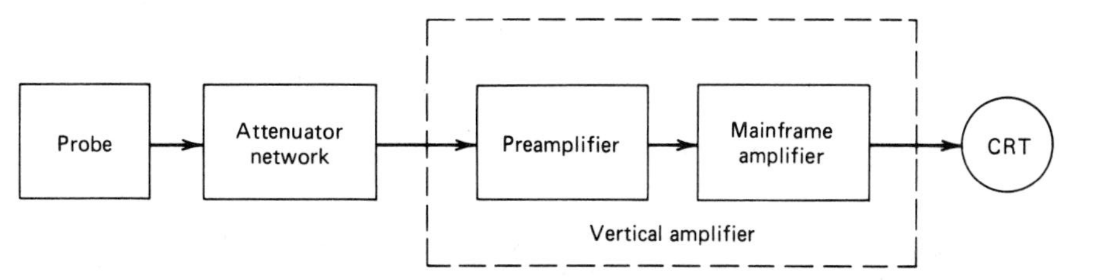

FIGURE 8-14 **Block diagram showing the vertical amplifier of a laboratory-quality oscilloscope.**

8-10 HORIZONTAL AMPLIFIER

The *horizontal amplifier* basically serves two purposes.

1. When the oscilloscope is being used in the ordinary mode of operation to display a signal applied to the vertical input, the horizontal amplifier will amplify the sweep generator output.
2. When the oscilloscope is being used in the $X-Y$ mode, the signal applied to the horizontal input terminal will be amplified by the horizontal amplifier.

When the oscilloscope is being used in its ordinary mode of operation, the gain and bandwidth requirements for the horizontal amplifier are not as stringent as those for the vertical amplifier. Although the vertical amplifier must be able faithfully to reproduce low-amplitude, high-frequency signals with fast rise times, the horizontal amplifier is required to provide only a faithful reproduction of the sweep signal, which has a relatively high amplitude and a slow rise time.

As with the vertical amplifier, the final stage of the horizontal amplifier is a push–pull amplifier, as shown in Fig. 8-15. The attenuator network given in the figure reduces, by voltage division, the amplitude of the horizontal input signal to a level equal to the sensitivity of the horizontal amplifier.

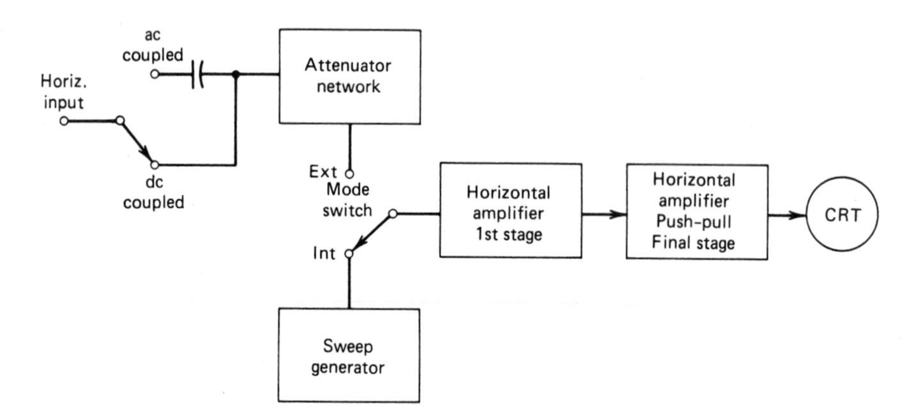

FIGURE 8-15 Block diagram of the horizontal amplifier of an oscilloscope.

8-11 SWEEP GENERATORS

Oscilloscopes are generally used to display a waveform that varies as a function of time. If the waveform is to be accurately reproduced, the beam must have a constant horizontal velocity. Since the beam velocity is a function of the deflecting voltage, the deflecting voltage must increase

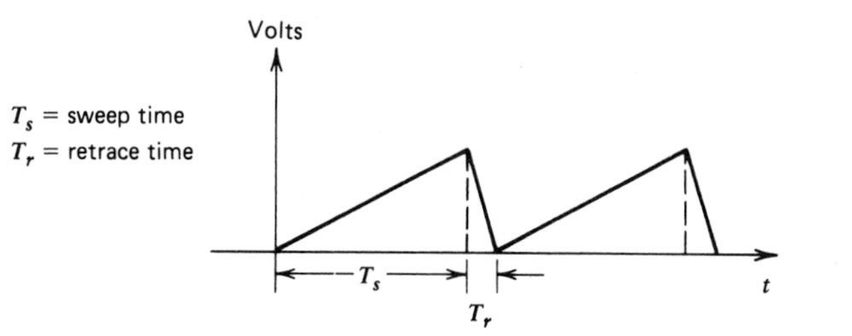

FIGURE 8-16 Typical sawtooth waveform applied to the horizontal deflection plates.

linearly with time. A voltage with this characteristic is called a *ramp voltage*. If the voltage decreases rapidly to zero with the waveform repeatedly reproduced, as shown in Fig. 8-16, the pattern is generally called a sawtooth waveform.

During the **sweep time**, T_s, the beam moves from left to right across the CRT screen. The beam is deflected to the right by the increasing amplitude of the ramp voltage and the fact that the positive voltage attracts the negative electrons. During the **retrace time**, T_r, the beam returns quickly to the left side of the screen. The control gird is generally "gated off," which blanks out the beam during retrace and prevents an undesirable retrace pattern from appearing on the screen.

Since signals of many different frequencies will be observed with the oscilloscope, the sweep rate must be adjustable. We can change the sweep rate in steps by switching different capacitors into the circuit. The front panel control for this adjustment is marked Time/Div or Sec/Div. The sweep rate can be adjusted in minor ways by making the resistor, *R*, in Fig. 8-17*a* a variable resistor.

The circuit shown in Fig. 8-17*a* is a simple sweep circuit in which the capacitor *C* charges through the resistor *R*. The capacitor discharges periodically through the transistor, Q_1, which causes the waveform shown in Fig. 8-17*b* to appear across the capacitor. The signal, V_i, which must be applied

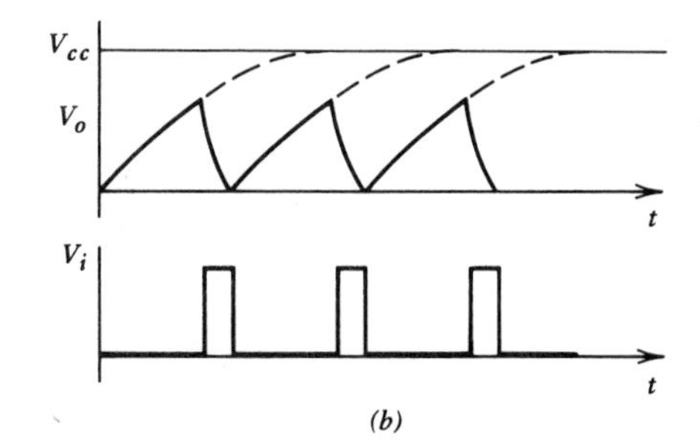

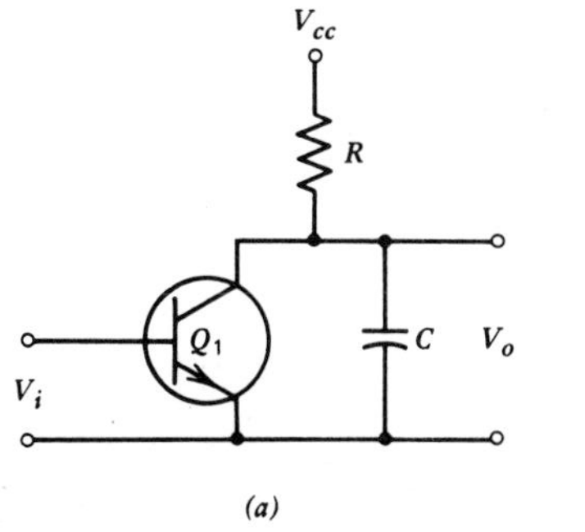

FIGURE 8-17 Simple sawtooth generator and associated waveforms.

to the base of the transistor to turn it "on" for short time intervals is also shown in Fig. 8-17*b*. When the transistor is turned completely "on," it presents a low-resistance discharge path through which the capacitor discharges quickly.

If the transistor is not turned "on," the capacitor will charge exponentially to the supply voltage V_{cc} according to the equation

$$V_o = V_{cc}(1 - \varepsilon^{-t/RC}) \tag{8-20}$$

where

V_o = instantaneous voltage across the capacitor at time t
V_{cc} = supply voltage
t = time of interest
R = value of series resistor in ohms
C = value of capacitor in farads
ε = constant having a value of 2.71828

EXAMPLE 8-3

A trigger pulse is applied to the sweep generator in Fig. 8-18 every 10 msec. Compute the amplitude of the voltage, V_o, across the capacitor when the trigger pulse is applied.

Solution

Using Eq. 8-20 we compute the voltage as

$$\begin{aligned} V_o &= V_{cc}(1 - \varepsilon^{-t/RC}) \\ &= (50\text{ V})\left[1 - \varepsilon^{-10\text{ msec}/(500\text{ k}\Omega \times 0.2\ \mu\text{F})}\right] \\ &= (50\text{ V})(1 - \varepsilon^{-0.1}) \\ &= (50\text{ V})(1 - 0.905) \\ &= (50\text{ V})(0.095) = 4.76\text{ V} \end{aligned}$$

Since V_o is less than 10% of V_{cc}, the charge curve should still be quite linear at this point.

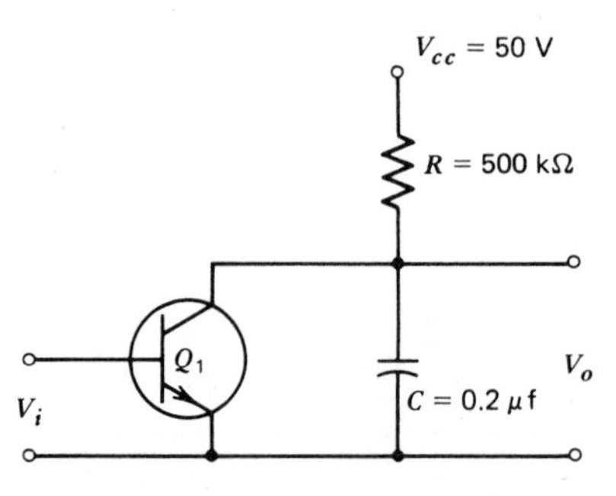

FIGURE 8-18 Circuit for Example 8-3.

When a trigger pulse V_i is applied to the transistor Q_1 in Fig. 8-17*a*, the capacitor C discharges through the resistance presented by the "on" transistor according to the equation

$$V_o' = V_o \varepsilon^{-t_r/R_1 C} \tag{8-21}$$

where

V_o' = voltage across capacitor C at time t_r during retrace
V_o = amplitude of the voltage across capacitor C at the start of discharge.
C = value of capacitor C, in farads
t_r = time of interest after start of retrace
R_1 = resistance from collector to emitter of the saturated transistor

EXAMPLE 8-4

An input pulse, V_i, of 5-nsec duration is applied to the circuit in Fig. 8-18 at the instant V_o reaches 4.76 V. What is the voltage across the capacitor after 50 μsec if the saturated transistor presents a resistance of 0.2 kΩ to the circuit?

Solution

Using Eq. 8-21, we find the voltage as

$$\begin{aligned} V_o' &= V_o \varepsilon^{-t_r/R_1 C} \\ &= (4.76\ \text{V})\left[\varepsilon^{-(50\times10^{-6}\ \text{sec})/(200\times0.2\times10^{-6})}\right] \\ &= (4.76\ \text{V})(\varepsilon^{-1.25}) \\ &= 1.36\ \text{V} \end{aligned}$$

The relationship between the current charging a capacitor and the voltage across the capacitor is given as

$$V_o = \frac{1}{C}\int_0^t i_c\, dt \tag{8-22}$$

If the current is constant, the solution of Eq. 8-22 is

$$V_o = \frac{It}{C} \tag{8-23}$$

Substituting V/R for I, we have

$$V_o = \frac{V_{cc}}{RC} t \tag{8-24}$$

Equation 8-24 is the equation of a straight line with a slope of V/RC. The voltage V_o would be a "linear ramp" if it followed this equation. Figure 8-19 compares the graph of the exponential function of Eq. 8-20 to the graph of the linear function of Eq. 8-24.

As can be seen by comparing the linear and exponential curves, the exponential function is fairly linear during the first 10% or 15% of the graph.

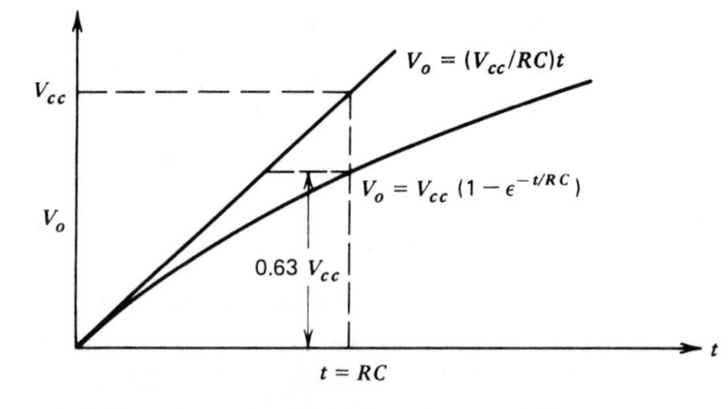

FIGURE 8-19 Graph comparing a capacitor charged by a constant-current source to one charged by a constant-voltage source.

Some inexpensive oscilloscopes use this type of *RC* circuit as a sweep generator with reasonably satisfactory results. However, linearity can be improved by using one of several other types of circuits as a sweep generator. The circuits most commonly used are

1. Some form of constant-current generator (see the discussion in Chapter 4) to charge a capacitor at a constant rate.
2. A bootstrap circuit in which the current is kept constant by maintaining a constant voltage across a charging resistor.
3. An operational amplifier in a circuit known by one of the following names: integrating amplifier, Miller integrator, Miller sweep generator.

The circuit shown in Fig. 8-20*a* utilizes an ordinary bipolar junction transistor, Q_1, connected in the common-base configuration to charge capacitor

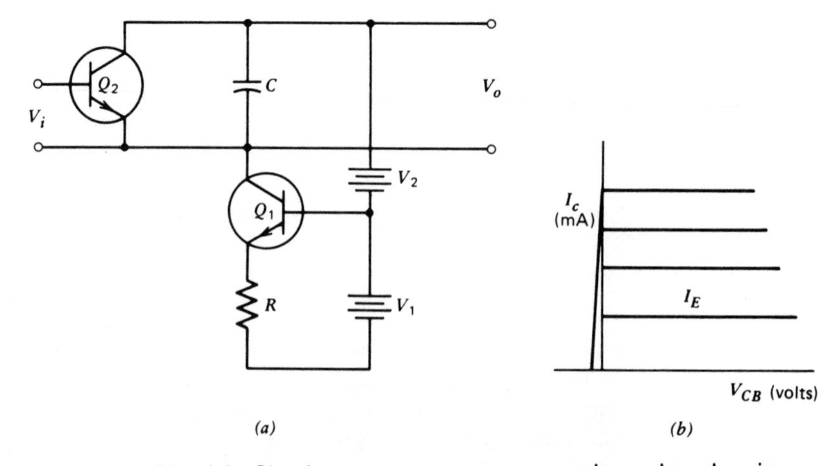

FIGURE 8-20 (*a*) Circuit to generate a ramp voltage by charging a capacitor at a constant rate. (*b*) The characteristic curves for transistor Q_1.

C at a constant rate. If the resistor, R has a large value of resistance, the emitter current will be nearly constant as will the collector current. Since the collector current charges the capacitor C, it will charge at a nearly constant rate. Typical common-base output characteristic curves are shown in Fig. 8-20*b*. Note that the curves are very nearly horizontal in the active region of the transistor. The collector-to-base voltage V_{CB} can change considerably without any appreciable effect on the collector current.

In order to have a repetitive sweep voltage, the capacitor must be discharged after it has reached its maximum voltage. This can be done by connecting transistor Q_2 across the capacitor as shown. At specific time intervals, a voltage pulse applied to the base of Q_2 turns the transistor "on" providing a discharge path for the capacitor.

8-12 VERTICAL INPUT AND SWEEP GENERATOR SIGNAL SYNCHRONIZATION

Most waveforms that we will have occasion to observe with an oscilloscope will be changing at a rate much faster than the eye can follow, perhaps many million times per second. If we are to be able to observe such rapid changes, the beam must retrace the same pattern repeatedly. If the pattern is retraced in such a manner that the pattern always occupies the same location on the screen, the eye will see a stationary display. The beam will retrace the same pattern at a rapid rate if the vertical input signal and the sweep generator signal are synchronized, which means that the vertical input signal must be equal to, or an exact multiple of, the sweep generator signal, as shown in Fig. 8-21. If the vertical input frequency is not exactly equal to, or an exact

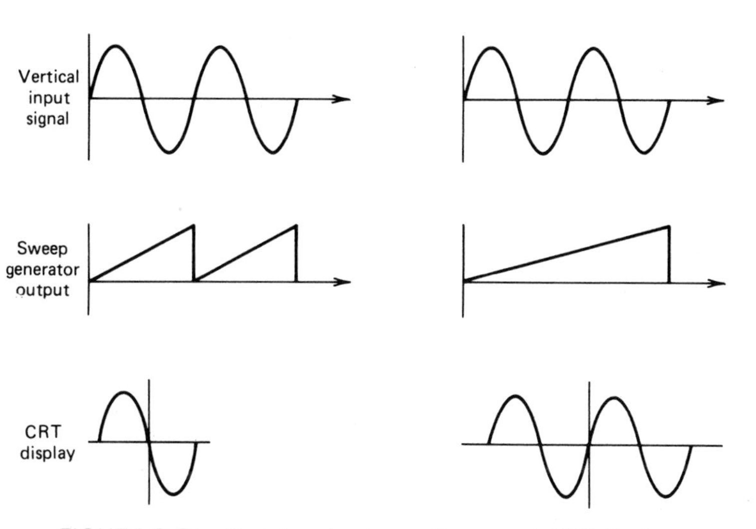

FIGURE 8-21 Synchronized waveforms and CRT display.

multiple of, the sawtooth frequency, the waveforms will not be synchronized and the display "walks" across the screen. If the pattern moves toward the right, the frequency of the sawtooth waveform is too high. Movement of the pattern toward the left indicates that the frequency of the sawtooth is too low.

Synchronization of the waveforms can be accomplished in two different ways. In very basic oscilloscopes the sweep generator is continuously changing and discharging a capacitor. One ramp voltage is followed immediately by another; hence, the sawtooth pattern appears. A sweep generator operating in this manner is said to be "free running." In order to present a stationary display on the screen, the sweep generator signal must be forced to run in synchronization with the vertical input signal. In basic oscilloscopes this is accomplished by carefully adjusting the sweep frequency to a value very close to the exact frequency of the vertical input signal, or a submultiple of this frequency. With both signals at the same frequency, an internal sync pulse will lock the sweep generator into the vertical input signal. This method of synchronization has some serious limitations when an attempt is made to observe low-amplitude signals. However, the most serious limitation is probably the inability of the instrument to maintain synchronization when the amplitude or frequency of the vertical signal is not constant, such as with voice or music signals.

These limitations are overcome by incorporating a trigger circuit into the oscilloscope as shown in Fig. 8-22. The trigger circuit may receive an input from one of three sources depending on the setting of the trigger selector switch. The input signal may come from an external source when the trigger selector switch is set to EXT, from a low-amplitude ac voltage at line

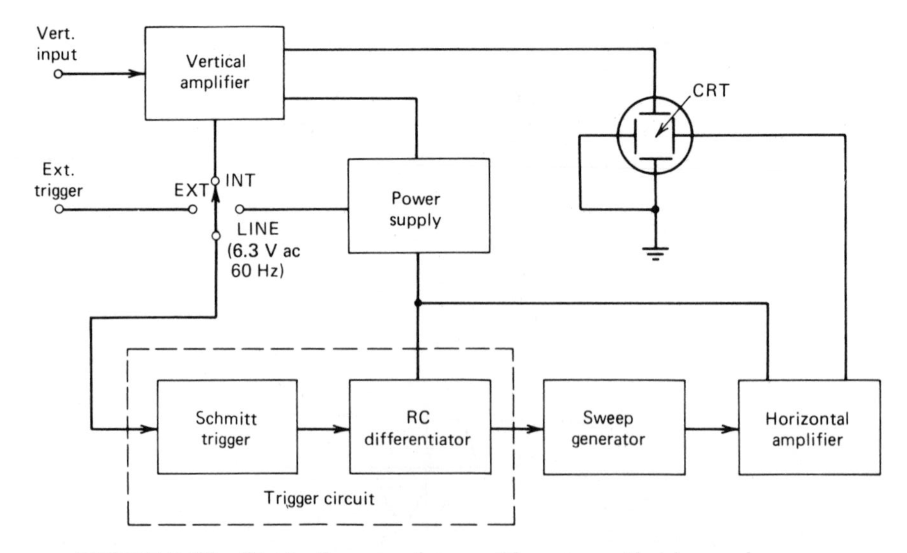

FIGURE 8-22 Block diagram of an oscilloscope with triggered sweep.

frequency when the switch is set to LINE, or from the vertical amplifier when the switch is set to INT. When set for internal triggering (INT), the trigger circuit receives its input from the vertical amplifier. When the vertical input signal that is being amplified by the vertical amplifier reaches a certain level, the trigger circuit provides a pulse to the sweep generator, thereby ensuring that the sweep generator output is synchronized with the signal that triggers it.

One type of circuit frequently used in the "trigger circuit" block of Fig. 8-22 is called a Schmitt trigger or a voltage-level detector. Basically, the Schmitt trigger compares an input voltage, in this case from the vertical amplifier, with a voltage at a point in the circuit. When the input voltage exceeds the voltage to which it is being compared, the circuit changes states, which means the output voltage goes to a high state. The point at which this occurs is called the *upper trigger point* (UTP). When the input voltage drops below a certain level called the *lower trigger point* (LTP), the Schmitt trigger output returns to its original level, and thus the output of the circuit is a square wave. Unless this square wave is of very short duration, it will not be suitable to trigger the sweep generator directly. It is common practice to apply the square wave to an *RC* circuit called a differentiation whose output is a short-duration spike suitable for triggering the sweep generator.

8-13 LABORATORY OSCILLOSCOPES

A high-quality, triggered-sweep oscilloscope with interchangeable, plug-in vertical amplifiers is highly versatile and can be made more so by incorporating into it additional features for special applications. Two of the more important features are discussed in the following paragraphs.

8-13.1 Dual-Trace Oscilloscope

The popularity of dual-trace oscilloscopes has increased tremendously in recent years. A dual trace is obtained by electronically switching the single electron beam. Figure 8-23 shows a block diagram of the two vertical input channels and the electronic switch that alternatively connects the two input channels to the vertical amplifier.

There are generally at least four nodes of operation with dual-trace oscilloscopes; they are labeled 1, 2, *alternate*, and *chopped*. When the oscilloscope is set to 1 or 2, only the input at that channel is displayed. In the alternate mode the inputs are displayed on alternate traces. Since the switching rate is synchronized with the sweep generator, switching occurs at the same rate as the output of the sweep generator. The alternate mode of operation is generally preferred when displaying relatively high-frequency signals. In the chopped mode electronic switching occurs at a rate completely independent of the sweep rate. Therefore, each display has portions missing during the time the other signal is being displayed. The chopped mode is

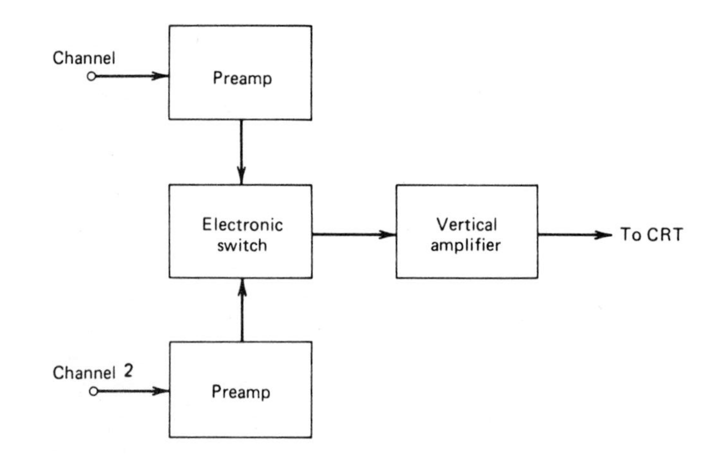

FIGURE 8-23 Block diagram of the input channels of a dual-trace oscilloscope.

normally used at low sweep rates when the alternate mode would provide a display with appreciable flicker.

The oscilloscope shown in Fig. 8-24 is a dual-trace laboratory-quality instrument with *storage* capability. The instrument is a Model 7633 which is manufactured by Tektronix, Inc. Industrial firms and research laboratories make extensive use of oscilloscopes like this one.

8-13.2 Delayed Sweep

Many laboratory-quality oscilloscopes include a **delayed-sweep** feature. This feature increases the versatility of the instrument by making it possible to magnify a selected portion of the undelayed sweep, measure waveform jitter or rise time, and check pulse-time modulation, as well as many other applications.

Delayed sweep is a technique that adds a precise amount of time between the trigger point and the beginning of the scope sweep. When the scope is being used in the delayed-sweep mode, the start of the delayed sweep can range from a few microseconds to perhaps 10 seconds or more. The delayed-sweep operation allows the instrument user to view a small segment of a waveform—for example, an oscillation or **ringing** that occurs during a small portion of a lower-frequency waveform.

Sometimes the delayed-sweep feature is used for convenience to allow triggering at some other point than at the leading or trailing edge. However, in some situations a measurement is possible only if the delayed-sweep feature is used. For example, suppose the part of a waveform that is to be measured is too far from the only available trigger point to permit a stable display on the CRT screen. The problem can be solved by using delayed sweep to trigger at the only available trigger point and then starting the sweep at the point of interest.

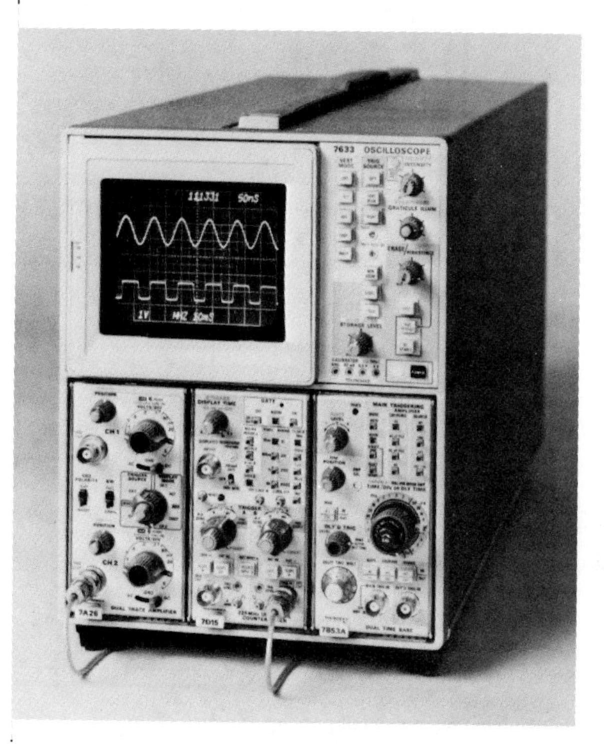

FIGURE 8-24 **Laboratory-quality dual-trace oscilloscope. (Courtesy Tektronix, Inc.)**

Although there are a few exceptions, delayed sweep is normally a feature of dual-time base oscilloscopes which have two completely separate sweep generators. One sweep functions as a main sweep, and the other serves as the delayed sweep. The main sweep is initiated by a trigger pulse at the leading edge of pulse 1 shown in Fig. 8-25. Suppose we wish to observe in detail a portion of the waveform near its trailing edge by expanding the waveform and using a higher sweep speed—for example, a sweep speed of 0.1 μsec/cm. As can be seen in Fig. 8-26, the portion of the waveform of interest to us is completely off the screen to the right when the sweep speed is set to 0.1 μsec/cm. To observe the portion of the waveform that is of interest at the higher sweep speed, we must use the delayed-sweep feature and reset the sweep speed to 5 μsec/cm.

Basically, the delayed sweep works as follows. The main sweep is initiated by a trigger pulse at time t_0 shown in Fig. 8-27. This time corresponds to the leading edge of pulse 1 in Fig. 8-26. The main sweep generator ramp, which is applied to a comparator along with a voltage from a *delay control* circuit, increases linearly until it trips the comparator at time t_1. When the comparator changes states, the delayed-sweep generator is triggered, which intensifies a portion of the original display. Adjusting the *delay time* should intensify the portion of the waveform that is of interest. The exact front panel adjustments

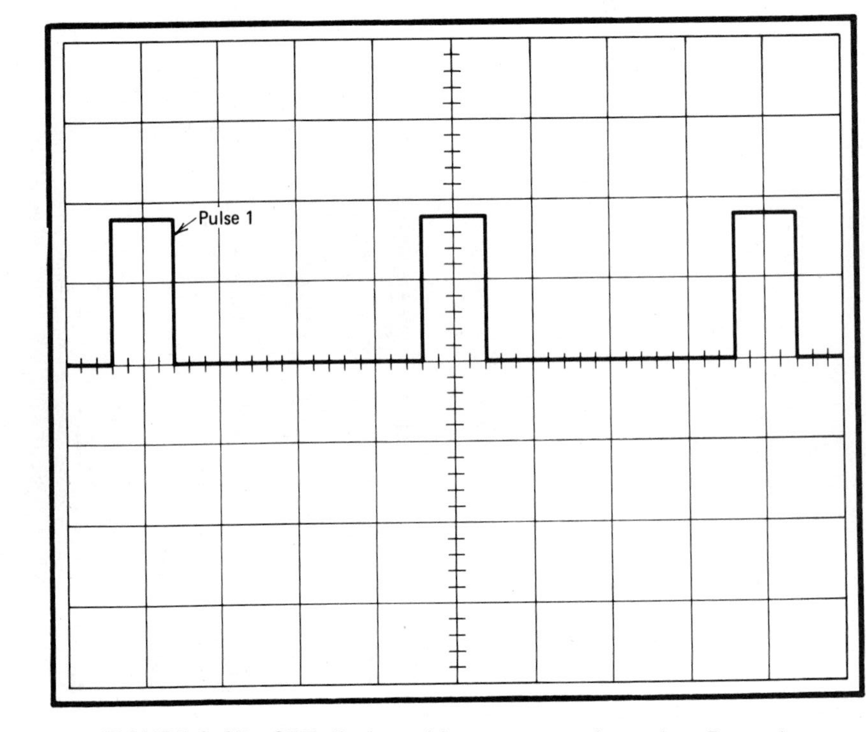

FIGURE 8-25 CRT display with sweep speed equal to 5 μsec/cm.

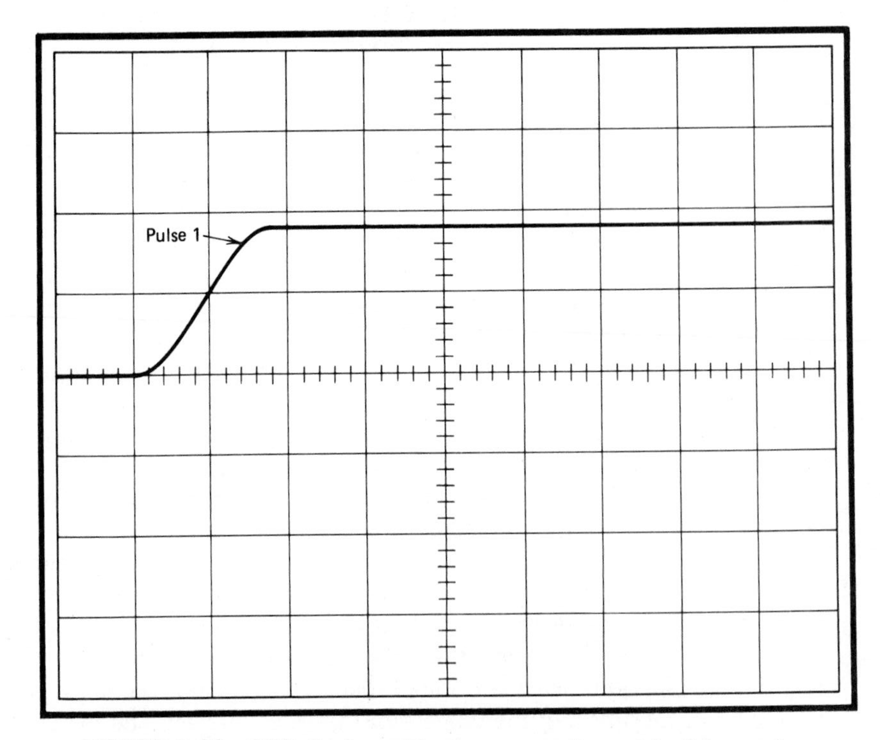

FIGURE 8-26 CRT display with sweep speed equal to 0.1 μsec/cm.

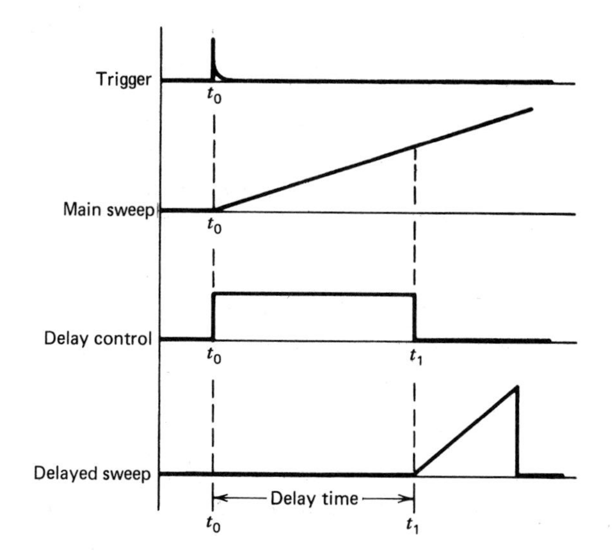

FIGURE 8-27 Delayed-sweep triggering waveforms.

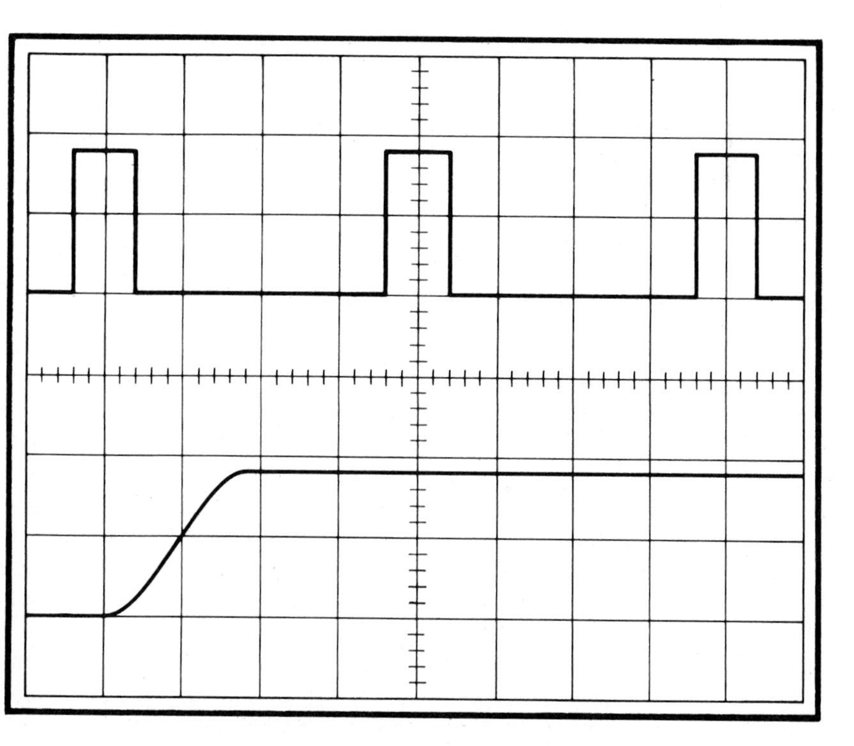

FIGURE 8-28 Delayed-sweep measurement.

that are necessary to display the expanded portion of the waveform on the CRT screen depend on the oscilloscope being used. Some oscilloscopes have an alternate *sweep separation* control so that two waveforms are displayed: the original waveform, with the portion of the waveform of interest intensified, and a second expanded waveform, as shown in Fig. 8-28. By adjusting the horizontal mode to channel *B*, we can now increase the sweep time so that we can view the portion of the waveform that is of interest.

8-14 STORAGE OSCILLOSCOPES

In many oscilloscope applications the limited persistence of the CRT phosphor makes real-time observation of one-time events nearly impossible. Although such events can be recorded photographically, this may prove to be expensive and time-consuming. The fairly recent development of the storage oscilloscope makes it possible to retain a CRT display for an extended period of time. After studying the stored display, we may then decide whether to take a photograph.

There are two types of storage oscilloscopes, one using a specially designed CRT and one using digital techniques. The storage CRT uses two electron guns: a *writing gun* which is the usual electron gun, and a *flood gun* which uniformly bombards the entire CRT screen with low-energy electrons. The phosphor particles struck by these low-energy electrons take on a low-level charge; however, nonenergized particles remain in a "no-charge" condition. When a trace is to be recorded, the writing gun is turned "on" and high-energy electrons strike the screen forming an image. The phosphor particles struck by the high-energy electrons possess considerable charge. Therefore, additional flood gun electrons are attracted, which sustains the image. The image is erased by grounding the phosphor screen, which removes excess charge.

8-15 SAMPLING OSCILLOSCOPES

The gain–bandwidth relationship of the vertical amplifier limits the frequency range of signals that can be displayed on the conventional oscilloscope. If signals with frequencies above the frequency that can be viewed with a conventional oscilloscope are repetitive, they can be effectively "slowed down" many thousands of times by a technique called *sampling*. Sampling is somewhat analogous to observing a rotating fan blade that is illuminated with a stroboscope. The rapidly rotating blade appears to be rotating slowly when observed through a narrow "time window," which samples the event at a rate slightly different from that at which the blade is rotating. The sampling technique used to observe very high-frequency repetitive electrical signals with an oscilloscope is shown in Fig. 8-29. Samples of the input waveform are taken from successive cycles, with one sample taken per cycle and each

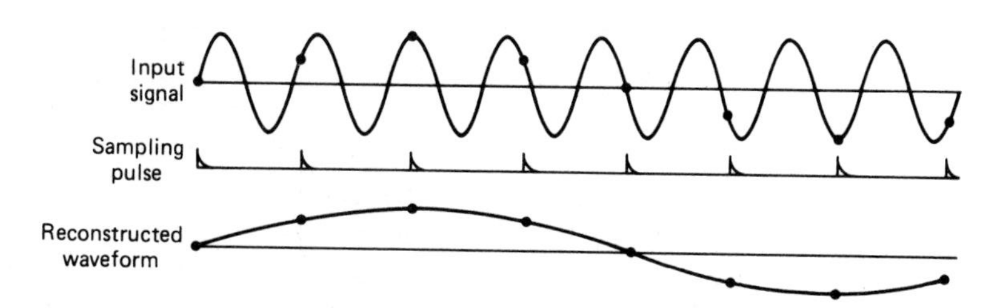

FIGURE 8-29 The principle of sampling.

sample slightly delayed with respect to the preceding sample. The reconstructed waveform, which is displayed on the CRT screen, is a composite waveform made up of the samples taken from successive cycles of the input signal.

8-16 DIGITAL STORAGE OSCILLOSCOPES

The conventional storage oscilloscopes discussed in the previous section have been used for a number of years to retain waveforms or transient signals. However, stored images gradually fade and cannot be recalled in this type of instrument.

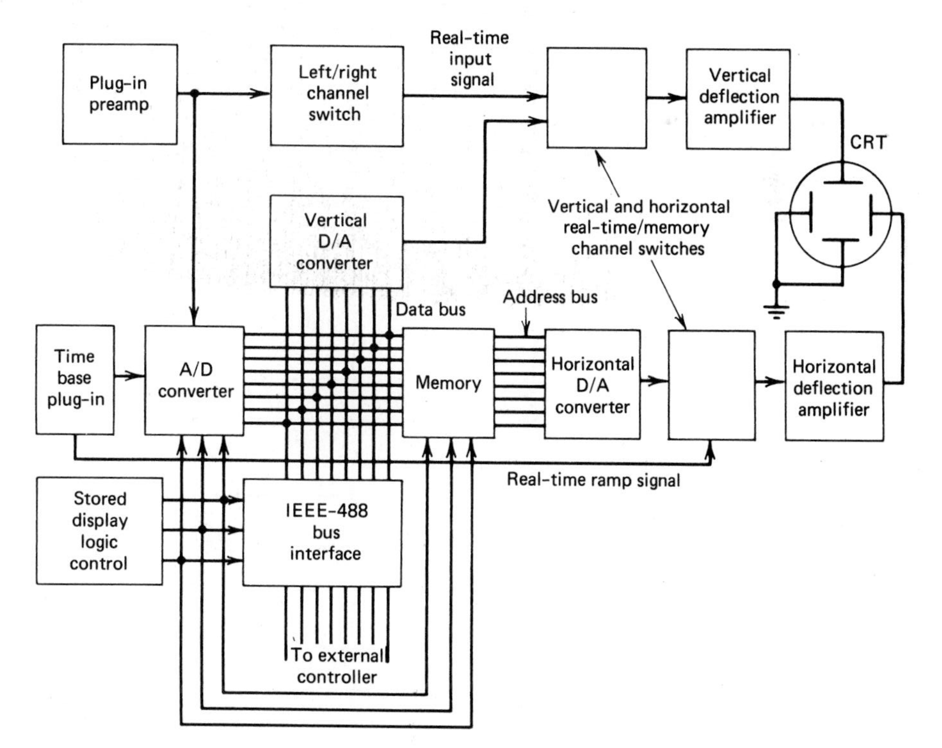

FIGURE 8-30 Block diagram of a digital storage oscilloscope.

Digital storage oscilloscopes provide a number of benefits when compared with their analog counterparts. For example, digital displays never fade or bloom. Of more significance is the fact that digital storage oscilloscopes can be interfaced with desktop calculators or microcomputers to obtain a variety of quick and accurate measurements.

Compared to the technique used in analog storage oscilloscopes to display waveforms for an extended period of time, the CRTs of the digital storage oscilloscopes are much simpler. Digital storage oscilloscopes use a conventional CRT with no flood guns. They incorporate memory circuitry to store information about the waveform. The waveform is then repeatedly read from memory to form the display on the CRT screen. Figure 8-30 gives a block diagram for a digital storage oscilloscope.

8-17 ATTENUATORS

The voltage at the input terminal of the vertical amplifier, which causes the beam to be deflected off the CRT screen, is quite low in amplitude. So that high-amplitude signals may be displayed, an **attenuator** network is placed between the vertical input terminal and the input terminal of the vertical amplifier. The term *attenuate* means to "reduce in size." The purpose of the attenuator is to reduce the amplitude of the vertical input signal before applying it to the vertical amplifier. The most basic attenuator is a simple resistive voltage divider such as the one shown in Fig. 8-31. With this circuit,

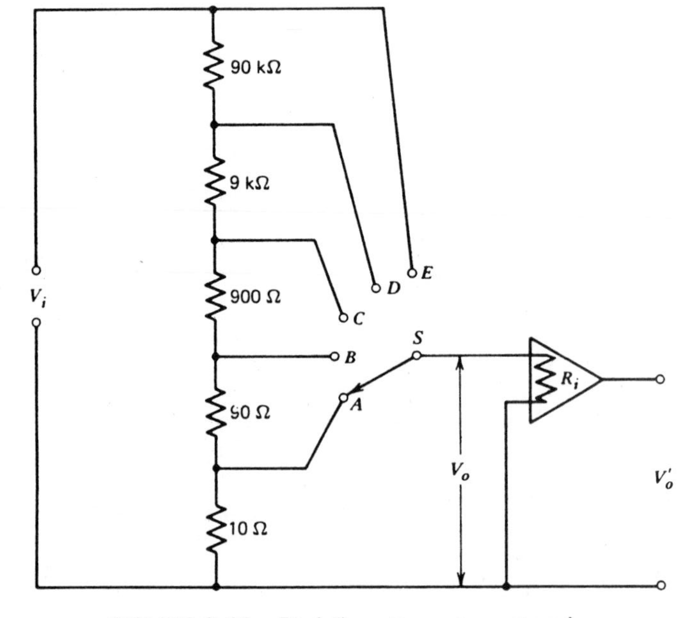

FIGURE 8-31 Resistive attenuator network.

the input voltage V_i will be attenuated by a factor of 10,000 with the switch S set to position A as shown. In positions B, C, and D, the attenuation factors will be 1000, 100, and 10, respectively. In switch position E there is no attenuation. In any switch position, the total input resistance, as seen by V_i, is 100 kΩ. The attenuation in any switch position can be determined from the ratio of the total resistance from the position of interest to ground to the total series resistance, written as

$$\frac{V_o}{V_i} = \frac{R}{R_t} \tag{8-25}$$

where

R = total resistance from the desired attenuator terminal to ground
R_t = total series resistance

Equation 8-25 is generally referred to as the voltage divider equation. The attenuation factor is the reciprocal of the voltage divider ratio.

EXAMPLE 8-5 If the switch in Fig. 8-31 is set to position D, what is the attenuation factor?

Solution Using Eq. 8-25, we obtain the voltage divider ratio as

$$\frac{V_o}{V_i} = \frac{R}{R_t}$$

$$= \frac{9\ \text{k}\Omega + 900\ \Omega + 90\ \Omega + 10\ \Omega}{100\ \text{k}\Omega} = 0.1$$

Therefore, the attenuation factor is

$$\text{Attenuation factor} = \frac{1}{0.1} = 10$$

The resistance values shown in Fig. 8-31 will provide the desired attenuation only if the input resistance R_i of the vertical amplifier is much greater than the attenuator resistance values. If the amplifier input resistance is not much larger than the attenuator resistance with which it is in parallel, appreciable error in the attenuation factor will be noted since the attenuator is "loaded down" by the amplifier.

EXAMPLE 8-6 If the switch in Fig. 8-31 is in position D, as in Example 8-5, and the input resistance of the amplifier R_i is 100 kΩ, what is the attenuation factor?

Solution Since the amplifier input resistance is in parallel with the attenuator resistance, this parallel combination must be used in the voltage divider equation.

The parallel resistance is

$$R_p = \frac{R_i R}{R_i + R}$$

$$= \frac{100\ \text{k}\Omega \times 10\ \text{k}\Omega}{100\ \text{k}\Omega + 10\ \text{k}\Omega} = 9.09\ \text{k}\Omega$$

Therefore,

$$\frac{V_o}{V_i} = \frac{9.09\ \text{k}\Omega}{100\ \text{k}\Omega} = 0.0909$$

The attenuation factor is now

$$\text{Attenuation factor} = \frac{1}{0.0909} = 11$$

As can be seen by comparing the results of Examples 8-5 and 8-6, the attenuation increased in response to the parallel effect of the input resistance of the amplifer.

Switch S in Fig. 8-31 is the rotary switch that is mounted on the front panel of an oscilloscope and labled Volts/Div, as was mentioned briefly in Section 8-9. Example 8-7 shows how to determine the value of the attenuating resistors to obtain desired attenuation factors.

EXAMPLE 8-7

An oscilloscope is to have an input resistance of 8 MΩ, a sensitivity of 50 mV, and attenuation factors of 4, 10, 40, 100, and 400. Compute the value of the attenuating resistors and the volts/division value corresponding to each attenuation factor. Assume R_i of the vertical amplifier is sufficiently high to be ignored.

Solution

The attenuating network required is shown in Fig. 8-32. When switch S is set to position A, as indicated, maximum voltage can be applied across the network because of the voltage divider action. Therefore, at this position the attenuation factor will be 400. The attenuation factor equals the reciprocal of the voltage divider ratio. For position A then

$$\text{Attenuation factor} = 400 = \frac{1}{V_o/V_i}$$

$$400 = \frac{V_i}{V_o}$$

Since V_o equals the value of the sensitivity (50 mV/div) because V_o is applied directly to the amplifier in each switch position, we can solve

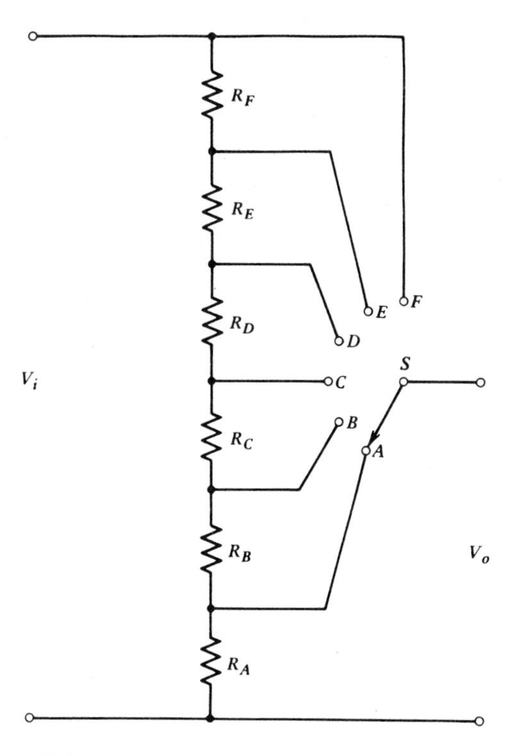

FIGURE 8-32 Circuit for Example 8-7.

for V_i as

$$V_i = 400V_o$$

$$= 400(50 \text{ mV/div}) = 20 \text{ V/div in position } A$$

From this result we see that we can multiply the attenuation factor by the sensitivity to obtain the value of the volts/division at each switch setting.

Position B: $V_i = 100(50 \text{ mV/div}) = 5 \text{ V/div}$

Position C: $V_i = 40(50 \text{ mV/div}) = 2 \text{ V/div}$

Position D: $V_i = 10(50 \text{ mV/div}) = 0.5 \text{ V/div}$

Position E: $V_i = 4(50 \text{ mV/div}) = 0.2 \text{ V/div}$

Position F: There is no attenuation or the attenuation factor is 1; therefore,

$$V_i = 1(50 \text{ mV/div}) = 50 \text{ mV/div}$$

Solving for the attenuating resistors, we obtain the following.

$$R_A = R_T(V_o/V_i) = (8\ \text{M}\Omega)(50\ \text{mV}/20\ \text{V}) = 20\ \text{k}\Omega$$

$$R_A + R_B = R_T(V_o/V_i) = (8\ \text{M}\Omega)(50\ \text{mV}/5\ \text{V}) = 80\ \text{k}\Omega$$

$$R_B = (R_A + R_B) - R_A = 80\ \text{k}\Omega - 20\ \text{k}\Omega = 60\ \text{k}\Omega$$

$$R_A + R_B + R_C = R_T(V_o/V_i) = (8\ \text{M}\Omega)(50\ \text{mV}/2\ \text{V}) = 200\ \text{k}\Omega$$

$$R_C = (R_A + R_B + R_C) - (R_A + R_B)$$
$$= 200\ \text{k}\Omega - 80\ \text{k}\Omega = 120\ \text{k}\Omega$$

$$R_A + R_B + R_C + R_D = R_T(V_o/V_i)$$
$$= (8\ \text{M}\Omega)(50\ \text{mV}/0.5\ \text{V}) = 800\ \text{k}\Omega$$

$$R_D = (R_A + R_B + R_C + R_D) - (R_A + R_B + R_C)$$
$$= 800\ \text{k}\Omega - 200\ \text{k}\Omega = 600\ \text{k}\Omega$$

$$R_A + R_B + R_C + R_D + R_E = V_o/V_i = (8\ \text{M}\Omega)(50\ \text{mV}/0.2\ \text{V})$$
$$= 2\ \text{M}\Omega$$

$$R_E = (R_A + R_B + R_C + R_D + R_E) - (R_A + R_B + R_C + R_D)$$
$$= 2\ \text{M}\Omega - 800\ \text{k}\Omega = 1.2\ \text{M}\Omega$$

$$R_F = R_T - (R_A + R_B + R_C + R_D + R_E) = 8\ \text{M}\Omega - 2\ \text{M}\Omega = 6\ \text{M}\Omega$$

The complete attenuator network is shown in Fig. 8-33.

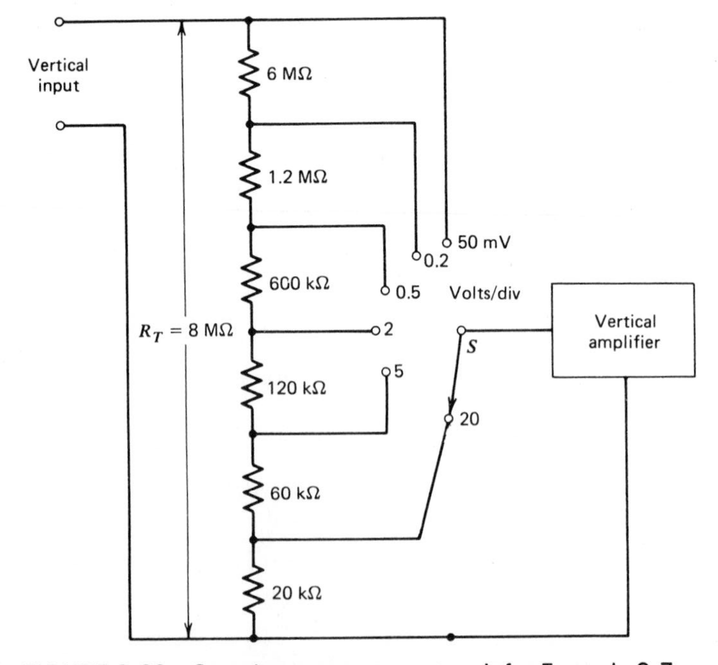

FIGURE 8-33 Complete attenuator network for Example 8-7.

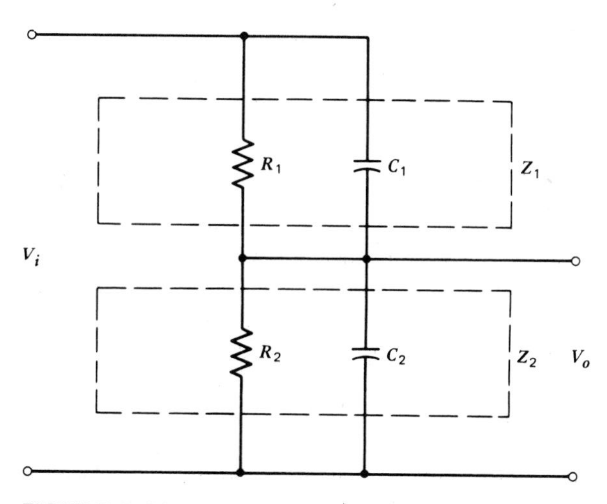

FIGURE 8-34 Attenuator network with associated stray capacitance.

If we were measuring dc voltages exclusively, our discussion of voltage divider attenuators might end here. However, oscilloscopes are used primarily to display ac waveforms. This introduces the problem of the stray capacitance of the attenuator network as shown in Fig. 8-34. For ac signals, the attenuation factor is actually related to the ratio of the impedances given by the voltage divider ratio as

$$\frac{V_o}{V_i}=\frac{Z_2}{Z_1+Z_2} \tag{8-26}$$

If the stray capacitance values are not correctly related to the resistance values, the voltage divider ratio will not have the same value at all frequencies.

EXAMPLE 8-8

Compare the output voltage of the voltage divider attenuator shown in Fig. 8-35 for a dc voltage and a 10-MHz ac signal.

Solution

For the dc voltage source

$$V_o=V_i\left(\frac{R_2}{R_1+R_2}\right)$$

$$=(1\text{ V})\left(\frac{100\text{ k}\Omega}{900\text{ k}\Omega+100\text{ k}\Omega}\right)=0.1\text{ V}$$

For the ac source, X_{C1} is approximately 3184 Ω and X_{C2} is approximately 1592 Ω. Since these values are much less than the value of the resistances

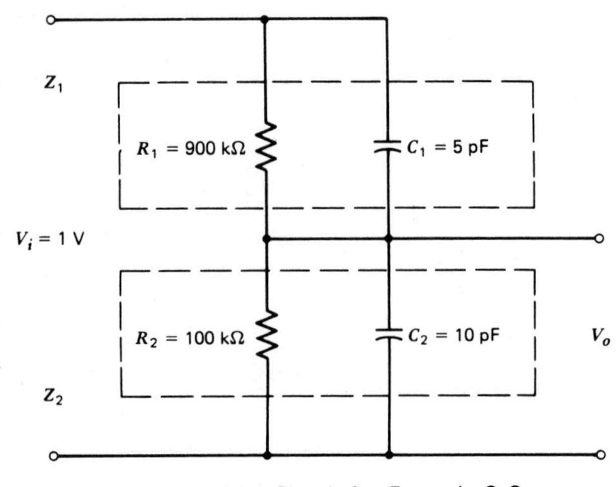

FIGURE 8-35 Circuit for Example 8-8.

in parallel with them, we can approximate the output voltage as

$$V_o = (1\text{ V})\left(\frac{X_{C2}}{X_{C1}+X_{C2}}\right)$$

$$= (1\text{ V})\left(\frac{1592\ \Omega}{3184\ \Omega + 1592\ \Omega}\right) = 0.33\text{ V}$$

Consider the attenuator network of Fig. 8-36. The resistance ratio of R_2 to R_1 is the reciprocal of the ratio of capacitors C_2 to C_1. When this is true, the voltage divider ratio will remain constant over a wide range of frequencies. Moreover, for a frequency-compensated voltage divider, the following relationships apply:

$$R_2 = \frac{R_1}{k-1} \tag{8-27}$$

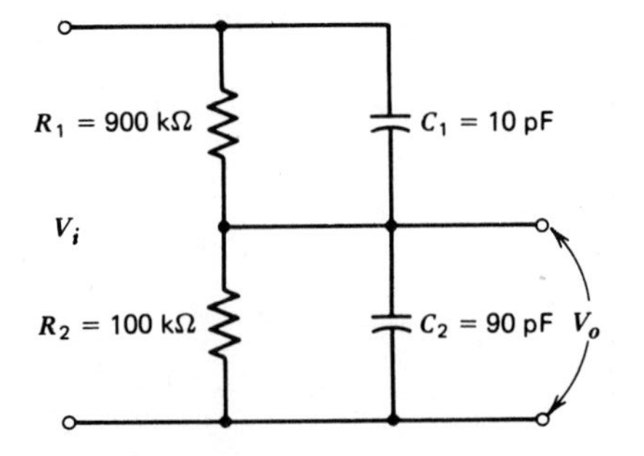

FIGURE 8-36 Compensated attenuator network.

and

$$C_2 = C_1(k-1) \tag{8-28}$$

where k is the voltage divider ratio, V_i/V_o.

8-18 HIGH-IMPEDANCE PROBES

External, high-impedance probes are used to increase the input resistance and reduce the effective input capacitance of an oscilloscope. A resistor and capacitor combination can be added to an oscilloscope as shown in Fig. 8-37, in effect, moving the input terminals from the front panel of the instrument to the end of the probe. Suppose that in Fig. 8-37 we wish to attenuate the input signal by a factor of 10. By using Eqs. 8-27 and 8-28, where $k = 10$, we can compute the value of resistor R_1 and capacitor C_1 as follows:

$$R_1 = R_2(k-1)$$
$$= (1\ \text{M}\Omega)(10-1) = 9\ \text{M}\Omega$$

and

$$C_1 = \frac{C_2}{k-1}$$
$$= \frac{30\ \text{pF}}{10-1} = 3.33\ \text{pF}$$

Note that the new input impedance R_i is the total resistance; therefore,

$$R_i = R_1 + R_2 = 10\ \text{M}\Omega$$

and

$$C_i = \frac{C_1 C_2}{C_1 + C_2} = 3\ pF$$

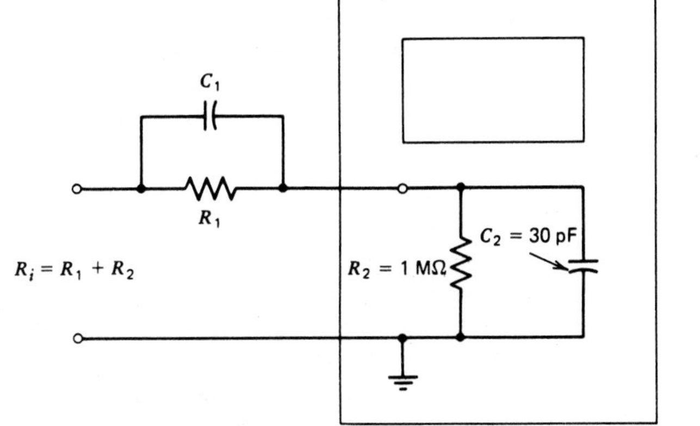

FIGURE 8-37 External high-impedance probe.

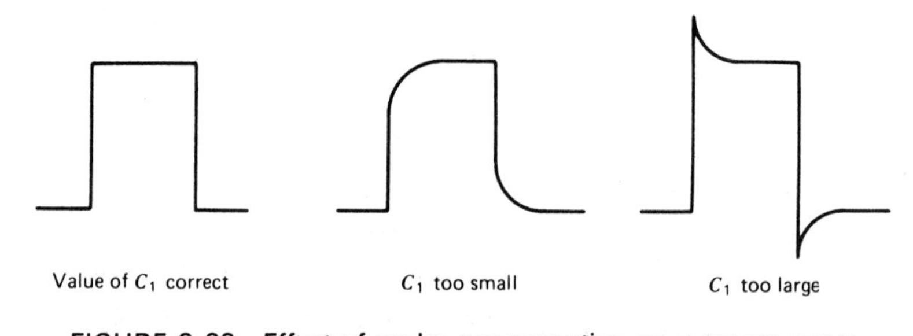

FIGURE 8-38 Effect of probe compensation on a square wave.

The input resistance has been increased by a factor of 10 while, at the same time, input capacitance is one-tenth its original value. The combination of R_1 and C_1 is called a $\times 10$ ("times ten") probe.

Capacitor C_1 is usually adjustable to compensate for differences in input capacitance between oscilloscopes. If the probe capacitance is adjusted to the wrong value, the oscilloscope will exhibit poor frequency response characteristics. Probe adjustment is usually checked by displaying a square wave on the CRT screen. If the probe is not properly compensated, the square wave will be adversely affected, as shown in Fig. 8-38. If the value of capacitor C_1 is too small, the leading edge of the square wave is rounded off; but if capacitor C_1 is too large in value, the leading edge of the square wave overshoots.

8-19 SPECIFICATIONS

A person purchasing an oscilloscope needs to consider specifications carefully. The intended use of an oscilloscope is a deciding factor in establishing the required specifications for an instrument. A particular oscilloscope represents several compromises which directly affect the specifications of the instrument and, therefore, in the final analysis, determine the kind of measurements that can be made with it. For example, in general-purpose bench use, specifications often represent a compromise between sensitivity and bandwidth, whereas for specialized high-frequency work, bandwidth might be increased at the expense of the sensitivity. Most major oscilloscope manufacturers offer several models of general-purpose oscilloscopes that differ primarily in bandwidth and sensitivity. These instruments are generally designed around a 5-inch CRT, which is itself a convenient compromise between the size of the instrument and the size of the display.

In the design of oscilloscopes certain principal-use requirements impose limitations that influence many of the specifications. For example, when portability is required, batteries become the principal source of electrical power, which limits both the voltage and current. These limitations affect the sensitivity, frequency response, and trace intensity.

Reading specifications is often confusing and perplexing. The reader should know the exact meaning of each parameter specified since companies

TABLE 8-2
Summary of Typical Oscilloscope Specifications

Subsystem	Typical Specification	Remarks
I. *Cathode-Ray Tube*		
Screen size	8 × 10 divisions	Divisions are typically not always 1 cm
Screen type	P31	Refer to Table 8-1 regarding type of phosphor
Accelerating voltage	7.5 kV	Ranges from about 1.5 kV depending on screen size and bandwidth
Graticule	Internal, illuminated	Internal graticule
Front panel controls	Focus, intensity, beam finder	
II. *Vertical Amplifier*		
Two identical amplifiers, channels *A* and *B*		
A. Modes of operation		
Channel *A* only		Signal amplified by channel *A* displayed alone
±Channel *B* only		Signal amplified by channel *B* alone, normally or inverted (±)
A and ±*B* alternate		Both signals displayed on alternate sweeps, with *B* normal or inverted
A − *B* only		One signal equal to signal *A* minus signal *B* displayed
A − *B* and ±*B* alternate		Two signals displayed alternately; *A* − *B* and ±*B*
B. Amplifiers		
Bandwidth	DC: 0 to 50 MHz	Input directly coupled to the amplifiers
Rise time	7 nsec	Signals with rise time faster than 7 nsec will be improperly displayed
Overshoot	Less than 2% at maximum sensitivity	
Deflection coefficient	1 mV/div to 10 V/div in 14 calibrated steps of sensitivity	Range of deflection; 10,000 can be displayed
	±3% tolerance	Accuracy of the display deflection
	1:2.5 uncalibrated	An uncalibrated gain control increases with maximum sensitivity to about 1:25,000

TABLE 8-2 (*Continued*)

Subsystem	Typical Specification	Remarks
Maximum input voltage	±400 V dc or +ac peak	Limited by the BNC input connector, ac coupling capacitor, and other components at the input
Input impedance	1 MΩ 1125 pF	Somewhat standard for most oscilloscopes
C. Calibration		
Calibration voltage	600 mV ± 1%	Usually a calibrated square wave
Frequency	2 kHz ± 1%	Can serve as a rough frequency standard
Calibration current	6 mA, tolerance ±2%	Used to calibrate current probes
III. *Horizontal Amplifiers*		
Frequency range	dc to 1 MHz	
Deflection coefficient	2 mV/div	For EXT X input when used in the $X-Y$ mode
Maximum input voltage	±400 V dc or +ac peak	Limited by input connector and coupling capacitors
Input impedance	1 MΩ 1125 pF	Somewhat standard
IV. *Time Base*		
Sweep speeds	1 sec/div to 50 nsec/div in 23 calibrated steps	
Variable time control	Uncalibrated between steps with maximum sweep speed of 1.25 sec/div	
Magnification	×10	Increases maximum sweep speed by a factor of 10
V. *Triggering*		
Sources	Internal or external line	Channel A or B
Modes	Automatic	The sweep generator is automatically triggered by any signal falling within certain voltage and frequency limits
	Triggered	The sweep generator will be activated by any signal whose amplitude and frequency fall within the capabilities of the instrument, regardless of when it occurs
	Single shot	The sweep generator will be swept once on the first signal that occurs after the controls have been reset.

TABLE 8-2 (*Continued*)

Subsystem	Typical Specification	Remarks
Slope	+ or −	Controls can be set to initiate triggering on either the positive or the negative slope
Sensitivity	0.3 div deflection	
Internal	From direct current to 25 MHz	
External	50 mV to 10 MHz 150 mV at 25 MHz	

Aspects of the Total System	Typical Specification	Remarks
Power Requirements		
Line voltage range	90 to 132 V ac 180 to 250 V ac	
Line frequency	46 to 60 Hz	
Maximum power consumption	215 W, 3.3 A at 90 V ac, 60 Hz	
Mechanical Data		
Height	150 mm	Including feet and handle
Width	320 mm	Including handle
Length	340 mm	Including front cover
Weight	4.8 kg	
Temperature		
Reference	23°C	
Nominal operating range	+5°C to +40°C	
Storage and transport range	−40°C to +70°C	

write specifications differently. For example, specifications given as a numerical value without tolerances are generally typical of an "average" instrument, whereas specifications expressed in numerical form with tolerances stated are guaranteed by the manufacturer.

Table 8-2 illustrates a typical set of specifications for a general-purpose, laboratory-quality oscilloscope.

8-20 APPLICATIONS

The range of uses for oscilloscopes varies from basic voltage measurements and waveform observations to highly specialized applications in all areas of science, engineering, and technology. We discuss a few applications here.

8-20.1 Voltage Measurements

The most direct voltage measurement made with an oscilloscope is the peak-to-peak value. The rms value of the voltage can easily be calculated from the peak-to-peak measurement if desired. To arrive at a voltage value from the CRT display, one must observe the setting of the vertical attenuator, expressed in volts/division, and the peak-to-peak deflection of the beam. The peak-to-peak value of voltage is then computed as

$$V_{p-p} = \left(\frac{\text{volts}}{\text{div}}\right)\left(\frac{\text{no. div}}{1}\right) \tag{8-29}$$

EXAMPLE 8-9 The waveform shown in Fig. 8-39 is observed on the screen of an oscilloscope. If the vertical attenuator is set to 0.5 volt/div, determine the peak-to-peak amplitude of the signal.

Solution Using Eq. 8-29, we can compute the peak-to-peak value of the voltage as

$$V_{p-p} = \left(\frac{\text{volts}}{\text{div}}\right)\left(\frac{\text{no. div}}{1}\right)$$

$$= \left(\frac{0.5\text{ V}}{\text{div}}\right)\left(\frac{3\text{ div}}{1}\right) = 1.5\text{ V}_{p-p}$$

8-20.2 Period and Frequency Measurements

The period and frequency of periodic signals are easily measured with an oscilloscope. The waveform must be displayed in such a manner that one complete cycle appears on the CRT screen. Accuracy is generally improved if the single cycle displayed fills as much of the horizontal distance across the screen as possible. The period is calculated as

$$T = \left(\frac{\text{time}}{\text{div}}\right)\left(\frac{\text{no. div}}{\text{cyc}}\right) \tag{8-30}$$

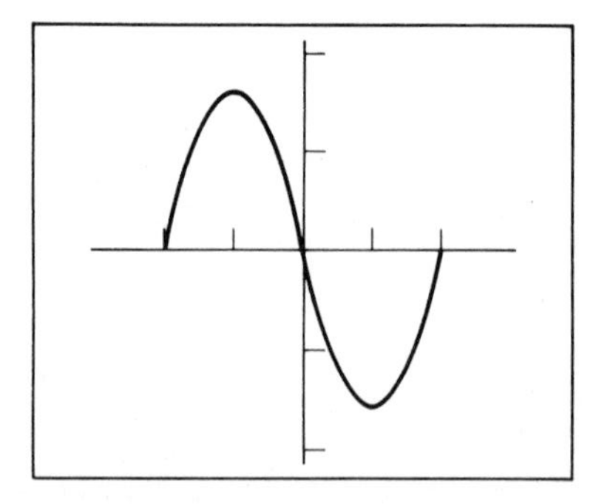

FIGURE 8-39 CRT display for Examples 8-9 and 8-10.

The frequency is then computed as the reciprocal of the period, or

$$f = \frac{1}{T} \tag{8-31}$$

EXAMPLE 8-10

If the time/division control is set to 2 μsec/div when the waveform in Fig. 8-39 is displayed on the CRT screen, determine the frequency of the signal.

Solution

The period of the signal is computed, using Eq. 8-30, as

$$T = \left(\frac{2\ \mu\text{sec}}{\text{div}}\right)\left(\frac{4\ \text{div}}{\text{cyc}}\right) = \frac{8\ \mu\text{sec}}{\text{cyc}}$$

Using Eq. 8-31, we then compute the frequency as

$$f = \frac{1}{T}$$

$$= \frac{1}{8\ \mu\text{sec/cyc}} = 125\ \text{kHz}$$

8-20.3 Determining Frequency with Lissajous Patterns

The oscilloscope can be used in the $X-Y$ mode to determine the frequency of a signal. The frequency is determined by applying the signal of unknown frequency to either the X or the Y input terminal and a signal of known frequency to the other input terminal. The pattern observed on the screen is called a **Lissajous figure**. The particular Lissajous pattern observed depends on the ratio of the two frequencies. This method has some limitations and is not widely used since low-cost digital frequency counters have come on the market. One limitation is that the ratio of the two frequencies must be set up with whole numbers in both the numerator and denominator. Another limitation is that 10:1 is about the maximum ratio of frequencies that can be used. At higher ratios the Lissajous pattern becomes so complex that a determination of the unknown frequency is very difficult.

If the frequencies of the signals applied to both the X and Y inputs are equal, that is, if their ratio is 1:1, and if the X and Y signals are out of phase by 90°, a circular pattern will be observed. A ratio of 2:1 produces a figure 8 pattern, as shown in Fig. 8-40. If the signal is applied to the horizontal input terminal, the figure 8 will be upright; if the vertical input signal is twice the horizontal input, the figure 8 pattern will be on its side. Ratios that are not equal to a whole number, such as 5:3, produce very complex patterns. The pattern shown in Fig. 8-41 is a 3:2 Lissajous figure with the vertical frequency higher. For frequency ratios greater than 10:1, the Lissajous pattern becomes too complex to be usable. In place of this technique, the ring Lissajous figure shown in Fig. 8-42 is used if the oscilloscope has a Z-axis input. The ratio of frequencies is determined by counting the dashes in the

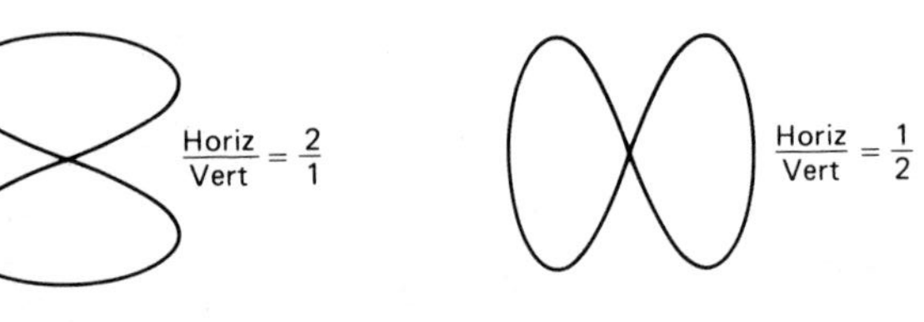

FIGURE 8-40 Lissajous patterns for 2:1 frequency ratios.

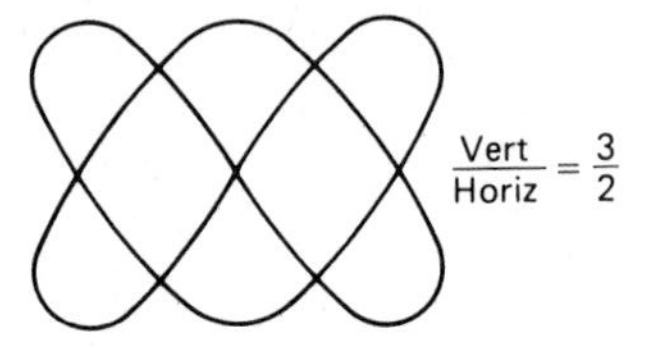

FIGURE 8-41 Lissajous pattern for 3:2 frequency ratio.

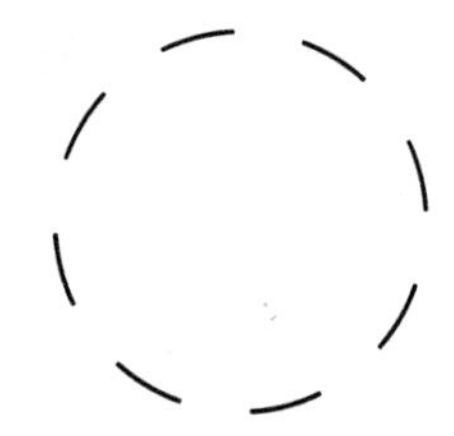

FIGURE 8-42 Ring Lissajous figure showing 8:1 frequency ratio.

ring. The ring pattern in Fig. 8-42 shows that the unknown frequency applied to the Z-axis input is eight times a known signal applied to the horizontal and vertical inputs to produce the circle.

8-20.4 Phase Angle Computation

Oscilloscopes can also be used in the $X-Y$ mode to determine the phase angle between two signals of the same frequency. The pattern displayed on the CRT screen may vary from a straight line with a positive slope, if the signals are in phase, to a straight line with a negative slope for signals 180° out of phase, as shown in Fig. 8-43. If the phase angle is any angle between 0° and 360° besides 180°, a circle or an ellipse, as shown in Fig. 8-44, will be displayed. The phase angle is easily determined from the ellipse. The ratio of the Y-axis intercept, represented as Y_1 in Fig. 8-44, and the maximum vertical

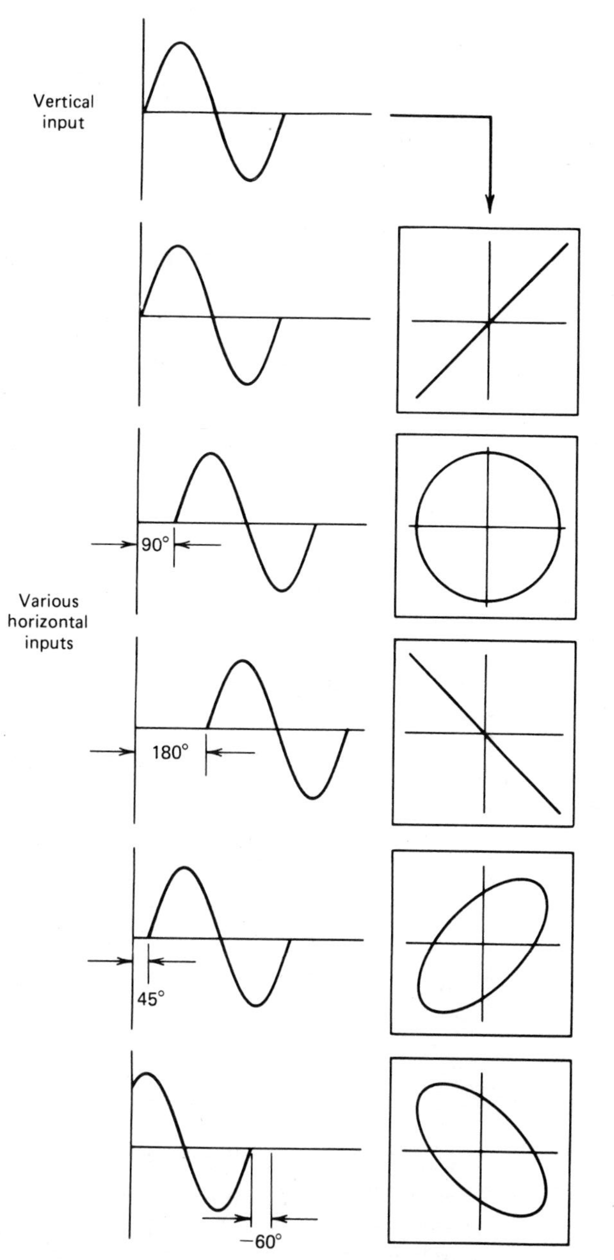

FIGURE 8-43 Lissajous patterns for selected phase angles.

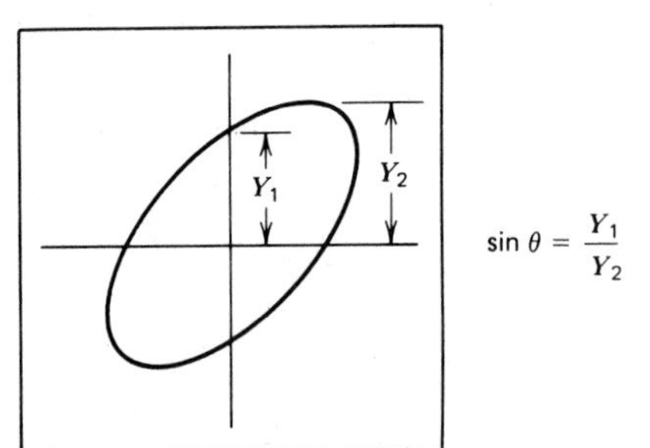

FIGURE 8-44 Evaluation of phase relationship.

deflection, Y_2, is equal to the sine of the phase angle; that is,

$$\sin\theta = \frac{Y_1}{Y_2} \tag{8-32}$$

where

θ = phase angle in degrees
Y_1 = Y-axis intercept
Y_2 = maximum vertical deflection

EXAMPLE 8-11 If, in Fig. 8-44, the distance Y_1 is 1.8 cm and Y_2 is 2.3 cm, what is the phase angle?

Solution Using Eq. 8-32, we can determine the phase angle as

$$\sin\theta = \frac{Y_1}{Y_2}$$

$$= \frac{1.8}{2.3} = 0.783$$

$$\theta = 51.5°$$

8-21 SUMMARY

The cathode-ray oscilloscope is a highly versatile laboratory instrument. The heart of the oscilloscope is the cathode-ray tube, which has a phosphor-coated face on which a reproduction of any signal applied to the input terminals of the instrument is displayed. The electronic circuitry in the instrument is required for operation of the CRT and for signal conditioning. The primary electronic circuits are the horizontal and vertical amplifiers, the sweep generator, including triggering circuitry, and the associated power supplies.

The sensitivity and the bandwidth of an oscilloscope are determined primarily by the vertical amplifier. The sweep generator produces a sawtooth voltage waveform which sweeps the electron beam horizontally. If the displayed pattern is to be a faithful reproduction of the input signal, the electron beam must be deflected equal horizontal distances per unit of time. Therefore the sawtooth waveform must be very linear.

Laboratory-quality oscilloscopes have provisions for triggering, delayed sweep, and other advanced features. In addition, many have dual-trace capabilities that allow two signals to be displayed simultaneously. Some oscilloscopes are also designed with storage capabilities and the ability to sample very high frequencies.

The oscilloscope finds applications in virtually every area of science, engineering, and technology. Special features such as sampling and storage capabilities enhance the usefulness of the instrument.

8-22 GLOSSARY

Attenuator: A fixed or variable device that is used to reduce the amplitude of an input or output signal to or from an electronic circuit.

Bandwidth: The range of frequencies over which a particular device or circuit is designed to operate within specified limits.

Collimate: To cause the electrons to follow straight and parallel paths.

Deflection sensitivity: The minimum voltage required to cause one division of vertical deflection

Delayed sweep: A technique to delay the sweep of the beam across the face of the CRT by a precise amount of time after the oscilloscope is triggered.

Fluorescent: Having the property of emitting light when bombarded with electrons.

Graticule: A scale on transparent material that is fitted to the face of a CRT for the purpose of measurement.

Horizontal amplifier: The circuit that provides amplification of the signal to be applied to the vertical deflection plates.

Lissajous figure: Patterns that form when applying periodic signals to the deflection plates of an oscilloscope. The two frequencies must make a ratio of whole numbers.

Phosphor: Any substance that becomes luminous through exposure to radiant energy or bombardment by atomic particles.

Retrace time: The time required for an electron beam to return to its original position on a CRT screen after being deflected to the right by a sawtooth waveform.

Ringing: Damped oscillations that may occur in improperly adjusted or poorly designed circuits.

Sensitivity: The ratio of required input signal to the amount of deflection. The property of an instrument that determines the scale factor.

Sweep generator: The circuit producing the waveform which in turn causes the electron beam to be deflected in the horizontal plane across the face of the CRT.

Sweep time: Time during which the beam is being swept from left to right on the CRT screen by the linearly increasing sawtooth voltage.

Thermionic emission: The evaporation of electrons from a heated surface.

8-23 REVIEW QUESTIONS

The following questions should be answered after a thorough study of the chapter. The purpose of the questions is to determine the reader's comprehension of the material.

1. What are the major components of a cathode-ray tube?
2. What does the term *phosphorescence* mean?
3. Describe the output waveform of the sweep generator and state why this type of waveform is needed.
4. What ratio is used to compute phase angle when observing a Lissajous figure?
5. Will the waveform displayed on the CRT screen of a sampling oscilloscope be at higher or lower frequency than the actual input signal?
6. The compensation of a probe is checked by applying a square wave. If the leading edge of the square wave becomes rounded, what is the value of the probe capacitance relative to what it should be?
7. What is the name of the pattern that is displayed when an oscilloscope is used in the $X-Y$ mode?
8. How can the input resistance of an oscilloscope be increased while simultaneously decreasing the input capacitance?

8-24 PROBLEMS

8-1 The deflection sensitivity of an oscilloscope is 35 V/cm. If the distance from the deflection plates to the CRT screen is 16 cm, the length of the deflection plates is 2.5 cm, and the distance between the deflection plates is 1.2 cm, what is the acceleration anode voltage?

8-2 What is the minimum bandwidth that an oscilloscope must have to be able to display, without distortion, a square wave with a rise time of 18 ns?

8-3 What is the amplitude of the voltage V_o in Fig. 8-45 after 20 msec?

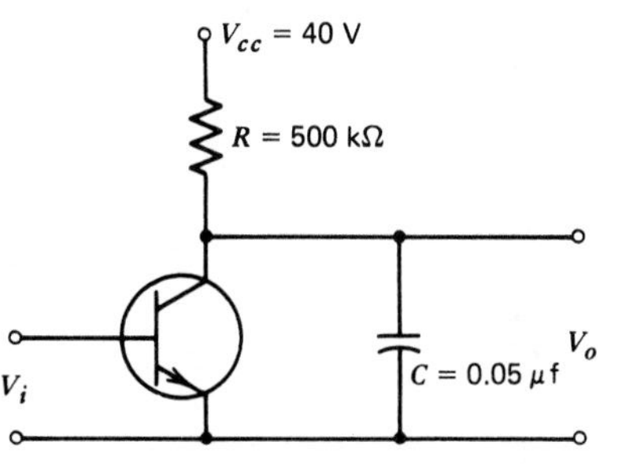

FIGURE 8-45 Circuit for Problem 8-3.

8-4 If the voltage across capacitor C in Fig. 8-45 is 20 V at the time an input pulse of 10-μsec duration is applied to the circuit, compute V_o at the end of the 10-μsec pulse. The transistor presents 250 **Ω** to the circuit.

8-5 Draw a block diagram of a basic oscilloscope, label each block, and briefly describe the function of each block.

8-6 Identify each of the items indicated in the drawing in Fig. 8-46.

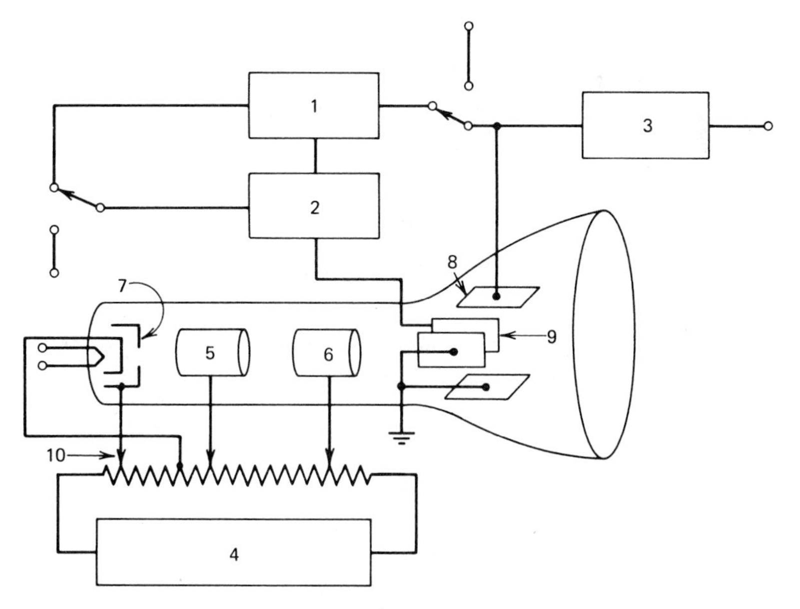

FIGURE 8-46 Drawing for Problem 8-6.

8-7 Compute the value of the resistors in the attenuator network shown in Fig. 8-47 to provide the volts/division values indicated for each position.

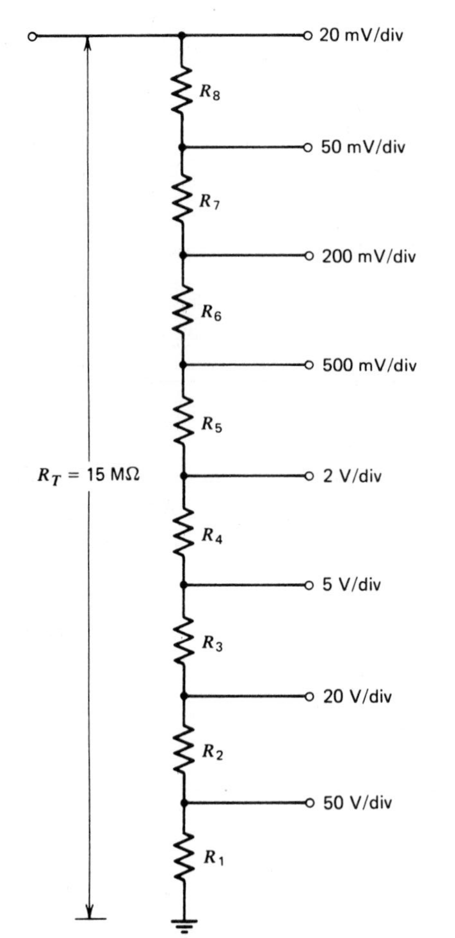

FIGURE 8-47 Attenuator network for Problem 8-7.

8-8 Compute the attenuation factor for each volts/division position in Problem 8-7.

8-9 The waveform shown in Fig. 8-48 is observed on a CRT screen. If the Time/Div switch is set to 10 μsec and the Volts/Div switch is set to

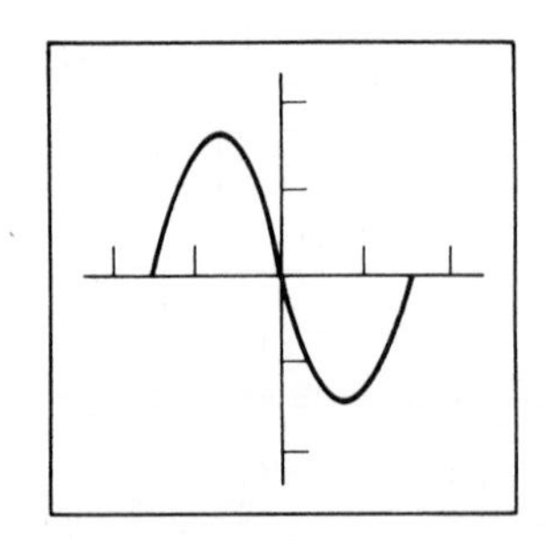

FIGURE 8-48 CRT display for Problem 8-9.

ermine the frequency and peak-to-peak amplitude of the

rain shown in Fig. 8-49*a* triggers a sweep generator, display shown in Fig. 8-49*b* to appear on the CRT screen. setting of the Time/Div switch?

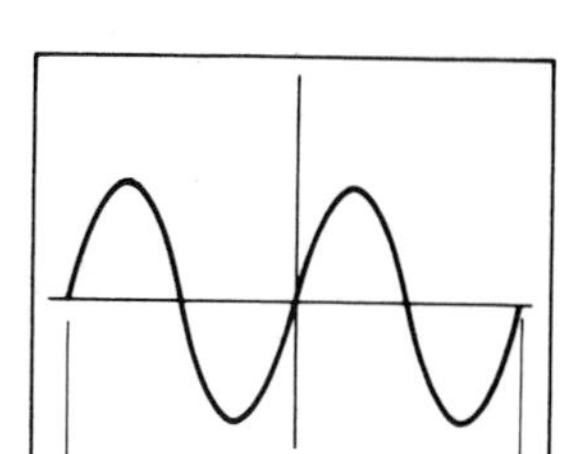
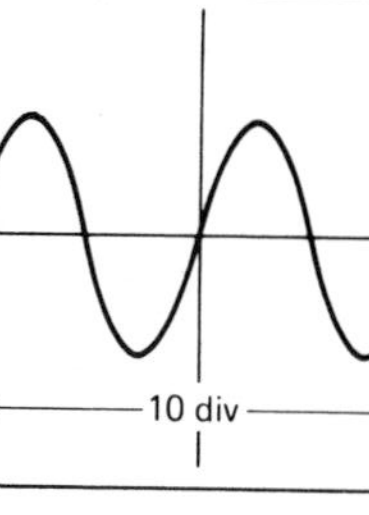

(a) *(b)*

Waveform and CRT display for Problem 8-10.

should C_1 have for V_o to be equal to $0.1V_i$ in the circuit g. 8-50?

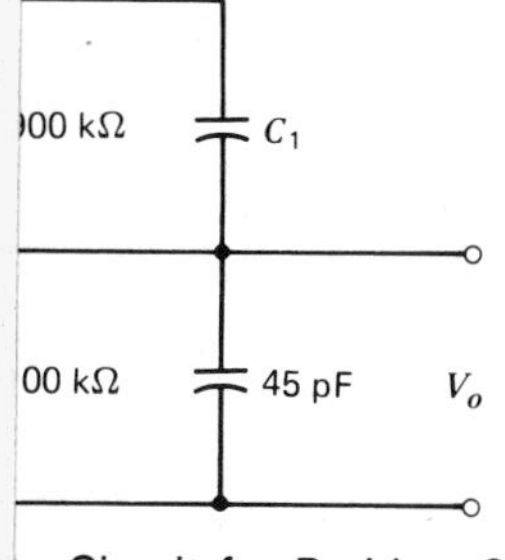

Circuit for Problem 8-11.

8-12 If one cycle of a 10-kHz sine wave fits exactly onto a CRT graticule that is ten divisions wide, what is the setting of the Time/Div switch?

8-13 If the vertical amplifier of an oscilloscope has a bandwidth of 15 MHz, what is the fastest rise time that an input may have to be displayed without distortion?

8-14 Two sine waves that are 90° out of phase are applied to the input terminals of an oscilloscope operating in the $X-Y$ mode. If the signal applied to the vertical input is twice the frequency of the horizontal input signal, sketch the waveform that will be observed on the CRT screen.

8-15 A high-impedance probe with 9-MΩ resistance and 4-pF capacitance is connected to an oscilloscope with an input resistance of 1 MΩ. If the effective capacitance decreased to 3.6 pF when the probe was connected, what is the capacitance of the oscilloscope alone?

8-16 If the horizontal frequency is 51 Hz, determine the vertical frequency for the Lissajous pattern shown in Fig. 8-51.

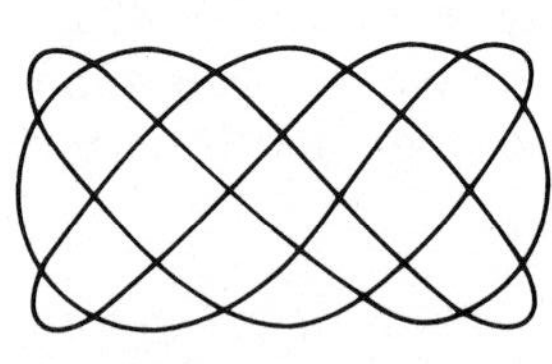

FIGURE 8-51 **Lissajous figure for Problem 8-16.**

8-17 Determine the phase shift between two sine waves, which is indicated by the pattern shown in Fig. 8-52.

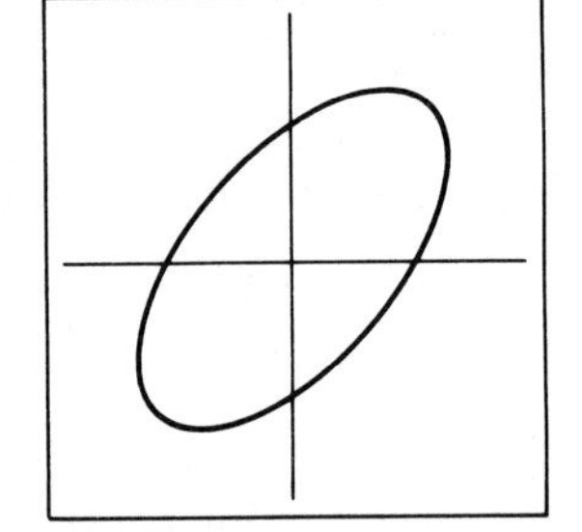

FIGURE 8-52 **CRT display for Problem 8-17.**

8-18 A sampling oscilloscope is being used to observe a 400-MHz sine wave. A sampling pulse occurs every 3 nsec. Draw five cycles of the 400-MHz signal and place a dot at the sampled point on each of the five cycles.

8-25 LABORATORY EXPERIMENTS

Laboratory experiments E18 through E22 apply the theory presented in Chapter 8. The purpose of the experiments is to provide students with hands-on experience so essential for a thorough understanding of concepts.

Experiments E18 through E21 require no special equipment and can easily be performed in any reasonably well-equipped electronics laboratory. Experiment E22 requires a cathode-ray table.

The contents of the laboratory report to be submitted by each student are listed at the end of each experimental procedure.

CHAPTER 9 Recording Instruments

9-1 Instructional Objectives

Chapter 9 discusses basic recording instruments. Although the basic theory of the operation of recorders is presented, the primary emphasis is on the applications of recorders. After completing the chapter you should be able to

1. Describe the recording instrument classified as a *strip-chart* recorder.
2. Describe the recording instrument classified as an $X-Y$ recorder.
3. List the typical frequency response for a strip-chart recorder.
4. Describe how automatic balance is achieved in a recording instrument.
5. List the three major systems of a strip-chart recorder.
6. List the three functions that a recorder may serve simultaneously in industrial applications.
7. Calculate the frequency of a recorded signal.
8. List a minimum of five specifications that should be evaluated when one is considering a recording instrument.
9. Describe the purpose of the error detector in a recorder.
10. Describe three applications for recording instruments.

9-2 INTRODUCTION

Recorders are instruments that provide a graphical record of the measurement of some physical event over time. A modern recording instrument is an electromechanical device. Its basic elements are a paper chart on which information is recorded, a writing instrument such as a stylus or a pen, and an interfacing apparatus between the measured parameter and the writing instrument.

Recording instruments may be broadly classified as

1. Instruments that record one or more parameters that change with respect to time. Such instruments are called *strip-chart* recorders. Strip-chart recorders may be self-balancing instruments or galvanometric recorders.
2. Instruments that record one or more dependent parameters that change with respect to some independent parameter. Such instruments are called *X–Y* recorders or function plotters.

Recorders that use mechanical writing instruments, such as a pen or a stylus, have a very limited frequency response, typically, from direct current to about 125 Hz. However, there is a wide and varied range of industrial applications for which this frequency response is completely adequate. Some instruments that are also generally classified as strip-chart recorders and record on photographic paper have a frequency response of a few thousand hertz. These instruments are also widely used.

This chapter discusses the characteristics and applications of different types of recording instruments.

9-3 SELF-BALANCING SYSTEM

Modern recording instruments with high-gain electronic amplifiers are usually *self-balancing* instruments. A wide range of self-balancing instruments, such as the instrument shown in Fig. 9-1, are commercially available from a substantial number of manufacturers.

Automatic balance, or "self-balance," is achieved by comparing the input signal with a reference voltage that is varied by a mechanical coupling from a **servo drive motor** at the output. Figure 9-2 shows a block diagram of a basic self-balancing system.

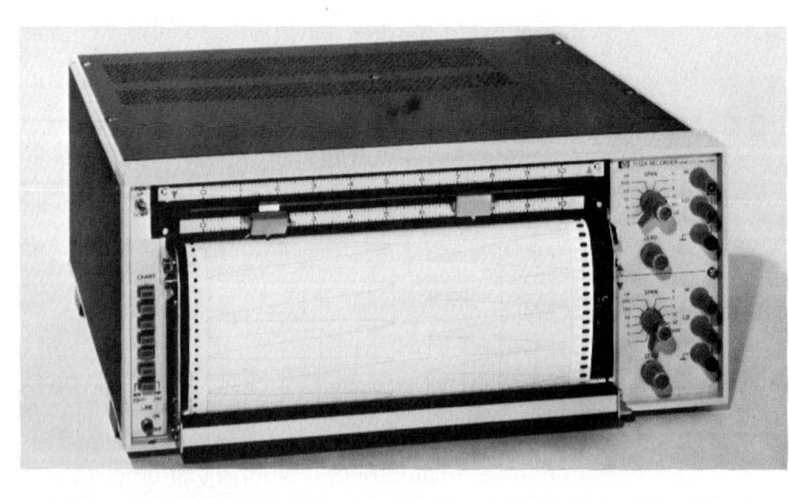

FIGURE 9-1 Self-balancing strip-chart recorder. (Courtesy Hewlett-Packard Company.)

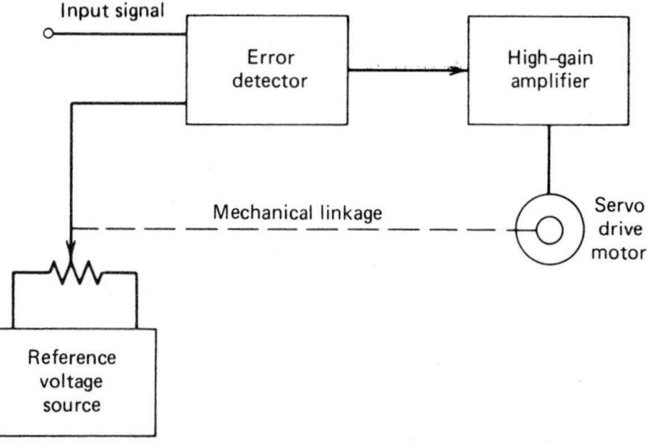

FIGURE 9-2 Block diagram of a self-balancing system.

If the input signal and the voltage at the movable arm of the potentiometer are equal, the output of the error detector will be zero. In this condition the instrument is balanced; therefore, the writing instrument is recording a straight line. When the voltages of the input signal and the movable arm differ, the drive motor turns. This action drives the pen in one direction or the other. The direction of the motion depends on the relative polarity of the input signal and the movable arm voltage. This can more readily be seen in Fig. 9-3.

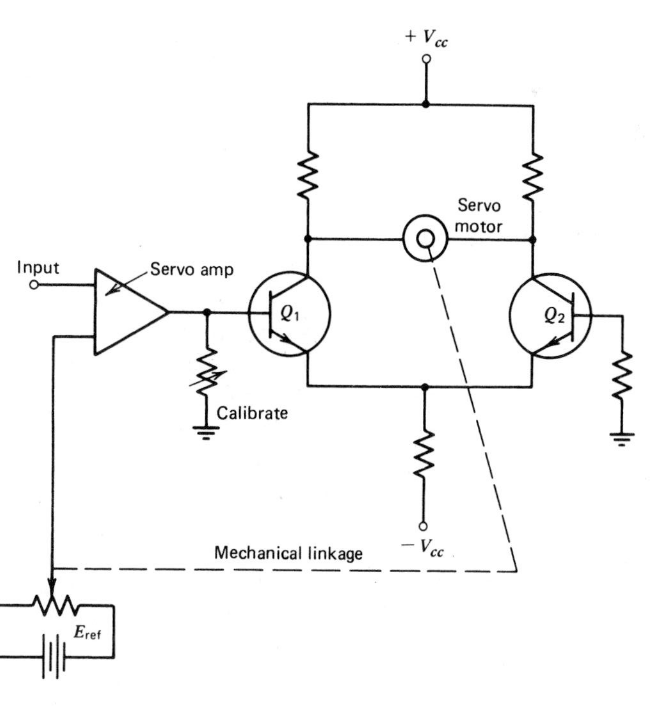

FIGURE 9-3 Basic schematic of a self-balancing system.

To establish a null condition (no error signal driving the motor) initially, we can adjust the calibrating resistor until the voltages on the base of both transistors, Q_1 and Q_2 in Fig. 9-3, are equal. Now, if an input signal is applied, there is a voltage difference at the inputs of the differential amplifier which is being used as an **error detector**. This difference is amplified and applied to the base of transistor Q_1. It is amplified by Q_2 and causes a voltage difference across the terminals of the motor. This difference in turn causes the motor, to which a writing instrument is attached, to rotate *proportionally* to the input signal. As the motor turns, the writing instrument reproduces the waveform of the input signal. The motor is also mechanically coupled back to the center tap of the potentiometer connected across the reference voltage source (Fig. 9-2). As the motor shaft turns, the movable arm of the potentiometer is mechanically positioned until the error voltage is reduced to zero.

Because of the principle of operation of the **null-seeking** electronic system driving a mechanical system, the writing instrument will always show some inability to respond exactly to the electronic signal. There will almost always be a small chart length, depending on the gain and the sensitivity of the system, over which the writing instrument will move without returning to its former point of balance. This is called the **dead-band,** which is defined as the minimum signal to which the recorder will respond. A significant proportion of this error is due to **backlash** in the mechanical system to which the writing instrument is attached.

As the writing instrument approaches the null position, it tends to **overshoot** the null point if it is recording a rapidly changing variable. The actual distance of overshooting back the null position is generally specified as a percentage of full-scale deflection. As an example, an overshoot of 1 mm on a 50-mm full-scale deflection instrument would be expressed as a 2% overshoot.

9-4 STRIP-CHART RECORDERS

Strip-chart recorders are instruments that provide a graphic record of a physical event varying with respect to time. A basic strip-chart recorder consists of a device for writing, such as a stylus, a chart on which the graphical record is recorded, and suitable circuitry for conditioning the signal to drive the chart at the desired rate. Figure 9-4 shows a block diagram for a basic strip-chart recorder. The output of the signal-conditioning circuit drives the reversible motor to which the stylus is attached. The movement of the stylus causes a reproduction of the input signal to be recorded on the chart. The rate of movement of the chart is controlled by the **chart speed** selector, which is a front panel control. The event marker allows the user to mark the edge of the chart to indicate the beginning or termination of some recording sequence.

A second type of strip-chart recorder is the *galvanometer* recorder. In this type of recorder the pen assembly is mounted at the end of the pointer of a

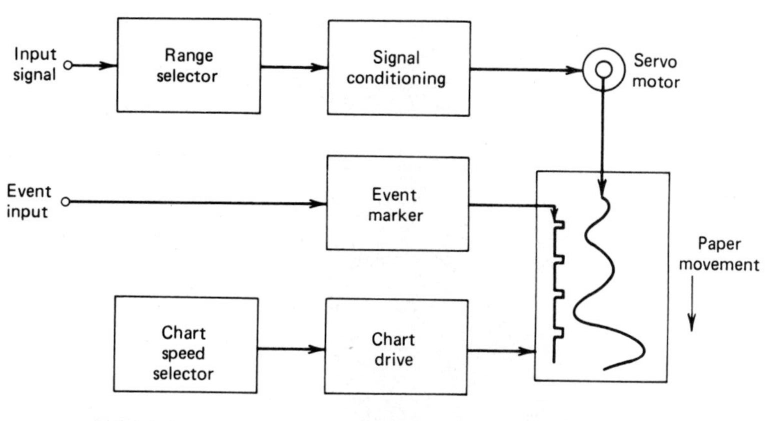

FIGURE 9-4 Basic strip-chart recorder block diagram.

rugged d'Arsonval-type movement, similar to the movement used in basic dc meters, as shown in Fig. 9-5. When a recording is being made with a galvanometer recorder, the restraining springs of the d'Arsonval movement, rather than a self-balancing signal, provide a counterforce to balance the force developed by the quantity being measured.

Although galvanometer recorders are not as sensitive as self-balancing systems, they have some other features that make them a very attractive choice for many applications. The most obvious advantages are their size and lower cost. Their compact size makes galvanometer movements ideal for use in multichannel recorders. They cost less because of their basic design, which provides less sensitivity.

Galvanometer recorders generally use a very slow drive system for the recording paper. Chart speeds between 1 and 10 in./hr are fairly typical. These instruments are widely used for recording changes or trends that occur over relatively long periods of time.

There is a wide variety of strip-chart recorders for fixed-installation industrial applications. These recorders generally serve the following three purposes simultaneously.

1. *Indicate*. To provide an immediate indication of the value of a parameter under observation.
2. *Record*. To provide a permanent graphic record of variation in value of the parameter under observation with respect to time.
3. *Control*. To provide control of the system being monitored.

There are literally thousands of scientific and industrial applications for strip-chart recorders—for example, in control rooms of utility companies, in oil refineries, in steel and paper mills, in chemical plants, and in many research facilities.

FIGURE 9-5 Galvanometer-type recorder. (Courtesy Astro-Med, West Warwick, R.I.)

9-5 CHART SPEED

Chart speed is a term used to express the rate at which the recording paper in a strip-chart recorder moves. Chart speed is usually expressed in inches per second or millimeters per second and is determined by a mechanical gear train. Many commercial strip-chart recorders have a wide range of chart speeds that are selected by switch. If the chart speed is known, the period of the recorded signal can be calculated as

$$\text{Period} = \frac{\text{Time}}{\text{Cycle}} = \frac{\text{Time base}}{\text{Chart speed}} \tag{9-1}$$

From Eq. 9-1 the frequency can easily be determined as

$$\text{Frequency} = \frac{1}{\text{Period}} \tag{9-2}$$

EXAMPLE 9-1

The chart speed of a recording instrument is 25 mm/sec. One cycle of the signal being recorded extends over 5 mm. (This is sometimes referred to as the time base.) What is the frequency of the signal?

Solution

$$\text{Period} = \frac{\text{Time}}{\text{Cycle}} = \frac{\dfrac{5\text{ mm}}{\text{cycle}}}{\dfrac{25\text{ mm}}{\text{sec}}}$$

$$= \frac{5\text{ mm}}{\text{cycle}} \times \frac{\text{sec}}{25\text{ mm}} = \frac{0.2\text{ sec}}{\text{cycle}}$$

$$\text{Frequency} = \frac{1}{\text{Period}} = \frac{1}{\dfrac{0.2\text{ sec}}{\text{cycle}}}$$

$$= \frac{\text{cycle}}{0.2\text{ sec}} = \frac{5\text{ cycles}}{\text{sec}}$$

9-6 *X–Y* RECORDERS

X–Y recorders, such as the one shown in Fig. 9-6, are instruments that provide a graphic record of the relationship between two variables. These instruments have along two axes a pair of servo systems, each of which drives a recording pen which is attached to a moving-arm mechanism. The graphic record appears on a fixed paper chart.

The recording paper used with the *X–Y* recorder may be held in place with clamps, or a differential pressure may be created, to hold the paper in place. In the differential pressure technique, which is frequently employed in more expensive recorders, a vacuum pump decreases the pressure beneath the paper. One advantage of *X–Y* recorders over strip-chart recorders is that almost any type of paper can be used.

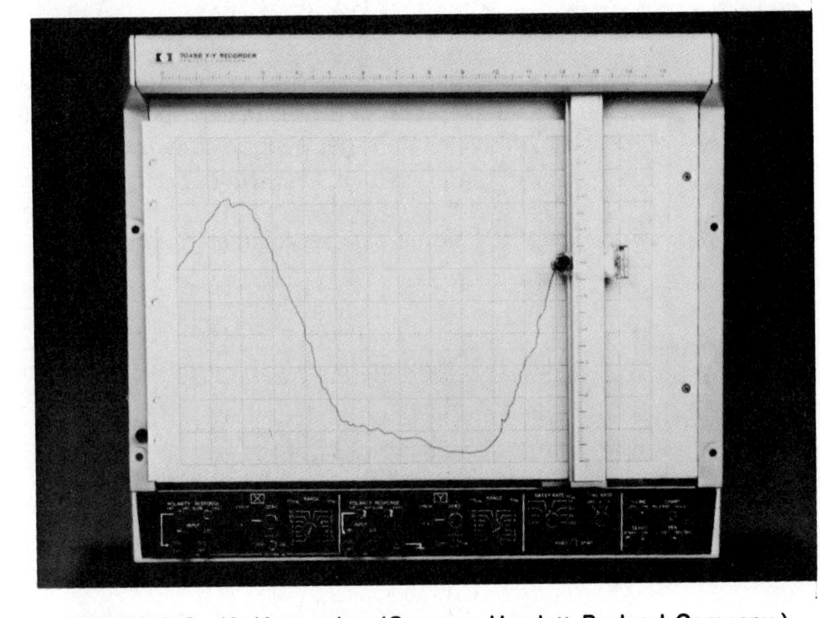

FIGURE 9-6 $X-Y$ recorder. (Courtesy Hewlett-Packard Company.)

An $X-Y$ recorder can be used to plot time-varying signals by applying a linear ramp signal to the X-channel input. Some $X-Y$ recorders contain a time base generator for recording one variable with respect to time. This technique is common in oscilloscopes (see Chapter 8).

$X-Y$ recorders also have some disadvantages compared to strip-chart recorders. They are generally considerably more expensive than a simple strip-chart recorder, and they cannot be used for continuous recording. Figure 9-7 shows a block diagram of a basic $X-Y$ recorder.

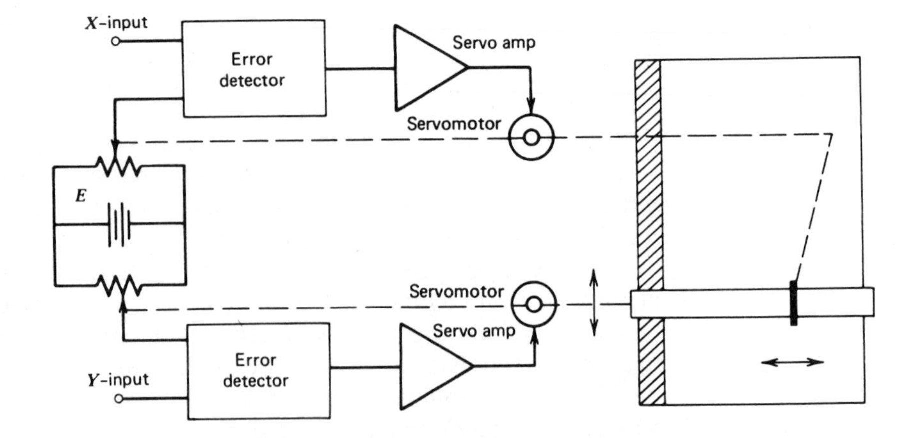

FIGURE 9-7 Basic $X-Y$ recorder block diagram.

The X-input signal drives the pen in the X-plane, and the Y-input signal drives the arm in the Y-plane. When an X-input signal is present, the pen drive motor continues to drive the pen until the error signal goes to zero. This is accomplished by the mechanical coupling between the pen drive motor and the movable arm of the potentiometer across the reference voltage. When the voltage at the movable arm equals the X-input voltage, the error signal equals zero and the pen drive motor stops turning. The same action occurs at the arm drive motor.

There are many applications for $X-Y$ recorders, particularly in laboratory work. The following are a few examples.

1. Plotting characteristic curves of vacuum tubes, transistors, and diodes.
2. Plotting speed–torque curves of electric motors.
3. Plotting resistance of electrical materials as a function of temperature.
4. Making physical and mechanical measurements such as pressure–volume, temperature–linear expansion, and stress–strain.

These or other applications for recorders are discussed in more detail at the end of this chapter.

9-7 SELECTING A RECORDER

A number of requirements should be considered when selecting a recording for a particular application. The primary consideration is the frequency of the waveform to be recorded. For signals in the frequency range from about 125 Hz to a few thousand hertz, an optical recorder is well suited. This is a strip-chart recorder that directs a light beam through an optical system and onto a photographic paper. The paper used in such instruments may need to be developed in a darkroom, or the paper may be light-sensitive and self-develop when exposed to light. Optical recorders may also be used at lower frequencies. However, the photographic paper is considerably more expensive than the paper in instruments using a pen or stylus.

For signals in the frequency range for approximately 50 to 125 Hz, a servo type of strip-chart recorder with preamplifier is most suitable. The preamplifier is recommended in this frequency range to provide additional energy to the pen or stylus at the rate required to record waveforms.

At lower frequencies of approximately 10 Hz or less, a servo-type recorder offers the user sensitivity, linearity, stability, mechanical ruggedness, and ample energy to drive the pen or stylus as well as a control device if the recorder is part of a control system.

9-8 RECORDER SPECIFICATIONS

Specifications for both strip-chart recorders and $X-Y$ recorders are extremely varied to meet the needs of thousands of industrial and scientific applications.

The following are general features and specifications that should be considered whenever one is examining a recorder.

1. The type of writing instrument desired, such as pen and ink, heated stylus, pressure stylus, or electrical discharge.
2. The type of recording chart desired, such as paper or wax-coated paper; or photographic film which is self-developing or requires darkroom processing.
3. Maximum amplification of the signal to be recorded.
4. Frequency response expressed in cycles per second (hertz).
5. Recording speed expressed in seconds for full-scale deflection.
6. Input signals: voltage or current.
7. Input impedance; should be several hundred thousand ohms for a general-purpose recorder.
8. Chart speed expressed in inches or centimeters per second, per minute, or per hour.
9. Input circuit; the input terminals should be floating (isolated from the ground).
10. Channel width expressed in inches, centimeters, or millimeters.
11. Number of recordings available from single channel to approximately 40 channels.
12. Drift expressed in inches or millimeters per hour.
13. Power requirements.
14. Weight and physical dimensions.

9-9 APPLICATIONS

The following applications are typical of the thousands of ways in which recorders are used in science and industry.

9-9.1 Temperature Recording

A strip-chart recorder may be used to provide a graphical record of temperature as a function of time. The two primary methods used for recording temperature are the thermocouple method and the resistance method.

The thermocouple method utilizes commercially available thermocouples that cover a wide range of temperatures. These serve very well as temperature-sensing elements and are readily adapted for use with strip-chart recorders. The circuit shown in Fig. 9-8 may be used to record the thermocouple voltage. A thermocouple, which is discussed in detail in Section 11-15, is made by joining two dissimilar metals. A small potential difference, which is proportional to the temperature, exists across the junction. This

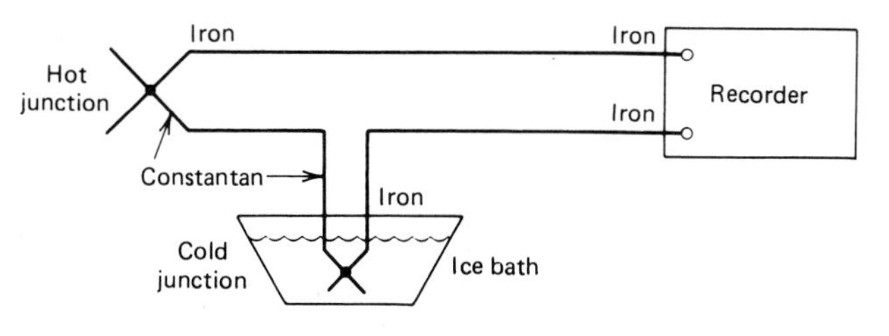

FIGURE 9-8 Circuit used to record thermocouple voltage.

potential difference has an almost linear relationship with the temperature and is very repeatable.

Elaborate tables of temperature versus potential difference have been developed for certain pairs of metals that are used for commercial thermocouples. These tables allow one to determine the "hot junction" temperature when the "cold junction" or the "reference junction" temperature is 0°C. The purpose of the ice bath is to maintain the cold-junction temperature at 0°C. Correction factors may be used with the tables if the reference junction is not at 0°C. However, it is more convenient to maintain the reference junction at 0°C or to develop an artificial electromotive force at the reference junction by using a circuit such as the one shown in Fig. 9-9. Thermocouple TC2 is maintained at 70°C by the oven. The emf developed by TC2 is balanced out by the voltage drop across resistor *R*, owing to current flow from the regulated power supply, until the total emf is equal to the emf of TC2 at 0°C. Notice that in Figs. 9-8 and 9-9 the two metal leads connected to the recorder terminals are of the same metal.

9-9.2 Providing a Record of Sound Level

Engineers frequently need to record over a period of time the sound level near highways, airports, hospitals, schools, or residences. This can be done with an ordinary microphone and a strip-chart recorder if the output signal from

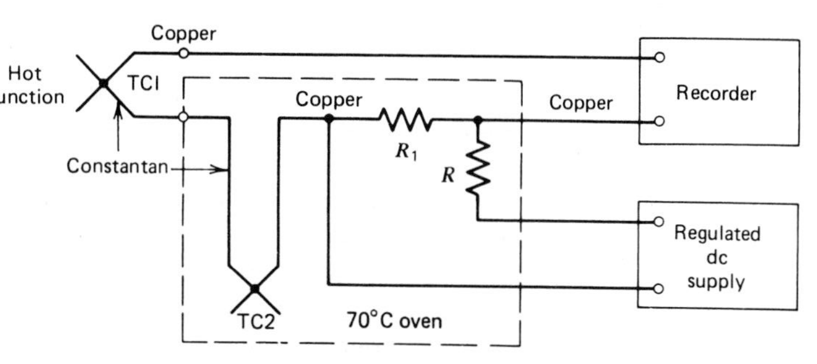

FIGURE 9-9 Circuit with artificial reference junction voltage used to measure thermocouple voltage.

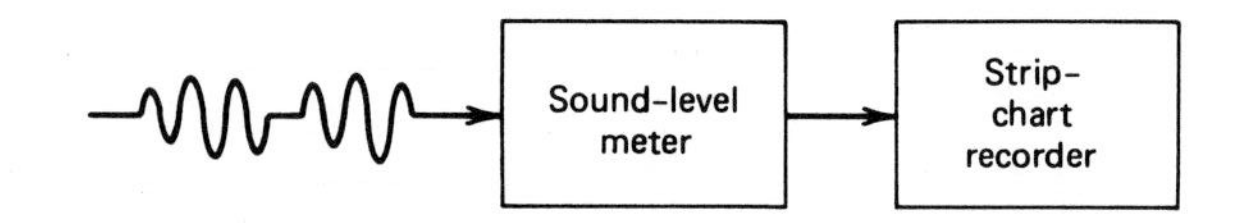

FIGURE 9-10 Setup for measuring the sound level as a function of time.

the microphone is of sufficient amplitude to drive the recorder. A better setup for obtaining sound-level data is to use a sound-level meter and a strip-chart recorder as shown in Fig. 9-10.

9-9.3 Recording Amplifier Drift

Transistor amplifiers are sensitive to temperature changes. Temperature changes cause the bias voltages on the transistor to change, generally by a small amount. As a result, the amplifier operates at a different point on its load line. This change in the point of operation is called *drift*. Since this is generally a slow process, a strip-chart recorder may be used to monitor and record the drift by connecting the recorder to the output of the amplifier, as shown in Fig. 9-11.

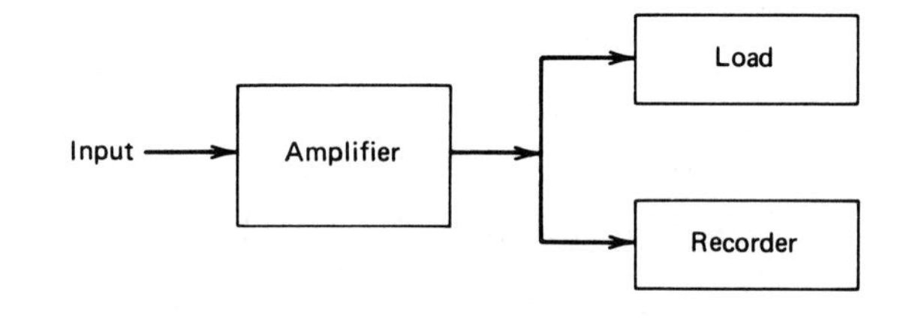

FIGURE 9-11 Circuit to record amplifier drift.

9-11 SUMMARY

Recorders provide a graphic record of variations of a measured quantity. These variations may occur with respect to time or with respect to another physical parameter. Recorders are electromechanical devices and therefore have a limited frequency response.

Modern self-balancing recorders utilize high-gain electronic amplifiers with mechanical feedback from the output to the input to achieve this very important self-balancing feature.

There is an almost unlimited range of industrial applications for recorders. For all applications, careful consideration should be given to which particular recorder is to be used.

9-11 GLOSSARY

Backlash: A form of mechanical hysteresis in which there is a lag between the application of a driving force and the response of what is being driven. The presence of backlash is most evident when the direction of motion is reversed.

Chart speed: The distance per second that the recording paper in a strip-chart recorder travels.

Dead band: The largest amplitude signal to which the instrument will not respond.

Error detector: A differential amplifier that provides an output proportional to the algebraic sum of the two input signals.

Null-seeking: A term used to describe the manner in which a self-balancing system approaches a null condition. The instrument "seeks" a reference voltage value that will cause the error signal to equal zero volts, thereby establishing a null condition.

Overshoot: The distance back to a null position as a percentage of full-scale deflection.

Servo drive motor: The component of a closed-loop system that converts an electrical error signal to a mechanical position.

9-12 REVIEW QUESTIONS

The following questions should be answered after a thorough study of the chapter. The purpose of the questions is to determine the reader's comprehension of the material.

1. What is the basic difference between a strip-chart recorder and an $X-Y$ recorder?
2. How is automatic balance achieved in a recorder?
3. What are the three major systems of a strip-chart recorder?
4. What are some of the specifications to be considered when examining a recording instrument?
5. What is the purpose of the error detector in a recorder?

9-13 PROBLEMS

9-1 What is the maximum frequency response of a recorder that uses a mechanical writing instrument?

9-2 How is automatic balance achieved in a recording instrument?

9-3 Define the following terms: dead band, overshoot, chart speed.

9-4 List the three functions of strip-chart recorders in fixed-installation industrial applications.

9-5 What is the approximate maximum frequency response of a galvanometer-type recorder?

9-6 The chart speed of a recording instrument is 10 mm/sec. If the time base of the recorded signal is 20 mm, what is the frequency of the recorded signal?

9-7 If the frequency of a signal to be recorded with a strip-chart recorder is 15 Hz, what chart speed must be used to record one complete cycle on 5 mm of the recording paper?

9-8 What is the primary disadvantage of optical recorders compared to a pen-type recorder?

9-9 List five features or specifications to be considered when purchasing a recorder.

9-10 What is one advantage of $X-Y$ recorders over strip-chart recorders?

9-11 How can an $X-Y$ recorder be used to record time-varying signals?

9-12 What is the primary application of galvanometer recorders?

9-13 How does the sensitivity of galvanometer recorders compare with that of self-balancing recorders?

9-14 LABORATORY EXPERIMENTS

Laboratory experiments E23 and E24 apply the theory presented in Chapter 9. The purpose of the experiments is to provide students with hands-on experience essential to obtaining a thorough understanding of the concepts involved.

A strip-chart recorder is required to perform the first experiment, and the second experiment requires an $X-Y$ recorder. No other special equipment is needed. The contents of the laboratory report to be submitted by each student are listed at the end of each experimental procedure.

CHAPTER 10 Signal Generators

10-1 Instructional Objectives

In Chapter 10 we discuss the basic theory, construction, and application of signal generators. We examine several types of instruments of various frequency ranges and output waveforms as well as applications for these signal-generating instruments. After completing the chapter you should be able to

1. Describe two types of circuits used in audio oscillators and list an advantage of each.
2. List the two conditions that must be satisfied for a circuit to sustain oscillation.
3. Describe why circuits used for radio-frequency (RF) oscillators are not practical for audio-frequency (AF) oscillators.
4. Define the terms *oscillator* and *generator*.
5. List three beneficial features of the Wien bridge oscillator.
6. Describe why sine-wave generating circuits are not generally used to generate the primary waveform in function generators.
7. Describe the function generator.
8. Describe the basic difference between a square-wave generator and a pulse generator.
9. Calculate basic oscillator and generator circuit parameters.
10. Define technical terms related to oscillators and generators.

10-2 INTRODUCTION

A signal source is a vital component at a test setup, whether at the end of a production line, on the service bench, or in the research laboratory. Signal sources have a variety of applications including checking stage gain, frequency response, and alignment in receivers and in a wide range of other electronics equipment.

Signal sources provide a variety of waveforms for testing electronic circuits, usually at low power. The various waveforms are generated by several different kinds of instruments, which range in complexity from simple fixed-frequency sine-wave oscillators to highly sophisticated instruments such as might be used in testing complex communications equipment.

Although terminology is not universal, the term **oscillator** is generally used for an instrument that provides only a sinusoidal output signal, and the term **generator** is applied to an instrument that provides several output waveforms, including sine wave, square wave, triangular wave, and pulse trains as well as amplitude **modulation** of the output signal.

Although we speak of oscillators as "generating" a signal, it should be emphasized that no energy is created; it is simply converted from a dc source into an ac energy at some specific frequency.

10-3 REQUIREMENTS FOR OSCILLATION

Basically, an oscillator is an amplifier with positive feedback. The gain equation for an amplifier with positive feedback is

$$A_f = \frac{A}{1 + A\beta} \qquad (10\text{-}1)$$

where

A_f = gain with feedback
A = open-loop gain
β = feedback factor, $\frac{V_i}{V_o}$

Feedback is provided by the feedback, or **phase-shift**, network shown in Fig. 10-1. The output signal of the amplifier is fed back to the amplifier's

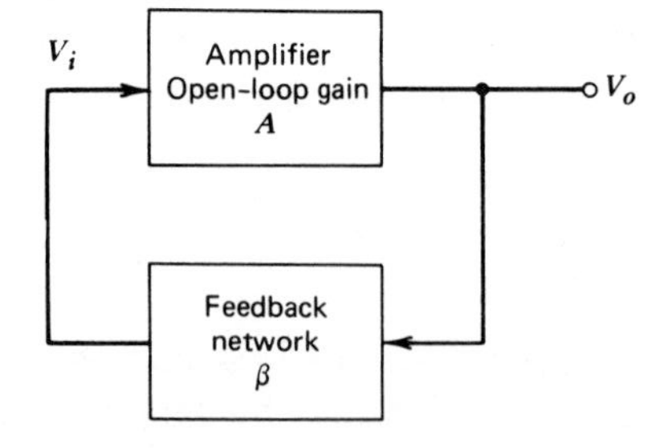

FIGURE 10-1 Closed-loop system consisting of amplifier with feedback.

input terminal through the phase-shift network, undergoing there a 180° shift in phase. The amplifier also causes a 180° phase shift. Therefore, the net phase shift in the circuit is zero, which is one condition that must be satisfied for sustained oscillation. A second requirement for oscillation is that the closed-loop gain, which is the product $A\beta$, must be equal to, or greater than, unity. These two conditions are known as the **Barkhausen criteria.**

It is not necessary to supply an input signal to initiate oscillation. A noise voltage or a **transient** is sufficient to start the process. It is usually desirable for the closed-loop gain to be greater than unity to ensure strong oscillations. However, if the gain is too high, the circuit operates at saturation and the output waveform will not be sinusoidal.

The phase-shift network consists of a combination of components with resistance and capacitance (*RC*) or inductance and capacitance (*LC*). A combination of inductance and capacitance is usually called a tuned circuit or a tank circuit. Oscillators that employ an *RC* combination have a distinct advantage over *LC*-type oscillators at low frequencies since the physical size of the required inductors is prohibitive.

10-4 AUDIO OSCILLATORS

An audio oscillator is useful for testing equipment that operates in the **audio-frequency** range. Such instruments always produce a sine-wave signal, variable in both amplitude and frequency, and usually provide a square-wave output as well. The maximum amplitude of the output waveform is typically on the order of 25 V_{rms}, whereas the range of frequencies covers at least the audio-frequency range from 20 Hz to 20 kHz. The most common output impedances for audio oscillators are 75 **Ω** and 600 **Ω**.

The two most common audio-oscillator circuits are the *Wien bridge oscillator* and the *phase-shift oscillator*, both of which employ *RC* feedback networks. The Wien bridge offers some very attractive features, including a straightforward design, a relatively pure sine-wave output, and a very stable frequency.

The Wien bridge oscillator is essentially a feedback amplifier in which the Wien bridge serves as the phase-shift network. The Wien bridge is an ac bridge, the balance of which is achieved at one particular frequency. The basic Wien bridge oscillator is shown in Fig. 10-2.

As can be seen, the Wien bridge oscillator consists of a Wien bridge and an operational amplifier represented by the triangular symbol. Operational amplifiers are integrated circuit amplifiers and have high-voltage gain, high input impedance, and low output impedance. The condition for balance for an ac bridge is

$$Z_1 Z_4 = Z_2 Z_3 \tag{10-2}$$

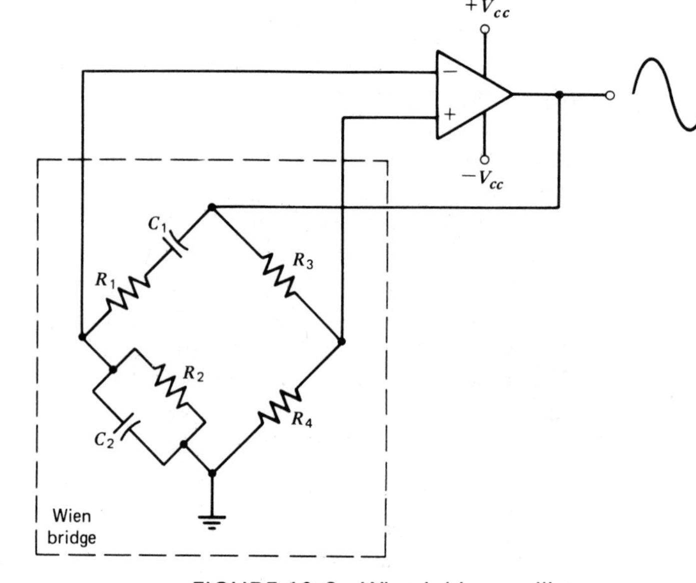

FIGURE 10-2 Wien bridge oscillator.

where

$$Z_1 = R_1 - j/\omega C_1$$

$$Z_2 = \frac{R_2(-j/\omega C_2)}{R_2 - j/\omega C_2} = \frac{-jR_2}{-j + R_2\omega C_2}$$

$$Z_3 = R_3$$

$$Z_4 = R_4$$

Substituting the appropriate expressions into Eq. 10-2 yields

$$\left(R_1 - \frac{j}{\omega C_1}\right)R_4 = \left(\frac{-jR_2}{-j + R_2\omega C_2}\right)R_3 \tag{10-3}$$

If the bridge is balanced, both the magnitude and phase angle of the impedances must be equal. These conditions are best satisfied by equating real terms and imaginary terms. Separating and equating the real terms in Eq. 10-3 yields

$$\frac{R_3}{R_4} = \frac{R_1}{R_2} + \frac{C_2}{C_1} \tag{10-4}$$

Separating and equating imaginary terms in Eq. 10-3 yields

$$\omega C_1 R_2 = \frac{1}{\omega C_2 R_1} \tag{10-5}$$

where $\omega = 2\pi f$. Substituting for ω in Eq. 10-5, we can obtain an expression for frequency which is

$$f = \frac{1}{2\pi(C_1 R_1 C_2 R_2)^{1/2}} \tag{10-6}$$

If $C_1 = C_2 = C$ and $R_1 = R_2 = R$, then Eq. 10-4 simplifies to yield

$$\frac{R_3}{R_4} = 2 \tag{10-7}$$

and from Eq. 10-6 we obtain

$$f = \frac{1}{2\pi RC} \tag{10-8}$$

where

f = frequency of oscillation of the circuit in Hertz
C = capacitance in farads
R = resistance in ohms

EXAMPLE 10-1 Determine the frequency of oscillation of the Wien bridge oscillator shown in Fig. 10-3 if $R = 6\ \text{k}\Omega$ and $C = 0.003\ \mu\text{F}$.

Solution Using Eq. 10-8, we compute the frequency as

$$f = \frac{1}{2\pi RC}$$

$$= \frac{1}{(2\pi)(6\ \text{k}\Omega)(0.003\ \mu\text{F})}$$

$$= 8.885\ \text{kHz}$$

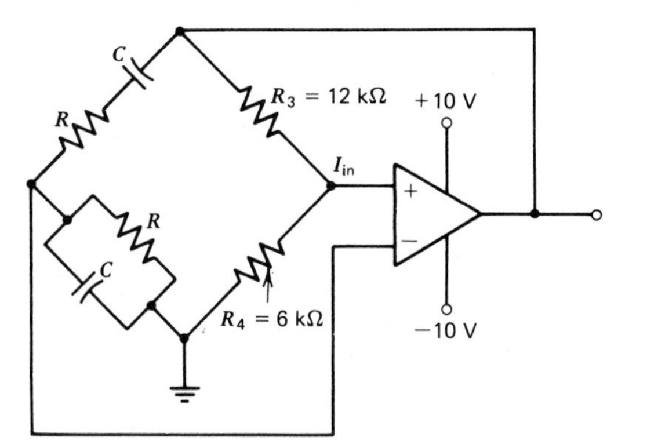

FIGURE 10-3 Wien bridge oscillator for Example 10-1.

It can be shown, by using ordinary ac circuit analysis techniques, that Eq. 10-2 is satisfied by the value of the components in the circuit in Fig. 10-3.

The design of a Wien bridge oscillator can be approached by selecting an operating frequency and level of current that will be acceptable through each arm of the bridge. The bridge currents are typically larger by at least a factor of 100 than the maximum input current to the amplifier, and the peak value of the sinusoidal output voltage is typically on the order of 90% of V_{cc}.

EXAMPLE 10-2

Design a Wien bridge oscillator around the following specifications:

- $f = 15$ kHz
- $V_{cc} = \pm 10$ V
- $I_{in} = 1\ \mu A$
- $I_{R4} = 100 I_{in}$

Solution

If the peak value of the sinusoidal waveform is 90% of V_{cc}, we can solve for the value of $R_3 + R_4$ as

$$R_3 + R_4 = \frac{0.9 V_{cc}}{100\ \mu\text{A}} = \frac{9\ \text{V}}{100\ \mu\text{A}} = 90\ \text{k}\Omega$$

Using Eq. 10-7, we can say that

$$R_3 = 2R_4$$

Therefore,

$$3R_4 = 90\ \text{k}\Omega$$

$$R_4 = 30\ \text{k}\Omega$$

and

$$R_3 = 2R_4 = 60\ \text{k}\Omega$$

We arrived at Eq. 10-7 by letting $R_1 = R_2 = R$. It is generally convenient to let $R_1 = R_2 = R_4$; therefore,

$$R_1 = R_2 = 30\ \text{k}\Omega$$

Using Eq. 10-8, we can now solve for the capacitance C as

$$C = \frac{1}{2\pi f R} = \frac{1}{(2\pi)(15\ \text{kHz})(30\ \text{k}\Omega)} = 354\ \text{pF}$$

The Wien bridge oscillator is widely used in audio oscillators because of its relatively small amount of **distortion,** excellent frequency stability, comparatively wide frequency range, and ease of changing frequency. A typical commercial Wien bridge oscillator can have a frequency range extending from 5 Hz to 500 kHz in **decade steps**.

The second audio-oscillator circuit of interest is the phase-shift oscillator. The phase-shift network for the phase-shift oscillator is an *RC* network made up of equal-value capacitors and resistors connected in cascade as shown in Fig. 10-4. Each of the three *RC* stages shown provides a 60° phase shift, with the total phase shift equal to the required 180°.

The phase-shift oscillator is analyzed by ignoring any minimal loading of the phase-shift network by the amplifier. By applying classical network analysis techniques, we can develop an expression for the *feedback factor* in terms of the phase-shift network components. The result is

$$\beta = \frac{V_i}{V_o} = \frac{1}{1 - \dfrac{5}{(\omega RC)^2} + j\left[\dfrac{1}{(\omega RC)^3} - \dfrac{6}{\omega RC}\right]} \tag{10-9}$$

If the phase shift of the feedback network satisfies the 180° phase-shift requirements, the imaginary components of Eq. 10-9 must be equal to zero, or

$$\frac{1}{(\omega RC)^3} - \frac{6}{\omega RC} = 0 \tag{10-10}$$

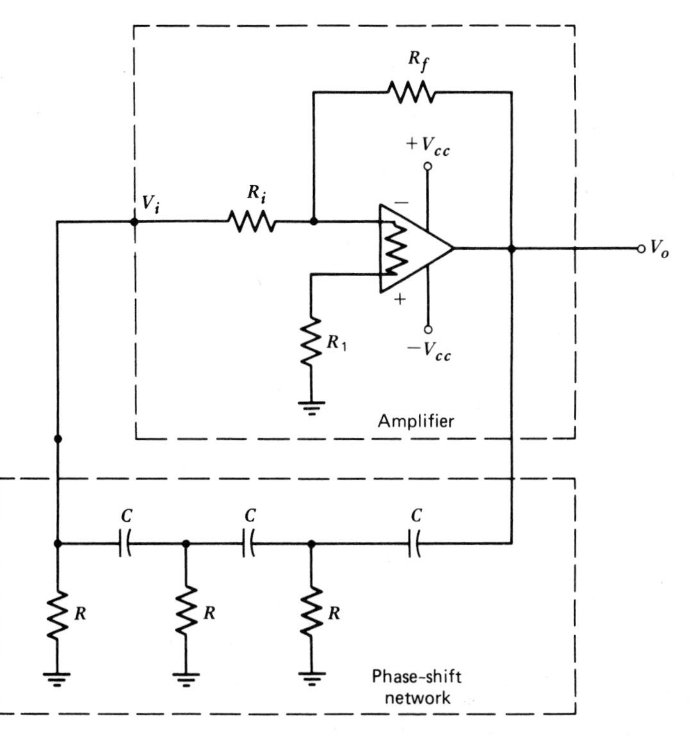

FIGURE 10-4 Basic phase-shift oscillator circuit.

The frequency of oscillation for the circuit can be determined by substituting $2\pi f$ for ω in Eq. 10-10 and solving for the frequency. The result is

$$f = \frac{1}{2\sqrt{6}\pi RC} \tag{10-11}$$

We can express Eq. 10-11 as

$$2\pi f = \frac{1}{\sqrt{6}RC} \tag{10-12}$$

or

$$\omega = \frac{1}{\sqrt{6}RC} \tag{10-13}$$

Substituting for ω in Eq. 10-9, we obtain

$$\beta = \frac{V_i}{V_o} = \frac{1}{1 - 5/(1/6) + j(6\sqrt{6} - 6\sqrt{6})} \tag{10-14}$$

or

$$\beta = \frac{V_i}{V_o} = \frac{1}{1 - 5 \times 6} = -\frac{1}{29} \tag{10-15}$$

Rewriting Eq. 10-15, we see that

$$V_o = -29V_i \tag{10-16}$$

which means that the gain of the amplifier must be at least 29 if the circuit is to sustain oscillation.

EXAMPLE 10-3 Determine the frequency of oscillation of a phase-shift oscillator with a three-section feedback network consisting of 13-Ω resistors and 100-μF capacitors.

Solution Using Eq. 10-11, we can compute the frequency of oscillation as

$$f = \frac{1}{2\pi\sqrt{6}RC}$$

$$= \frac{1}{(2\pi\sqrt{6})(13\ \Omega)(100\ \mu\text{F})} = 50\ \text{Hz}$$

The phase-shift oscillator is useful for noncritical applications, particularly at medium and low frequencies, even down to 1 Hz, because of its simplicity. However, its frequency stability is not as good as that of the Wien bridge oscillator, distortion is greater, and changing frequency is inconvenient because the value of each capacitor must be adjusted. The choice of an oscillator circuit to operate in the audio-frequency range is determined by the particular application.

10-5 RADIO-FREQUENCY OSCILLATORS

Radio-frequency (RF) oscillators must satisfy the same basic criteria for oscillation as was discussed in Section 10-4 for audio oscillators. That is, the Barkhausen criteria must be satisfied. The phase-shift network for RF oscillators is an inductance–capacitance (*LC*) network. This *LC* combination, which is generally referred to as a tank circuit, acts as a filter to pass the desired oscillating frequency and block all other frequencies. The tank circuit is designed to be resonant at the desired frequency of oscillation. An *LC* circuit is said to be resonant when the inductive and capacitive reactances are equal, that is, when

$$X_L = X_C \qquad (10\text{-}17)$$

or when

$$2\pi f L = \frac{1}{2\pi f C} \qquad (10\text{-}18)$$

Solving Eq. 10-18 for the frequency f, we obtain an expression for the frequency of oscillation of an RF oscillator, which is

$$f = \frac{1}{2\pi (LC)^{1/2}} \qquad (10\text{-}19)$$

where

f = frequency of oscillation
L = total inductance of the phase-shift network
C = total capacitance of the phase-shift network

There are a number of standard RF oscillator circuits in use; the most popular are the Colpitts oscillator and the Hartley oscillator shown in Fig. 10-5.

As can be seen, the phase-shift network contains a tapped inductor consisting of sections L_1 and L_2 and an adjustable capacitor to vary the frequency of oscillation. The feedback factor β, is given as

$$\beta = \frac{-L_1}{L_2} \qquad (10\text{-}20)$$

The negative sign means there must be a 180° phase shift across the amplifier. This is accomplished by connecting the amplifier in the inverting configuration as shown in Fig. 10-5. As we stated earlier, the circuit must satisfy the Barkhausen criterion which states that $A\beta \geq 1$ to sustain oscillation. This may be written as

$$A \geq \frac{1}{\beta} \qquad (10\text{-}21)$$

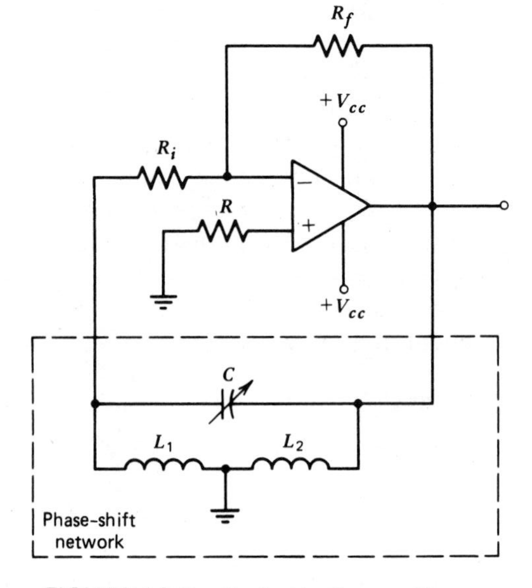

FIGURE 10-5 Basic Hartley oscillator.

Substituting Eq. 10-20 into Eq. 10-21 gives us

$$A \geq -\frac{L_2}{L_1} \tag{10-22}$$

Equation 10-22 states that the gain of the amplifier must be greater than or equal to the ratio of L_1 to L_2 to sustain oscillation. Using the equation for the gain of an inverting amplifier yields

$$A = -\frac{R_f}{R_i} \tag{10-23}$$

We can determine the value of either R_f or R_i given the value of the other.

EXAMPLE 10-4

Determine the frequency of oscillation and the minimum value of R_f to sustain oscillation for the Hartley oscillator shown in Fig. 10-6.

Solution

The frequency of oscillation is determined from Eq. 10-19 as

$$f = \frac{1}{2\pi[(L_1 + L_2)C]^{1/2}}$$

$$= \frac{1}{2\pi[(280\ \mu\text{H})(0.001\ \mu\text{F})]^{1/2}}$$

$$= \frac{1}{(2\pi)(5.29 \times 10^{-7})} = 300\ \text{Hz}$$

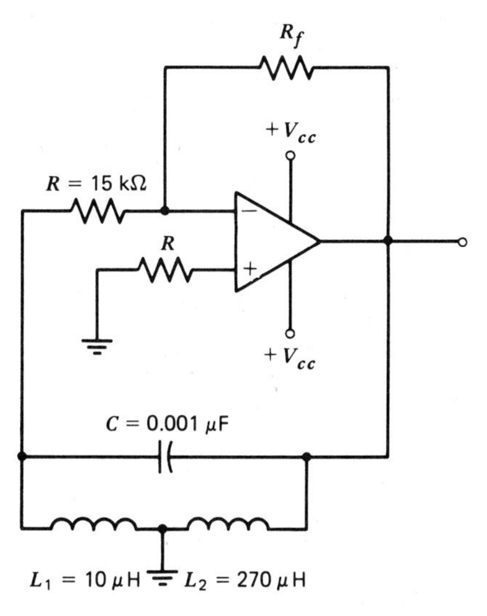

FIGURE 10-6 Hartley oscillator circuit for Example 10-4.

The minimum gain of the amplifier is computed using Eq. 10-22 as

$$A_{min} = -\frac{L_2}{L_1} = \frac{270\,\mu H}{10\,\mu H} = -27$$

Using Eq. 10-23, we can compute the value of the feedback resistor R_f as

$$R_f = AR_{in}$$
$$= (27)(15\,\mathbf{k\Omega}) = 405\,\mathbf{k\Omega}$$

10-6 RADIO-FREQUENCY GENERATORS

Radio-frequency (RF) generators are designed to provide an output signal over a wide range of frequencies from approximately 30 kHz to nearly 3000 MHz. The term *generator* is generally used for an instrument that is capable of providing a modulated output signal.

Laboratory-quality RF generators contain a precision output attenuator network that permits selection of output voltages from approximately 1 μV to nearly 3 V in precise steps. The output impedance of RF generators is generally 50 **Ω**. Few generators cover the entire RF spectrum, but many cover a very wide range of frequencies. A frequency range exceeding 100 MHz is fairly commonplace in RF generators.

A block diagram for a basic RF signal generator is shown in Fig. 10-7. The circuit is that of a very stable, multiple-band, RF oscillator. The frequency

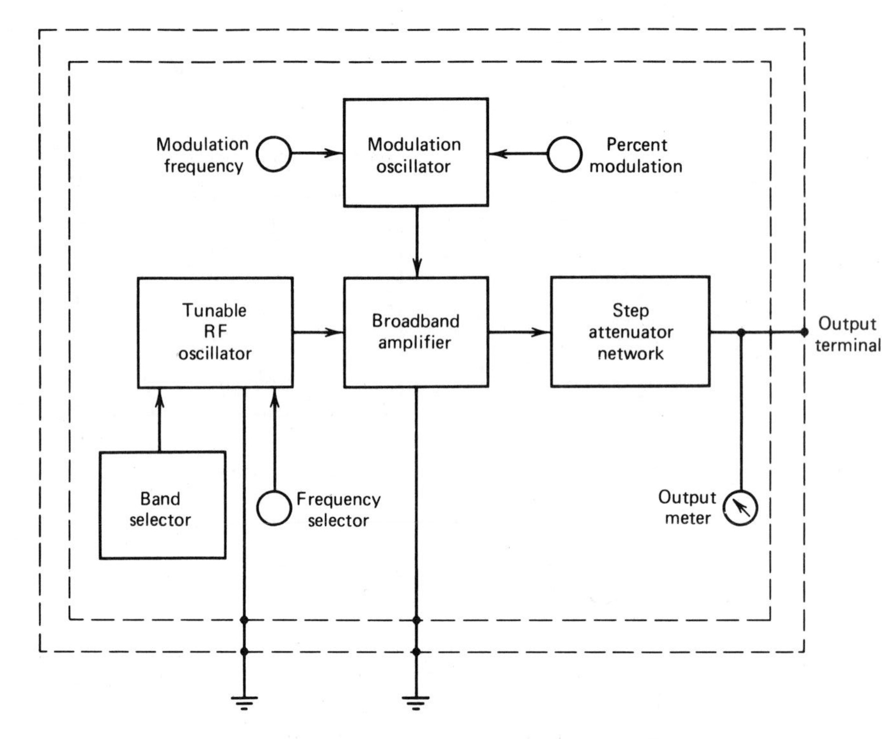

FIGURE 10-7 Basic RF signal generator.

range is selected with the band selector. The exact frequency is selected with the vernier frequency selector. The frequency stability of the generator is limited by the stability of the *LC* oscillator circuit. Therefore, considerable attention should be directed toward design of the master oscillator. The minimally distorted sinusoidal output of the oscillator is applied to a broadband amplifier that amplifies the signal and provides buffering between the oscillator and any load connected to the output terminal. If so desired, the RF signal is modulated at the amplifier by the modulation oscillator.

Both the modulation frequency and the percentage of modulation can be adjusted by a vernier control. The amplifier output is applied to a step attenuation network. The output of the attenuator is monitored by the output meter, which indicates the signal level for the user. The entire instrument is contained in a shielded cabinet. Many laboratory-quality RF generators provide shielding for the oscillator plus shielding for the entire instrument.

Figure 10-8 shows a Hewlett-Packard Model 8640A RF signal generator. This laboratory-quality instrument covers the frequency range from 500 kHz to 512 MHz in ten bands. The calibrated and metered output is adjustable from 0.013 μV to 2 V (-145 to $+19$ db). The instrument provides internal AM and FM modulation as well as external AM, FM, and pulse modulation. The metered and calibrated modulated signal is variable in frequency from 20 Hz to 600 kHz.

FIGURE 10-8 Laboratory-quality RF signal generator. (Courtesy Hewlett-Packard Company.)

10-7 FUNCTION GENERATORS

Function generators, which are very important and versatile instruments, provide a variety of output waveforms over a wide frequency range. The most common output waveforms are sine, square, triangular, ramp, and pulse. The frequency range generally extends from a fraction of a hertz to at least several hundred kilohertz.

Since a function generator provides sine, square, and triangular wave outputs, any of these may be the primary waveform generated by the instrument. This primary waveform can then be applied to the proper circuitry to generate the remaining waveforms. For example, the primary waveform may be a sine wave generated with the *RC* or *LC* oscillator circuit. However, because of difficulties with amplitude and frequency stability, particularly at very low frequencies, oscillators with a sine wave as the primary output are generally not used. Figure 10-9 shows a schematic diagram of one of several alternative approaches that can be used in a basic function generator. The primary waveform in the circuit shown is a square wave. This waveform is chosen because some circuits generating square waves are simpler and offer significantly better amplitude and frequency stability than do circuits generating sine waves.

The first stage, A_1, which is a voltage *comparator*, generates a square-wave output. The output of A_1 is driven to saturation; therefore, the square wave is either at $+V_{cc}$ or $-V_{cc}$. The second stage, A_2, is an integrator which generates a triangular output. The square wave is applied to a square-to-sine wave converter that filters out the odd harmonics making up the square wave while passing on only the fundamental sine wave.

The operation of the circuit can be analyzed by starting at the output of the comparator, which is at either $+V_{cc}$ or $-V_{cc}$. Consider V_{01} to be at $-V_{cc}$. The

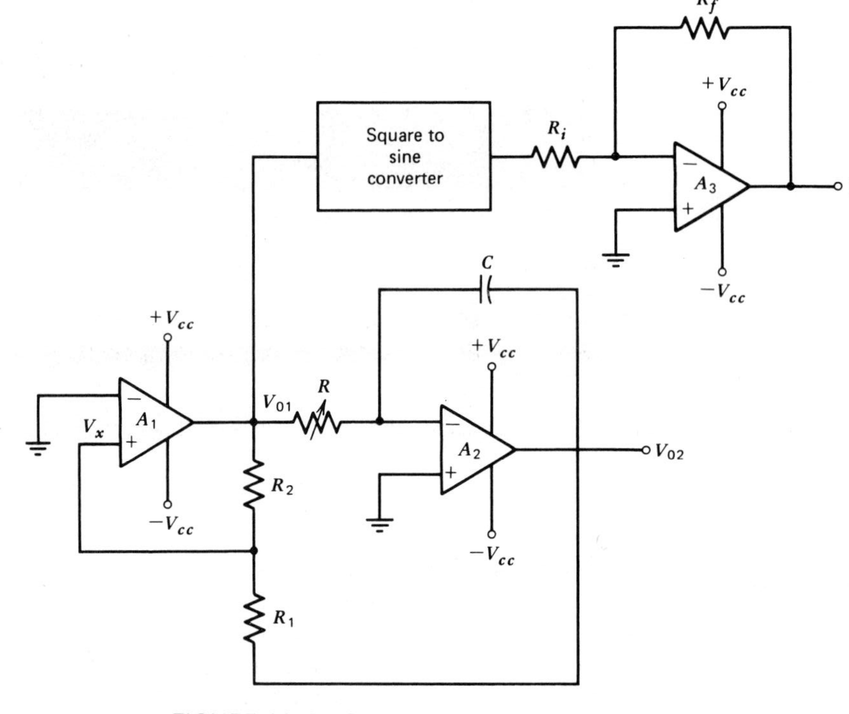

FIGURE 10-9 Circuit for a basic function generator.

voltage V_{01} will remain at $-V_{cc}$ until the voltage at the inverting input of A_1 exceeds the voltage at the noninverting input, which in this case is at zero volts. The noninverting input voltage, V_x, is due, in part, to the voltage V_{01} and, in part, to the voltage V_{02}, according to the expression

$$V_x = -V_{cc}\frac{R_1}{R_1+R_2} + V_{02}\frac{R_2}{R_1+R_2} \qquad (10\text{-}24)$$

The output V_{01} changes states when $V_x = 0$; therefore, we can say

$$0 = -V_{cc}\frac{R_1}{R_1+R_2} + V_{02}\frac{R_2}{R_1+R_2} \qquad (10\text{-}25)$$

which simplifies to

$$V_{02}R_2 = V_{cc}R_1 \qquad (10\text{-}26)$$

From Eq. 10-26 we can determine the maximum amplitude of the triangular output, V_{02}, which is expressed as

$$V_{02} = V_{cc}\frac{R_1}{R_2} \qquad (10\text{-}27)$$

When the output voltage V_{02} reaches the amplitude given by Eq. 10-27, the output of the comparator changes states and the triangular wave begins to decrease linearly. Since the output is symmetrical about 0 V, Eq. 10-27 also expresses the minimum value of V_{02} at which switching occurs. The waveforms at V_x, V_{01}, and V_{02} are shown in Fig. 10-10 for the situation in which $R_1 = R_2$.

The frequency of the circuit is controlled by the RC time constant of the integrator. To obtain an expression for the frequency, we begin with the expression relating capacitor current, charge, and time of change:

$$q = i_c t \tag{10-28}$$

The rate of charge of the capacitor is

$$dq = i_c \, dt \tag{10-29}$$

which can be written as

$$i_c = \frac{dq}{dt} \tag{10-30}$$

As the capacitor charges, the relationship between charge, capacitance, and voltage across the capacitor plates is

$$q = CV_{02} \tag{10-31}$$

Substituting Eq. 10-31 into Eq. 10-30 yields

$$i_c = C\frac{d(V_{02})}{dt} \tag{10-32}$$

Since the input resistance of the operational amplifier is very high, the current

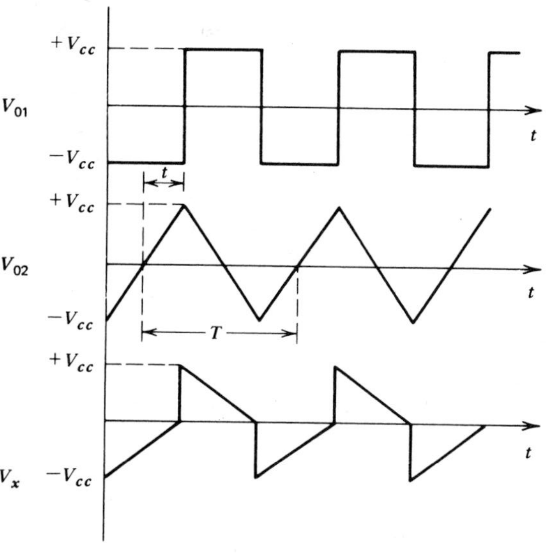

FIGURE 10-10 Waveforms for the function generator of Fig. 10-9.

through resistor R is approximately equal to the charging current of the capacitor. Therefore, we can write

$$i_R = C\frac{d(V_{02})}{dt} \tag{10-33}$$

In addition, since the voltage gain of the operational amplifier is very high, the voltage at the input to the amplifier is very nearly zero. Therefore

$$i_R = \frac{V_{01} - 0}{R} \tag{10-34}$$

Substituting Eq. 10-34 into Eq. 10-33, we obtain

$$d(V_{02}) = \frac{1}{RC}V_{01}\,dt \tag{10-35}$$

Integrating both sides of Eq. 10-35, we obtain

$$V_{02} = \frac{1}{RC}\int V_{01}\,dt = \frac{V_{01}}{RC}t \tag{10-36}$$

Substituting Eq. 10-27 into Eq. 10-36 yields

$$V_{cc}\frac{R_1}{R_2} = \frac{V_{01}}{RC}t \tag{10-37}$$

Since $V_{01} = V_{cc}$, Eq. 10-37 simplifies to

$$t = RC\frac{R_1}{R_2} \tag{10-38}$$

The development of Eq. 10-38 began with Eq. 10-28, which allows us to compute the charge on a capacitor after a period of time t. Equation 10-28 is valid only if the initial charge and, therefore, the initial voltages on the capacitor are zero. Therefore, the time t in Eq. 10-38 is the time for the capacitor to charge from 0 V until switching occurs, which is at one-fourth cycle as shown in Fig. 10-10. Since $t = \frac{1}{4}T$, Eq. 10-38 becomes

$$T = 4RC\left(\frac{R_1}{R_2}\right) \tag{10-39}$$

The frequency, which is the reciprocal of the period, is now expressed as

$$f = \frac{1}{4RC}\left(\frac{R_2}{R_1}\right) \tag{10-40}$$

EXAMPLE 10-5

Compute the frequency and the peak amplitude of the triangular output of the circuit shown in Fig. 10-11.

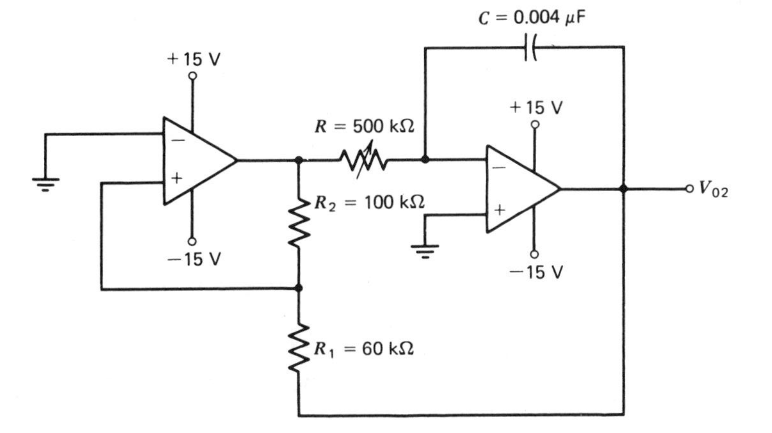

FIGURE 10-11 Function generator for Example 10-5.

Solution

The output frequency can be computed using Eq. 10-40 as

$$f=\frac{1}{4RC}\left(\frac{R_2}{R_1}\right)$$

$$=\frac{1}{4(500\ \text{k}\Omega)(0.004\ \mu\text{F})}\left(\frac{100\ \text{k}\Omega}{60\ \text{k}\Omega}\right)$$

$$=208\ \text{Hz}$$

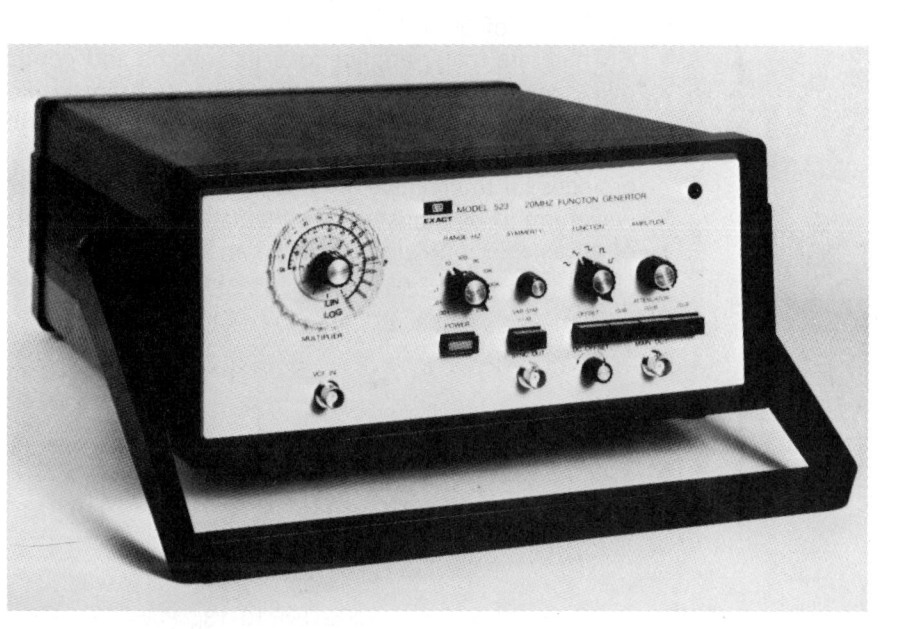

FIGURE 10-12 Laboratory quality function generator. (Courtesy Exact Electronics.)

The amplitude of the triangular waveform can be computed from Eq. 10-27 as

$$V_{02} = V_{cc}\frac{R_1}{R_2}$$

$$= (15\text{ V})\left(\frac{60\text{ k}\Omega}{100\text{ k}\Omega}\right) = 9\text{ V}$$

Figure 10-12 shows an Exact Electronic Model 528 function generator. This laboratory-quality instrument generates sine, square, triangle, ramp, and pulse waveforms over the frequency range from 0.001 Hz to 20 MHz. The output voltage is 30 V peak to peak in an open circuit and 15 V peak to peak across a 50-**Ω** load.

10-8 PULSE GENERATORS

Pulse generators are instruments that produce a rectangular waveform similar to a square wave but with a different duty cycle. **Duty cycle** is defined as the ratio of the pulse width to the pulse period, expressed in percent, or

$$\text{Duty cycle} = \frac{\text{Pulse width}}{\text{Pulse period}} \times 100\%$$

The duty cycle of a square wave is 50%, whereas the duty cycle of a pulse is generally from approximately 5% to 95%. Figure 10-13 shows a pulse with a duty cycle of 30%.

There are many applications in systems and component testing in which pulse generators, used in conjunction with an oscilloscope, are very useful. The most basic pulse generator is the astable **multivibrator** which generates symmetrical square waves with variable pulse repetition frequency (PRF) and provides for the relative lengths of the positive and negative excursions of the waveform to be adjusted. A schematic for a basic pulse generator, consisting of an astable multivibrator and a monostable multivibrator, is shown in Fig. 10-14. The output of the astable multivibrator is a square wave. The square

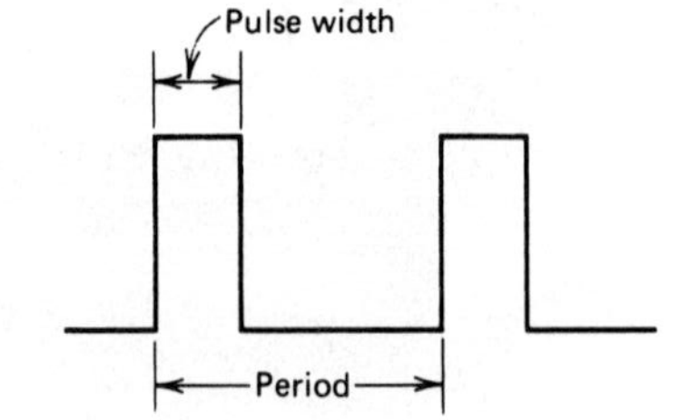

FIGURE 10-13 Pulse with 30% duty cycle.

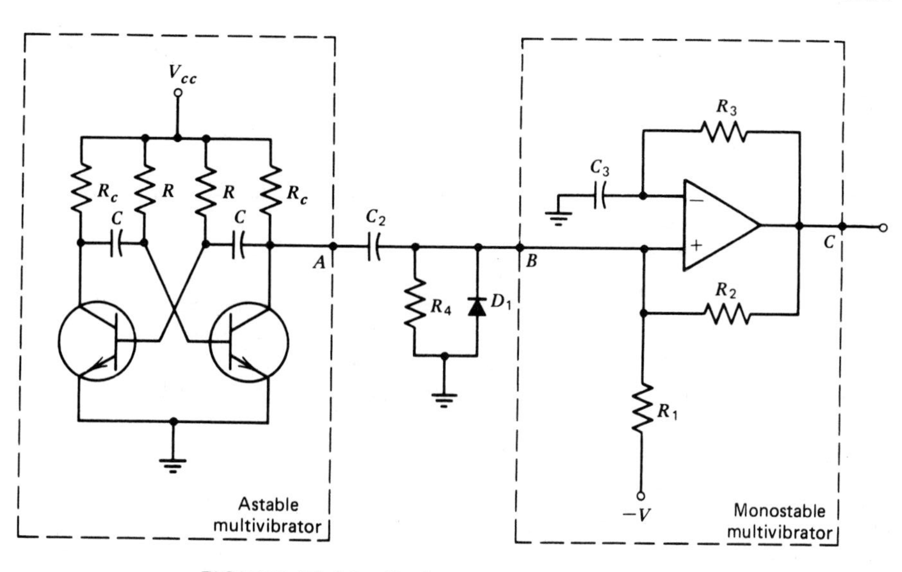

FIGURE 10-14 Basic pulse generator circuit.

wave will be symmetrical if both capacitors, C, and both resistors, R, have the same values.

The frequency, which is given as

$$f = \frac{0.693}{RC}$$

can be changed, while maintaining the symmetry of the waveform, by changing both resistors or both capacitors. The duty cycle of the square wave can be varied by changing the value of one capacitor or one resistor. Common practice would be to use several switch-selectable capacitors to provide different ranges of frequency and to provide for vernier control by making one of the resistors variable.

The purpose of the "one-shot," or monostable multivibrator, is to provide a short-duration output pulse each time a pulse is applied to its input. The monostable multivibrator has one stable state and one unstable state. The circuit operates in its stable state until an input pulse drives it into its unstable state, thus producing an output pulse. The duration of the output pulse is computed as

$$t = R_3 C_3 \ln\left(\frac{R_1 + R_2}{R_3}\right)$$

If resistors R_1 and R_2 are equal in value, as they frequently are, then the expression simplifies to

$$t = 0.693 R_3 C_3$$

The circuit $C_2 R_4$ is called a *differentiator*. Its purpose is to provide a short-duration "spike" to trigger the monostable multivibrator. The purpose

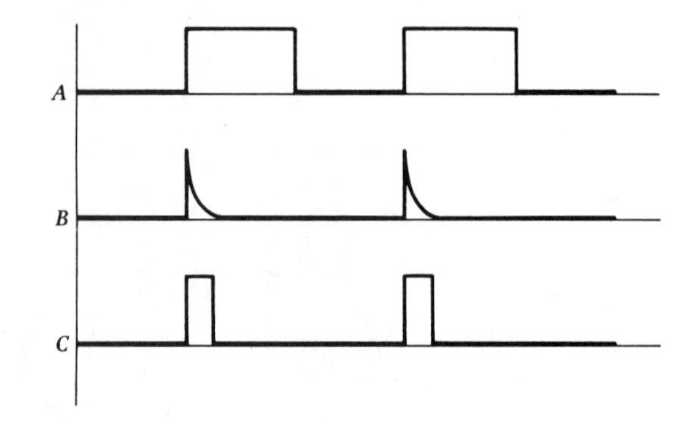

FIGURE 10-15 Waveforms for the pulse generator in Fig. 10-14.

of diode D_1 is to bypass to ground the negative spike from the differentiator. The waveforms at points A, B, and C are shown in Fig. 10-15.

EXAMPLE 10-6

Determine the frequency, pulse width, and duty cycle of the circuit shown in Fig. 10-16.

Solution

The frequency of the square wave generated by the astable multivibrator is computed as

$$f = \frac{0.693}{RC}$$

$$= \frac{0.693}{75\ \text{k}\Omega \times 0.01\ \mu\text{F}} = 924\ \text{Hz}$$

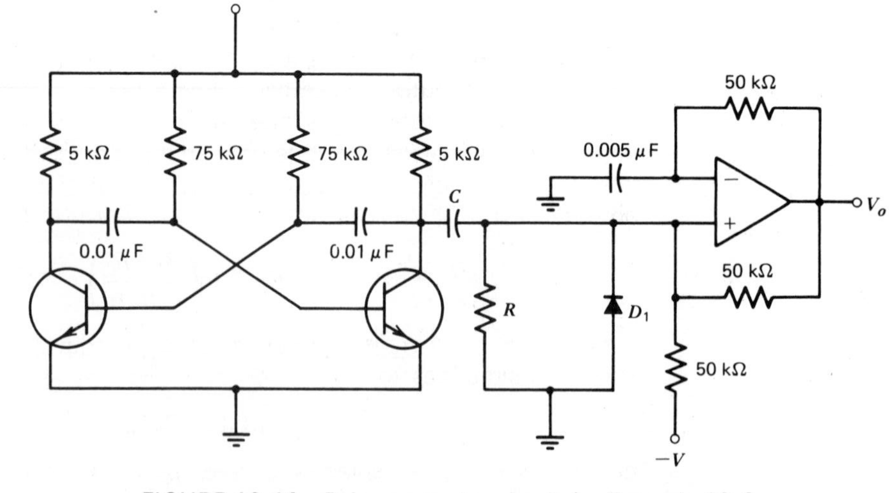

FIGURE 10-16 Pulse generator circuit for Example 10-6.

The pulse width is computed as

$$t = 0.69 R_3 C_3$$
$$= (0.69)(50\ \text{k}\Omega)(0.005\ \mu\text{F}) = 0.173\ \text{msec}$$

To compute the duty cycle, we must first compute the pulse period which is

$$T = \frac{1}{f} = \frac{1}{924\ \text{Hz}} = 1.08\ \text{msec}$$

We can now compute the duty cycle as

$$\text{Duty cycle} = \frac{\text{Pulse width}}{\text{Pulse period}} \times 100\%$$
$$= \frac{0.173\ \text{msec}}{1.08\ \text{msec}} \times 100\%$$
$$= 16\%$$

Figure 10-17 shows a laboratory-quality pulse generator. The instrument shown is a Hewlett-Packard Model 8005B, which is designed to the following partial list of specifications.

- Pulse repetition rate: 0.3 Hz to 20 MHz, five ranges.
- Pulse width: 25 nsec to 3 sec in five ranges.
- Maximum duty cycle: greater than 80% from 0.3 Hz to 1 MHz; greater than 50% form 1 to 10 MHz.
- Output impedance: 50 Ω.
- Amplitude continuously variable from 0.03 to 10 V.

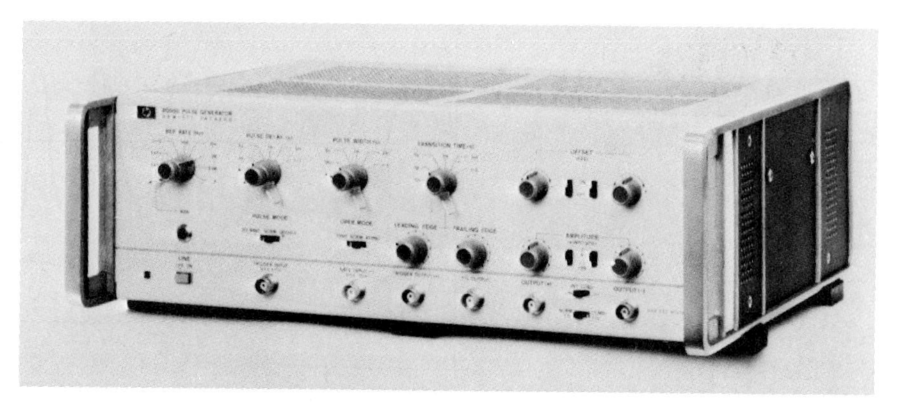

FIGURE 10-17 Laboratory-quality pulse generator. (Courtesy Hewlett-Packard Company.)

The instrument, which is simple to operate, has flexible output parameters and can be triggered by an external pulse as well as internally.

10-9 SWEEP-FREQUENCY GENERATORS

Sweep-frequency generators are instruments that provide a sine wave in the RF range, the frequency of which can be varied smoothly and continuously over an entire frequency band. These instruments are very useful in applications in which the characteristics of a device must be determined over a wide, continuous range of frequencies. Replacing laborious and time-consuming point-by-point measurements with sweep measurements increases the speed and convenience of broadband testing. Sweeping continuously over an entire frequency band also eliminates the chance of missing important information between frequency points.

The output signal of a sweep-frequency generator is a frequency-modulated waveform. The RF carrier may be modulated by many different kinds of waveforms in the audio-frequency range. Figure 10-18 shows an RF signal modulated by an audio-frequency ramp voltage.

Basically, a sweep-frequency generator is nothing more than an RF oscillator that incorporates the additional circuitry necessary to vary the output signal continuously and smoothly over a wide frequency range. A basic block diagram for a sweep-frequency generator is shown in Fig. 10-19. The ramp, or sawtooth generator, is the same circuit used in oscilloscopes to sweep the electron beam horizontally, and the RF oscillator is one of the standard RF oscillator circuits discussed in Section 10-5.

The circuit also incorporates a **voltage-controlled oscillator** (VCO). Perhaps the simplest VCO is an astable multivibrator with the base resistors connected to a variable voltage. In Fig. 10-17 the variable voltage connected

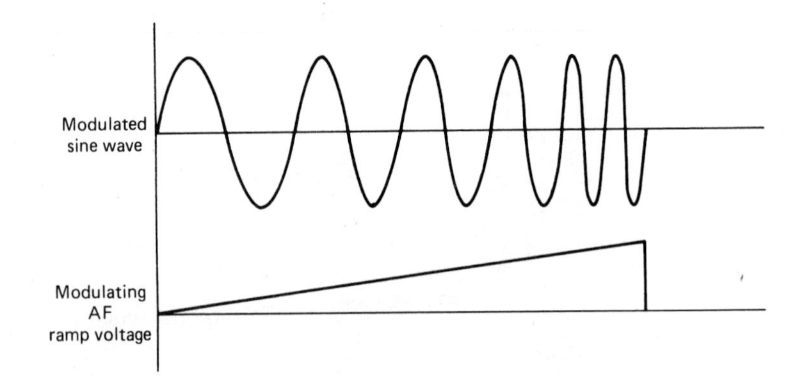

FIGURE 10-18 Waveforms associated with a voltage-controlled oscillator (VCO), showing an RF signal modulated by an audio-frequency ramp voltage.

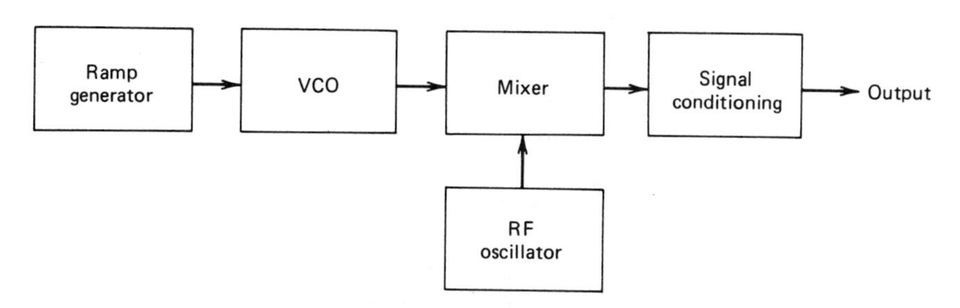

FIGURE 10-19 Block diagram of a basic sweep-frequency generator.

to the base resistors within the VCO is the ramp voltage from the ramp generator. A basic mixer circuit can be configurated by applying the VCO output to the base and the RF oscillator output to the emitter of a transistor amplifier. The frequency of the output signal is the difference in the two frequencies applied to the mixer stage.

A typical laboratory-quality sweep-frequency generator is shown in Fig. 10-20. The instrument is a Model 184 sweep-frequency generator, manufactured by Wavetek. The principal specifications for the instrument are as follows:

- Output waveforms: sine, triangle, square, ± pulse.
- Symmetry: variable from 1:19 to 19:1.
- Operating frequency range: 0.0001 Hz to 5 MHz in ten ranges.
- Amplitude: 0 to 20 V_{p-p} into open circuit, 10 V_{p-p} into 50 Ω.
- Operating modes: triggered, gated, sweep.
- Sweep rates: from 100 sec to 1 msec.

FIGURE 10-20 Laboratory-quality sweep-frequency generator. (Courtesy Wavetek.)

10-10 APPLICATIONS

The range of applications for the various types of signal generators discussed in this chapter is wide and varied. One of the most frequent applications of AF oscillators and RF generators is as a signal substitute in inoperative amplifiers or communications equipment. For example, an RF signal generator can be used to substitute for a suspect local oscillator in a radio or television receiver. Figure 10-21 shows how an RF generator may be used to determine the frequency response of a radio-frequency or an intermediate-frequency amplifier stage in a radio or television receiver.

Pulse generators with fast rise times are widely used in the development of digital circuitry. They can test for diode and transistor switching rates and propagation delays. Pulse generators are also used as modulators for **klystrons** and other RF sources to obtain high peak power while average power dissipation is low.

Other applications of pulse generators are to detect faults in transmission cables and telephone lines, to provide a stimulus to living tissue in physiological and biological research, to drive lasers, and to stimulate data transmission signals.

A primary application of sweep-frequency generators is the display of the response curve for the various stages of radio or television receivers.

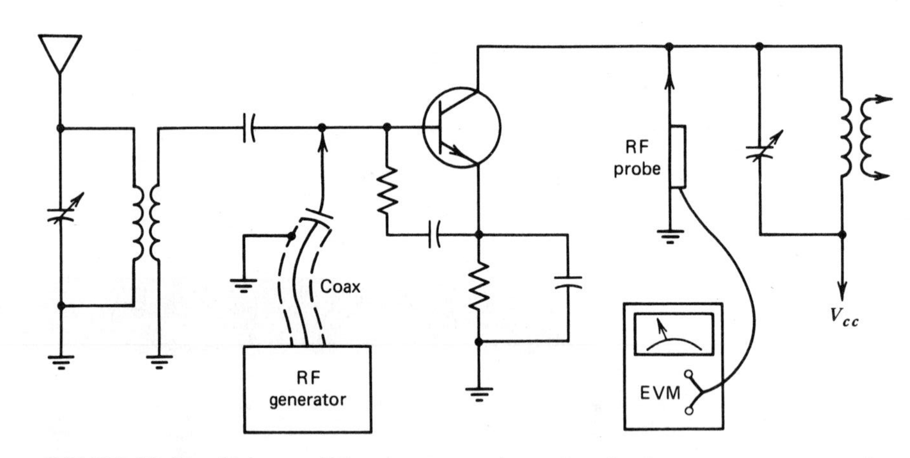

FIGURE 10-21 Using an RF generator to determine the frequency response of an RF or IF amplifier stage.

10-11 SUMMARY

Signal-generating instruments include audio-frequency oscillators, radio-frequency generators, function generators, pulse generators, and sweep-frequency generators.

Audio oscillators produce a sine-wave signal over the range of frequencies from 20 Hz to at least 20 kHz. Most audio oscillators provide a square-wave output as well. The most widely used circuit in audio oscillators is the Wien bridge oscillator.

Radio-frequency generators produce a sine-wave signal that can be modulated by an audio-frequency signal. Radio-frequency oscillators operate in the frequency range from 30 kHz to 3000 MHz. Although few instruments cover the entire frequency, a frequency range exceeding 100 MHz is fairly commonplace.

Function generators provide sine, square, and triangular waveforms. Many provide ramp and pulse outputs as well. The frequency range of these versatile instruments goes from a fraction of a hertz to several hundred kilohertz or even several megahertz.

Pulse generators produce a rectangular output waveform variable in both frequency and duty cycle. Typically, the frequency range is from a fraction of a hertz to several megahertz, and the duty cycle can be varied from approximately 5% to 95%.

Sweep-frequency generators provide an RF sine-wave output that can be varied smoothly and continuously over an entire frequency band.

10-12 GLOSSARY

Audio frequency (AF): A frequency corresponding to an audible sound wave. The extreme limits of audible frequencies vary with the individual and range from about 20 to 20,000 Hz.

Barkhausen criteria: Conditions that must be satisfied if a circuit is to sustain oscillation.

Decade steps: Increasing the value of a parameter in steps, the value of each step being ten times greater than that of the previous step.

Distortion: An unwanted change in a waveform generally caused by nonlinearity of an active device, nonuniform response at different frequencies, or a nonproportional relationship between phase shift and frequency.

Duty cycle: The ratio of the pulse width to the period expressed as a percentage.

Generator: A circuit or instrument that develops a varying output waveform when energized by a dc source. Provision is made for modulating the output waveform.

Klystron: A type of vacuum tube used to generate or amplify microwave signals.

Modulation: The process in which the amplitude, frequency, or phase of a carrier wave is varied with time in accordance with the waveform of an intelligence signal.

Multivibrator: A two-stage amplifier with positive feedback; one active device or the other is at saturation, thereby generating a square-wave output.

Oscillator: A circuit or instrument that incorporates positive feedback to cause self-sustained oscillation, thereby providing an unmodulated sine-wave output whose frequency is determined by the value of the circuit components.

Phase shift: A change in the phase relationship between two periodic quantities.

Radio-frequency (RF): The part of the frequency spectrum between audio sound and infrared light (about 20 kHz to 10 MHz).

Transient: A temporary component of current existing in a circuit during adjustment to a load change, source voltage change, or line impulse.

Voltage-controlled oscillator (VCO): An instrument that produces an output whose frequency is dependent on the amplitude of a input voltage.

10-13 REVIEW QUESTIONS

The following questions should be answered after a thorough study of the chapter. The purpose of these questions is to determine the reader's comprehension of the material.

1. What are the two types of circuits used in audio oscillators?
2. What is the basic difference between an oscillator and a generator?
3. What two conditions must be satisfied for a circuit to sustain oscillation?
4. Why are radio-frequency oscillator circuits not practical for audio-frequency oscillators?
5. What is a function generator?
6. What is the basic difference between a square-wave generator and a pulse generator?
7. What is an astable multivibrator?
8. What is a "one-shot" or monostable multivibrator?

10-14 PROBLEMS

10-1 Determine the gain with feedback A_f for the amplifier in Fig. 10-22.

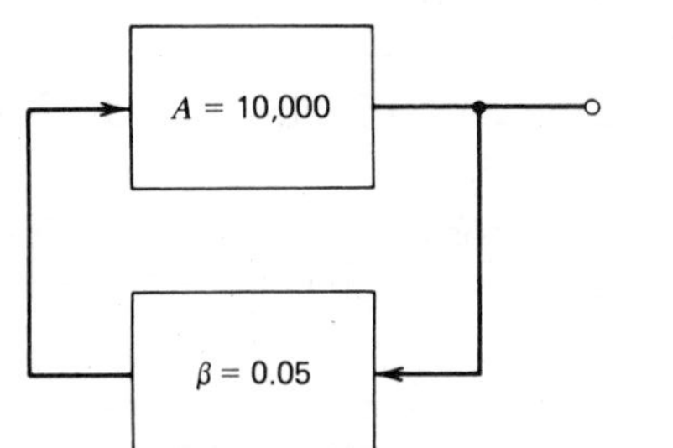

FIGURE 10-22 Circuit for Problem 10-1.

10-2 Describe the requirements that must be satisfied for a circuit to sustain oscillation.

10-3 Determine the values of C_1 and C_2, that will cause the circuit in Fig. 10-23 to oscillate at 12 kHz.

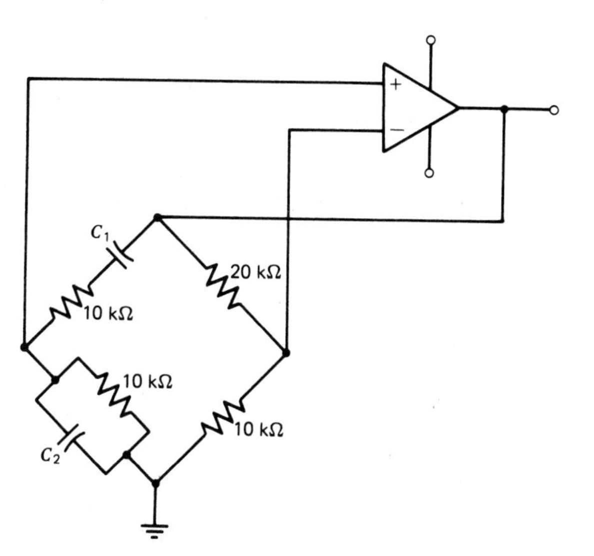

FIGURE 10-23 **Circuit for Problem 10-3.**

10-4 Determine the frequency of oscillation of a phase-shift oscillator with a three-section feedback network consisting of 20-**Ω** resistors and 0.0015-μF capacitors.

10-5 Determine the value of L_2 in the circuit shown in Fig. 10-24 if the frequency of oscillation is to be 100 kHz.

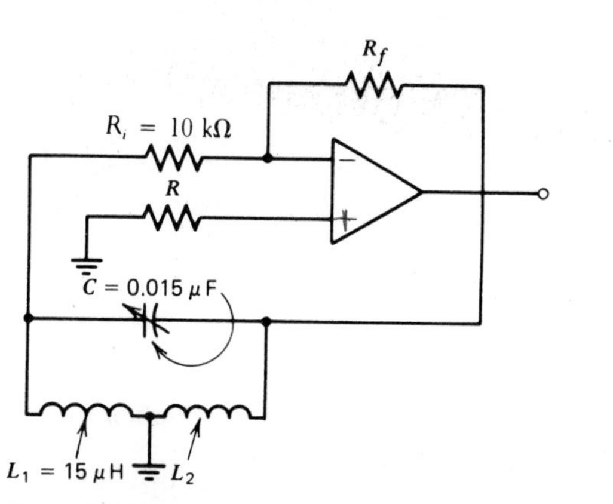

FIGURE 10-24 **Circuit for Problem 10-5.**

10-6 Determine the minimum value of R_f in Fig. 10-24 to sustain oscillation if L_2 equals 125 μH.

10-7 Determine the frequency of the triangular output of the circuit shown in Fig. 10-25.

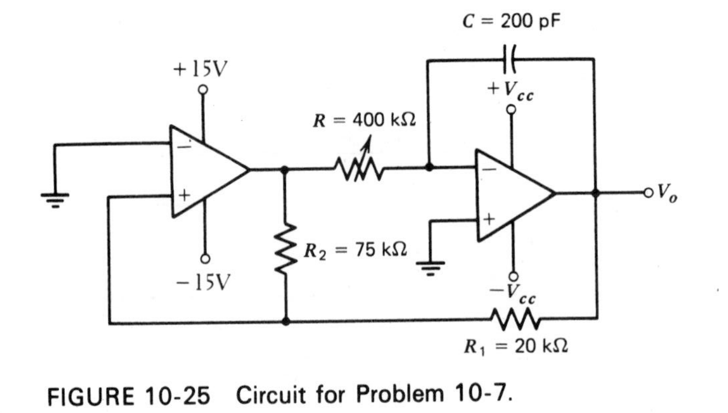

FIGURE 10-25 Circuit for Problem 10-7.

10-8 Determine the peak-to-peak amplitude of the output signal V_o for the circuit of Fig. 10-25.

10-9 Determine the value of R in the circuit of Fig. 10-25 if the frequency of the output is to be 5 kHz.

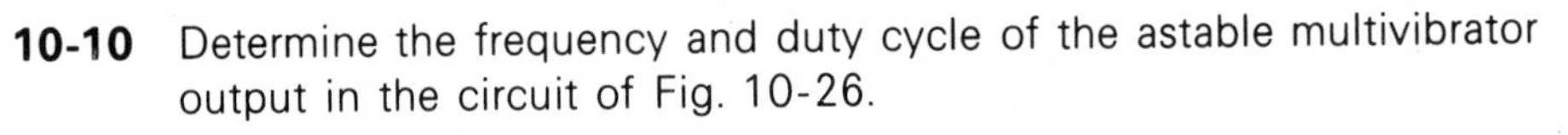

10-10 Determine the frequency and duty cycle of the astable multivibrator output in the circuit of Fig. 10-26.

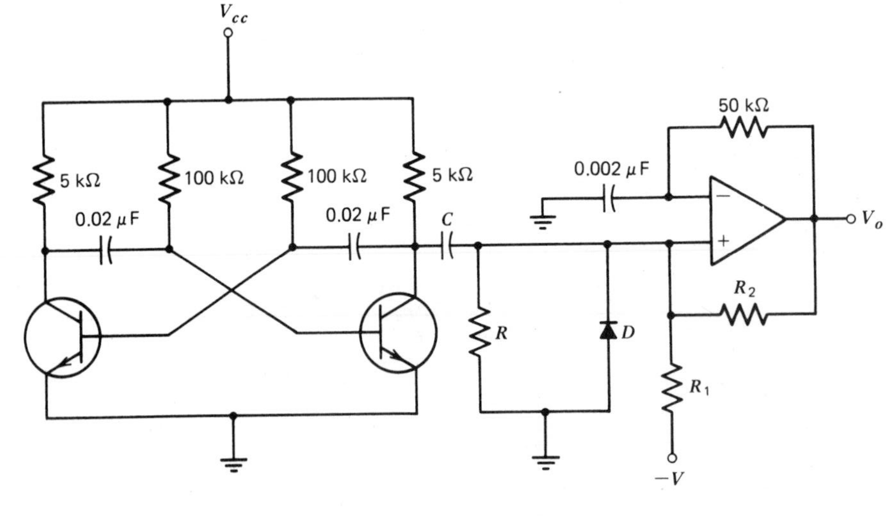

FIGURE 10-26 Circuit for Problem 10-10.

10-11 Determine the pulse width of the output pulse V_o in the circuit of Fig. 10-26.

10-12 Design a Wien bridge oscillator around the following specifications:

- $f = 10$ kHz
- $V_{cc} = \pm 15$V
- $I_{in} = 1\ \mu$A
- $I_{R4} = 100\ I_{in}$

10-15 LABORATORY EXPERIMENTS

Experiments E25 and E26 apply the theoretical material presented in Chapter 10. The purpose of the experiments is to provide hands-on experience to reinforce the theory.

Experiment E25 requires a sweep-frequency generator and $X-Y$ recorder. If this equipment is not available, this experiment will have to be skipped. The equipment for Experiment E26 should be available in any general electronics laboratory.

The contents of the laboratory report to be submitted by each student are included at the end of each experimental procedure.

CHAPTER 11 Transducers

11-1 Instructional Objectives

In this chapter we will examine transducers in order to gain an awareness of what they can do. We will discuss their operations, characteristics, and functions. After completing Chapter 11 you should be able to

1. Define a transducer.
2. Explain the operation and function of each of the following:
   (a) The resistive position transducer.
   (b) The strain gauge.
   (c) The capacitive transducer.
   (d) The inductive transducer.
   (e) The linear variable differential transformer.
   (f) The piezoelectric transducer.
   (g) The thermocouple.
   (h) The thermistor.
   (i) The ultrasonic temperature transducer.
   (j) The photomultiplier tube.
   (k) The photocell.
   (l) The photovoltaic cell.
   (m) The semiconductor photodiode.
   (n) The phototransistor.
3. Explain the limitations of certain transducers.
4. Explain the applications of certain transducers.

11-2 INTRODUCTION

An important function in the industrial and scientific use of electronics is the measurement of physical parameters such as position, temperature, force, pressure, and the rate of flow of fluids. Without transducers, advances in the application of control and computation would have been impossible. Transducers have become convenient, economical, and highly efficient in operation because they convert the various physical quantities into related electrical values which are readily used for measuring, amplifying, transmitting, and control. These electrical values enhance accuracy and facilitate permanent recording of the information. For control systems, transducers put information immediately into the form in which the electronic devices and circuits can most readily use it.

11-3 DEFINITION OF A TRANSDUCER

In general terms, a **transducer** is any device that converts energy in one form to energy in another. However, in its applied usage, the term *transducer* refers to rather specialized devices. The majority either convert electrical energy to mechanical displacement or convert some *nonelectrical* physical quantity, such as temperature, sound, or light, to an electrical signal. Since this is a text in electronic instrumentation, the second conversion is of primary interest.

The functions of a transducer are (1) to *sense* the presence, magnitude, change in, and frequency of some **measurand** and (2) to *provide* an electrical output that, when appropriately processed and applied to a readout device, gives accurate quantitative data about the measurand (Fig. 11-1). The term *measurand* refers to the quantity, property, or condition which the transducer translates to an electrical signal.

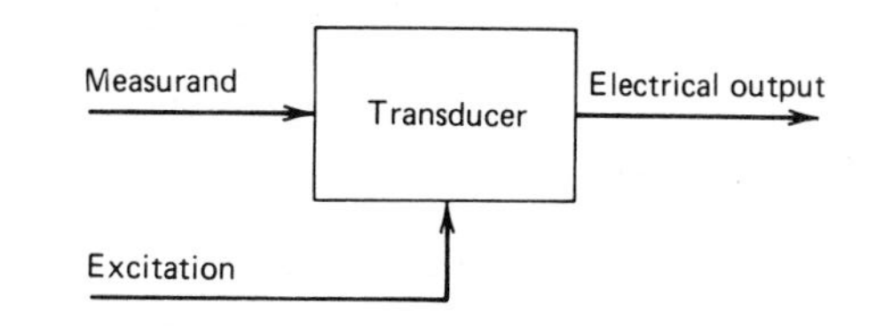

FIGURE 11-1 Block diagram of a transducer.

11-4 CLASSIFICATION OF TRANSDUCERS

Perhaps the most satisfactory way to classify transducers is by the electrical principle involved in their operation. Transducers can also be classified according to their application, based primarily on the physical quantity, property, or condition that is measured. A sharp distinction between, and classification of, types of transducers is difficult. Table 11-1 shows a classification of transducers according to their electrical operating principles. The first part of the table lists the *passive* transducers. These transducers require an external power, and their output is a measure of some variation, such as

TABLE 11-1
Types of Transducers[a]

Classes and Examples	Nature of the Device	Quantities Measured or Typical Applications
	Externally Powered Transducers (Passive)	
Variable resistance		
Slide-wire resistor	Slider or contact varies the resistance in a potentiometer, rheostat, or bridge circuit	Dimension, displacement
Resistance strain gauge	Resistance wire, foil, or semiconductor stress	Strain, force, torque, changed by pressure
Resistance thermometer	Wire or thermistor with large temperature coefficient of resistivity	Temperature and temperature effects, radiant heat
Hot-wire meter	Electrically heated wire exposed in gas stream	Flow rate, turbulence, gas density, vacuum
Resistance hygrometer	Resistivity of conductive stripe changed by moisture	Relative humidity
Thermistor radiometer	Radiation focused on thermistor bolometer	Missile and satellite tracking
Contact thickness gauge	Resistance between contacts depends on material and thickness	Sheet thickness, liquid level
Photoconductive cell	Resistance of cell as circuit element varied by incident radiation	Relays sensitive to light or to infrared
Photoemissive and photomultiplier tubes	Radiation causes electron emission and current (amplification available)	photosensitive relays (with amplification)
Ionization gauge	Electron flow induced by ionization	Radiation and particle counting, vacuum
Variable inductance		
Air-gap gauge	Self-inductance or mutual inductance changed by varying the magnetic path	Thickness, displacement, pressure
Reluctance pickup	Reluctance of magnetic circuit varied by position of material	Position, displacement, phonograph pickup, vibration, pressure
Eddy-current gauge	Inductance of ac coil varied by proximity of an eddy-current plate	Thickness, displacement
Differential transformer	Transformer with differential secondaries and movable magnetic core	Displacement, position, pressure, force
Magnetostriction gauge	Magnetic properties varied by pressure and stress	Sound, pressure, force
Hall-effect pickup	Magnetic field interacts with current through semiconductor to produce voltage at right angle	Field strength, current

TABLE 11-1 (*Continued*)

Classes and Examples	Nature of the Device	Quantities Measured or Typical Applications
	Externally Powered Transducers (Passive)	
Variable capacitance		
Adjustable capacitor	Capacitance between electrodes varied by spacing or area	Displacement, pressure
Condenser microphone	Capacitance between diaphragm and fixed electrode varied by sound pressure	Speech and music, noise vibration
Dielectric gauge	Capacitance varied by changes in dielectric	Liquid level, thickness
	Self-Generating Transducers	
Moving-coil generator	Relative movement of magnet and coil varies output voltage	Vibration velocity, speed of displacement
Thermocouple and thermopile	Pairs of dissimilar metals or semiconductors at different temperatures	Temperature difference, radiation, heat flow
Piezoelectric pickup	Quartz or other crystal mounted in compression, or by bending or twisting	Vibration, acceleration, heat flow
Photovoltaic cell	Layer-built semiconductor cell or transistor generates voltage from light	Exposure meters, light meters, solar batteries

[a]For specific data on 1250 models of transducers, see "ISA Transducer Compendium" published by the Instrument Society of America.

resistance or capacitance. In the second category are the *self-generating* transducers. These transducers do not require an external power, and they produce an analog voltage or current when stimulated by some physical form of energy.

11-5 SELECTING A TRANSDUCER

The transducer, or sensor, as it is sometimes called, has to be physically compatible with its intended application. In the selection of a transducer, there are approximately eight areas of concentration.

1. *Operating range*. The transducer should maintain range requirements and good resolution.
2. *Sensitivity*. The transducer must be sensitive enough to allow sufficient output.

3. *Frequency response and resonant frequency*. Is the transducer flat over the needed range? Will the resonant frequency be excited?
4. *Environmental compatibility*. Do the temperature range of the transducer, its corrosive fluids, the pressures, shocks, and interactions it is subject to, its size and mounting restrictions make it inapplicable?
5. *Minimum sensitivity*. The transducer must be minimally sensitive to expected stimuli other than the measurand.
6. *Accuracy*. The transducer may be subject to repeatability and calibration errors as well as errors expected owing to sensitivity to other stimuli.
7. *Usage and ruggedness*. The ruggedness both of mechanical and electrical intensities of the transducer versus its size and weight must be considered. Who will be installing and using the transducer?
8. *Electrical*. What length and type of cable is required? What are the signal-to-noise ratios when combined with amplifiers and frequency-response limitations.

We will discuss the most common type of transducers, emphasizing the descriptive story of what the transducers do, where they are used, and the basic measurement principles associated with them.

The most common transducers are those in which a force is applied to the system; the transducer converts this applied force into displacement. These transducers are generally classified as

- Capacitive.
- Differential transformer.
- Inductive.
- Piezoelectric.
- Piezoresistive—the strain gauge.
- Photoelectric.
- Signal converters.
- Potentiometric.
- Thermocouple.

We will discuss some of these transducers as well as others in the sections that follow.

11-6 RESISTIVE POSITION TRANSDUCERS

The principle of the resistive position transducer is that the physical variable under measurement causes a resistance change in the sensing element. A common requirement in industrial measurement and control work is to be able to sense the position of an object, or the distance it has moved.

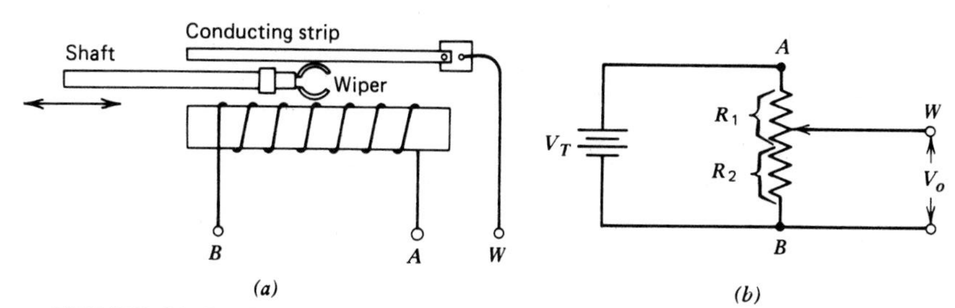

FIGURE 11-2 **Resistive positive transducer, or displacement transducer.**

One type of displacement transducer uses a resistance element with a sliding contact or wiper linked to the object being monitored. Thus, the resistance between the slider and one end of the resistance element depends on the position of the object. Figure 11-2*a* shows the construction of this type of transducer.

Figure 11-2*b* shows a typical method of use. The output voltage depends on the wiper position and therefore is a function of the shaft position. This voltage may be applied to a voltmeter calibrated in inches for visual display.

Typical commercial units provide a choice of maximum shaft strokes from an inch or less to 5 feet or more. Deviation from linearity of the resistance-versus-distance specification can be as low as 0.1% to 1.0%.

Consider Fig. 11-2*b*. If the circuit is unloaded, the output voltage V_o is a certain fraction of V_T, depending on the position of the wiper:

$$\frac{V_o}{V_T} = \frac{R_2}{R_1 + R_2} \tag{11-1}$$

In its application to resistive position sensors, this equation shows that the output voltage is directly *proportional* to the position of the wiper, if the resistance of the transducer is distributed uniformly along the length of travel of the wiper, that is, if the element is perfectly *linear*.

EXAMPLE 11-1

A displacement transducer with a shaft stroke of 3.0 in. is applied in the circuit of Fig. 11-2*b*. The total resistance of the potentiometer is 5 kΩ, and the applied voltage $V_T = 5.0$ V. When the wiper is 0.9 in. from *B*, what is the value of the output voltage V_o?

Solution

$$R_2 = \frac{0.9\text{ in.}}{3.0\text{ in.}} \times 5000\ \Omega = 1500\ \Omega$$

$$V_o = \frac{R_2}{R_T} V_T = \frac{1500\ \Omega}{5000\ \Omega} \times 5.0\text{ V} = 1.5\text{ V}$$

EXAMPLE 11-2

A resistive position transducer with a resistance of 5000 Ω and a shaft stroke of 5.0 in. is used in the arrangement of Fig. 11-3. Potentiometer R_3R_4 is also 5000 Ω, and $V_T = 5.0$ V. The initial position to be used as a reference point is such that $R_1 = R_2$ (i.e., the shaft is at midstroke). At the start of the test,

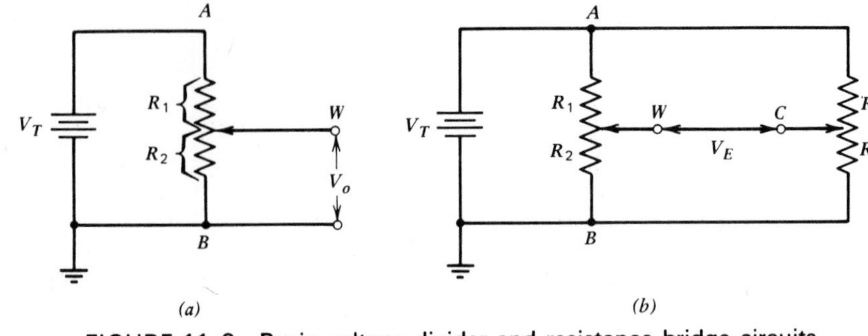

FIGURE 11-3 Basic voltage divider and resistance bridge circuits.

potentiometer R_3R_4 is adjusted so that the bridge is balanced ($V_E = 0$). Assuming that the object being monitored will move a maximum distance of 0.5 in. toward A, what will the new value of V_E be?

Solution

If the wiper moves 0.5 in. toward A from midstroke, it will be 3.0 in. from B.

$$R_2 = \frac{3.0\text{ in.}}{5.0\text{ in.}} = 5000\ \Omega = 300\ \Omega$$

$$V_E = V_{R2} = V_{R4} - \frac{R_2}{R_1 + R_2} V_T - \frac{R_4}{R_3 + R_4} V_T \tag{11-2}$$

$$= \left(\frac{3000\ \Omega}{5000\ \Omega}\right)(5\text{ V}) - \left(\frac{2500\ \Omega}{5000\ \Omega}\right)(5\text{ V}) = 0.5\text{ V}$$

This answer is a measure of the distance and direction that the object has traveled.

11-7 STRAIN GAUGE

The **strain gauge** is an example of a passive transducer that uses electrical resistance variation in wires to sense the strain produced by a force on the wires. It is a very versatile detector and transducer for measuring weight, pressure, mechanical force, or displacement.

The construction of a bonded strain gauge (Fig. 11-4) shows a fine-wire element looped back and forth on a mounting plate, which is usually cemented to the member undergoing stress. A tensile stress tends to elongate the wire and thereby increase its length and decrease its cross-sectional area. The combined effect is an increase in resistance as seen from Eq. 11-3,

$$R = \frac{\rho L}{A} \tag{11-3}$$

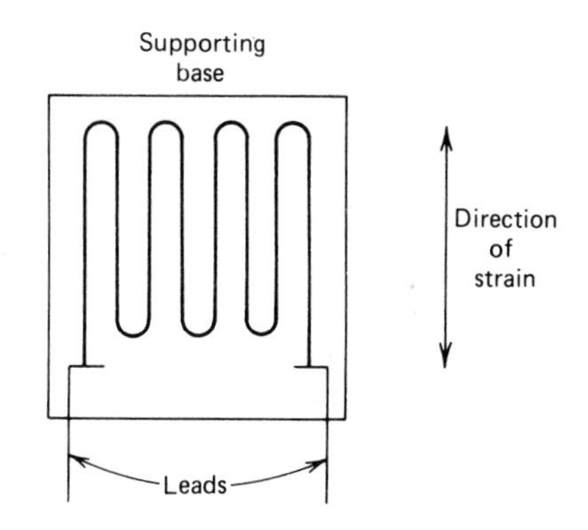

FIGURE 11-4 Resistive strain gauges; wire construction.

where

ρ = the specific resistance of the conductor material in ohm meters (product)
L = the length of the conductor in meters
A = the area of the conductor in square meters

As a consequence of strain, two physical qualities are of particular interest: (1) the change in gauge *resistance* and (2) the change in *length*. The relationship between these two variables expressed as a ratio is called the **gauge factor**, K, expressed mathematically as

$$K = \frac{\Delta R/R}{\Delta L/L} \tag{11-4}$$

where

K = the gauge factor
R = the initial resistance in ohms (without strain)
ΔR = the change in initial resistance in ohms
L = the initial length in meters (without strain)
ΔL = the change in initial length in meters

Note that the term $\Delta L/L$ in the denominator is the same as the unit strain G. Therefore, Eq. 11-4 can be written as

$$K = \frac{\Delta R/R}{G} \tag{11-5}$$

Robert Hooke pointed out in the seventeenth century that for many common materials, there is a constant ratio between stress and strain. **Stress** is defined as the internal force per unit area. The stress equation is

$$S = \frac{F}{A} \tag{11-6}$$

where

S = The stress in kilograms per square meter
F = the force in kilograms
A = the area in square meters

The constant of proportionality between stress and strain for a linear stress–strain curve is known as the **modulus of elasticity** of the material, E, or Young's modulus. Hooke's law is written as

$$E = \frac{S}{G} \tag{11-7}$$

where

E = Young's modulus in kilograms per square meter
S = the stress in kilograms per square meter
G = the strain (no units)

For strain gauge applications, a *high degree of sensitivity* is very desirable. A high gauge factor means a relatively large resistance change for a given **strain.** Such a change is more easily measured than a small resistance change. Relatively small changes in strain can be sensed, as shown in Example 11-3.

EXAMPLE 11-3

A resistant strain gauge with a gauge factor of 2 is fastened to a steel member, which is subjected to a strain of 1×10^{-6}. If the original resistance value of the gauge is 130 **Ω**, calculate the change in resistance.

Solution

$$K = \frac{\Delta R/R}{\Delta L/L} = \frac{\Delta R/R}{G} \tag{11-7}$$

$$\Delta R = KGR = (2)(1 \times 10^{-6})(130\ \Omega) = 260\ \mu\Omega$$

EXAMPLE 11-4

A round steel bar, 0.02 m in diameter and 0.40 m in length, is subjected to a tensile force of 33,000 kg, where $E = 2 \times 10^{10}$ kg/m². Calculate the elongation, ΔL, in meters.

Solution

$$A = \pi\left(\frac{D}{2}\right)^2 = \pi\left(\frac{0.02\ \text{m}}{2}\right)^2 = 3.14 \times 10^{-4}\ \text{m}^2$$

$$E = \frac{S}{G} = \frac{F/A}{\Delta L/L}$$

$$\Delta L = \frac{FL}{AE} = \frac{33{,}000\ \text{kg} \times 0.40\ \text{m}}{(3.14 \times 10^{-4}\ \text{m}^2)(2 \times 10^{10}\ \text{kg/m}^2)}$$

$$= 2.1 \times 10^{-3}\ \text{m}$$

Metallic strain gauges are formed from thin resistance wire or etched from thin sheets of metal foil. Wire gauges are generally small in size, are subjected to minimal leakage, and can be used in high-temperature applications. Foil elements are somewhat larger in size and are more stable than wire gauges. They can be used under conditions of extreme temperature and under prolonged loading, and they dissipate self-induced heat easily.

Semiconductor strain gauges are often used in high-output transducers as load cells. These gauges are extremely sensitive, with gauge factors from 50 to 200. They are, however, affected by temperature fluctuations and often behave in a nonlinear manner.

The strain gauge is generally used as one arm of a bridge. The simple arrangement shown in Fig. 11-5*a* can be employed when temperature variations are not sufficient to affect accuracy significantly, or in applications for which great accuracy is not required. However, since gauge resistance is affected by temperature, any change of temperature will cause a change in the bridge balance conditions. This effect can cause an error in the strain measurement. Thus, when temperature variation is significant, or when unusual accuracy is required, an arrangement such as that illustrated in Fig. 11-5*b* may be used. Here two gauges of the same type are mounted on the item being tested, close enough together that both are subjected to the same temperature. Consequently, the temperature will cause the same change of resistance in the two, and the bridge balance will not be affected by the temperature. However, one of the two gauges is mounted so that its sensitive direction is at right angles to the direction of the strain. The resistance of this *dummy gauge* is not affected by the deformation of the material. Therefore, it acts like a passive resistance (such as R_3 of Fig. 11-5*a*) with regard to the strain measurement. Since only one gauge responds to the strain, the strain causes bridge unbalance just as in the case of the single gauge.

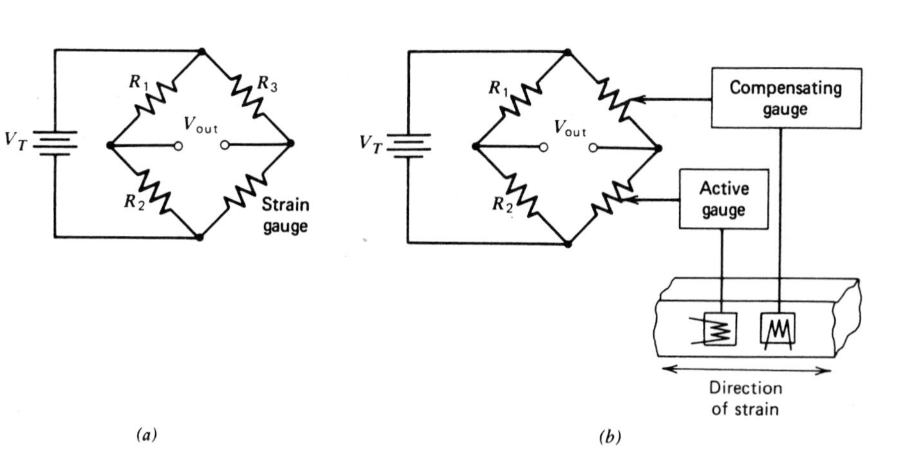

FIGURE 11-5 Basic gauge bridge circuits.

11-8 DISPLACEMENT TRANSDUCERS

Most displacement transducers sense displacement by means of a sensing shaft, which is mechanically connected to the point or object whose displacement is to be measured. The mechanical elements that are used to convert the applied force into a displacement are called force-summing devices.

Among the various types of displacement transducers, classified primarily on the basis of their transduction principle, three groups are used most frequently. (1) *Reluctive transducers* are used in ac measuring circuits; this group includes the reluctance bridge, the differential transformer, and some other types. (2) *Potentiometric transducers* are used in dc systems. (3) *Digital output transducers* are used when very close accuracy of measurements is required. A number of additional types are commercially available but are used less frequently. A few designs are not yet sufficiently developed for widespread use.

11-9 CAPACITIVE TRANSDUCERS

The capacitance of a parallel-plate capacitor is given by

$$C = \frac{kA\varepsilon_o}{d} \text{ (farads)} \qquad (11\text{-}8)$$

where

k = dielectric constant
A = the area of the plate, in square meters
$\varepsilon_o = 8.854 \times 10^{-12}$, in farads per meter
d = the plate spacing in meters

Since the capacitance is inversely proportional to the spacing of the parallel plates, any variation in d causes a corresponding variation in the capacitance.

Figure 11-6 shows several forms of capacitive transducers. In Fig. 11-6*a* we see a rotary plate capacitor, one that is not unlike the variable capacitor used to tune radio transmitters and receivers. The capacitance of this unit is proportional to the amount of area on the fixed plate that is convered, that is, "shaded" by the moving plate. This type of transducer will give signals proportional to curvilinear displacement or angular velocity.

A rectilinear capacitance transducer is shown in Fig. 11-6*b*, and it consists of a fixed cylinder and a moving cylinder. These pieces are configured so that the moving piece fits inside the fixed piece but is insulated from it.

Figure 11-6*c* shows a transducer that varies the spacing between surfaces, that is, the thin diaphragm. The dielectric is either air or vacuum. Such devices are often used as capacitance microphones.

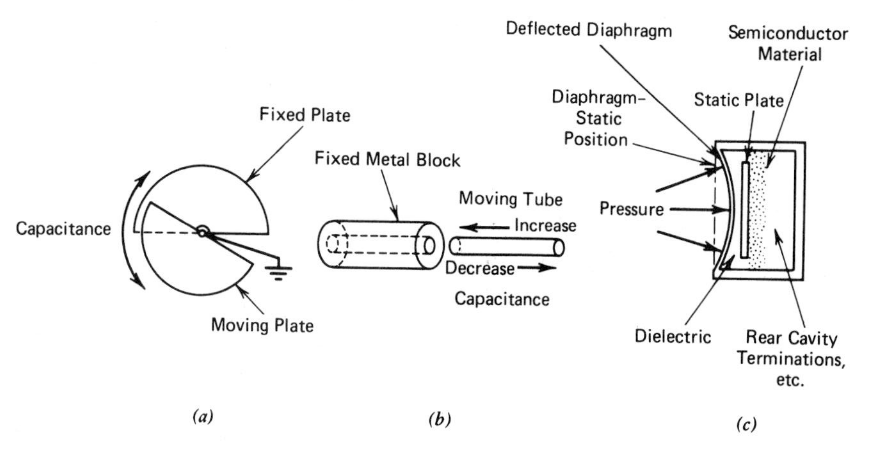

FIGURE 11-6 **Capacitance transducers. (*Source*: Harry Thomas, *Biomedical Instrumentation and Measurements*, Fig. 1-6, p. 13, Reston Publishing Company, Reston, Va.)**

EXAMPLE 11-5

An electrode–diaphragm pressure transducer has plates whose area is 5×10^{-3} m^2 and whose distance between plates is 1×10^{-3} m. Calculate its capacitance if it measures air pressure. The dielectric constant of air is $k = 1$.

Solution

$$C = \frac{kA\varepsilon_o}{d}$$

$$= \frac{(1)(5 \times 10^{-3}\ \text{m}^2)(8.854 \times 10^{-12}\ \text{F/m})}{1 \times 10^{-3}\ \text{m}}$$

$$= 44.25\ \text{pF}$$

Capacitance transducers can be used in several ways. One method is to use the varying capacitance to frequency-modulate an RF oscillator. This method is the one employed with capacitance microphones (like Fig. 11-6*c*). Another method is to use the capacitance transducer in an ac bridge circuit.

The capacitance transducer has excellent frequency response and can measure both static and dynamic phenomena. Its disadvantages are sensitivity to temperature variations and the possibility of erratic or distorted signals owing to long lead length.

11-10 INDUCTIVE TRANSDUCERS

Inductive transducers may be either the self-generating or the passive type. The self-generating type utilizes the basic electrical generator principle, that when there is relative motion between a conductor and magnetic field, a voltage is induced in the conductor (generator action). This relative motion between field and conductor is supplied by changes in the measurand.

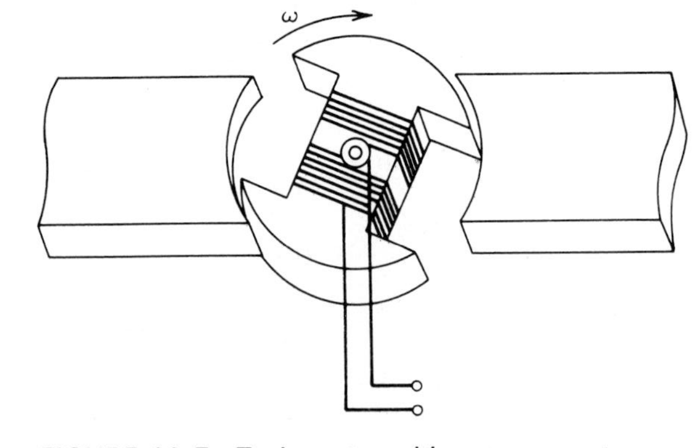

FIGURE 11-7 Tachometer with a permanent magnet stator. (The device is a dc generator.)

A **tachometer** is an inductive transducer that directly converts speed or velocity into an electrical signal. In one technique, the object whose angular velocity is to be measured is directly coupled to the rotor of a dc generator, as shown in Fig. 11-7. The coupling turns the rotating armature between the poles of a permanent magnet, thereby inducing a voltage in the windings of the rotor. The voltage developed may approach 10 mV per revolution per minute (rpm) and can be fed directly into a dc voltmeter calibrated in rpm units. Alternatively, the rotating armature may be simply a permanent magnet. The coils are wound around the fixed poles as shown in Fig. 11-8. This configuration provides an alternating signal, which has certain advantages over a dc voltage in that noise and ripple signals can be filtered more readily before further signal amplification.

One obvious application of this transducer is in the field of frequency determination, when the tachometer is attached directly to the frequency generator. For more exact frequency determination, a digital counting system is used.

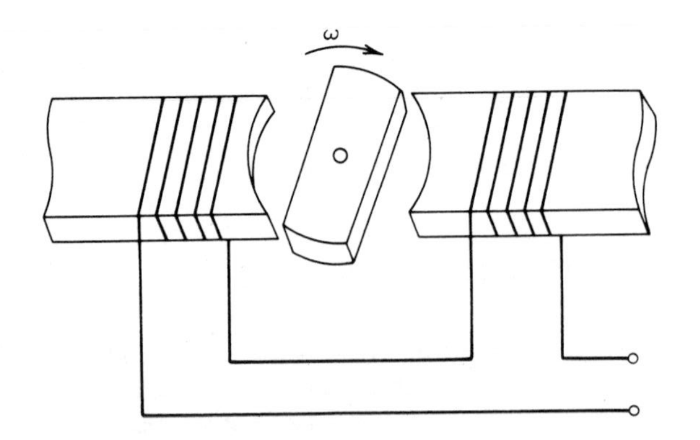

FIGURE 11-8 Tachometer with a permanent magnet rotor.

Flow velocities are also measured by inductive transducers. This method finds extensive use in systems that cannot be opened to the atmosphere. The measurement of liquids containing suspended solids such as sewage and the feed to paper mills presented considerable problems until the advent of the electromagnetic flowmeter. As its name implies, it can be used to measure the flow of any flowing material that is electrically conductive. The meter can be regarded as a section of pipe that is lined with an insulating material. Two saddle coils are arranged opposite each other, and electrodes diametrically opposed are arranged flush with the inside of the lining. If the coils are energized, the moving liquid, as a length of conductor, cuts the lines of force and generates an electromotive force that is picked up by the electrodes. By suitable circuitry and amplification, an electrical signal proportionate to flow can be obtained.

Figure 11-9 gives a general idea of the construction of such a meter. The principle on which the meter operates is that of the dc generator. The generator rotor is replaced by the pipe between two magnetic poles. As the fluid flows through the magnetic field, an electromotive force is induced in it and can be picked up by the electrodes. This can be expressed mathematically as

$$E = Blv \tag{11-9}$$

where

E = emf, volts
B = field strength, webers per square meter (teslas)
l = conductor length, meters
v = velocity of conductor, meters per second

In the case of the electromagnetic flowmeter, the flowing liquid represents the conductor and the internal diameter corresponds to the length l. If the

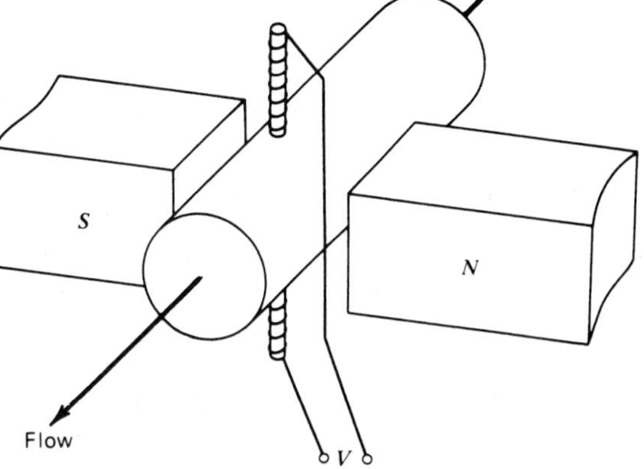

FIGURE 11-9 Inductive input transducer for measuring flow.

field strength B is maintained constant, the only variable is v; hence, the electromotive force is proportional to the velocity.

The electromagnetic flowmeter has the advantage of causing no drop in the pressure of the fluid and having a very large range. It is not suitable for low velocities, the smallest range possible being 2 ft/sec for full scale or 1 ft/sec with the sacrifice of some accuracy. It is limited to the measurement of fluids having a conductivity in excess of 0.1 siemens/m. Of great importance is the fact that the readings are unaffected by variations in viscosity, density, temperature, pressure, or conductivity. The electrodes must, of course, be kept clean, and this can present a problem in the measurement of sewage, in which grease, a poor electrical conductor, may be present. (For certain designs for measuring a very small amount of flow, the conductivity of the liquid may have to be as high as 0.5 siemens/m.)

EXAMPLE 11-6 Given an inductive input transducer (Fig. 11-9) for flow measurement, the diameter of the conduit is 0.0125 m. If a flow velocity of 10 m/sec produces a magnetic field strength of 0.2 T, find the electromotive force generated.

Solution

$$E = Blv$$

$$= (0.2\ \mathrm{W/m^2})(0.0125\ \mathrm{m})(10\ \mathrm{m/sec})$$

$$= 0.025\ \mathrm{V}$$

11-11 VARIABLE INDUCTANCE TRANSDUCERS

Passive inductive transducers require an external source of power. The action of the transducer is principally one of modulating the excitation signal.

The differential transformer is a passive inductive transformer. It is also known as the linear variable differential transformer (LVDT) and is shown constructively in Fig. 11-10*a*. It consists basically of a primary winding and two secondary windings, wound over a hollow tube and positioned so that the primary is between two secondaries.

An iron core slides within the tube and therefore affects the magnetic coupling between the primary and the two secondaries. When the core is in the center, the voltage induced in the two secondaries is equal. When the core is moved in one direction from center, the voltage induced in one winding is increased and that in the other is decreased. Movement in the opposite direction reverses this effect.

In the schematic diagram shown in Fig. 11-10*b*, the windings are connected "series opposing." That is, the polarities of V_1 and V_2 oppose each other as we trace through the circuit from terminal A to terminal B. Consequently, when the core is in the center so that $V_1 = V_2$, there is no voltage output, $V_o = 0$.

When the core is away from center toward S_1, V_1 is greater than V_2 and the

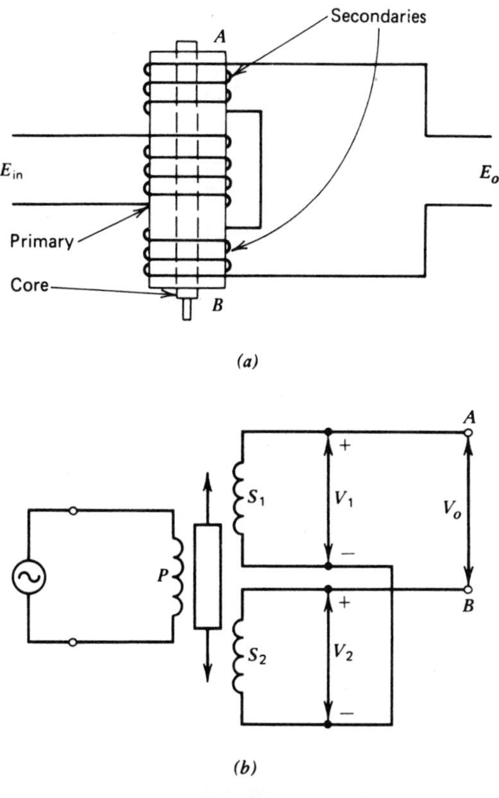

FIGURE 11-10 The linear variable differential transformer. (*a*) Construction. (*b*) Schematic diagram.

output voltage V_o will have the polarity of V_1. When the core is away from center toward S_2, V_2 is greater than V_1 and the output will have the polarity of V_2. That is, the output ac voltage *inverts* as the core passes the center position. The farther the core moves from center, the greater the *difference* in value between V_1 and V_2, and consequently the greater the value of V_o. Thus, the amplitude of V_o is a function of the distance the core has moved, and the *polarity* or *phase* indicates which direction it has moved (Fig. 11-11). If the core is attached to a moving object, the LVDT output voltage can be a measure of the position of the object.

One advantage of the LVDT over the inductive bridge-type transducer is that it produces a higher output voltage for small changes in core position. Several commercial models that produce 50 to 300 mV/mm are available. This means that a 1-mm displacement of the core can produce a voltage output of 300 mV.

LVDTs are available with ranges from as low as ± 0.05 in. to as high as ± 25 in., and they are sensitive enough to be used to measure displacement of well below 0.001 in. They can be obtained for operation at temperatures as

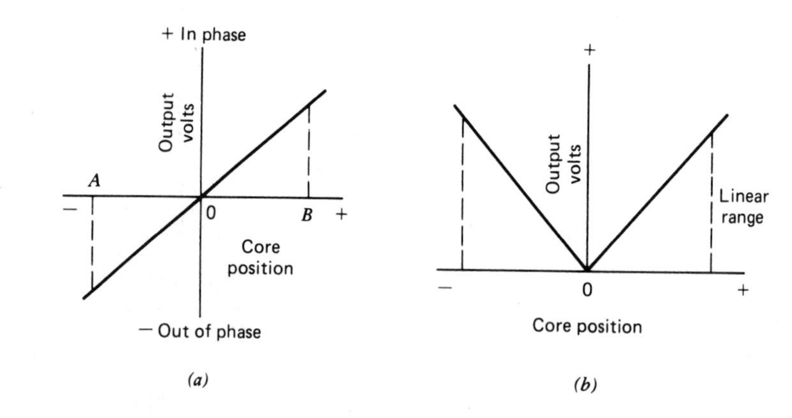

FIGURE 11-11 Output voltage. (*a*) Phase relationship. (*b*) Absolute magnitude.

low as −265°C and as high as +600°C, and they are also available in radiation-resistance designs for operation in nuclear reactors.

Typical applications that illustrate the capabilities of LVDTs include controls for jet engines in close proximity to exhaust gases and measuring roll positions and the thickness of materials in hot-stripe or hot-slab steel mills.

EXAMPLE 11-7

An ac LVDT has the following data: input 6.3 V, output 5.2 V, range ±0.50 in. Determine

(a) The plot of output voltage versus core position for a core movement going from +0.45 to −0.03 in.

(b) The output voltage when the core is −0.25 in. from center.

Solution

(a) A core displacement of 0.5 in. produces 5.2 V. Therefore, a 0.45-in.

core movement produces $\dfrac{(0.45)(5.2)}{0.5} = 4.68$ V. Similarly, a −0.30-in.

core movement produces $\dfrac{(-0.3)(-5.2)}{-0.5} = -3.12$ V.

(b) A core movement of −0.25 in. produces $\dfrac{(-0.25)(-5.2)}{-0.5} = -2.6$ V.

11-12 PIEZOELECTRIC TRANSDUCERS

When a mechanical pressure is applied to a crystal of the **Rochelle salt**, **quartz**, or **tourmaline** type, a displacement of the crystals causes a potential difference to occur. This property is used in piezoelectric transducers; in these tranducers a crystal is placed between a solid base and force-summing member, as shown in Fig. 11-12. Externally applied forces exert pressure to

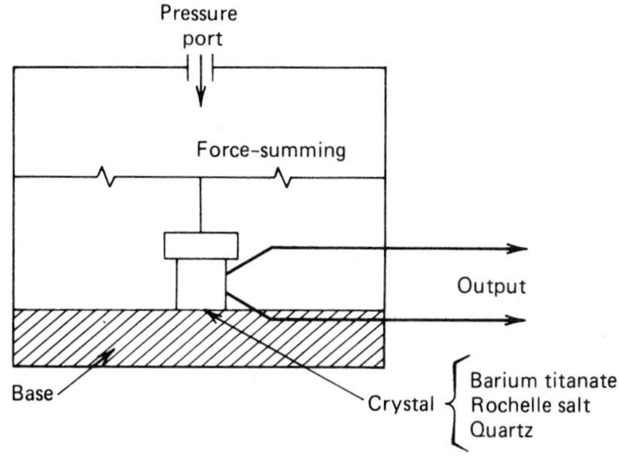

FIGURE 11-12 Elements of a piezoelectric transducer.

the top of the crystal. This produces an electromotive force across the crystal proportional to the magnitude of the applied pressure.

For a piezoelectric element under pressure, part of the energy will be converted to an electric potential that will appear on opposite faces of the element, analogous to the charge on the plates of a capacitor. The rest of the applied energy is converted to mechanical energy, analogous to that of a compressed spring. When the pressure is removed, the piezoelectric element will return to its original shape and also lose its electric charge. From these relationships the following formulas have been derived for the coupling coefficient k.

$$k = \frac{\text{Mechanical energy converted to electrical energy}}{\text{Applied mechanical energy}} \qquad (11\text{-}10)$$

$$k = \frac{\text{Electrical energy converted to mechanical energy}}{\text{Input electrical energy}} \qquad (11\text{-}11)$$

An alternating voltage applied to a crystal causes it to vibrate at its natural resonance frequency. Since the frequency is a very stable quantity, piezoelectric crystals are used principally in high-frequency accelerometers. The output voltage is typically on the order of 1 to 30 mV per gram of acceleration. The device needs no external power source and is therefore self-generating. The principal disadvantage of this transducer is that voltage will be generated only as long as the pressure applied to the piezoelectric element is changing.

EXAMPLE 11-8

A certain crystal has a coupling coefficient of 0.32. How much electrical energy must be applied to produce an output of 1 in.-oz of mechanical energy?

Solution

$$1 \text{ in.-oz} = 1 \text{ in.-oz} \times \frac{1 \text{ ft}}{12 \text{ in.}} \times \frac{1 \text{ lb}}{16 \text{ oz}} \times \frac{1.356 \text{ J}}{1 \text{ ft-lb}}$$

$$= 7.06 \times 10^{-3} \text{ J}$$

$$\text{Electrical energy} = \frac{\text{Electrical energy converted to mechanical energy}}{k}$$

$$= \frac{7.1 \times 10^{-3} \text{ J}}{0.32} = 22.19 \text{ mJ}$$

11-13 TEMPERATURE TRANSDUCERS

Temperature transducers can be divided into four main categories.

1. Resistance temperature detectors (RTD).
2. Thermocouples.
3. Thermistors.
4. Ultrasonic transducers.

11-14 RESISTANCE TEMPERATURE DETECTORS

Detectors of resistance temperatures commonly employ platinum, nickel, or resistance wire elements, whose resistance variation with temperature has a high intrinsic accuracy. They are available in many configurations and sizes and as shielded or open units for both immersion and surface applications. The relationship between temperature and resistance of conductors can be calculated from the equation

$$R = R_0(1 + \alpha\,\Delta T) \qquad (11\text{-}12)$$

where

R = the resistance of the conductor at temperature t (°C)
R_0 = the resistance at the reference temperature, usually 20°C
α = the temperature coefficient of resistance
ΔT = the difference between the operating and the reference temperature

EXAMPLE 11-9

A platinum resistance thermometer has a resistance of 150 Ω at 20°C. Calculate its resistance at 50°C ($\alpha_{20} = 0.00392$).

Solution

$$R = R_0(1 + \alpha\,\Delta T)$$

$$= 150\,\Omega[1 + 0.00392(50 - 20)\,°\text{C}]$$

$$= 167.64\,\Omega$$

11-15 THERMOCOUPLES

One of the most commonly used methods of measuring temperature in science and industry depends on the **thermocouple** effect. When a pair of wires made of different metals are joined together at one end, a temperature *difference* between this end and the other end of the wires produces a voltage between the wires (Fig. 11-13). The magnitude of this voltage depends on the materials used for the wires and the amount of temperature difference between the joined ends and the other ends.

The junction of the two wires of the thermocouple is called the *sensing* junction. In normal use this junction is placed in or on the material being tested, and the other ends of the wire are connected to the voltage-measuring equipment.

Since the temperature difference between this sensing junction and the other ends is the critical factor, the other ends are either kept at a constant *reference* temperature or, when the cost of the equipment is very low, simply maintained at room temperature. When the other ends are kept at room temperature, the temperature is monitored and the thermocouple output voltage readings are corrected for any changes in room temperature. Because the temperature at this end of the thermocouple wires is a reference temperature, the junction here with the equipment terminals or with other connecting wires is known as the *reference* junction. It is also quite often referred to as the *cold* junction. Because the thermocouple is frequently used for measuring high temperatures, the reference junction in such cases is indeed the colder of the two junctions.

Since any junction of dissimilar metals will produce some thermocouple voltage, the wires and any metal terminals between the sensing junction and the rest of the equipment must be carefully controlled. Usually this means that the wires between the sensing junction and the reference junction are of specific materials provided by the thermocouple supplier, and the wires from the reference junction to the measuring equipment are copper.

Thermocouples are made from a number of difference metals or metal alloys covering a wide range of temperatures from as low as −270°C(−418°F)

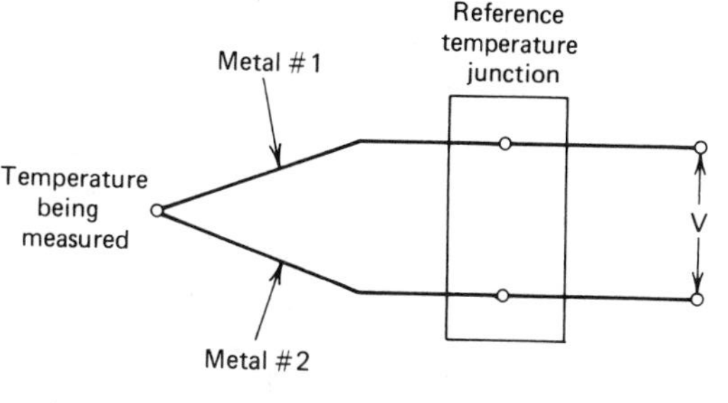

FIGURE 11-13 Schematic representation of a thermocouple assembly.

to as high as 2700°C (about 5000°F). They may be obtained in simple uninsulated wire form, in insulated wire form, or within protective sheaths or probes. Sheath diameters as small as 0.25 mm are available. For applications in which small size is not essential and more physical protection is needed, some suppliers manufacture thermocouple *wells*. These are metal castings with threaded bushings, intended to be mounted so that they extend into the container (pipe, tank, etc.) that holds the material whose temperature is to be monitored. Figure 11-14 shows several typical thermocouple assemblies.

The magnitude of the thermal emf depends on the wire materials used and on the temperature difference between the junctions. Figure 11-15 shows the thermal emfs for some common thermocouple materials. The values shown are based on a reference temperature of 32°F. The effective emf of the thermocouple is given as

$$E = c(T_1 - T_2) + k(T_1^2 - T_2^2) \qquad (11\text{-}13)$$

where

c and k = constants of the thermocouple materials
T_1 = the temperature of the "*hot*" junction
T_2 = the temperature of the "cold" or "reference" junction

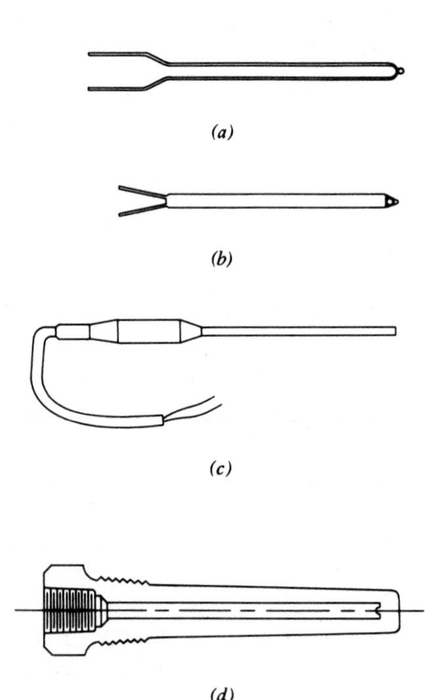

FIGURE 11-14 Thermocouples and thermocouple assemblies. (*a*) Uninsulated thermocouple. (*b*) Insulated thermocouple. (*c*) Probe assembly. (*d*) Thermocouple well.

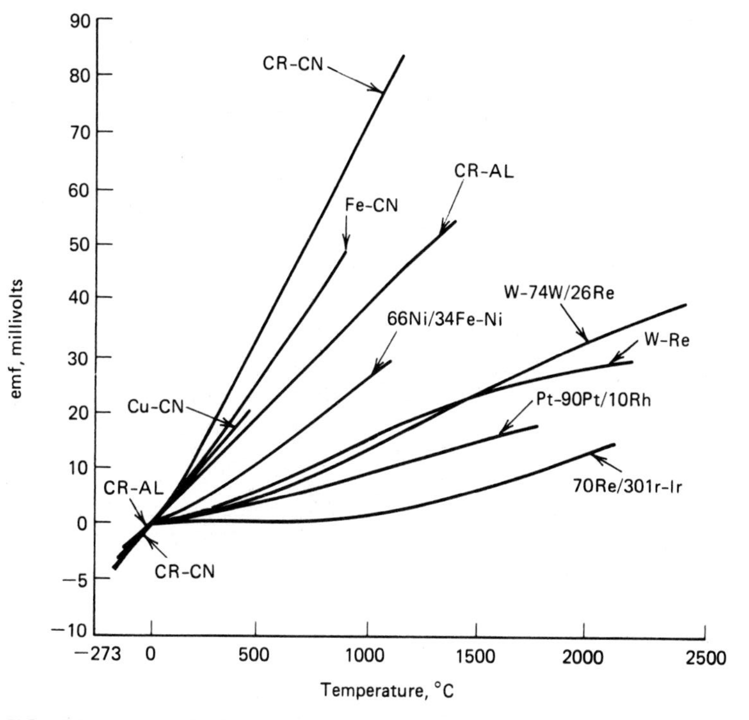

FIGURE 11-15 Calibration curves for several thermocouple combinations.

EXAMPLE 11-10

During experiments with a copper–constantan thermocouple it was found that $c = 3.75 \times 10^{-2}$ mV/°C and $k = 4.50 \times 10^{-5}$ mV/°C². If $T_1 = 100$°C and the cold junction T_2 is kept in ice, compute the resultant electromotive force.

Solution

$$E = c(T_1 - T_2) + k(T_1^2 - T_2^2)$$

$$= 3.75 \times 10^{-2}\,\frac{\text{mV}}{°\text{C}}\,(100°\text{C} - 0°\text{C})$$

$$+ 4.50 \times 10^{-5}\,\frac{\text{mV}}{°\text{C}^2}\,(100^2 - 0^2)°\text{C}^2$$

$$= 3.75\text{ mV} + 0.45\text{ mV} = 4.20\text{ mV}$$

The temperature ranges covered by thermocouples make them appropriate for use in industrial furnaces as well as for measurements in the **cryogenic** range. Cryogenic temperatures are very low, within a few degrees of absolute zero, much below those prevailing in industry, in scientific research, and for medical instrumentation. In engineering, thermocouples are used to monitor the operating temperatures of electrical and mechanical equipment. In industrial processes they monitor the temperatures of liquids and gases in storage and flowing in pipes and ducts. In medical work, the extremely small-size probes that are possible permit their use in measuring internal body temperatures.

11-16 THERMISTORS

The electrical resistance of most materials changes with the temperature. By choosing materials that are very sensitive to temperature, we can make devices that are useful in temperature control circuits as well as in temperature measurement. A **thermistor** is a semiconductor made by sintering mixtures of metallic oxide, such as oxides of manganese, nickel, cobalt, copper, and uranium.

Thermistors have a *negative* temperature coefficient. That is, their resistance decreases as their temperature rises. Resistance at 25°C for typical commercial units ranges from the vicinity of 100 **Ω** to over 10 M**Ω**. A graph showing resistance versus temperature for a family of thermistors is given in Fig. 11-16. The resistance value marked at the bottom end of each curve is the value at 25°C. In addition to the choice of resistance values, choices of

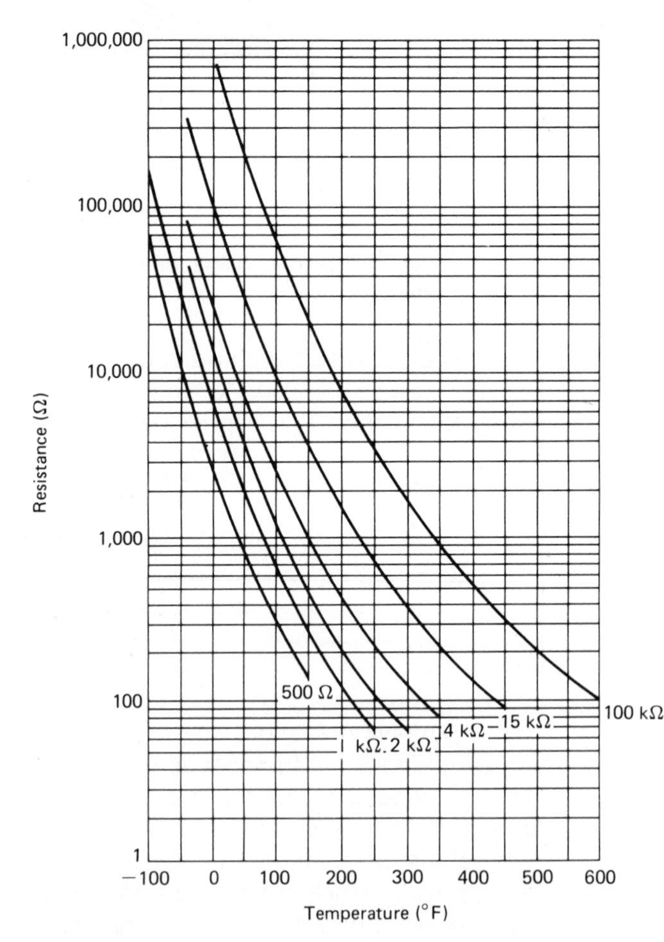

FIGURE 11-16 Typical thermistor resistance-versus-temperature curves. (Courtesy Fenwal Electronics, Framingham, Mass.)

power rating, physical size and shape, resistance tolerance, and thermal time constant are also available.

The smallest thermistors are made in the form of beads. Some are as small as 0.15 mm (about 0.006 in.) in diameter. These may come glass-coated or sealed in the tips of solid-glass probes. The probes are used in measuring the temperatures of liquids. Where greater power-dissipating ability is needed, thermistors may be obtained in disk, washer, or rod form.

Thermistors can be connected in series–parallel arrangements for applications requiring greater power-handling capability. High-resistance units find application in measurements that employ wires or cables with small quantities of lead. Thermistors are chemically stable and can be used in nuclear environments. Their wide range of characteristics also permits them to be used in limiting and regulation circuits, as time delays, for the integration of power pulses, and as memory units.

Typical thermistor configurations are shown in Fig. 11-17, and the electrical symbol of the device is depicted in the same figure.

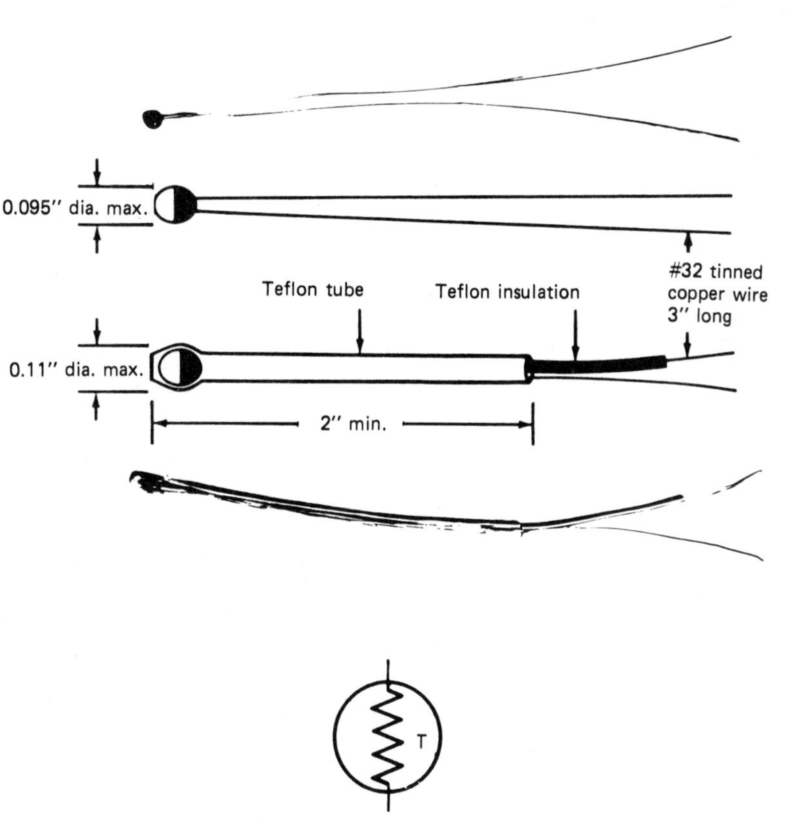

FIGURE 11-17 Thermistor configuration and the electrical symbol for a thermistor. (Courtesy Yellow Springs Instrument Company, Yellow Springs, Ohio.)

A thermistor in one leg of a Wheatstone bridge circuit will provide precise temperature information. In most applications accuracy is limited only by the readout device.

Thermistors are nonlinear over a temperature range, although units today are available with a better than 0.2% linearity over a temperature range of 0°C to 100°C. The typical sensitivity of a thermistor is approximately 3 mV/°C at 200°C.

EXAMPLE 11-11 The circuit of Fig. 11-18 is to be used for temperature measurement. The thermistor is a 4-kΩ type identified in Fig. 11-16. The meter is a 50-mA ammeter with a resistance of 3 Ω, R_c is set to 17 Ω, and the supply voltage V_T is 15 V. What will the meter readings at 77°F and at 150°F be?

Solution The graph for the 4-kΩ thermistor in Fig. 11-16 shows that its resistance at 77°F is 4 kΩ. Therefore, the current at 77°F is

$$I = \frac{V_T}{R_T} = \frac{15\text{ V}}{4000\ \Omega + 17\ \Omega + 3\ \Omega} = 3.73\text{ mA}$$

At 150°F the graph shows the thermistor resistance to be 950 Ω. The meter reading at this temperature, therefore, should be

$$I' = \frac{V_T}{R'_T} = \frac{15\text{ V}}{950\ \Omega + 17\ \Omega + 3\ \Omega} = 15.5\text{ mA}$$

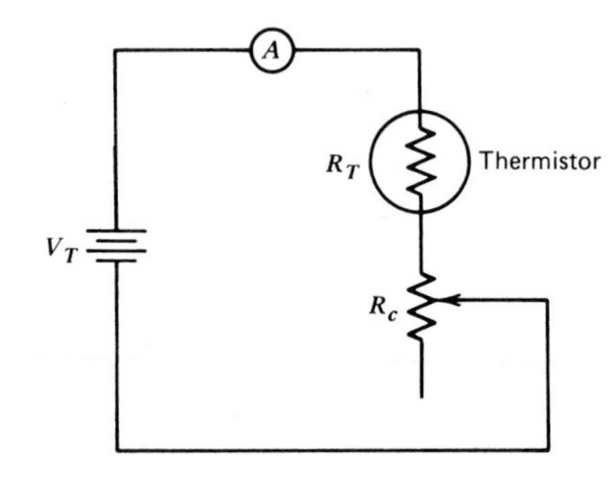

FIGURE 11-18 Basic thermistor circuit for measuring.

11-17 ULTRASONIC TEMPERATURE TRANSDUCERS

Ultrasonics, which are sound vibrations above 20,000 Hz, can be useful when we are concerned with rapid temperature fluctuations, temperature extremes, limited access, nuclear, and other severe environmental conditions and when we must measure the temperature distribution inside solid bodies. The need to measure simultaneously the distribution of parameters other than temperature (e.g., flow) may also justify an ultrasonic approach. Ultrasonics also offers possibilities of *remote* sensing and sometimes can prevent penetration of the system (nonintrusive).

Apart from profiling, ultrasonic thermometer sensors permit the measurement of an extremely wide range of temperatures, from cryogenic to **plasma** levels, response times in microseconds to milliseconds, millidegree resolution,

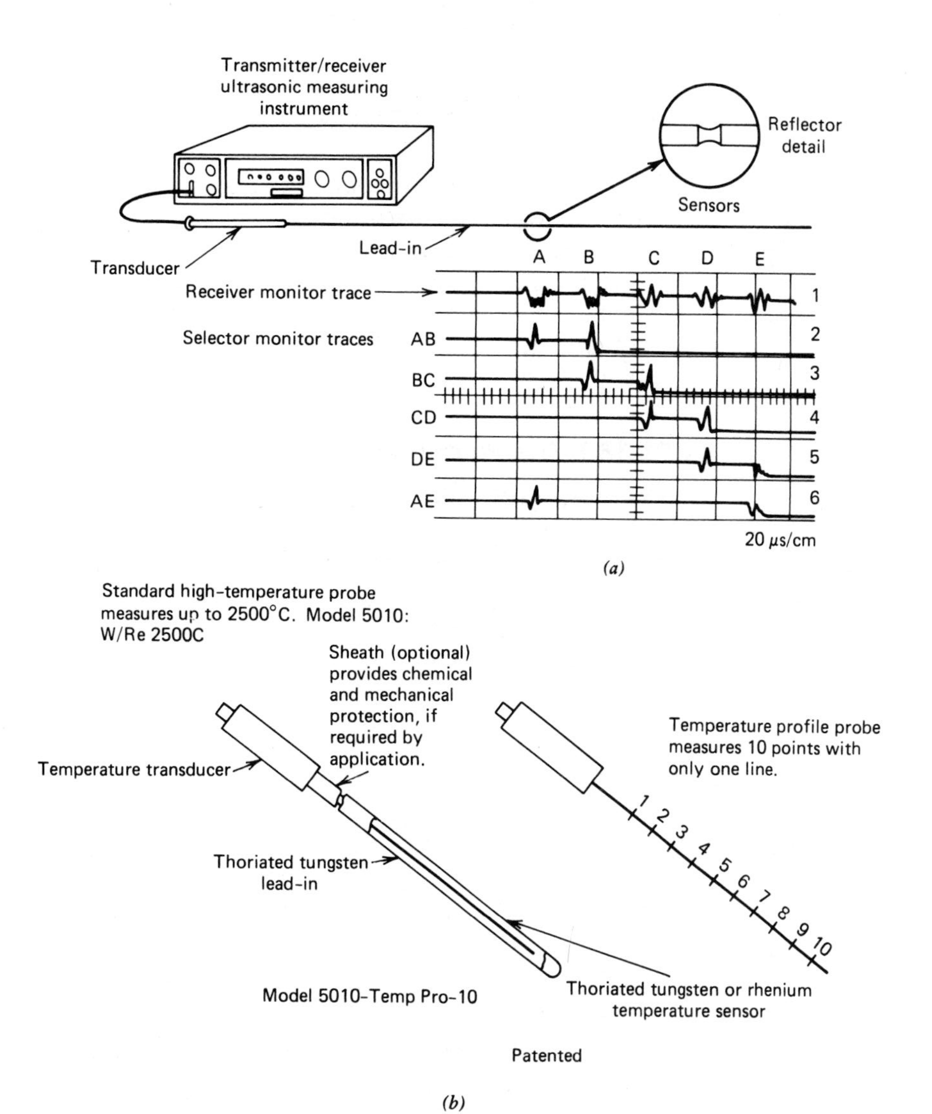

FIGURE 11-19 **(*a*) Schematic and oscillogram illustrating ultrasonic temperature profiling. (*b*) Ultrasonic thermometer. (Courtesy Parametics, Waltham, Mass.)**

a greater choice of materials for sensors, operation in extreme nuclear or corrosive environments, an averaging capability over a defined path, and the remote location of transducer and electronics. Naturally, not all these features are available simultaneously.

With regard to profiling, ultrasonics permits us to obtain from two to ten or more temperatures with a single transmission line. This feature minimizes the perturbation of the region in question, simplifies installation, and provides reliable, accurate data at a reasonable price per point. A schematic and oscillogram illustrating ultrasonic temperature profiling and a thermometer fabricated by Panametrics of Waltham, Massachusetts, are shown in Fig. 11-19.

11-18 PHOTOELECTRIC TRANSDUCERS

A **photoelectric transducer** can be categorized as **photoemissive**, **photoconductive**, or **photovoltaic**. In photoemissive devices, radiation falling on a cathode causes electrons to be emitted from the cathode surface. In photoconductive devices, the resistance of a material is changed when it is illuminated. Photovoltaic cells generate an output voltage proportional to radiation intensity. The incident radiation may be infrared, ultraviolet, gamma rays, or X rays as well as visible light.

11-18.1 The Photomultiplier Tube

The **photomultiplier tube** consists of an evacuated glass envelope containing a photocathode, an anode, and several additional electrodes called *dynodes*, each at a higher voltage. Figure 11-20 illustrates the principle of the photomultiplier. Electrons emitted by the cathode are attracted to the first anode. Here a phenomenon known as *secondary emission* takes place. When electrons moving at a high velocity strike an appropriate material, the material

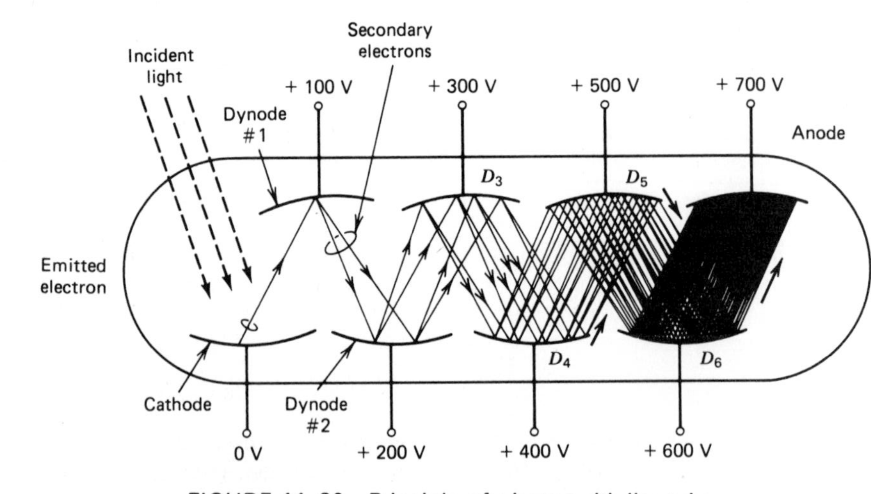

FIGURE 11-20 Principle of photomultiplier tube.

emits a greater number of electrons than it was struck with. In this device the high velocity is achieved by using a high voltage between the first anode and the cathode. The electrons emitted by the first anode are then attracted to the second anode, where the same thing takes place again. Each anode is at a higher voltage, in order to achieve the requisite electron velocity each time. Thus, secondary emission and a resulting "electron multiplication" occur at each step, with an overall increase in electron flow that may be very great. Amplification of the original current by as much as 10^5 to 10^9 is common. Luminous sensitivities range from 1 A per **lumen** or less, to over 2000 A per lumen. Typical anode current ratings are 100 μA minimum to 1 mA maximum. The extreme luminous sensitivity possible with these devices is illustrated by the fact that with a sensitivity of 100 A per lumen, only 10^{-5} lumen is needed to produce a 1-mA output current.

Magnetic fields affect the gain of the photomultiplier because some electrons may be deflected from their normal path between stages and therefore never reach a dynode or, eventually, the anode. In scintillation-counting applications this effect may be disturbing, and mu-metal magnetic shields are often placed around the photomultiplier tube.

11-18.2 Photoconductive Cells or Photocells

Another photoelectric effect that has proved very useful is the photoconductive effect, which is used in photoconductive cells or **photocells**. In this type of device, the electrical resistance of the material varies with the amount of light striking it.

A typical form of construction is shown in Fig. 11-21*a*. The photoconductive material, typically cadmium sulfide, cadmium selenide, or cadmium sulfoselenide, is deposited in a zigzag pattern, to obtain a desired resistance

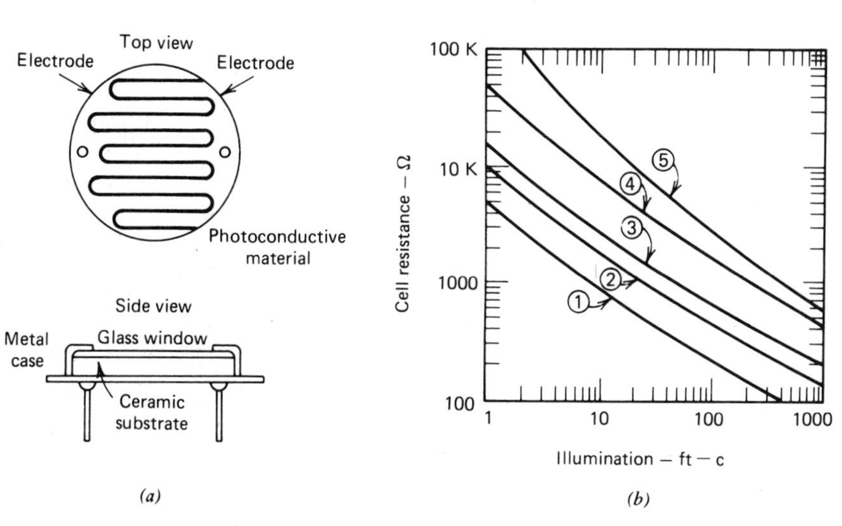

FIGURE 11-21 The photoconductive cell. (*a*) Construction. (*b*) Typical curves of resistance versus illumination.

value and power rating. The material separates two metal-coated areas acting as electrodes, all on an insulating base such as ceramic. The assembly is enclosed in a metal case with a glass window over the photoconductive material. Photocells of this type are made in a range of sizes, having diameters of one-eighth inch to over one inch. The small sizes are suitable where space is critical, for example, in equipment for reading punched cards and similar applications. However, the very small units have very low power dissipation ratings.

A typical control circuit utilizing a photoconductive cell is illustrated in Fig. 11-22. The potentiometer is used to make adjustments to compensate for manufacturing tolerances in photocell sensitivity and relay-operating sensitivity. When the photocell has the appropriate light shining on it, its resistance will be low and the current through the relay will consequently be high enough to operate the relay. When the light is interrupted, the resistance will rise, causing the relay current to decrease enough to deenergize the relay.

EXAMPLE 11-12

The relay of Fig. 11-23*a* is to be controlled by a photoconductive cell with the characteristics shown in Fig. 11-23*b*. The circuit delivers 10 mA at a 30-V setting when the cell is illuminated with about 400 lm/m^2. The circuit becomes deenergized when the cell is dark. Calculate

(a) The required series resistance.

(b) The level of the dark current.

Solution

(a) The cell's resistance at 400 lm/m^2 $\approx$ 1 kΩ

$$I = \frac{30\text{ V}}{R_1 + R_{\text{cell}}}$$

$$R_1 = \frac{30\text{ V}}{I} - R_{\text{cell}}$$

$$= \frac{30\text{ V}}{10\text{ mA}} - 1\text{ k}\Omega = 2\text{ k}\Omega$$

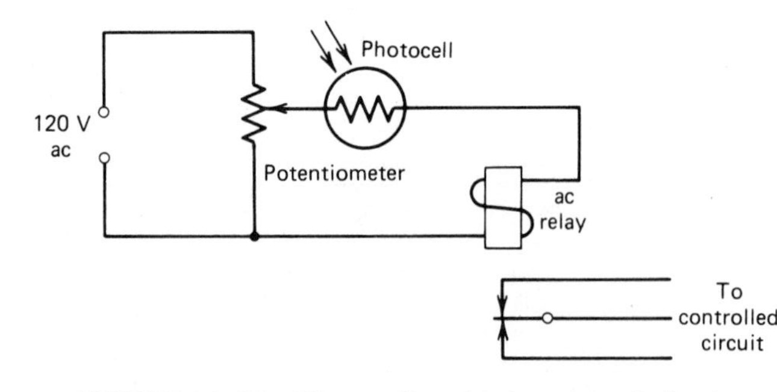

FIGURE 11-22 Photocell and relay control circuit.

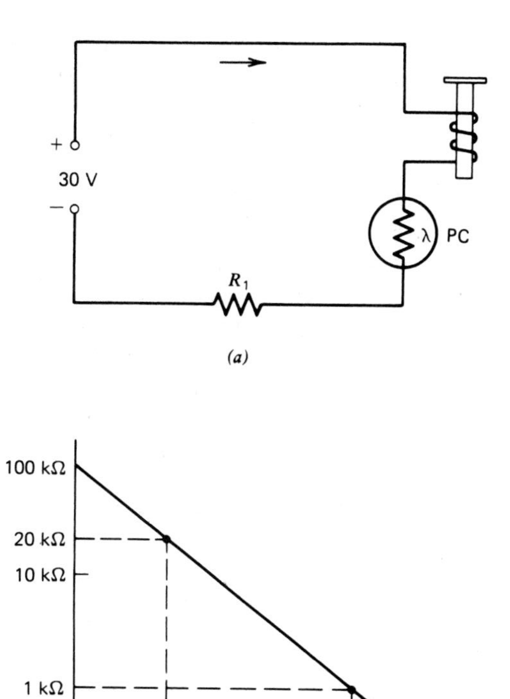

FIGURE 11-23 **(*a*) Relay control by a photoconductive (PC) cell and (*b*) PC cell illumination characteristics.**

(b) The cell's dark resistance $\approx 100\ \text{k}\Omega$

$$\text{Dark current} \approx \frac{30\ \text{V}}{2 \times 10^3\ \Omega + 100 \times 10^3\ \Omega} = 0.3\ \text{mA}$$

Counting objects, as on the converyor belt of a production line, would be a typical use for this kind of circuit. A light source is focused in a beam passing across the path traversed by the objects to be counted and shining on the photocell. Each time an object passes, it interrupts the beam and the relay actuates a counter. The same techniques may be used for sorting objects, if there is sufficient difference in the height of the objects. With the focused light beam directed at a height at which the larger objects intercept the beam, the photocell output can be used to actuate a gate or turnstile to direct the larger objects into a different path, or to perform some other desired function.

11-18.3 The Photovoltaic Cell

The **photovoltaic cell**, or "**solar cell**," as it is sometimes called, will produce an electrical current when connected to a load. Both silicon (Si) and selenium (Se) types are known.

Photovoltaic cells may be used in a number of applications. Multiple-unit silicon photovoltaic devices may be used for sensing light as a means of reading punched cards in the data processing industry. Gold-doped germanium cells with controlled spectral responses act as photovoltaic devices in the infrared region of the spectrum and may be used as infrared detectors.

11-19 THE SEMICONDUCTOR PHOTODIODE

A reverse-biased semiconductor diode passes only a very small leakage current (a fraction of 1 μA in typical small silicon diodes) if the junction is not exposed to light. Under illumination, however, the current rises almost in

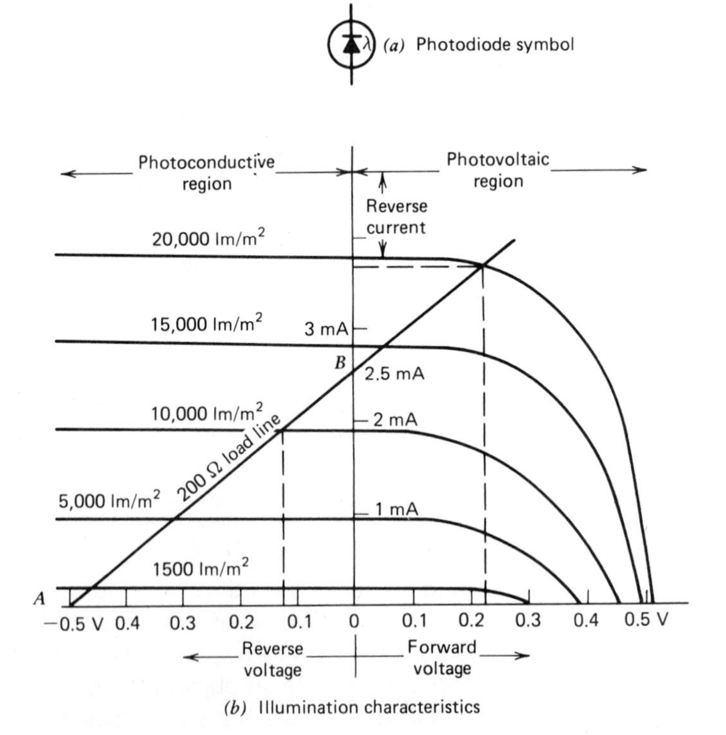

FIGURE 11-24 The symbol and typical illumination characteristics for the silicon photodiode.

direct proportion to the light intensity. Thus, the **photodiode** can be used in applications similar to those of the photoconductive cell.

When the device operates with a reverse voltage applied, it functions as a photoconductive device. When operating without the reverse voltage, it functions as a photovoltaic device. It is also possible to arrange for a photodiode to change from the photoconductive mode to the photovoltaic mode. However, the photodiode has a very important advantage over the photoconductive cell. Its response time is much faster, so that it may be used in applications in which the light fluctuations occur at quite high frequencies. The photoconductive cell is useful only at very low frequencies.

Figure 11-24 shows the symbol and typical illumination characteristics for the silicon photodiode.

EXAMPLE 11-13

A photodiode with the illumination characteristics shown in Fig. 11-24 is connected in series with a 200-**Ω** resistance and a 0.5-V supply. The supply polarity reverse-biases the device. Draw the dc load line for the circuit and determine the diode currents and voltages at 1500, 10,000, and 20,000 lm/m² illumination.

Solution

The circuit is shown in Fig. 11-25. When $I_D = 0$, $V_{R1} = I_D R_1 = 0$ V. Plot point A on Fig. 11-25 at $I_D = 0$ and $V_D = -0.5$ V. When $V_D = 0$, $V_{R1} = E_S$,

$$I_D = \frac{V_{R1}}{R_1} = \frac{-0.5\ \text{V}}{200\ \Omega} = -2.5\ \text{mA}$$

Plot point B at $I_D = -2.5$ mA and $V_D = 0$ V. Draw the load line through A and B. From the load line we get

At 1500 lm/m², $I_D = -0.2$ mA and $V_D = -\ 0.45$ V

At 10,000 lm/m², $I_D = 1.9$ mA and $V_D = -0.12$ V

At 20,000 lm/m², $I_D = -3.6$ mA and $V_S = 0.22\ V$

Note that the polarity of V_D changes from negative to positive at the highest level of illumination.

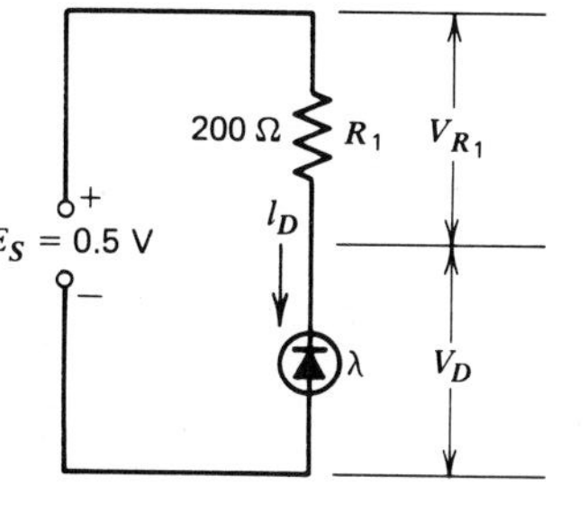

FIGURE 11-25 Photodiode with load resistance.

11-20 THE PHOTOTRANSISTOR

The sensitivity of the photodiode can be increased as much as 100 times by adding a junction—which makes it an *n-p-n* device. A simple representation of the construction is shown in Fig. 11-26, which also represents several of the circuit symbols that have been used for this device. Illumination of the central region causes the release of electron–hole pairs here. This lowers the barrier potential across both junctions, causing an increase in the flow of electrons from the left-hand region into the center region, and on to the right-hand region.

For a given amount of illumination on a very small area, the **phototransistor** provides a much larger output current than that available from a photodiode. That is, the phototransistor is the more sensitive of the two.

The circuit symbol and typical output characteristics of the phototransistor are shown in Fig. 11-27. Arrays of phototransistors and low-current photodiodes are widely used as photodetectors for such applications as punched card and tap readouts. Although phototransistors have the advantage of being more sensitive than photodiodes, the photodiode has faster switching times.

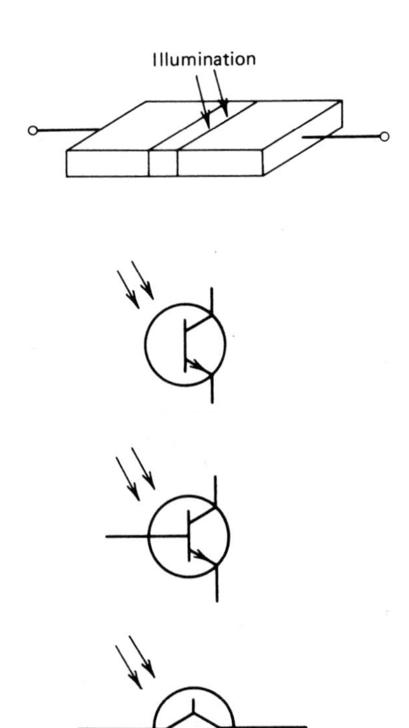

FIGURE 11-26 The phototransistor.

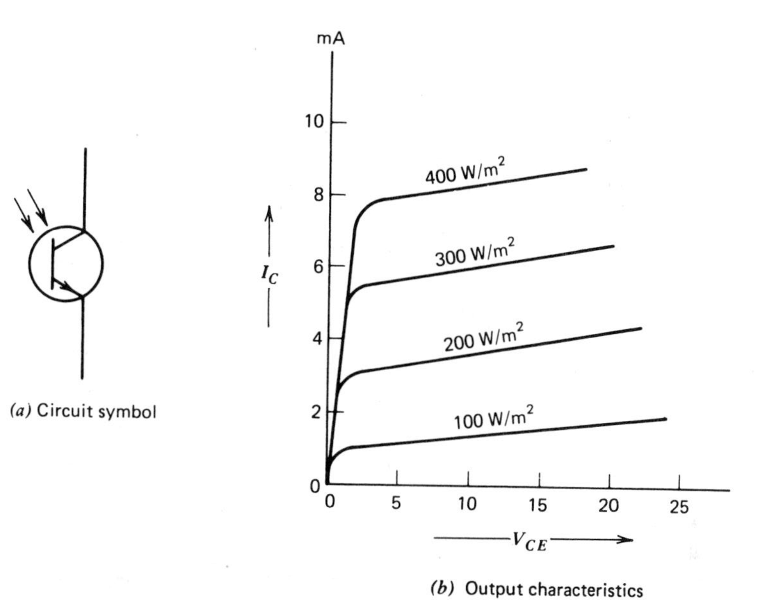

(*a*) Circuit symbol

(*b*) Output characteristics

FIGURE 11-27 The circuit symbol and output characteristics for the phototransistor.

EXAMPLE 11-14

A phototransistor with the characteristics shown in Fig. 11-27 has a supply voltage of 20 V and a collector load resistance of 2 k**Ω**. Determine the output voltage when the illumination level is

(a) Zero.

(b) 200 W/m^2.

(c) 400 W/m^2.

Solution

The circuit and dc load line are shown in Fig. 11-28. The load line is drawn in the usual way. From the intersection of the load line and the characteristic curve, we obtain, when the illumination level is zero,

$$\text{Output voltage } (V_{CE}) = 20 \text{ V}$$

At an illumination level of 200 W/m^2,

$$\text{Output voltage} \approx 12.5 \text{ V}$$

And an illumination level of 400 W/m^2,

$$\text{Output voltage} \approx 4 \text{ V}$$

One circuit utilizing phototransistors is shown in Fig. 11-29. The light incident on the phototransistor causes its current to increase and thus increases the *IR* drop across the 50-K**Ω** resistor and the input to the transistor driving the relay. This raises the current to the operational value.

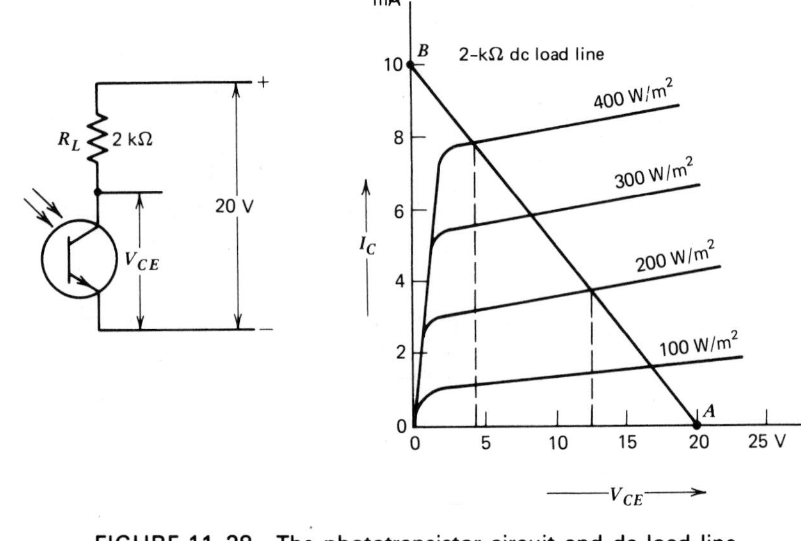

FIGURE 11-28 The phototransistor circuit and dc load line.

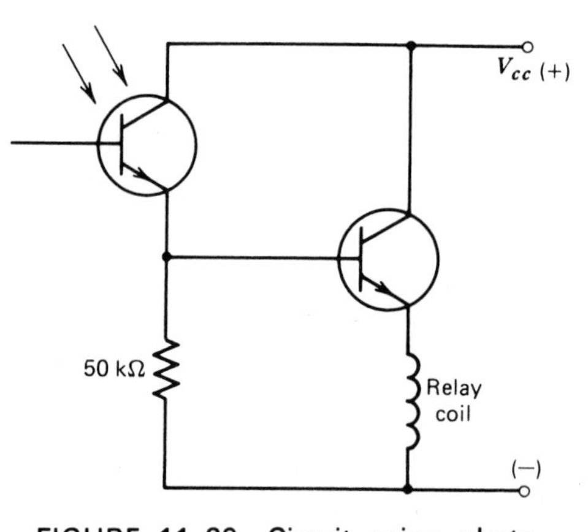

FIGURE 11-29 Circuit using phototransistors.

11-21 SUMMARY

We have examined various transducers with emphasis on the measurements that can be made with them. Some theory has been introduced. Each of the transducers can be examined in an experimental lab. Without transducers, electronics could not have attained the important role it plays in real-life situations. Our electronic instruments would be nothing more than laboratory or production phenomena. With transducers we can use electronic instrumentation to measure, modify, and improve the technological world we have created.

Integrated temperature transducers are now available with output operational amplifiers. Such transducers have a sensitivity per degree centrigrade of 10 to 200 mV and are very linear. They are available from the National Semiconductor Corporation of Santa Clara, California.

Ion-sensitive field-effect transistors are now used as transducers in measuring bioelectronic potentials.

One further use of transducer systems is in the field of telemetry. Telemetering is transmitting the readings of measuring instruments to a remote location for recording. In this system, a transducer is fed to a conversion system which is modified for telephone transmission. At the receiving end the transmitted signal is converted by a conversion scheme to an analog or digital signal.

11-22 GLOSSARY

Cryogenic: Pertaining to temperatures that are very low, within a few degrees of absolute zero, much below those prevailing in industry, scientific research, and medical instrumentation.

Gauge factor: The measure of the amount of resistance change for a given strain.

Lumen: A light unit equal to the light emitted within a unit solid angle from a uniform point source with the intensity of one candle.

Measurand: The quantity, property, or condition that the transducer translates to any electrical signal.

Modulus of elasticity: Synonymous with Young's modulus. A constant that expresses the ratio of unit stress to unit deformation for all values within the proportional limit of the material.

Photocell: The term applied to both photoconductive and photovoltaic devices.

Photoconductive: Having a high resistance in the dark and a low resistance when exposed to light.

Photodiode (semiconductor): A semiconductor diode built with glass or clear plastic over its junction area, so that the diode current varies with the intensity of the light striking it.

Photoelectric transducer: A light-sensitive device that converts light energy into electric energy.

Photoemissive: Capable of emitting electrons when under the influence of light or other radiant energy.

Photomultiplier tube: A vacuum tube with a photosensitive cathode and an anode. With appropriate voltage applied between anode and cathode, light striking the cathode is converted to electrical current. The term refers only to photoemissive, vacuum-tube devices.

Phototransistor: A bipolar junction transistor built so that illumination of the base region causes the transistor current to vary with light intensity.

Photovoltaic cell: Capable of generating a voltage when exposed to visible or other light radiation.

Plasma: A mixture of charged-particle gases.

Rochelle salt, tourmaline, quartz: Natural materials that have piezoelectric properties.

Solar cell: See *Photovoltaic cell.*

Strain: The elastic deformation produced in a solid under stress.

Strain gauge: A measuring element for converting force, pressure, tension, and so on, into an electrical signal.

Stress The internal force per unit area.

Tachometer: A device for indicating speed of rotation.

Thermistor: A semiconductor device whose resistance decreases as the temperature increases.

Thermocouple: A pair of dissimilar conductors joined together so that an electromotive force is developed by the thermoelectric effects when the two junctions are at different temperatures.

Transducer: A device capable of converting one form of energy or signal to another.

Ultrasonics (supersonics): The study of sound in the frequency range above 20 kHz/sec.

11-23 REVIEW QUESTIONS

These questions test your knowledge of the subject matter. If you cannot answer any particular question, look through the text for the section discussing the subject.

1. State the function of a transducer.
2. Why are transducers important in electronic instrumentation?
3. List five physical characteristics (parameters) that transducers measure.
4. Describe a strain gauge.
5. List three types of temperature transducers and describe the uses of each.
6. Draw an electrical circuit using a thermistor.
7. Explain how to use a potentiometric transducer.
8. Describe the operation of a piezoelectric transducer.
9. Name the five different types of photoelectric transducers.
10. Give the difference between a self-generating and passive inductive transducer.

11-24 PROBLEMS

11-1 A resistive displacement transducer with a shaft stroke of 1 in. is applied to the circuit of Fig. 11-2*b*. The applied voltage $V_T = 10$ V. What is the displacement indicated by each of the voltage readings: 3.0 V; 5.0 V; 8.0 V?

11-2 A resistive displacement transducer with a shaft stroke of 5 in. is used in the circuit of Fig. 11-2*b*. The voltage supply and connections are the same as those in Problem 11-1. What displacements do the following voltages indicate: 2 V; 4 V; 6 V?

11-3 The circuit of Fig. 11-3 is being used, with $V_T = 5.0$ V. If the resistance of the potentiometer (i.e., $R_1 + R_2$) is 1000 **Ω** and the wiper is set so $R_2 = 300$ **Ω**, what is the value of V_o?

11-4 Using the same potentiometer as in Problem 11-3, but with $V_T = 3.5$ V, what is the value of V_o when the potentiometer is set so $R_2 = 200$ **Ω**?

11-5 A displacement transducer with a shaft stroke of 2.0 in. is used in the circuit of Fig. 11-3. The total resistance $(R_1 + R_2)$ is 400 **Ω** and $V_T = 4.0$ V. The wiper is 1.5 in. from *B*. (a) What is the value of R_2? (b) What is the value of V_o?

11-6 With the same transducer and circuit as in Problem 11-5, what will V_o be when the wiper is 0.7 in. from *B*?

11-7 A strain gauge having a gauge factor of 4 is used in testing a machine. If the gauge resistance is 100 **Ω**, and the strain of the gauge is 2×10^{-5}, how much will the resistance of the strain gauge change?

11-8 A strain gauge has a gauge factor of 4. If the strain gauge is attached to a metal bar that stretches from 10 to 10.2 in., what is the percentage change in resistance? If the unstrained value of resistance is 120 **Ω**, what is the value after strain?

11-9 A strain gauge has a resistance of 120 **Ω** unstrained and a gauge factor of −12. What is the resistance value if the strain is 1%?

11-10 A strain gauge with a resistance of 250 **Ω** undergoes a change of 0.150 **Ω** during a test. The strain is 1.5×10^{-4}. What is the gauge factor?

11-11 A strain gauge with a gauge factor of 4 has a resistance of 500 **Ω**. It is to be used in a test in which the strain to be measured may be as low as 5×10^{-6}. What will the change in gauge resistance be?

11-12 Calculate the gauge factor if a 1.5-mm-diameter conductor that is 24 mm long changes length by 1 mm and diameter by 0.02 mm under a compression force. $\left[\textit{Hint}: K = 1 + 2\left(\frac{\Delta d/d}{\Delta L/L}\right) \right]$

11-13 A parallel-plate capacitor has plates that are 4×10^{-3} m^2 in area. The distance between plates is 5×10^{-4} m. Calculate the capacitance if the dielectric is

(a) Ceramic ($k = 1000$)

(b) Oil ($k = 4$).

11-14 Given a parallel-plate capacitor, what is the relationship on capacitance:

(a) If the plate area is doubled?

(b) If the distance between plates is decreased by one-half?

(c) If the dielectric could be changed to one with a lower dielectric constant?

11-15 We are given an inductive transducer (Fig. 11-9) for measuring flow. The diameter of the conduit is 0.02 m. If a flow velocity of 5.0 m/sec produces a magnetic field strength of 0.1 T, what electromotive force is generated?

11-16 We are given an inductive transducer (Fig. 11-9). The diameter of the conduit is 0.02 m and the magnetic field strength is 0.5 T. Find the flow velocity when an electromotive force of 0.03 V is recorded.

11-17 An ac LVDT with a secondary voltage of 5 V has a range of ± 1 in. Find

(a) The output voltage when the core is -0.75 in. from center.

(b) The plot of output voltage versus core position for a core movement going from $+0.75$ in. to -0.4 in.

11-18 Given the same LVDT as in Problem 11-17, find

(a) The core movement from center when the output voltage is -3 V.

(b) The plot of core position versus output voltage for output voltages varying from $+3$ V to -4.5 V.

11-19 A certain crystal has a coupling coefficient of 0.25. How much mechanical energy (in joules) must be applied to produce an output of 50 mJ of electrical energy?

11-20 Given the same crystal as Problem 11-19, find the electrical energy that must be applied to produce an output of 0.5 in.-oz of mechanical energy.

11-21 A platinum resistance thermometer has a resistance of 100 Ω at 25°C. What is the temperature when the resistance is 150 Ω ($\alpha = 0.00392$)?

11-22 Given the same thermometer as in Problem 11-21, find the resistance at 65°C.

11-23 A certain thermocouple provides outputs in the vicinity of 0.05 mV/°C with a constant k equal to 5×10^{-5} mV/°C and $T_1 = 80$°C; the cold junction is kept in ice. Compute the resultant electromotive force.

11-24 Using the same thermocouple as in Problem 11-23, find the temperature of the hot junction if the output voltage is 5 mV. (*Hint*: Use an approximate equation by deleting the second term.)

11-25 The circuit of Fig. 11-18 is being used for temperature measurement. A "15-kΩ" thermistor (Fig. 11-16) is used. The meter is a 100-μA ammeter with a resistance of 1800 **Ω**, R_c is set to 8 **KΩ**, and the supply voltage V_T is 10 V. What will the meter read at 100°F and at 400°F?

11-26 In Problem 11-25 a 150-mV voltage is applied to the thermistor in series with a relay. What temperature causes the relay to be energized? (The energizing current is 1 mA.)

11-27 We are given the circuit of Fig. 11-23 using a 10-V supply voltage. A current of 5 mA flows when the cell is illuminated with about 400 lm/m^2, and deenergized when the cell is dark. Find

(a) The required series resistance R_1.

(b) The level of dark current.

11-28 The input voltage of Problem 11-27 is decreased to 10 V. Find

(a) The current flowing at 400 lm/m^2.

(b) The level of dark current.

11-29 Given the photodiode circuit (Fig. 11-25) and photodiode characteristic curve with the load line (Fig. 11-24), at an illumination of 5000 lm/m^2, find

(a) The diode current.

(b) The IR drop across R_1.

(c) The diode drop.

11-30 We are given the same photodiode and characteristic curve of Problem 11-29. The photodiode is forward-biased at an illumination of 20,000 lm/m^2. Find.

(a) The diode current.

(b) The IR drop across R_1.

(c) The diode drop.

(d) The resistance of the photodiode.

11-31 Given the circuit and characteristic curve of Fig. 11-28, at an illumination level of 300 W/m^2, find

(a) V_{CE}
(b) V_{RL}
(c) I_C

11-32 Using the same circuit and characteristic curve of Problem 11-31, find

(a) The current at 100 W/m^2.
(b) The current at 400 W/m^2.
(c) The change of resistance (ΔR_{CE}) across the transistor at the two illuminations.

11-25 LABORATORY EXPERIMENTS

Experiments E27 and E28 apply the theory that has been presented in Chapter 11. The purpose of the experiments is to provide hands-on experience to reinforce the theory.

Both experiments suggest the use of feedback equipment; however, both can be modified to use available equipment. The contents of the laboratory report to be submitted by each student are included at the end of each experimental procedure.

CHAPTER 12 Noise

12-1 Instructional Objectives

Chapter 12 discusses noise. Noise is defined, and sources of noise are discussed, as are means by which noise may enter a system and techniques for measuring and eliminating noise. After completing the chapter you should be able to

1. Discuss noise sources.
2. Describe how noise enters a system.
3. Define the following terms.
   (a) Noise.
   (b) Signal-to-noise ratio.
   (c) Noise factor.
   (d) Noise figure.
4. Calculate noise factors.
5. Calculate noise figures.
6. Calculate signal-to-noise ratios.
7. Discuss and evaluate methods for reducing noise.

12-2 INTRODUCTION

Any spurious current or voltage extraneous to the current or voltage of interest in an electrical or electronic circuit is called **noise**. Noise may be generated external to a particular circuit of interest and enter the circuit in various ways, or it may be generated within the circuit of interest.

The effect of noise in a circuit may range from an annoying "hum" in the speaker of a radio receiver owing to 60-Hz line-frequency noise, to the transmission of incorrect data in telemetry or data communications, to life-threatening situations caused by incorrect interpretation of waveforms related to vital organs in biomedical electronics.

This chapter discusses both internally and externally generated noise, noise measurement, and techniques to reduce noise.

12-3 SOURCES OF NOISE

Noise can be defined as any deviation from an expected value or as an unwanted value or signal superimposed on the value or signal of interest. The extent to which noise becomes important in the measurement process depends on the relative amplitudes or magnitudes of the unwanted values to the signal of interest. If the unwanted value is small compared to the signal level, then the signal-to-noise ratio is large and the noise becomes unimportant.

The ratio of desired signal to undersired noise is called the **signal-to-noise ratio** and is written as

$$\frac{s}{n} = \frac{[\text{Expected signal value (volts)}]^2}{[\text{Unwanted or deviated value (volts)}]^2} = \frac{\text{Signal power}}{\text{Noise power}} \tag{12-1}$$

In any measurement process, it is desirable to establish and maintain a large signal-to-noise relationship either by increasing the signal level without increasing the noise or by decreasing, with some technique, the noise level.

EXAMPLE 12-1

A signal voltage of 10 μV is amplified by a transistorized amplifier so that its output voltage is 100 mV. Superimposed on the output signal voltage is a noise voltage of 15 μV which was generated in the amplifier. What is the signal-to-noise ratio of the output signal?

Solution

$$\frac{s}{n} = \frac{(100 \times 10^{-3}\ \text{V})^2}{(15 \times 10^{-6}\ \text{V})^2} = \frac{1 \times 10^{-2}}{2.25 \times 10^{-10}} = 4.45 \times 10^7$$

In Example 12-1 the signal is 4.45×10^7 times the noise. The noise component was generated within the amplifier. Noise signals may develop in circuits by at least three different mechanisms. The three mechanisms may be identified by considering the amplifier shown in block form in Fig. 12-1.

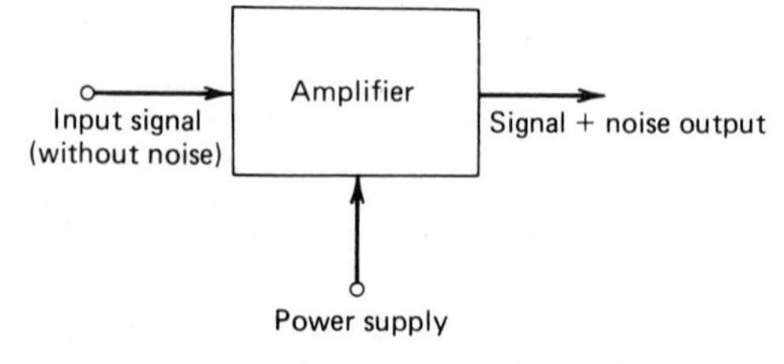

FIGURE 12-1 Mechanisms by which noise may develop in a circuit.

Assume an input signal that contains no noise. The amplifier has a source of energy, the power supply. The output signal is now some multiple of the input signal plus a noise signal now present. One possible source of the noise is in the amplifying systems (resistors, capacitors, transistors, etc.). The noise in this case is generated within the amplifier and is called **generated noise**.

The energy source or power supply could be the source of noise by virtue of the presence of spikes, ripple, or random deviations that are carried or conducted into the amplifier by the power wiring. This type of noise is called **conducted noise** and requires a metallic conducting medium such as ac power leads or instrumentation wiring.

Another source of the noise could be of the electromagnetic variety in which undesirable signals are radiated into the interior of the amplifier. Such noise could arise from nearby electric or magnetic fields or disturbances. A common source of **radiated noise** consists of electromagnetic impulses radiated from the ignition wiring of spark plugs and generating noise in a radio receiver. In this particular example, a combination of events may occur. The noise is radiated onto the radio antenna, which is then conducted into the active electronic circuitry. In the same system, alternator or generator noise may be conducted into the active circuits through the power wiring.

12-4 GENERATION OF NOISE

A known or standard noise source is a required item of test equipment for evaluating the noise characteristics of an amplifier. The method of noise measurement is to measure the noise output of the device (amplifier) under test with a known source of noise injected into the input compared with the noise output of the device with no noise injected at the input.

One common noise source is a temperature-limited diode, that is, a diode whose current is a function of temperature, an ordinary thermionic vacuum diode, for example. Another common example of a noise source is a gas discharge tube. Neon and argon are representative gases. The type of gas used depends on the frequency band of noise required. The noise generated by such devices is random and of a wide spectrum of frequencies.

12-5 MEASUREMENT OF NOISE

To discuss noise measurement parameters quantitatively, we need to establish some relevant expressions for the various parameters. The most easily quantifiable noise is the thermal noise generated within a conductor. This noise is called **Johnson noise** after its discoverer, John B. Johnson, an American physicist.

The noise power generated in a conductor, P_n, is jointly proportional to the absolute temperature of the conductor and to the frequency bandwidth which the noise occupies. Stated mathematically,

$$P_n \propto T \, \Delta B$$

where

P_n is the noise power
T is the absolute temperature in degrees Kelvin,
ΔB is the frequency bandwidth.

With the introduction of a proportionality constant, the proportion becomes the equation

$$P_n = KT \, \Delta B \tag{12-2}$$

where K is the constant of proportionality known as Boltzmann's constant and has the value of 1.38×10^{-23} J/°K · cyc. A second quantity of importance can be defined as the power spectrum density, which is the noise per unit of frequency bandwidth,

$$S_n \equiv \frac{P_n}{\Delta B} = \frac{KT \, \Delta B}{\Delta B} = KT \tag{12-3}$$

where S_n is the power spectrum density. This quantity, in watts per hertz or joules per cycle, is an expression of the noise energy generated per cycle of vibratory motion in the generating conductor.

The noise generation system can now be described as an energy or voltage source in series with a resistor, as shown in Fig. 12-2 where R_n is the equivalent noise resistance in which the noise is being generated and E_n is the noise voltage.

If the noise generator is now connected to an external load resistor, R_L, noise energy will be transferred to the load. This situation is shown in Fig. 12-3.

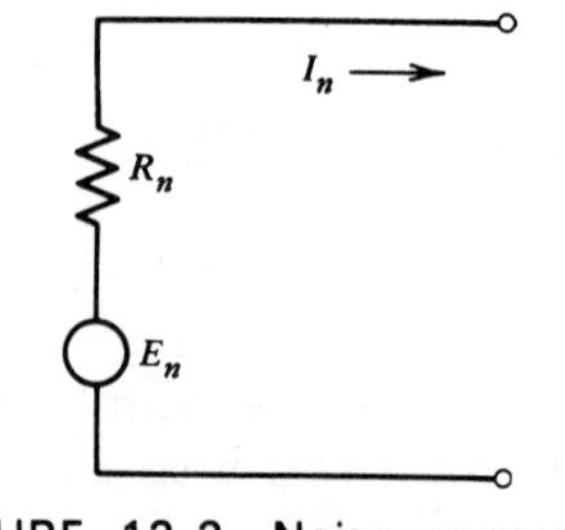

FIGURE 12-2 Noise generator in series with a resistor.

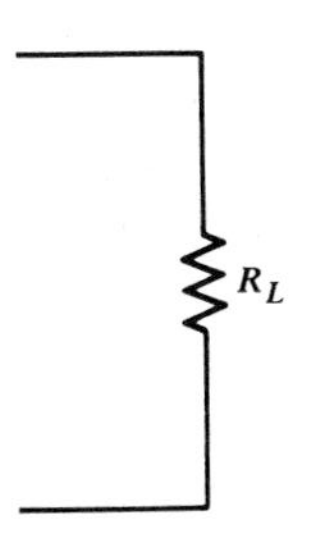

FIGURE 12-3 Circuit in which noise energy is transferred from the source to the load.

The noise power delivered to, and dissipated in, resistor R_L is given by

$$P_{nL} = I_n^2 R_L \tag{12-4}$$

The noise current I_n is given by

$$I_n = \frac{E_n}{R_n + R_L} \tag{12-5}$$

Equation 12-4 then becomes

$$P_{nL} = \frac{E_n^2 R_L}{(R_n + R_L)^2} \tag{12-6}$$

If we assume the maximum power transfer condition is met, that is, $R_n = R_L$, Eq. 12-6 then becomes

$$P_{nL} = \frac{E_n^2}{4R_n}$$

or

$$E_n^2 = 4R_n P_{nL} \tag{12-7}$$

Since $P_{nL} = KT\,\Delta B$, we can write Eq. 12-7 as

$$E_n^2 = 4KTR_n\,\Delta B \tag{12-8}$$

Equation 12-8 quantifies the noise voltage in terms of the absolute temperature of the conductor or resistor.

EXAMPLE 12-2

A voltmeter has an input resistance of 15 MΩ. How much voltage will it generate for each cycle of frequency bandwidth?

Solution

The noise voltage can be computed, using Eq. 12-9, as

$$E_n^2 = \left(\frac{4}{1}\right)\left(\frac{1.38 \times 10^{-23}\ \text{J}}{°\text{K} \cdot \text{cyc}}\right)\left(\frac{290\ °\text{K}}{1}\right)\left(\frac{15 \times 10^6\ \Omega}{1}\right)\left(\frac{1\ \text{cyc}}{\text{sec}}\right)$$

$$= \frac{(4)(1.38 \times 10^{-23}\ \text{J})(2.9 \times 10^2)(15 \times 10^6\ \Omega)}{\text{sec}}$$

$$= 24 \times 10^{-14}\ \text{V}^2$$

$$E_n = 4.9 \times 10^{-7}\ \text{V} = 0.49 \times 10^{-6}\ \text{V} = 0.49\ \mu\text{V}$$

If the voltmeter considered in Example 12-2 is connected to the input terminals of an amplifier to measure the input signal voltage, the voltmeter would inject a noise signal of 0.49 μV for each cycle of bandwidth rating of the amplifier. The voltmeter would then become a source of man-made noise to be reckoned with.

For resistors connected in series, an equivalent circuit shown in Fig. 12-4 may be drawn. The total noise is the sum of the noises in the individual elements,

$$E_n^2(T) = E_n^2(1) + E_n^2(2)$$

$$= 4KR_1T\ \Delta B + 4KR_2T\ \Delta B$$

$$= 4KR_sT\ \Delta B$$

where $E_n(T)$ is the total noise voltage and R_s is the series equivalent or R_1 and R_2. The equation can be expanded to quantify any number of resistors in series.

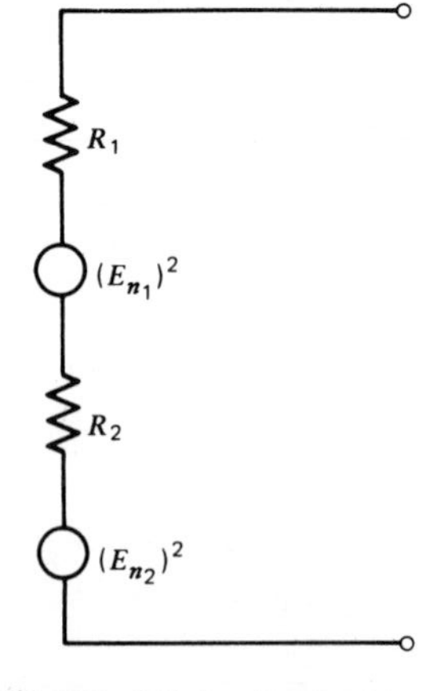

FIGURE 12-4 Equivalent circuit for series-connected resistors.

EXAMPLE 12-3 A high-input impedance voltmeter has a 15-MΩ input impedance which is composed of three 5-MΩ resistors in series. Calculate the total noise voltage generated in the resistors at room temperature (290°K) over a 100-kHz bandwidth.

Solution

$$E_n^2 = E_n^2(1) + E_n^2(2) + E_n^2(3)$$

$$E_n = \sqrt{E_n^2(1) + E_n^2(2) + E_n^2(3)}$$

$$= \sqrt{4KTR_1\,\Delta B + 4KTR_2\,\Delta B + 4KTR_3\,\Delta B}$$

$$= \sqrt{(3)(4)KTR_1\,\Delta B}$$

$$E_n^2 = \sqrt{\left(\frac{12}{1}\right)\left(\frac{290°\text{K}}{1}\right)\left(\frac{1.38 \times 10^{23}\ \text{J}}{\text{cyc} \cdot °\text{K}}\right)\left(\frac{5 \times 10^6\ \Omega}{1}\right)\left(\frac{100 \times 10^3\ \text{cyc}}{\text{sec}}\right)}$$

$$= 155\ \mu\text{V}$$

Example 12-3 illustrates the fact that the *squares* of the voltage are additive.

For resistors in parallel, the relevant equation for the noise voltage is

$$E_n^2 = 4KTR_p\,\Delta B \qquad (12\text{-}9)$$

where R_p is the equivalent resistance of the parallel combination.

Since most measuring instruments have resistive elements in the input circuit or the output circuit, the instrument user must recognize the noise-producing possibility of each test instrument.

The definition of the signal-to-noise ratio, which was given in Eq. 12-1, is

$$\frac{s}{n} = \frac{\text{Signal power}}{\text{Noise power}} \qquad (12\text{-}1)$$

Since signal power is V_s^2/R and noise power is V_n^2/R, Eq. 12-1 becomes

$$\frac{s}{n} = \frac{V_s^2}{V_n^2}$$

This equation represents one of the methods of quantitative measurements of noise.

EXAMPLE 12-4 (a) An amplifier has a signal voltage level of 3 μV and a noise voltage level of 1 μV. What is the signal-to-noise ratio at the input?

Solution

$$\frac{s}{n} = \frac{V_s^2}{V_n^2} = \frac{(3\ \mu\text{V})^2}{(1\ \mu\text{V})^2} = \frac{9\ \mu^2\text{V}^2}{1\ \mu^2\text{V}^2} = 9$$

(b) If the voltage gain of the amplifier is 20, what is the s/n ratio at the output?

Solution

$$\frac{s}{n} = \frac{V_o^2}{V_{on}^2} = \frac{(60\ \mu\text{V})^2}{(20\ \mu\text{V})^2} = \frac{3600\ \mu^2\text{V}^2}{400\ \mu^2\text{V}^2} = 9$$

(c) If the amplifier adds 5 μV of noise, what is the s/n ratio at the output?

Solution

$$\frac{s}{n}=\frac{(60\ \mu\text{V})^2}{(25\ \mu\text{V})^2}=\frac{3600\ \mu^2\text{V}^2}{625\ \mu^2\text{V}^2}=5.76$$

Another quantitative measurement of noise is the measurement of the **noise factor**, F. The noise factor, F, is defined as

$$F=\frac{s/n \text{ at input}}{s/n \text{ at output}} \tag{12-10}$$

Noise factor measurements are important in that they are a measure of the noise added to a signal in a network or amplifier stage. In Example 12-4, the noise factor can be computed as

$$F=\frac{9}{5.76}=1.54$$

If the noise factor is expressed in decibels, it is known as the **noise figure** where

$$\text{Noise figure} \equiv 10 \log_{10} F \tag{12-11}$$

EXAMPLE 12-5

An amplifier has a s/n ratio of 10 at the input and a s/n ratio of 1 at the output. Calculate the noise figure.

Solution

$$nf=10\log_{10}\frac{10}{1}=10\log_{10}10$$

$$=(10)(1)=10\text{ db}$$

The measurement of *noise figure* is the most meaningful measurement for amplifiers, transistors, and vacuum tubes since it is a measure of the noise generated within the device. Noise figure is expressed in decibels and requires a known source of noise for the measurement. The noise figure from Eq. 12-11 is

$$nf=10\log\left(\frac{\text{Noise voltage at output with noise at input}}{\text{Noise voltage at output with no noise at input}}\right)$$

$$=10\log\frac{V_n}{V_o}$$

where V_n is the output noise voltage with a noise source injected into the input and V_o is the output noise voltage without noise at the input. The difference between the two represents the noise added by the device under test.

EXAMPLE 12-6

A transistor output noise voltage is measured at 15 μV when a standardized noise source is at the input. With the noise source removed, a noise voltage of 5 μV is measured. What is the noise figure?

Solution

$$nf = 10 \log \frac{15\ \mu\text{V}}{5\ \mu\text{V}} = 10 \log 3$$

$$= (10)(0.477) = 4.77 \text{ db}$$

12-6 NOISE REDUCTION TECHNIQUES

In eliminating noise or reducing noise signals to acceptable levels, we must first identify how the noise enters the system, whether it is generated, conducted, or radiated, and then take possible solution steps.

Internally generated noise comes from several mechanisms within elements. One source is the simple carbon composition resistor. The conductive portion of the resistor consists of a regularly arrayed group of atoms that maintain the same general position in the conductor. These atoms contribute charge carriers or conduction electrons for the mechanism of current flow. Although the atoms maintain their general position, each is in rapid vibratory motion because of temperature or thermal effects. The vibratory motion of the atoms is transferred to the conduction electrons, thereby modulating the current with an unwanted or noisy component. Since the noise is temperature-dependent, it increases with the internal I^2R heating or with an increase in the ambient environmental temperature. Noise generated by this mechanism is called *Johnson noise*.

The vibration within the resistor atoms covers a wide frequency range, and therefore the generated noise consists of a wide spectrum of frequencies. This wideband noise is sometimes referred to as *white noise* because it does contain such a wide range of frequency components. The noise generated by a jet aircraft is mostly white noise, and the audible effect of white noise in a radio receiver is a hiss. Since thermal agitation is the source of the noise, reducing resistor temperature is a possible means of reducing the noise. Special film and glass substrate resistors, designed to minimize the noise generation, are available. Since the noise generation is internal, shielding is ineffective. Selective filtering is ineffective in reducing the noise amplitude because it is wideband.

A second type of noise is generated internally by short-time electrical events within an active device such as a transistor. As the charges cross junctions within the semiconductor devices, they make the transition from one energy level to another, resulting in an acceleration. Accelerating charges generate electromagnetic disturbances, which makes a contribution to the noise generated within the system. Since the duration of the acceleration period is short, the resulting frequency spectrum of generated noise is once again wideband. Little can be done to reduce the noise since environmental

TABLE 12-1
Noise Suppression Techniques

Noise Source	Noise Suppression Techniques
Electric motors and generators	• Use a well-shielded housing. • Bond housing to ground. • Use bypass capacitors at brushes. • Use a feedthrough capacitor at the armature terminal. • Shield terminals and interconnecting wiring. • Keep brushes in good condition.
RF generators	• Use special multiple-shield enclosures. • Bypass and filter all lines entering or leaving the shield enclosures. • Use resonant traps for specific frequencies.
Relays, controllers, switching devices	• Shunt relay or switch contacts with a capacitor to reduce current surges. In general, a current-limiting resistor should be placed in series with the capacitor to prevent deterioration of the switching contacts. • Enclose the switching device in a shield.
Dc–dc converters	• Shield the unit. • Use low-pass filters on all leads passing through the shield. (At minimum, use feedthrough capacitors.)
Electromechanical vibrators	• Shield the vibrator. • Use feedthrough capacitors or, if necessary, more elaborate filters for power leads passing through the shields. All filter components should be within the shield, with leads passing through the shield via feedthrough capacitors.
Vibrator-type dc voltage regulators	• Shield the regulator. • Locate the regulator as close to the generator as possible. • Shield leads between regulator and generator. • Bypass input dc lead inside the shield with a capacitor, preferably a feedthrough capacitor.
Mechanically induced noise	• Support cables to reduce movement. • Use low-noise cables. • Use vibration and shock-damping mounts. • Select low-noise devices and handpick for lowest noise.
Arc and gaseous-discharge devices	• Install bypass capacitors on lines. • Use special conductive coatings over glass. • Use shield cases. • Substitute incandescent lights for fluorescents.
Ignition noise	• Place the resistor (10 kΩ) in a high-voltage lead near the coil, or use resistive ignition leads. • Use shielded internal-resistor spark plugs. • Shield the coil. • Use shielded ignition wires. • Use bypass capacitors on the dc lines into the coil and distributor.

conditions do not affect the noise. The same type of noise is generated as electrons pass through the various electrode areas of a vacuum tube because electric fields vary between the electrodes. This type of noise is called **shot noise**. Selective filtering is effective to some extent in reducing shot noise.

Changing electric fields within the region between capacitor plates and changing magnetic fields surrounding inductors, as well as effective inductances owing to interconnecting leads within a circuit, cause spurious and unwanted signals to be present in electrical equipment. Since these fields change on an orderly basis, a filter tuned to a specific frequency, a selective filter, may be utilized to reduce the unwanted signal.

One of the most frequent noise components is the 60 cycles per second or some harmonic of 60 cycles per second whose source is the power supply frequency. This particular signal can be conducted into systems by the power supply line, and then transferred to most other elements in the system. In addition to being conducted into the system, the 60-cycle noise may be radiated into the system from adjacent power leads and from transformer and ballast fields. The reduction of 60-cycle noise is possible but difficult. First, the mode of entry, radiated or conducted, must be determined, and then adequate steps must be taken for the reduction. If entry is made by conduction, appropriate filters may be placed in leads to trap out the noise. If the noise is radiated, shielding will be necessary to reduce the interfering signal. The shielding problem can become highly involved because shielding of test leads, shielding of power supply leads, and even complete shielding of the entire circuit may become necessary.

One particularly troublesome source of noise is the induction of noise into a test circuit by the interaction between various items of test equipment necessary to perform a test. Voltmeters, ammeters, signal generators, and so on, can be sources of noise, which at times must be recognized and reckoned with.

In most instances, techniques for reducing noise are most successful if sensitive measuring instruments are physically removed as far as possible from the noise source. Table 12-1 lists techniques for noise suppression for some specific examples.

12-7 APPLICATIONS

In communications work, particularly in the microwave region, and in radar, the weakest signal that can be detected with instruments is usually determined by the amount of noise generated within the receiving system. If the noise generated within the receiving system can be reduced, this provides an increase in the signal-to-noise ratio at the output of the receiver. This is equivalent to a corresponding increase in the signal received. However, from a performance standpoint, an increase in the signal-to-noise ratio achieved by reducing the noise in the receiver is more economical than increasing the power of the transmitter. A signal source is generally desirable when analyzing or evaluating an amplifying system. This is true even of noise sources such as the one discussed in the following paragraphs.

12-7.1 Broadband Noise Source

There are many ways of generating noise for use as a noise source. The method selected depends primarily on the frequency range of interest. A very flexible broadband noise source can be constructed using a specially designed thermionic diode called a *noise diode*. The circuit is shown in Fig. 12-5. The noise, expressed in decibels, is

$$n_{dB} = 10 \log(20 I_d Z_o + 1)$$

where

I_d = saturated current of the noise diode (35 mA maximum for 5722)
Z_o = characteristic impedance of cable

As can be seen, a coaxial cable is connected directly across the diode and terminated with a resistor equal to its characteristic impedance. The inductor, L, may be used to increase the frequency range by offsetting the effects of the interelectrode capacitance of the diode. The frequency range of the noise generated is from about 1 kHz to 100 MHz.

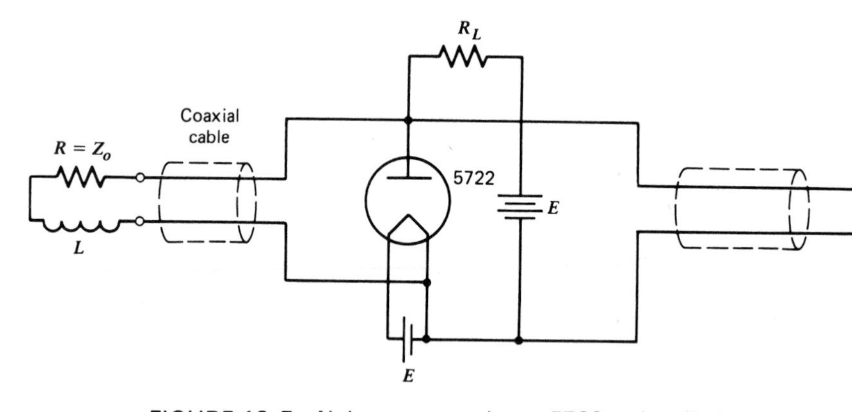

FIGURE 12-5 Noise source using a 5722 noise diode.

12-8 SUMMARY

Noise that may be defined as an unwanted signal superimposed on the signal of interest may be generated external to a particular circuit of interest, or it may be generated internally. Many electrical disturbances cause noise to be present in electrical circuits; therefore, noise is classified in different ways. It is generally classified as generated, conducted, or radiated. Noise generated within a circuit of interest is called generated noise, whereas noise that has its origin outside the circuit is classified as either conducted or radiated noise.

Certainly, a prime consideration in any circuit is maintaining a large signal-to-noise ratio, either by increasing the signal level without increasing the noise or by decreasing the noise level while maintaining a constant signal level. Noise reduction may be achieved in many instances by proper shielding or filtering, depending on the source.

12-9 GLOSSARY

Conducted noise: Noise carried into a circuit through a metallic conductor such as ac power leads or instrument wiring.

Generated noise: Noise generated within a circuit. The term *generated noise* is a broad heading and includes Johnson noise and shot noise.

Johnson noise: Noise caused by the thermal agitation of the free electrons carrying current, thereby modulating the current; also called *thermal noise*.

Noise: A term that refers to any unwanted electrical signal tending to interfere with the signal of interest. The term *noise* originally referred to fluctuating signals in the audio-frequency range that produced a static or hissing sound when connected to a speaker. Today the term refers to any spontaneous signal fluctuations regardless of the frequency.

Noise factor: The ratio of signal-to-noise at the input to signal-to-noise at the output of a circuit.

Noise figure: The ratio of signal-to-noise at the input to signal-to-noise at the output of a circuit expressed in decibels.

Radiated noise: Noise of an electromagnetic origin that is radiated into a circuit.

Shot noise: Noise resulting from the random emission of electrons across a *p-n* junction.

Signal-to-noise ratio: The ratio of desired signal to undesired noise.

12-10 REVIEW QUESTIONS

The following questions should be answered after a thorough study of the chapter. The purpose of the questions is to determine the reader's comprehension of the material.

1. What is noise?

2. How is noise generated?

3. How is noise measured?

4. In what units are noise measurements expressed?

5. How does noise enter electrical or electronic systems?

6. What type of noise is easiest to eliminate?

12-11 PROBLEMS

12-1 Plot the power spectrum density versus temperature for temperatures from 0°K up to 1000°K in 100-°K steps.

12-2 Calculate the noise power available in 100 kHz of bandwidth at a room temperature of 290°K.

12-3 Calculate the noise voltage across a 100-kΩ resistor at a room temperature of 290°K for a bandwidth of 100 kHz.

12-4 Calculate the noise voltage across a 500-kΩ resistor at a temperature equal to the boiling point of water if the bandwidth is 200 kHz.

12-5 Calculate the power spectrum density from a welding torch whose temperature is 7000°C.

12-6 An amplifier whose bandwidth is 100 kHz has a noise power spectrum density input of 7×10^{-21} J/cyc. If the input resistance is 50 kΩ, and the amplifier has a gain of 100, what is the noise output voltage?

12-7 A resistor, at room temperature, has a noise voltage of 2 μV for a 50-kHz bandwidth. Calculate the temperature at which the noise voltage is 20% of its value at room temperature.

12-8 For the resistor in Problem 12-7, calculate the temperature at which the noise voltage is double its room temperature value.

12-9 An amplifier has an input s/n of 15 and an output s/n of 5. Calculate the noise factor.

12-10 An amplifier has an input signal voltage of 25 μV and a noise voltage of 3 μV. Calculate the signal-to-noise ratio.

12-11 A low-pass filter has an input s/n of 20. The input signal voltage is measured to be 3 mV. Calculate the noise voltage.

12-12 An amplifier has a noise figure of 30 dB. If the input s/n is 100, what is the output s/n?

12-13 Calculate the noise factor in Problem 12-12.

12-14 In order to measure the noise figure of an amplifier, a noise voltage that is 10% of the signal voltage is applied at the input of the amplifier. The noise voltage at the output of the amplifier is measured as 20% of the signal voltage. Calculate the noise figure of the amplifier.

12-12 LABORATORY EXPERIMENTS

Experiments E29 and E30 apply the theory presented in Chapter 12. The purpose of the experiments is to provide students with hands-on experience, which is essential for a thorough understanding of the concepts involved.

The experiments require no special equipment; therefore, the required equipment should be found in any electronics laboratory. The contents of the laboratory report to be submitted by each student are listed at the end of each experimental procedure.

CHAPTER 13

Digital Instruments

13-1 Instructional Objectives

This chapter discusses the basic theory of digital techniques applicable to electronic test instruments. We consider several approaches for converting analog voltages to digital signals and for counting circuits, and we discuss the increasing use of microprocessors in instruments. Several applications are also discussed. After completing the chapter you should be able to

1. List five advantages of digital instruments over analog instruments.
2. Describe the difference between a digital instrument and a digital readout instrument.
3. Describe what the term *overrange* means and what a half-digit is.
4. List three types of analog-to-digital converters and list an advantage of each type.
5. Describe how flip-flops are used to count.
6. Describe how a binary counter can be made to count in the decimal system.
7. List five modes of operation of an electronic counter.
8. Describe what is meant by the term *intelligent instruments*.
9. Describe the purpose of the IEEE 488 bus.

13-2 INTRODUCTION

Few areas of technology have advanced as rapidly in recent years as digital electronics. Virtually every phase of electronics, including measuring instruments, have experienced tremendous change. A very substantial percentage of electronic instruments of recent design incorporate some digital circuitry.

Digital instruments offer several very attractive advantages over analog instruments, including greater speed, increased accuracy and resolution, reduction in user errors, and the ability to provide automatic measurement in systems applications.

Because we live in an analog world, physical parameters to be measured with digital instruments must first be converted to a digital format to facilitate measurement. Therefore, this chapter will include a discussion of analog-to-digital conversion techniques as well as basic digital techniques used in instruments.

13-3 DIGITAL INSTRUMENTS VERSUS DIGITAL READOUT INSTRUMENTS

A clear distinction should be made between a digital instrument and an instrument with digital readout. A digital instrument is one in which the circuitry required to obtain a measurement is of digital design. A digital readout instrument is one in which the measuring circuitry is of analog design and only the indicating device is of digital design.

An analog instrument with digital readout is generally no more accurate than the same analog instrument with analog readout. However, the digital display is unambiguous and can be read more quickly. In this respect, the digital display may be more desirable, though no more accurate. A block diagram of both a digital instrument and an instrument with digital readout is shown in Fig. 13-1.

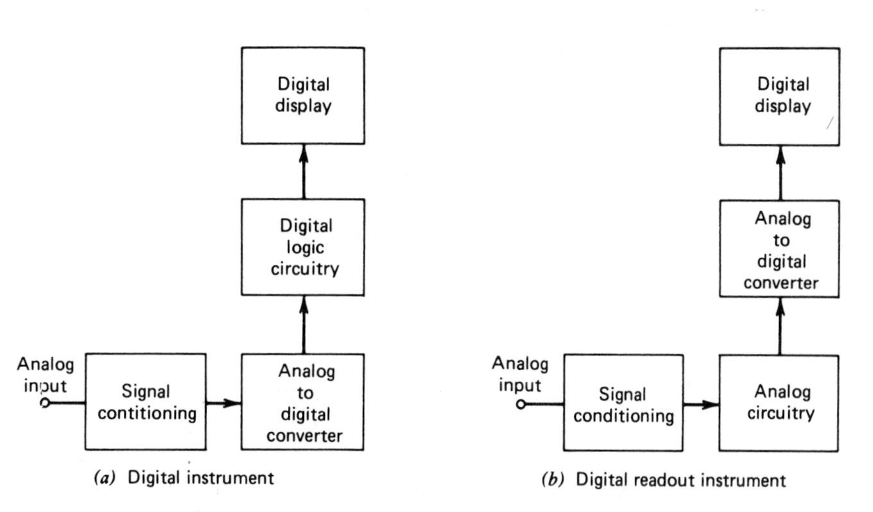

FIGURE 13-1 Block diagrams of a digital instrument and a digital readout instrument.

13-4 COMPARISON OF DIGITAL AND ANALOG METERS

Digital instruments use logic circuits and techniques to carry out measurements or to process data. Basically, any digital instrument may be viewed as an arrangement of logic gates that change states at very high speeds in the process of making a measurement. Because of the rapidly expanding use of digital techniques in measuring instruments, a comparison of factors affecting error in measurement when using analog and digital instruments is in order. Although several of the factors discussed in the remaining paragraphs of this section are pertinent to several kinds of digital instruments, the factors are described with respect to digital multimeters.

13-4.1 Readability

The single most significant factor favoring digital instruments over their analog counterparts is the readability of the measurement result because of the digital readout. When applying an analog instrument, the user must function as an analog-to-digital converter. The user must also read the analog scale properly, possess some skill in interpolation, be able to use mirrored scales, and, in general, have a "good eye." The process is prone to error and is time-consuming.

13-4.2 Accuracy

Accuracy is a second major advantage of digital instruments over analog instruments. In general, low-cost digital multimeters are more accurate than comparably priced VOMs and EVMs by a factor of 10, that is, 0.1% versus 1%. However, the accuracy of most digital multimeters is related to the frequency of calibration. Therefore, not all the digits of a digital readout are necessarily meaningful. Most conservative manufacturers design their instrument to read within one digit at normal laboratory temperature, $\pm 10°C$ for a calibration interval of 30 to 90 days. Typical specifications may state: $\pm 0.1\%$ of the reading in the temperature range from 10°C to 30°C for a period of 90 days. A person who is considering purchasing a digital multimeter or any other digital instrument should read the manufacturer's specifications carefully regarding operating temperature ranges and the required frequency of calibration.

13-4.3 Resolution

Digital instruments provide better resolution than their analog counterparts. In particular, digital multimeters yield an order-of-magnitude improvement in resolution over those of comparably priced VOMs and EVMs. The greater resolution of digital multimeters reduces the number of ranges required to cover the voltage range from 1 to 1000 V. Digital multimeters cover the same voltage range with three scales.

13-4.4 Sample Speed

Most digital multimeters contain a built-in triggering circuit. The trigger source, which may be fixed or variable, generally triggers at a rate that is independent of the response time of the analog circuits within the instrument. The triggering rate, and thus the sampling rate, of most digital multimeters is from three to ten samples per second. Although sampling at any of these rates is fine for dc measurements, with ac measurements the ac converter may take 1 or 2 seconds to respond. This means that the instrument user must wait for several samples before obtaining a stable reading. Thus, the speed of digital multimeters is dependent on the response time of the analog input circuitry and the sampling rate of the digital circuitry.

13-4.5 Digits Displayed and Overranging

Digital multimeters are often classified according to the number of full digits displayed. In addition to the full digits, an **overrange** digit is generally added to allow the user to read beyond full scale. This overrange digit is often called a "one-half" or a "partial" digit since it can display only digits 0 and 1. Therefore, the instrument specification states "X and $\frac{1}{2}$ digits" where the X is the number of full digits and the $\frac{1}{2}$ digit is the overrange digit.

Overranging greatly extends the usefulness of digital multimeters by maintaining resolution up to, and beyond, full scale. For example, if a voltage to be measured changes from 9.999 to 10.024 V, a four-digit instrument without overranging could measure the original voltage value as 9.999 V. To measure the new value of the voltage would require a range change with a resulting measurement of 10.02 V. The additional change of 0.004 V would not be displayed. With overranging, the second measurement would be displayed as 10.024 V with no loss of resolution. The half-digit that is used in overranging is the left-most digit in the display. Overranging is generally expressed as a percentage of full scale. For example, a $3\frac{1}{2}$ digit instrument with 20% overranging can display voltages up to 1.199 V on its 1-V range.

13-5 ANALOG-TO-DIGITAL CONVERTERS

Digital instruments, particularly digital multimeters, are used to measure analog parameters. Therefore, it is necessary to convert the analog signal to an equivalent digital signal by using an **analog-to-digital converter**. The three conversion techniques generally used are single-slope, dual-slope, and voltage-to-frequency conversion. The vast majority of laboratory-quality digital multimeters of recent vintage use dual-slope conversion. However, the other techniques are used often enough to warrant discussion.

13-5.1 Single-Slope Converters

Single-slope converters are at present used in low-cost instruments. The fundamental concept of single-slope converters is to make a linear conversion of unknown voltage to time. Conversion to time is chosen because digital counting circuits can then be used to display the time in a digital format.

Obtaining a relationship between voltage and time is quite straightforward; however, obtaining a linear relationship is somewhat more involved. If a constant-voltage source is used to charge a capacitor through a resistor, the voltage across the capacitor increases with time and approaches the value of the voltage source. The time required for the voltage across the capacitor to reach any particular value depends on the value of the voltage source, the resistance, and the capacitance. However, the relationship between capacitor voltage and time is exponential rather than linear. If the constant-voltage source is replaced with a constant-current source, the capacitor charges at a rate that is linear with respect to time. Figure 13-2 shows the capacitor voltage-versus-time curves when charging from a constant-voltage source and a constant-current source.

An operational amplifier integrator circuit is more widely used for obtaining a linear voltage–time relationship than the method just discussed. This type of circuit is used in most modern analog-to-digital converters to obtain a linear voltage–time relationship.

A circuit for a single-slope, analog-to-digital converter, which includes the integrator in Fig. 13-3, is shown in Fig. 13-4. If the circuit is to operate satisfactorily, V_{ref} must be a negative, well-regulated, dc voltage source, and V_x must be a fairly stable, positive, dc voltage.

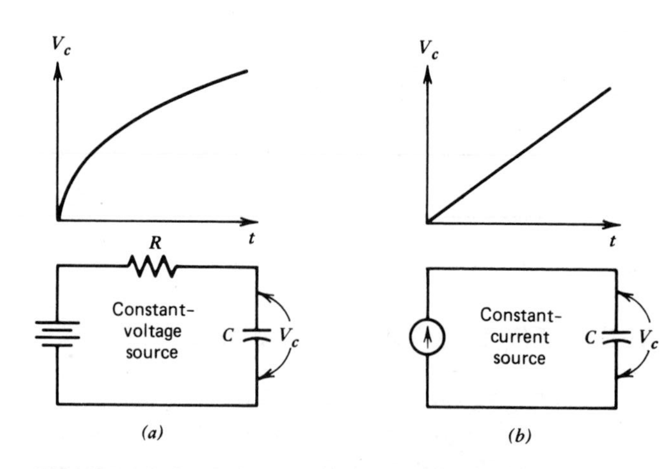

FIGURE 13-2 Voltage–time relationship for a charging capacitor. (*a*) An exponential voltage rise across a capacitor charged from a constant-voltage source. (*b*) A linear voltage rise across a capacitor charged from a constant-current source.

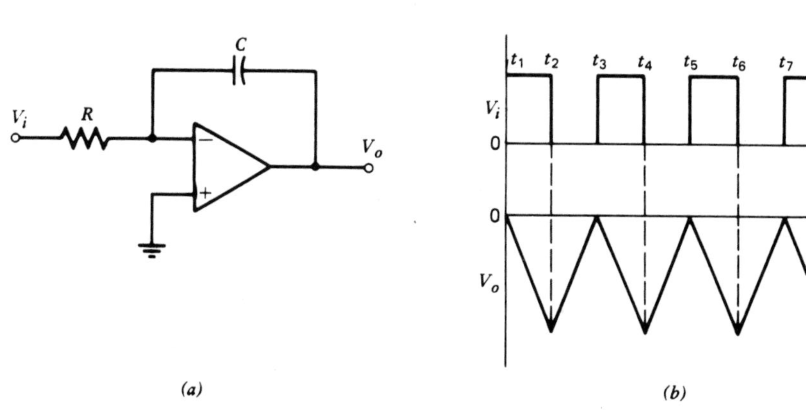

FIGURE 13-3 Op-amp integrator and associated waveform. (*a*) Schematic. (*b*) Input square wave to output with linear voltage-time relationship.

An unknown voltage is measured as follows.

1. The test probes are connected to the unknown voltage V_x. This places a positive voltage on the noninverting input of the op-amp in the comparator circuit.
2. The main gate control generates a positive pulse, which opens switch S on its leading edge and causes the top input of the AND gate to be "high" or positive.
3. When switch S opens, the integrator circuit capacitor, C, begins to charge linearly from zero in a positive direction. The increasing output signal from the integrator is applied to the inverting input of the comparator.
4. The output of the comparator is positive owing to the unknown voltage V_x, which is applied to its noninverting input. The positive comparator output causes the middle input of the AND gate to be "high."
5. The pulse train from the clock is applied to the third input of the AND gate. Since the other two inputs of the AND gate are high, the AND gate output is a series of clock pulses that are counted by the binary counter.
6. When capacitor C charges to a voltage level slightly higher (less than 1 mV higher) than V_x, the output of the comparator abruptly switches to zero. Since this removes one input from the AND gate, it stops passing clock pulses into the counter.
7. The count stored in the binary counter is directly proportional to the unknown voltage V_x and is indicated on the digital readout device as the value of the unknown voltage.
8. After a short time interval, which is determined by the main gate control, the output of the main gate control goes low. The leading edge of this pulse causes switch S to close, which discharges capacitor C. This disables the gate until the start of the next cycle.

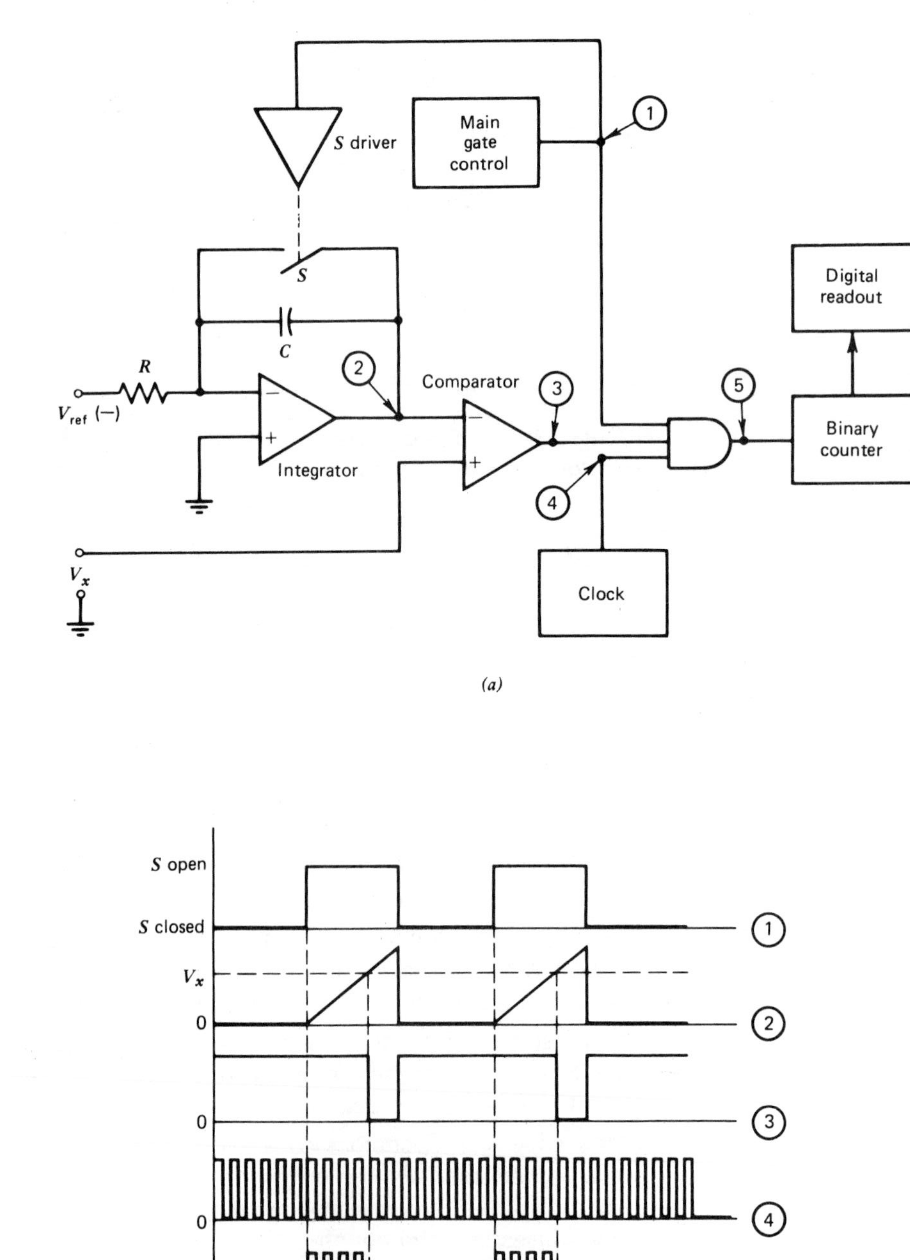

FIGURE 13-4 Circuit and timing diagram for a single-slope analog-to-digital converter.

Although the simplicity of the circuit makes it attractive, its simplicity is also responsible for its limitations. The most serious limitations are the following.

1. It can measure voltages of only one polarity.
2. Additional circuitry is required for overrange conditions.
3. The circuit is susceptible to oscillator frequency drift.
4. The circuit is susceptible to drift in the constant-current source.
5. Accuracy depends on the stability of the capacitor.
6. Accuracy depends on the stability of the differential voltage that trips the comparator.
7. The converter is very susceptible to noise on the analog voltage.

13-5.2 Dual-Slope Converters

Dual-slope converters overcome most of the limitations of single-slope converters, improving in particular long-term accuracy. Dual-slope converters use the linear charge concept of the single-slope converter; however, several improvements are incorporated to improve long-term stability.

Although dual-slope converters also use a capacitor charged by a constant-current source to provide a voltage-to-time conversion, as the name implies, the dual-slope converter not only charges the capacitor, but also discharges it during a measurement cycle. This charge–discharge cycle tends to reduce significantly the long-term drift and stability problems associated with single-slope converters.

Extensive efforts have been made with dual-slope converters to reduce the accuracy and stability requirements placed on components and yet improve the accuracy and stability of the converter. The result is a converter that uses low-cost components and yet is very accurate (usually about 0.1%) over extended periods of time, through considerable variations in temperature, and even when subject to a great amount of input noise. A block diagram of a basic dual-slope converter is shown in Fig. 13-5.

The dual-slope A/D converter is designed around an **integrator** consisting of the operational amplifier A_1, resistor R, and capacitor C, plus the voltage comparator A_2, high-speed electronic switches S_1 and S_2, and the logic control circuit. The remaining circuitry is associated with the actual counting circuit. The output of the voltage comparator is low if the output of the integrator is zero, but it will switch states to a high output when the integrator output rises to approximately 1 mV above ground potential.

The initial step in the operation of the converter is the momentary closure of switch S_2 and setting switch S_1 to position A, which is done by the logic control circuit. With switch S_1 in position A, the integrator is connected to the input voltage, which causes the voltage at the output of the integrator to

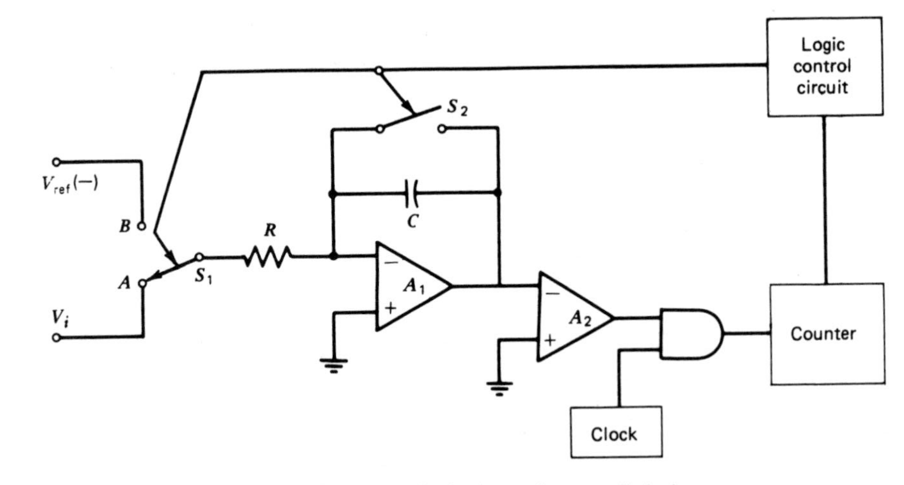

FIGURE 13-5 Basic dual-slope analog-to-digital converter.

begin rising linearly with respect to time according to the expression

$$V_A = \frac{1}{RC}\int_{t_1}^{t_2} V_i\,dt \qquad (13\text{-}1)$$

As soon as the voltage at the output of the integrator, V_A, rises a few millivolts, the voltage comparator A_2 changes states, setting its output high. This enables the AND gate to pass clock pulses to the counter circuit, which counts the pulses until it overflows at time T_2, shown in Fig. 13-6.

The amplitude of voltage V_A at the instant the counter overflows is determined by the following factors.

1. The value of R and C in the integrator. These values should remain constant over a conversion cycle since the components are of fixed value.
2. The time required for the counter to fill and provide an overrange pulse. This is determined by the frequency of the oscillator, which is assumed to be stable during the conversion cycle.
3. The value of the unknown voltage, V_i.

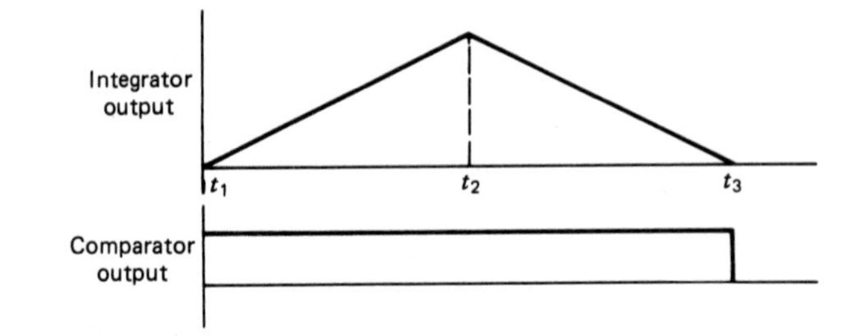

FIGURE 13-6 Integrator and comparator output waveforms for the circuit in Fig. 13-5.

As soon as the counter overflows, a "carry" pulse from the counter causes the control logic section to set switch S_1 to position B. With S_1 in position B, the reference voltage V_{ref} is connected to the input of the integrator, which causes the integrator capacitor C to discharge at a constant rate. During the period of discharge, from t_2 to t_3, the voltage V_A at the output of the comparator is given by the expression

$$V_A = \frac{1}{RC}\int_{t_2}^{t_3} V_{ref}\,dt \qquad (13\text{-}2)$$

As can be seen from Eq. 13-2, the rate of discharge of the integrator capacitor is determined by the value of the resistor R, the capacitor C, and the reference voltage V_{ref}, which must have good short-term and long-term stability. When the output of the integrator reaches zero, at t_3 in Fig. 13-6, the comparator changes states, setting its output "low," as shown in Fig. 13-6, which disables the counter. The count registered by the counter at this time is directly proportional to the ratio of the input voltage to the reference voltage. This proportional relationship can be developed mathematically by starting with Eq. 13-1, which is restated here,

$$V_A = \frac{1}{RC}\int_{t_1}^{t_2} V_i\,dt \qquad (13\text{-}1)$$

If the capacitor charges linearly, Eq. 13-1 can be simplified as

$$V_A = V_i\frac{T}{RC} = V_i\frac{t_2 - t_1}{RC} \qquad (13\text{-}3)$$

If the capacitor discharges at a linear rate, Eq. 13-2 can be simplified as

$$V_A = V_{ref}\frac{T}{RC} = V_{ref}\frac{t_3 - t_2}{RC} \qquad (13\text{-}4)$$

Since the right sides of both Eqs. 13-3 and 13-4 are equal to V_A, they can be set equal to each other as

$$V_i\frac{t_2 - t_1}{RC} = V_{ref}\frac{t_3 - t_2}{RC} \qquad (13\text{-}5)$$

Rewriting Eq. 13-5, we obtain

$$t_3 - t_2 = (t_2 - t_1)\frac{V_i}{V_{ref}} \qquad (13\text{-}6)$$

which shows that the count stored during the time interval $t_3 - t_2$ is directly proportional to the ratio of the input voltage to the reference voltage.

EXAMPLE 13-1

An integrator contains a 100-kΩ resistor and a 1-μF capacitor. If the voltage applied to the integrator input is 1 V, what voltage will be present at the output of the integrator after 1 sec?

Solution

Using Eq. 13-3, we compute the integrator output as

$$V_A = V_i \frac{t_2 - t_1}{RC}$$

$$= 1\text{ V} \times \frac{1\text{ sec}}{(1 \times 10^5\ \Omega)(1\ \mu\text{F})} = 10\text{ V}$$

Table 1-1 demonstrates that the results are dimensionally correct.

EXAMPLE 13-2

If the reference voltage applied to the integrator at time t_2 in Example 13-1 is 5 V in amplitude, what is the time interval from t_2 to t_3?

Solution

Using Eq. 13-6, we compute the time interval as

$$t_3 - t_2 = (t_2 - t_1)\frac{V_i}{V_{ref}}$$

$$= 1\text{ sec} \times \frac{1\text{ V}}{5\text{ V}} = 0.2\text{ sec}$$

13-5.3 Voltage-to-Frequency Converters

Voltage-to-frequency converters, as their name implies, convert an input voltage to a periodic waveform whose frequency is directly proportional to the input voltage. Voltage-to-frequency converters are very linear, wide-range, voltage-controlled oscillators (VCO). The basic concept of voltage-to-frequency conversion, for the purpose of measurement, is demonstrated in Fig. 13-7. The output signal from the VCO is applied to one input of the two-input AND gate. The second input to the AND gate is a gating pulse. During the time that both input signals are present, the output of the AND gate is identical to the VCO output. If there is a linear relationship between the VCO input voltage and the output frequency, the AND gate output can be applied to a digital counter to provide an indication of the VCO input voltage.

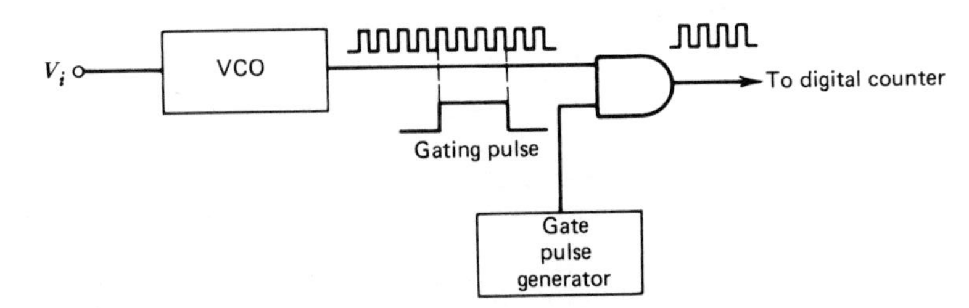

FIGURE 13-7 Block diagram of a basic voltage-to-frequency converter.

EXAMPLE 13-3

The relationship between the input voltage V_i and the output frequency f for the VCO in Fig. 13-7 is given as

$$V_i = \frac{f}{50}$$

If 530 pulses are passed by the AND gate during a 0.1-sec gating pulse, what is the amplitude of V_i?

Solution

The VCO output frequency is

$$f = \frac{\text{Pulses}}{\text{Gate duration}} = \frac{530 \text{ pulses}}{0.1 \text{ sec}} = 5300 \text{ Hz}$$

The voltage is, therefore,

$$V_i = \frac{f}{50} = \frac{5300}{50} = 106 \text{ V}$$

The basic circuit of Fig. 13-7 has limited usefulness, primarily because of the nonlinearity of the VCO. The block diagram shown in Fig. 13-8 is a more serviceable voltage-to-frequency converter. The basic circuit consists of an integrator, a voltage comparator, a pulse generator, and voltage reference source. When the unknown voltage is applied to the integrator, its output voltage begins to increase at a rate proportional to the amplitude of the input voltage. When the amplitude of the voltage at the output of the integrator exceeds the amplitude of the reference voltage, the comparator output changes states. This voltage change at the output of the comparator causes a pulse out of the pulse generator, which discharges the integrator capacitor and resets the comparator, after which a new ramp is initiated. A

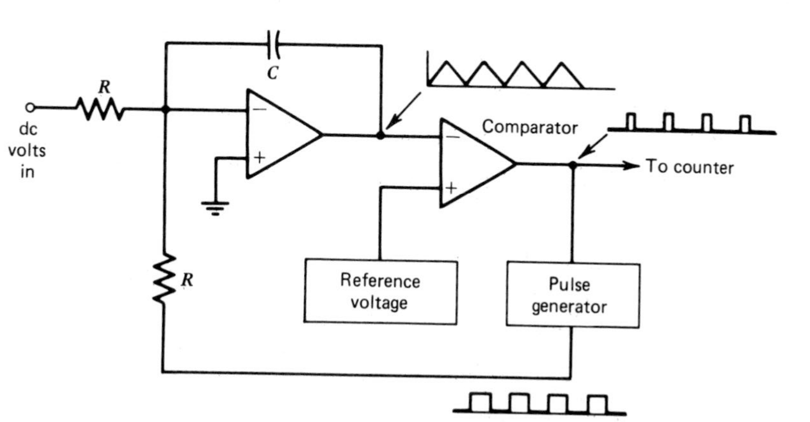

FIGURE 13-8 Voltage-to-frequency converter that uses an integrator.

short-duration pulse appears at the comparator output, at a frequency that is proportional to the input signal level. The number of pulses per unit of time can be counted with a digital counter, thereby completing the analog-to-digital conversion.

The primary advantages and limitations of voltage-to-frequency converters are as follows.

Advantages

1. Voltage-to frequency converters show good 60-Hz noise rejection without noise filters, which would reduce sampling speed.
2. The circuit is easily adapted to a digital counter.
3. The circuit requires no special overranging circuit.

Limitations

Accuracy is limited by

1. Stability of the integrating time constant.
2. Stability and accuracy of the comparator switching point.
3. Stability and accuracy of the reference voltage source.

13-6 COUNTING CIRCUITS

Many digital instruments provide a digital display of numbers that have been entered in a digital counter. These numbers may represent the frequency of a periodic signal, the digital equivalent of an analog signal, or the value of a passive component.

Counting circuits are a vital part of virtually all digital instruments and will therefore be discussed here in some detail.

13-6.1 The Binary Counter

A **binary counter** counts in the binary number system. In the binary system only two symbols are used for counting: 0 and 1. These two symbols are used in combinations to count beyond 1. In the familiar decimal, or base 10, number system, digits farther removed to the left from the decimal point carry greater "weight." We call the columns "units, tens, hundreds," and so on. The binary, or base 2, number system is much the same. The weight of the columns going left from the point is $2^0, 2^1, 2^2$ and so on. As the binary numbers develop, the symbol in the first column to the left of the point changes every 2^0 counts. Since 2^0 equals 1, the first column alternates between 0 and 1. The symbol in the second column changes every 2^1 counts. Therefore, the count pattern in the second column is an alternating pattern of two 0's and two 1's. The symbol in the third column alternates every fourth

TABLE 13-1
Decimal Numbers and Their Binary Equivalent

Binary Numbers	Decimal Numbers
0000	0
0001	1
0010	2
0011	3
0100	4
0101	5
0110	6
0111	7
1000	8
1001	9
1010	10

count, in the fourth column every eighth count, and so on. You should detect this pattern in the binary numbers presented in Table 13-1; these numbers represent the binary equivalent of the decimal numbers 0 through 10.

Binary counters are made up of circuits called multivibrators connected in cascade. Multivibrators are amplifier circuits with positive feedback that operate in only two states: either "on" in saturation or completely "off." Although there are three types of multivibrators, only the bistable multivibrator, or flip-flop, is of interest in this discussion. This circuit has two stable states and remains in one state or the other until triggered. Operating in this manner, the circuit lends itself ideally to counting in the binary system. Several flip-flops may be cascaded, as shown in Fig. 13-9, to allow large numbers of pulses to be counted.

An indicator, such as a lamp, may be connected to the output of each flip-flop. When an output voltage is present, the lamp is turned "on," indicating a pulse is stored in the flip-flop. The circuit counts pulses in the following manner. All flip-flops are initially "reset" so that the indicator lamps are turned "off." After one pulse is applied, flip-flop 1 changes states, which

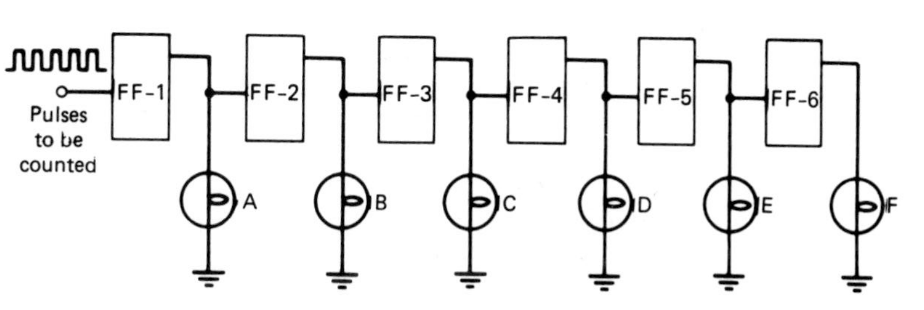

FIGURE 13-9 A binary counter using flip-flops.

turns lamp A "off." In addition flip-flop 2 changes states, which turns lamp B "on." When a pulse train is applied to the counter, flip-flop 1 changes states on every pulse, flip-flop 2 changes on every second pulse, flip-flop 3 changes on every fourth pulse, and so on. The progression equals $(2)^{n-1}$ where n equals the number of the flip-flop. Therefore, we can see that flip-flop 6 will change states on every thirty-second pulse because $(2)^{6-1}$ equals 32.

EXAMPLE 13-4

Which indicating lamps will be "on" after ten pulses are applied to the counter shown in Fig. 13.9?

Solution

The following sequence shows the lamps that are "on" after each pulse (1 indicates the lamp is "on").

Pulse	"On" lamps
	D C B A
0	0 0 0 0
1	0 0 0 1
2	0 0 1 0
3	0 0 1 1
4	0 1 0 0
5	0 1 0 1
6	0 1 1 0
7	0 1 1 1
8	1 0 0 0
9	1 0 0 1
10	1 0 1 0

We can see the pattern for the "on" lamps, which was just discussed. To determine which lamps are "on" without writing down the counting sequence, list the numbers raised to the second power that total to the desired number. The number 10 is $2^2 + 2^1$ or $8 + 2$. The number 8 is lamp D and 2 is lamp B.

EXAMPLE 13-5

Which indicating lamps will be "on" after 13 pulses are applied to the counter shown in Fig. 13-9?

Solution

The number 13 equals $2^3 + 2^2 + 2^0$ or $13 = 8 + 4 + 1$. Therefore, lamps D, C, and A will be "on."

Although the largest number that can be counted by a binary counter of n flip-flops is 2^n, the last pulse resets all flip-flop outputs. Therefore, in order to obtain a positive indication (all lamps "on"), the largest count that can be stored is 2^{n-1}.

13-6.2 The Decade Counter

A **decade counter** is a circuit of flip-flops in cascade that counts in the base 10 number system. This means that the counter progresses through a sequence of ten distinct counts. Three flip-flops in cascade progress through eight distinct states (the binary numbers 000 through 111), and four flip-flops in cascade progress through 16 distinct states (the binary numbers 0000 through 1111). Eight distinct states is too few, whereas 16 is too many for a decade counter. This problem can be overcome by using four flip-flops in cascade and by resetting the output of each flip-flop to 0 after the desired ten counts. Figure 13-10 shows four flip-flops in cascade. After ten pulses have been applied to the circuit, the B and D outputs will be high (equal to binary 1). Therefore, the output in binary will be 1010. We wish to reset the outputs to 0000 and repeat this counting sequence. This is the purpose of the AND gate. When B and D are equal to 1, the output from the AND gate resets all flip-flops to 0.

The waveforms in Fig. 13-11 show the pulse train applied to the trigger input of the decade counter of Fig. 13-10 and the output waveforms for each flip-flop. If we wish to drive some sort of indicating devices, such as lamps,

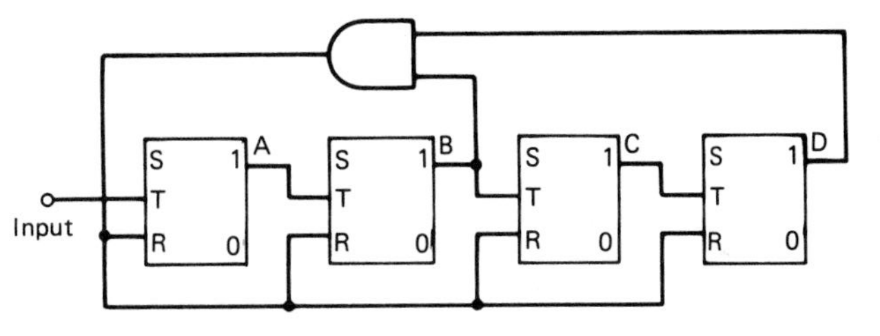

FIGURE 13-10 A decade counter.

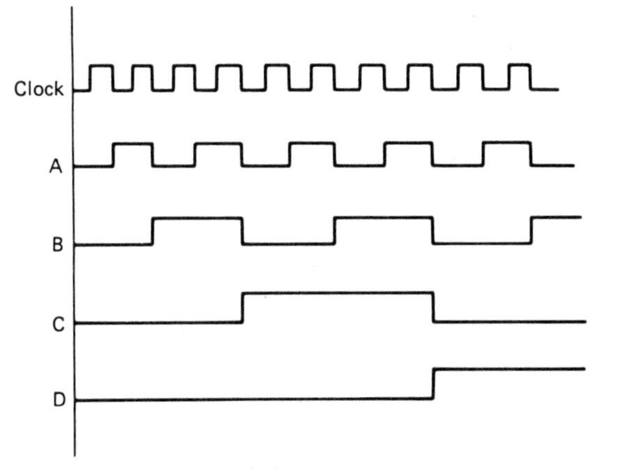

FIGURE 13-11 Waveforms for the decade counter of Fig. 13-10.

the signals must first be encoded into base 10 numbers. This can be accomplished with AND gates. The outputs of the flip-flops are applied to the AND gates. The AND gate inputs are a unique set of conditions that occur only once during the ten trigger pulses; therefore, each of the ten lights is "on" for only one particular pulse. This allows us to determine at a glance how many pulses have been counted. Before counting any pulses, we should check that the "0" light is "on." If the flip-flop outputs $\bar{A}$, $\bar{B}$, $\bar{C}$, and $\bar{D}$ are applied to the four-input AND gate shown in Fig. 13-12, the zero lamp will be turned "on."

EXAMPLE 13-6

Which four inputs should be applied to a four-input AND gate so that the output of the gate is high after six trigger pulses?

Solution

After six pulses, the flip-flop outputs equal binary 6, which is 0110. If all four AND gate inputs must be high, then the inputs can be $\bar{A}BC\bar{D}$. (If A and D and 0, then $\bar{A}$ and $\bar{D}$ are 1.)

The waveforms in Fig. 13-11 show that a flip-flop acts as a frequency divider. A flip-flop divides its input signal by 2. Frequency division, as well as counting upward, is an important characteristic of flip-flops.

Large-scale integration (LSI) has made it possible to incorporate an entire decade counter–divider circuit with binary-to-decimal encoding into one or

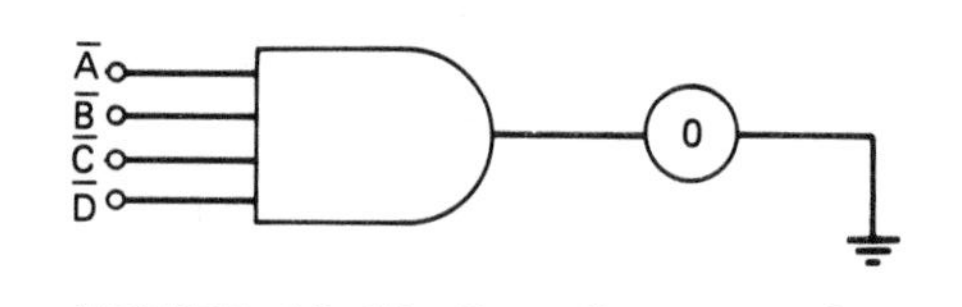

FIGURE 13-12 Decoding gate for zero lamp indicating 0 counts.

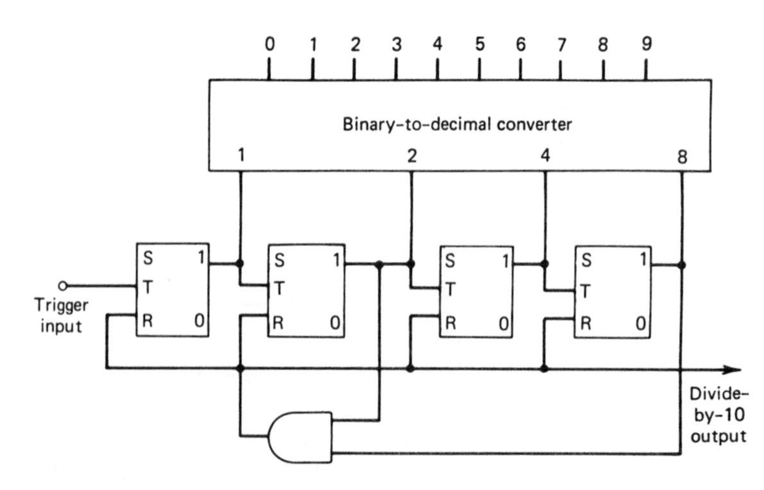

FIGURE 13-13 Block diagram of a decade counter–divider with base 10 readout.

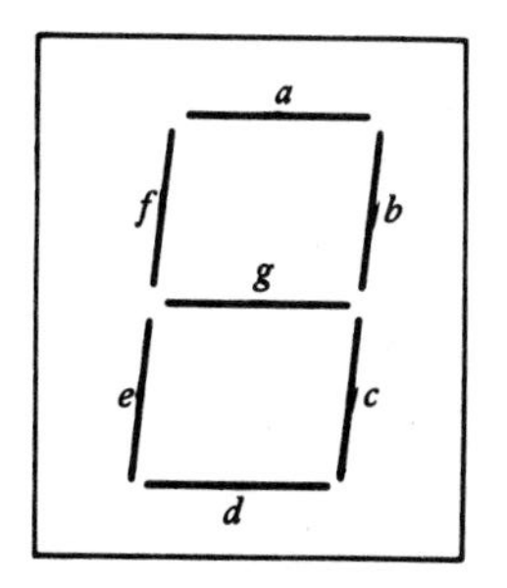

FIGURE 13-14 Seven-segment readout.

two integrated circuits (ICs). The binary-to-decimal encoding is generally accomplished with a binary-to-decimal converter. The block diagram of a decade counter–divider with decimal output is shown in Fig. 13-13.

If each of the ten outputs from the binary-to-decimal converter goes to a different indicating device, we have a rather cumbersome circuit and an output display that is confusing to interpret. A more desirable output device is a seven-segment digital readout, shown in Fig. 13-14.

When this type of readout device is used with a decade counter, the interfacing circuitry between the cascaded flip-flops and the seven-segment display is called a BCD (binary-coded decimal) to seven-segment decoder. The block diagram for a decade counter with digital readout is shown in Fig. 13-15.

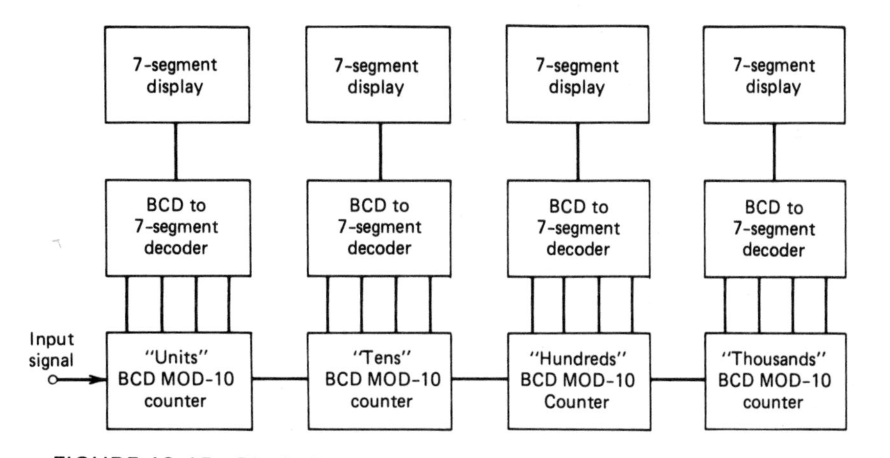

FIGURE 13-15 Block diagram of a decade counter with digital readout.

13-7 ELECTRONIC COUNTERS

The decade counters discussed in Section 13-6 are easily incorporated into a commercial test instrument called an electronic counter or universal counter. Considered alone, a decade counter behaves as a *totalizer* by totaling pulses applied to it during the time interval a gate pulse is present. However, this is but one of several modes of operation for most commercial electronic counters. Typical modes of operation are the *totalizing, frequency, period, ratio, time-interval,* and *averaging* modes.

13-7.1 The Totalizing Mode

In the totalizing mode, input pulses are totalized (counted) by the decade-counting units as long as switch S_1, shown in Fig. 13-16, is closed. If the pulse count exceeds the capacity of the decade counters, the overflow indicator is activated and the counter starts counting again from zero. If the overflow indicator is "on," the indicated count is ignored since it is incorrect.

13-7.2 The Frequency Mode

If the time interval in which pulses are being totalized is accurately controlled, the counter is operating in the frequency mode. Accurate control of the time interval is achieved by applying a rectangular pulse of known duration to the AND gate in Fig. 13-16 in place of the dc voltage source. This technique is referred to as "gating the counter." A block diagram of an electronic counter operating in the frequency mode is shown in Fig. 13-17. Commercial electronic counters use a more stable clock than an astable multivibrator. However, it is used here for the purpose of illustration because it simplifies the total circuit.

The frequency of the input signal is computed as

$$f = \frac{N}{t} \qquad (13\text{-}7)$$

where

f = frequency of the input signal
N = pulses counted
t = duration of gate pulse

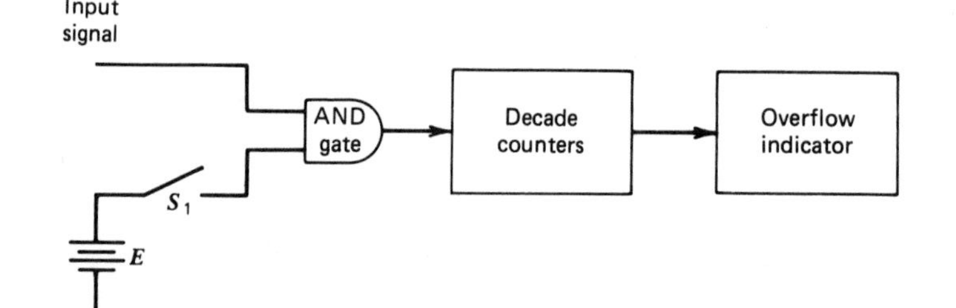

FIGURE 13-16 Block diagram of the totaling mode of an electronic counter.

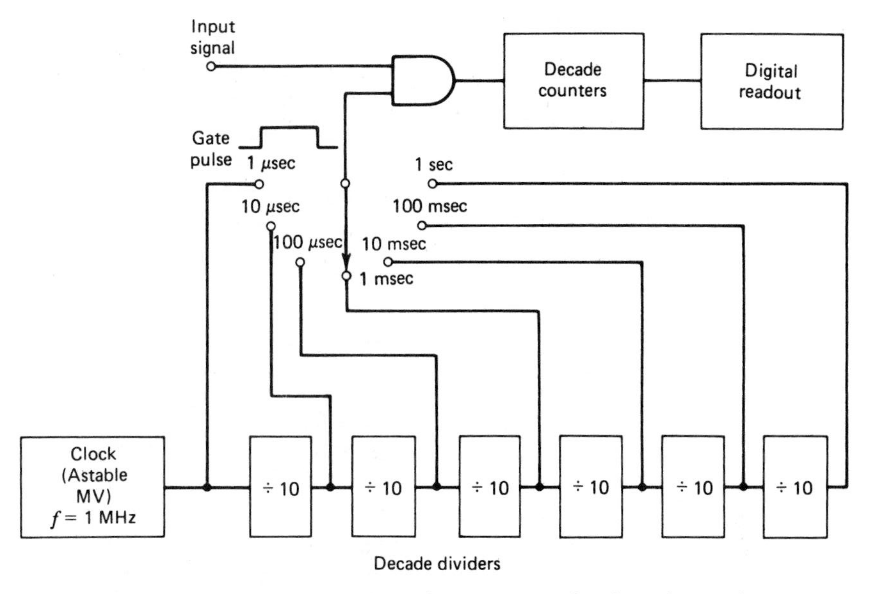

FIGURE 13-17 Block diagram of the frequency mode of an electronic counter.

13-7.3 The Period Mode

In some applications it may be desirable to measure the period of a signal rather than its frequency. Because the period is the reciprocal of the frequency, the period can easily be measured by using the input signal as a gating pulse and counting the clock pulses as shown in Fig. 13-18. The period of the input signal is determined from the number of pulses of known frequency, or known time duration, that are stored in the counter during one cycle of the input signal.

The period is computed as

$$T = \frac{N}{f} \tag{13-8}$$

where

T = period of the input signal
N = pulses counted
f = frequency of the clock

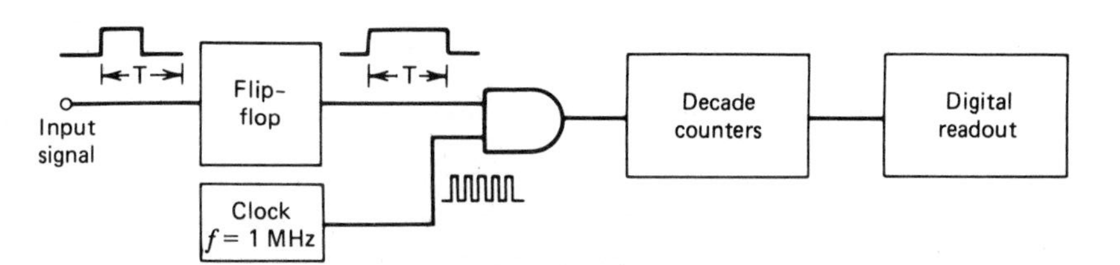

FIGURE 13-18 Block diagram of the period mode of an electronic counter.

13-7.4 The Ratio Mode

The ratio mode of operation simply displays the numerical value of the ratio of the frequencies of two signals. The lower-frequency signal is used in place of the clock to provide a gate pulse. The number of cycles of the higher-frequency signal, which are stored in the decade counters during the presence of the externally generated gate pulse, is read directly as the ratio of the frequencies. A basic circuit for the ratio mode of operation is shown in Fig. 13-19.

13-7.5 The Time-Interval Mode

The time-interval mode of operation measures the elapsed time between two events. The measurement can be carried out using the circuit of Fig. 13-20. As can be seen in the drawing, the gate is controlled by two independent inputs: the START input, which opens the gate, and the STOP input, which closes the gate. During the time interval between the START signal and the STOP signal, clock pulses accumulate in the register, thus indicating the time interval between the start and completion of an event.

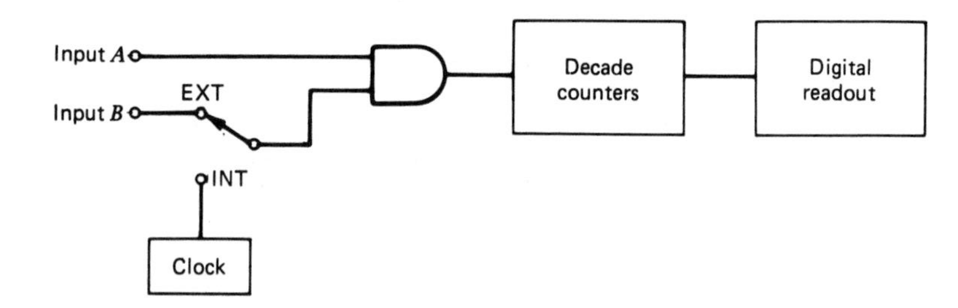

FIGURE 13-19 Block diagram of the ratio mode of an electronic counter.

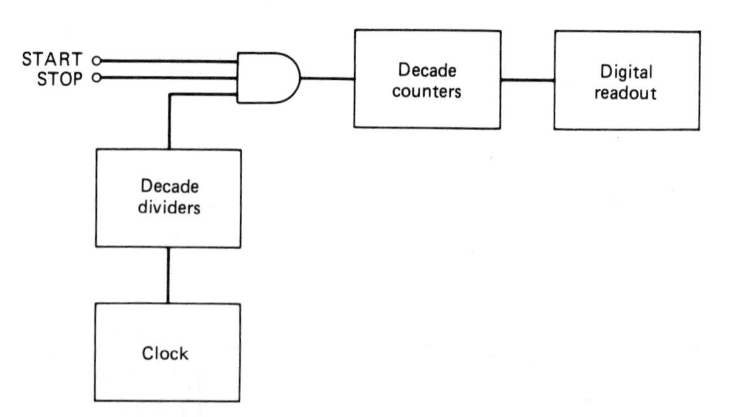

FIGURE 13-20 Block diagram of the time-interval mode of an electronic counter.

13-7.6 The Averaging Mode

It is sometimes desirable, when measuring frequency, period, or time interval, to increase accuracy and resolution by obtaining average measurements over several cycles, periods, or time intervals. This method is often referred to as multiple-period averaging.

13-8 COUNTER ERRORS

The primary sources of measurement error for an electronic counter are generally categorized as

- ± 1 count error.
- Time base error.
- Trigger error.
- Systematic error.

When a measurement is made with an electronic counter, a ± 1 count ambiguity can exist in the least significant digit. This ambiguity is generally the result of noncoherence between the internal clock signal and the input signal. The error caused by this ambiguity is in absolute terms: ± 1 count for the total accumulated count. Therefore, as the total accumulated count increases, the precentage of error attributable to ± 1 count error decreases.

Any error that is a result of a difference between the actual master clock frequency and its nominal frequency is directly translated into a measurement error. This error is called a time base error and is generally given as a dimensionless number expressed as so many parts per million.

Trigger error is a random error caused by noise on the input signal or noise from within the counter. The primary effect of noise is to cause the gate to be open for an incorrect period of time.

A systematic error is related to the instrument itself. Such things as the quality and age of the instrument affect amplifier rise times and triggering levels, causing what are called systematic errors.

13-9 COMMERCIAL ELECTRONIC COUNTERS

The instrument shown in Fig. 13-21 is a typical commercial electronic counter. The instrument is a Global Specialities Model 5001, which is a general-purpose, moderately priced counter–timer. The five-function instrument has a maximum frequency response of 10 MHz at input A, 2 MHz at input B, a sensitivity of 20 mV, and 1-MΩ input impedance. The two input channels have full signal conditioning, including attenuators, slope selection, variable triggering levels, and variable delays between measurements.

FIGURE 13-21 **Typical commercial electronic counter. (Courtesy Global Specialities Corporation.)**

13-10 DIGITAL MULTIMETERS

A basic digital multimeter (DMM) is made up of one of several types of analog-to-digital converters, including the three types discussed earlier in this chapter, and circuitry for counting, also discussed earlier. A block diagram of a basic digital multimeter is shown in Fig. 13-22. The attenuator networks are no different from those discussed in Chapter 8. Many commercial digital multimeters use the same attenuator network for both ac and dc voltage measurements.

The current-to-voltage converter shown in block form in Fig. 13-22 can be implemented with the circuit shown in Fig. 13-23. The current to be measured is applied to the summing junction Σ_i at the input of the operational amplifier. Since the current at the input of the amplifier is near zero because of the very high input impedance of the amplifier, the current I_R is very nearly equal to I_i. The current I_R causes a voltage drop, which is proportional to the current, to be developed across one of the resistors. This voltage drop is the analog input to the A/D converter, thereby providing a reading that is directly proportional to the unknown current.

Resistance is measured by passing a known current, from a constant-current source, through an unknown resistor. The voltage drop across the resistor is applied to the A/D converter, thereby providing an indication of the value of the unknown resistor.

EXAMPLE 13-7

A digital voltmeter that uses a dual-slope A/D converter is shown in Fig. 13-24. The reference voltage is 1 V, the clock frequency is 1 kHz, and the counter is a three-decade counter. The integrating capacitor changes to 5 V in the time interval from t_1 and t_2 (a 1-sec interval) in Fig. 13-6 and discharges at a rate of 10 V/sec. What is the input voltage if the counter has stored 1500 pulses?

Solution

During the 1-sec period of charge on the capacitor, the counter stores 1000 pulses. Capacitor discharge requires only 0.5 sec, since it is discharging at a rate of 10 V/sec. During this 0.5 sec the counter stores an additional 500

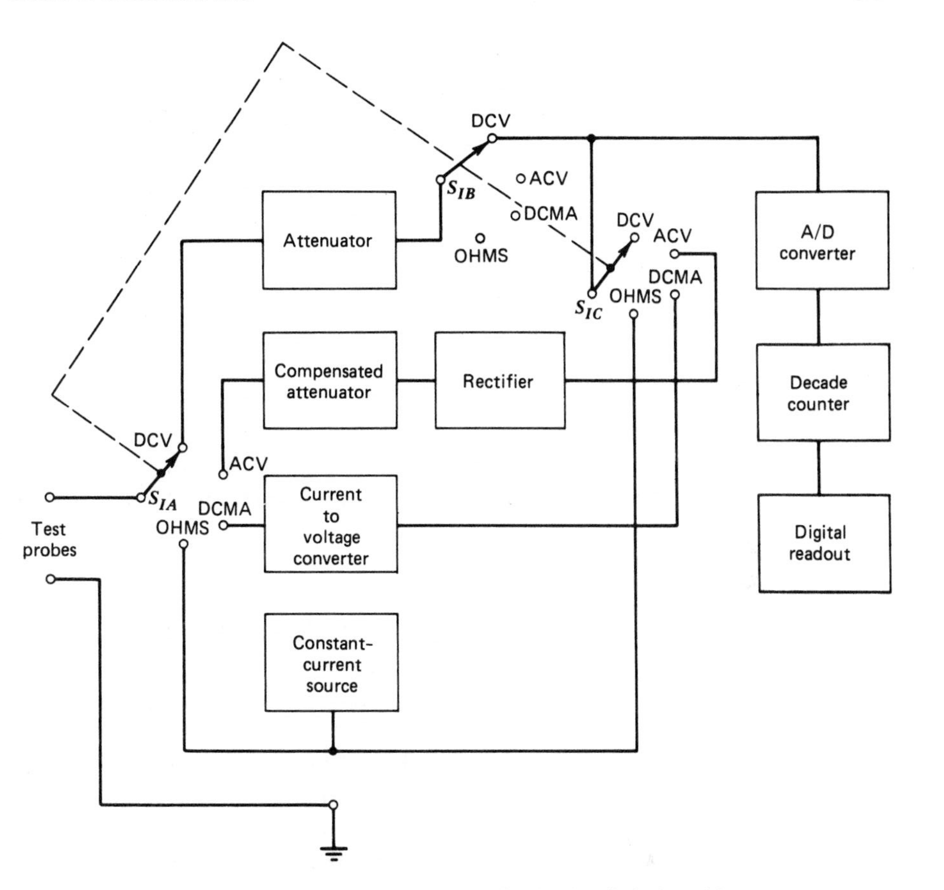

FIGURE 13-22 Block diagram of a basic digital multimeter.

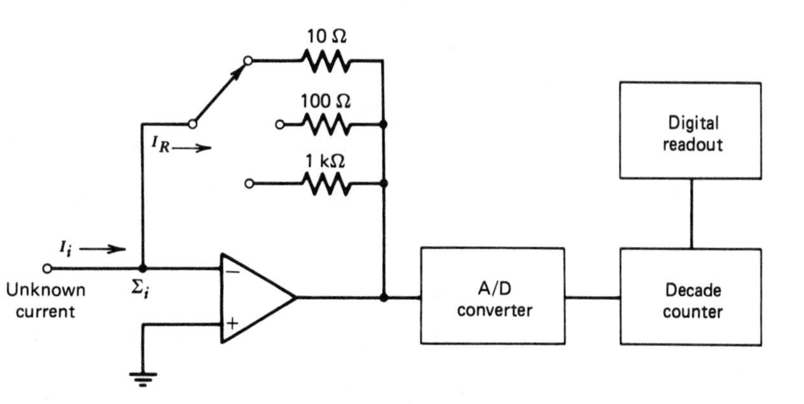

FIGURE 13-23 Block diagram of a basic current-to-voltage converter.

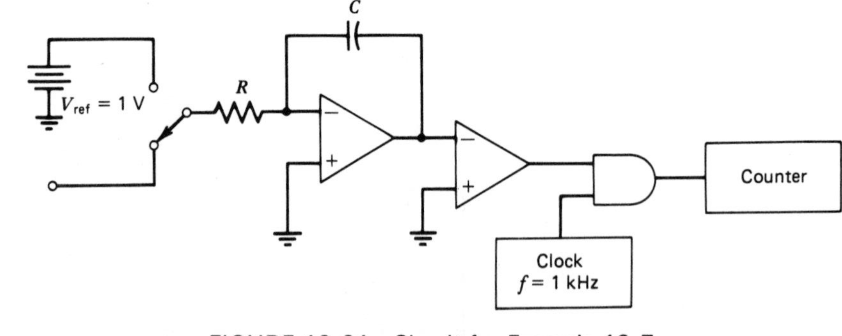

FIGURE 13-24 Circuit for Example 13-7.

pulses, or a total of 1500 pulses. Since the counter consists of three decades, the count registered will be 500, which represents an input voltage of 0.5 V.

13-11 COMMERCIAL DIGITAL MULTIMETERS

Digital multimeters are comparative newcomers for measurement outside of research laboratories. Until fairly recently, because of their cost, their use was limited primarily to the laboratory. However, significant advances in integrated circuit technology have permitted price reductions to the point that DMMs are price-competitive with analog instruments. The number of makes and models of DMMs from which we can select is almost staggering. The instrument shown in Fig. 13-25 is a typical commercial digital multimeter. The instrument, a Model 8050A, manufactured by the John Fluke Manufacturing Company, is a microprocessor-controlled, $4\frac{1}{2}$-digit DMM. The instrument is capable of measuring dc voltage, ac voltage, direct current, alternating current, resistance, conductance, and decibels. Some of the major specifications are listed below:

DC Voltage

- Five ranges from ± 200 mV to ± 1000 V.
- Resolution—10 μV on the lowest range.
- Accuracy—$\pm 0.03\%$ of reading + two digits.

AC Voltage

- Five ranges from 200 mV to 750 V.
- Resolution—10 μV on the lowest range.
- Accuracy—frequency-dependent, best accuracy of 0.5% + 10.
- Digits between 45 Hz and 1 kHz on all ranges.

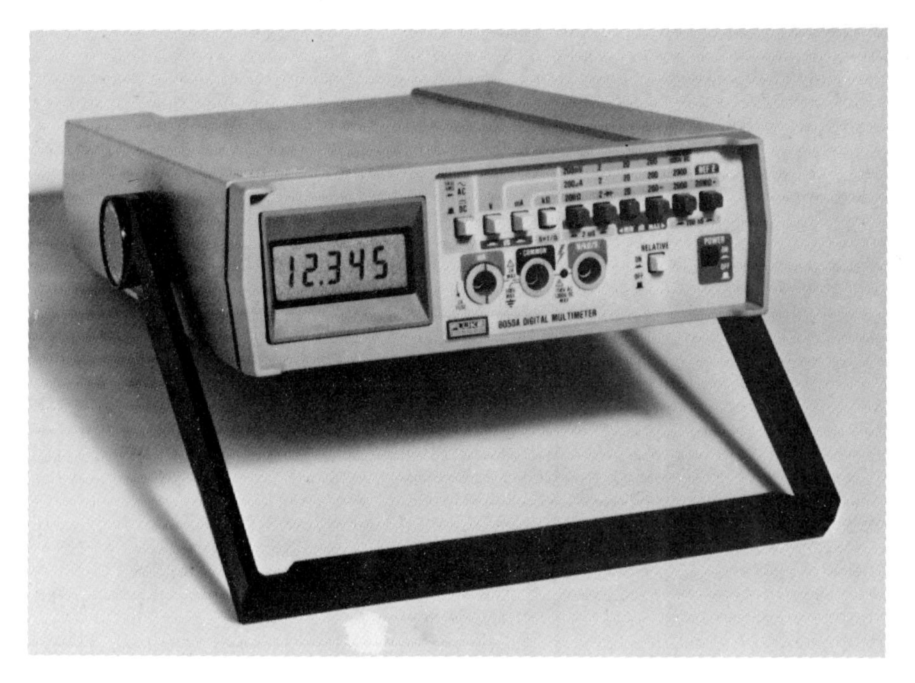

FIGURE 13-25 Typical commercial digital multimeter. (Courtesy John Fluke Manufacturing Company, Inc.)

DC Current

- Five ranges from $\pm 200\ \mu A$ to ± 2000 mA.
- Resolution—$\pm 0.01\ \mu A$ on the lowest range.
- Accuracy—$\pm 0.3\%$ of reading + two digits.

AC Current

- Five ranges from 200 μA to 2000 mA.
- Accuracy—frequency-dependent, best accuracy of $\pm 1\%$ + ten digits between 45 Hz on all ranges.
- Two kilohertz on all ranges.

Resistance

- Six ranges from 200 Ω to 20 MΩ.
- Accuracy—$\pm 0.1\%$ of reading + two digits + 0.02 Ω on lowest range.

13-12 MICROPROCESSOR-BASED INSTRUMENTS

The digital instruments discussed to this point in Chapter 13 are designed around digital logic circuits without memory circuitry. As was stated in Chapter 5, now that microprocessors can be made an integral part of

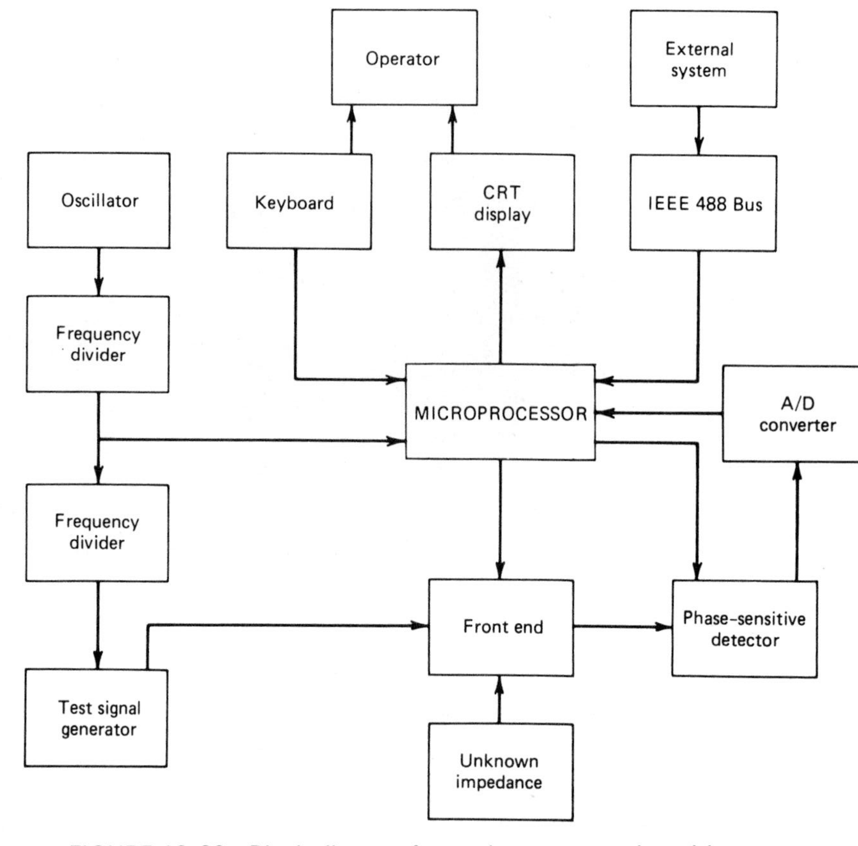

FIGURE 13-26 Block diagram for a microprocessor-based instrument.

measuring instruments, a whole new class of instruments called *intelligent instruments* have become possible.

Figure 13-26 shows a block diagram of a microprocessor-based impedance-measuring instrument. The operator makes interface with the instrument via the IEEE 488 bus to allow control by, or to make measurements available to, a larger, external computer system. The timing clock signal and the ac test signal are provided by frequency division of the oscillator signal. The "front-end" circuit applies the test signal to the unknown impedance and a standard impedance and provides an output signal, proportional to the voltage across each, to the phase-sensitive detector. Signal transfer is controlled by the microprocessor. The phase-sensitive detector, which is also controlled by the microprocessor, converts the ac input of the impedances in vector form to a dc output. The A/D converter provides the digital data which the microprocessor uses to compute the value of the unknown impedance. This value is then displayed on the CRT or outputted to the IEEE 488 bus.

13-13 THE IEEE 488 BUS

The IEEE 488 bus provides digital interfacing between programmable instruments. There are many instrumentation systems in which interactive instruments, under the command of a central controller, provide superior error-free results compared to those of conventional manually operated systems. Problems such as impedance mismatch, obtaining cables with the proper connectors, and logic-level compatibility are also eliminated by designing the system around bus-compatible instruments.

The basic structure of the IEEE 488 bus, which interfaces with interactive instruments, is shown in Fig. 13-27. Every device in the system must be able to perform at least one of the roles of *talker*, *listener*, or *controller*. A talker can send data to other devices via the bus. Some devices, such as programmable instruments, can both *listen* and *talk*. In the listen mode the device may receive an instruction to make a particular measurement, and in the talk mode it may send its measurement. A controller *manages* the operation of the bus system. It controls data gathering and transfer by designating which devices are to talk or listen, as well as controlling specific actions within other devices.

The IEEE first adopted the 488 bus standard in 1975. As an indication of its acceptance and potential, over 100 companies worldwide manufacture bus-compatible instruments. Tremendous progress has been made in providing bus-compatible instruments, but some problems remain, most of them related to software. Because there is no standard software, users may encounter problems when attempting to interface instruments from different manufacturers.

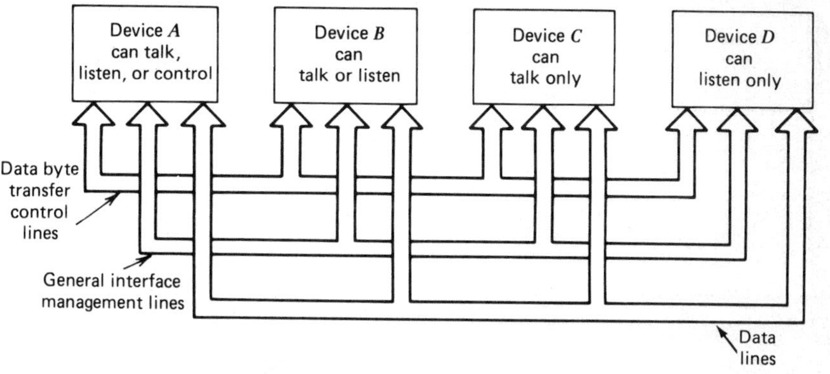

FIGURE 13-27 Block diagram of devices interfaced with the IEEE-488 bus.

13-14 APPLICATIONS

Electronic counters find many applications in research and development laboratories, in standards laboratories, on service benches, and in the everyday operation of many electronic installations.

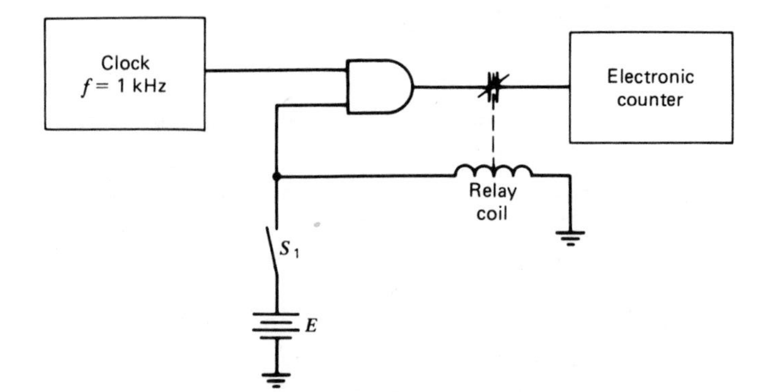

FIGURE 13-28 Checking for the operating delay of a relay.

Counters are used in communications to measure the carrier frequency, in digital systems to measure the clock frequency and other digital signals, and in many other applications. Figure 13-28 shows how relay delay time can be measured with an electronic counter. At the instant the switch S_1 is closed, the counter begins to register count. At the same instant the relay coil is energized; however, some delay is encountered before normally closed relay contacts open. If the frequency of the clock is adjusted to 1 kHz, the number of counts registered by the counter is the number of milliseconds of delay associated with the relay. Electronic counters also find many applications in time-interval measurements. Such things as pulse width, rise time, and phase shift can be measured. To measure phase shift, the technician measures the period of the reference waveform, shown in Fig. 13-29. The time interval, t_ϕ, is then measured, and the phase shift, in degrees, is computed using the expression

$$\text{Phase shift} = \frac{t_\phi}{T} \times 360 \qquad (13\text{-}9)$$

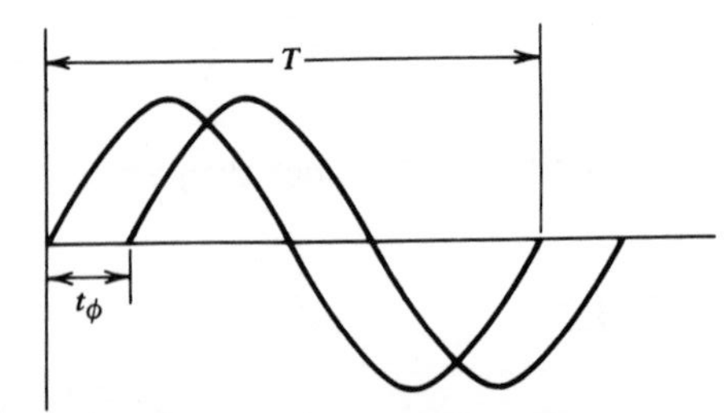

FIGURE 13-29 Phase-shift measurement with an electronic counter.

13-15 SUMMARY

Digital instruments, which are relative newcomers on the measurements scene, offer some significant advantages to the user. Because of recent technological accomplishments and accompanying cost reductions, digital instruments are cost-competitive with analog instruments and are therefore experiencing widespread industrial use.

At this time, a large percentage of the digital instruments in use are electronic counters and digital multimeters. However, increasing numbers of instruments are incorporating digital techniques.

13-16 GLOSSARY

Analog-to-digital converter: An electronic circuit that accepts an analog voltage input and provides a digital output signal that is equivalent to the analog input.

Binary counter: A circuit consisting of digital logic blocks arranged to count in the binary, or base 2, number system.

Decade counter: A circuit consisting of digital logic blocks arranged to count in the decimal, or base 10, number system. A decade counter is a binary counter modified to count base in 10.

Digital instruments: Instruments that utilize digital logic circuits, which are primarily counting circuits, for measuring.

Integrator: A circuit containing a resistor and capacitor in series with the output taken across the capacitor. The output voltage represents a continuous tally of how much voltage has, in a sense, been accumulated by the capacitor.

Overranging: A technique used with digital voltmeters to permit a voltage that exceeds a particular range by a nominal amount to be measured without changing ranges.

13-17 REVIEW QUESTIONS

The following questions should be answered after a thorough study of the chapter. The purpose of the questions is to determine the reader's comprehension of the material.

1. What are some advantages of digital instruments over analog instruments?
2. What does the term *half-digit* mean?
3. What is the most widely used type of analog-to-digital converter in digital voltmeters?
4. What are some advantages and disadvantages of the voltage-to-frequency analog-to-digital converter?
5. How can a binary counter be modified to count in the decimal system?

6. What are the three modes of operation of electronic counters?

7. What are three types of error associated with electronic counters?

8. What is the purpose of the IEEE 4888 bus?

13-18 PROBLEMS

13-1 An integrator consists of a 150-kΩ resistor and a 0.8-μF capacitor. If the voltage applied to the integrator is 0.4 V, what voltage will be present at the output of the integrator after 1.2 sec?

13-2 An input voltage of 0.6 V is applied to an integrator. After 1.5 sec the output voltage is 9 V. If the circuit contains a 60-kΩ resistor, what is the value of the capacitor?

13-3 The ratio of the output frequency to input voltage for the VCO in Fig. 13-30 is 80, disregarding units. If 480 pulses are stored in the counter during a 0.1-sec gating pulse, what is the amplitude of V_i?

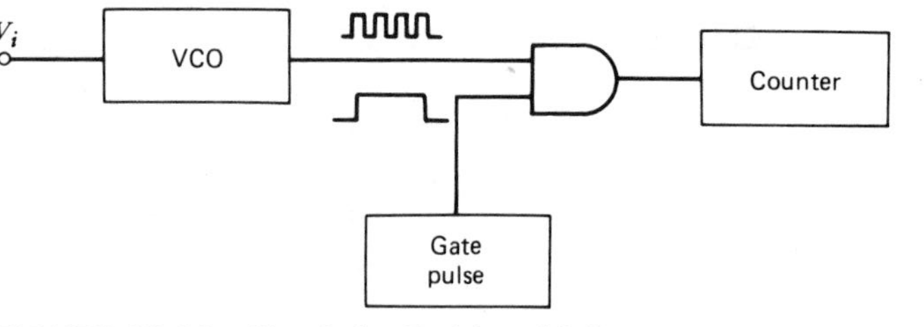

FIGURE 13-30 Circuit for Problem 13-3.

13-4 The ratio of input voltage to output frequency for the VCO in Fig. 13-30 is 0.2. If 405 pulses are stored by the counter during one gate pulse, what is the duration of the gate pulse?

13-5 Which indicating lamps will be on after nine pulses are applied to the counter shown in Fig. 13-9?

13-6 Output signals from a decade counter are applied to four-input AND gates for decoding. Inputs $\bar{A}B\bar{C}D$ are applied to one AND gate. What count indication does this represent?

13-7 An electronic counter is being used in its *period* mode. If 500 cycles of a 2-MHz clock are counted during the period of the input signal, what is the frequency of the input signal?

13-8 What is the period of a signal that causes the decade counter in Fig. 13-17 to register a count of 4275 when the gate pulse is 1 msec in duration?

13-9 The DVM shown in Fig. 13-24 uses a 1-V reference voltage, a 1-kHz clock, and a three-decade counter. The integrating capacitor charges to 6 V in 1 sec (t_1 to t_2 in Fig. 13-6) and discharges at a rate of 10 V/sec. What is the input voltage if the counter has stored 1600 pulses?

13-10 A constant current of 1 mA is passed through an unknown resistor to determine its resistance. The voltage drop across the resistor is applied to a voltage-to-frequency A/D converter with a voltage and frequency constant of 0.0025. If the output from the A/D converter causes 1580 pulses to be stored by the counter during a 0.1-sec gate pulse, what is the value of the resistor?

13-19 LABORATORY EXPERIMENTS

Experiments E31 and E32 apply the theory that has been presented in Chapter 13. The purpose of the experiments is to provide hands-on experience to reinforce the theory.

Both experiments use standard components and equipment, which should be available in any electronics laboratory. The contents of the laboratory report to be submitted by each student are included at the end of each experimental procedure.

CHAPTER 14

Troubleshooting with Instruments

14-1 Instructional Objectives

This chapter will discuss some practical troubleshooting techniques. It offers coverage of digital troubleshooting. After completing Chapter 14 you should be able to

1. Explain the generalized troubleshooting procedure.
2. Define signal tracing.
3. Define signal injection.
4. List six basic types of signal-flow paths.
5. Explain basic digital troubleshooting using a logic probe, pulser, clip, current tracer, and comparator.

14-2 INTRODUCTION

Solid-state troubleshooting may involve widely different kinds of equipment from radio receivers to hi-fi stereo units and industrial electronic installations to highly complex digital equipment. However, a troubleshooting procedure always starts with a preliminary analysis of symptoms of trouble from which various possibilities of malfunction are deduced. These are considered in order of probability, and various quick checks are usually made to eliminate or verify initial deductions.

After a malfunction has been confirmed, it may or may not be possible to proceed directly with repair activities. In many situations tests and measurement must be made to isolate the defective component or device, or to identify the source of malfunction. Although electronic instruments are indispensable in troubleshooting procedures, measurements must be meaningful if they are to guide the technician to the equipment fault. Measurements are meaningful only to the extent that the technician understands circuit actions and the significance of measured values.

14-3 GENERALIZED TROUBLESHOOTING

A summary of a generalized troubleshooting procedure is presented in Fig. 14-1. Unless the troubleshooter is fully familiar with the equipment, service data (or least a schematic diagram) will be practically indispensable. For example, standard service data for a television receiver provides a complete schematic diagram with dc voltages, resistor values, capacitor values, and operating waveforms. Service notes and instructions are given for receiver disassembly. Sometimes a check chart for device failures is provided. Signal flowcharts are included for the more elaborate type of receivers.

With reference to Fig. 14-1, quick checks include replacement of suspected modules, transistor turn-off and turn-on tests, bridging of suspected open components, and so on. Notice that there is not a sharp dividing line between quick checks and preliminary troubleshooting procedures. For example, a technician may make a current variation test of a "dead" radio receiver. This type of test can be classed as either a quick check or as a preliminary troubleshooting technique. Again, a technician may couple a normally operating radio receiver to a "dead" receiver to check for the possibility of an inoperative oscillator. This kind of test can be regarded as a quick check or as a preliminary troubleshooting technique.

Numerous test instruments are used in various areas of electronic servicing. The most basic instrument is the multimeter, used to make measurements of dc voltage, resistance, ac voltage, and dc current values. The amplitude-modulation (AM) signal generator ranks next in importance. It is used to make signal-injection tests and to align tuned circuits. Most technicians consider the oscilloscope to be the third-ranking instrument. Oscilloscopes are used for waveform analysis of signal tracing and peak-to-peak voltage measurements. Peak-to-peak voltage values can also be measured by many transistor multimeters. Service-type digital voltmeters are usually limited to rms (sine-wave) measurements of ac voltages, however. Semiconductor testers are also generally regarded as basic service instruments. Most technicians prefer an instrument that provides both in-circuit and out-of-circuit tests of both bipolar and unipolar transistors.

14-4 FUNCTIONAL BLOCK DIAGRAM

A functional **block diagram** is an overall representation of the functional units within the equipment as well as of the signal-flow paths between them. Figure 14-2 shows a typical functional block diagram for an AM transceiver set composed of six functional units. Thus, each unit may consist of a variety of circuits or stages, each performing its own major electronic function. For example, the transmitter unit may contain a radio-frequency (RF) oscillator circuit, an RF voltage amplifier circuit, and several RF power amplifier circuits, but its major electronic function is to produce an RF carrier wave having enough power to "transmit" the signal.

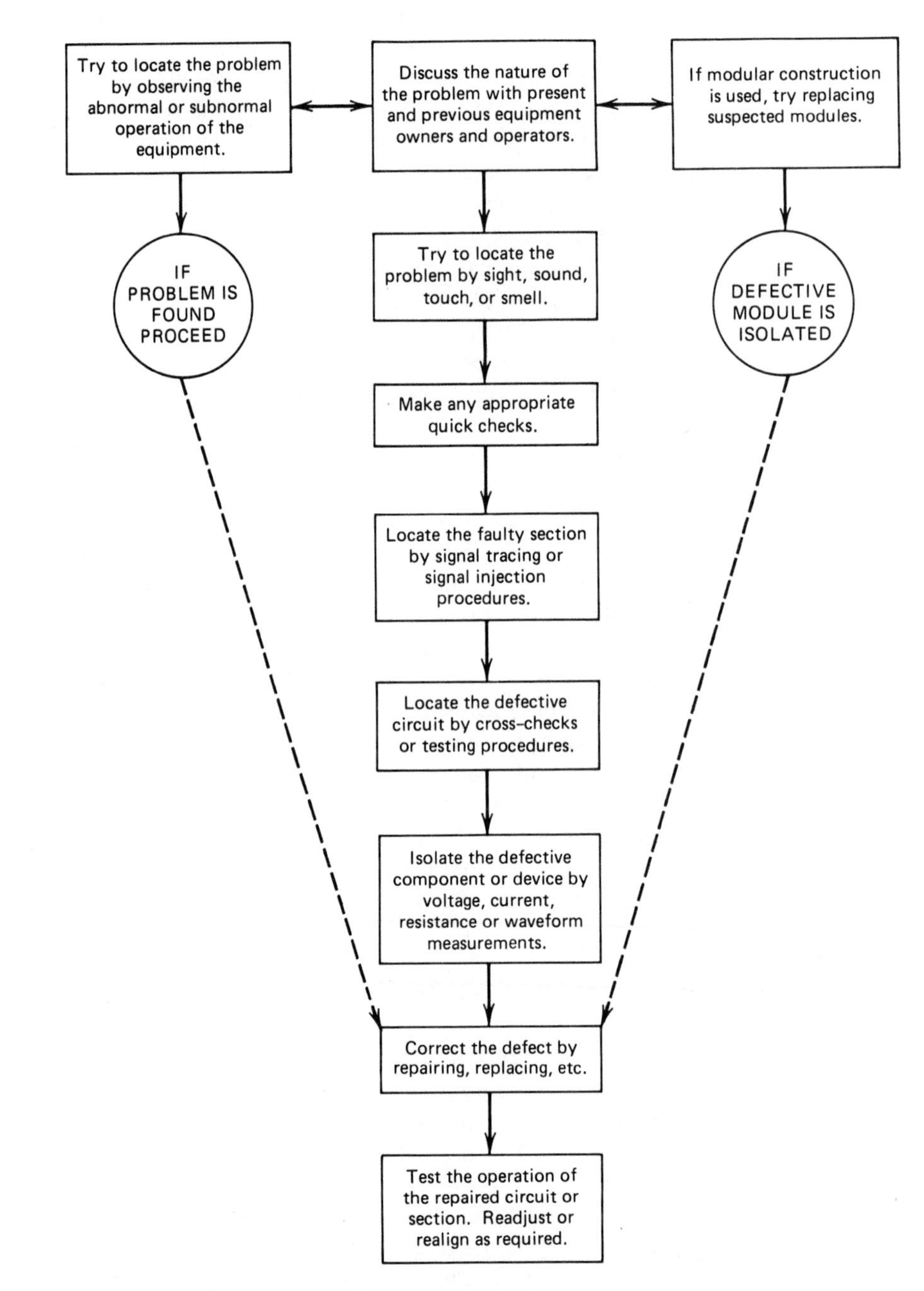

FIGURE 14-1 Generalized troubleshooting procedure.

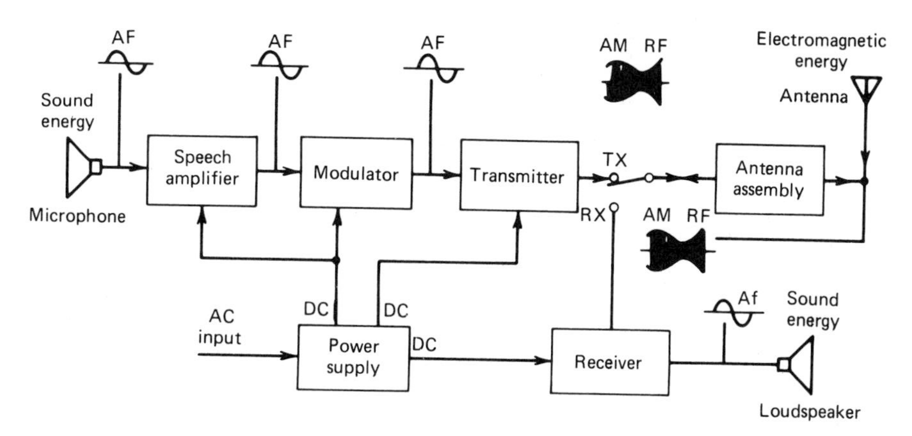

FIGURE 14-2 **Functional block diagram of an amplitude-modulated transceiver set.**

The connecting lines between the various functional blocks (units) represent important signal-flow connections, but the diagram does not necessarily indicate where these connections can be found in the actual equipment circuitry. Sometimes additional information such as frequencies and input and output waveforms may be included to show how far each type of signal progresses through the equipment.

Block diagrams provide a pictorial guide for isolating the trouble. On simple equipment, such as that covered only by a data sheet rather than a technical manual, the entire assembly will be represented by one servicing block diagram. Circuits within the functional units are enclosed by a dashed line, as are all circuits comprising the circuit groups within the unit.

Waveforms are given at several points, usually at the input and output of each circuit. Test points are identified by both letter and number. Generally, test points that are useful in localizing faulty functional units are numbered. Test points helpful in isolating faulty circuit groups or individual circuits are lettered. Notice that the physical location of the circuit groups within the equipment has no relation to their representation on the servicing block diagram.

14-5 SIGNAL-TRACING PROCEDURES

Signal tracing with an oscilloscope is a comparatively simple and effective method for locating defective devices or components. The amplitude of a displayed waveform can be quickly measured. Signal-tracing methods require that the operator identify the basic type of signal path to be tested. As shown in Fig. 14-3, six basic types of signal-flow paths are encountered in electronic

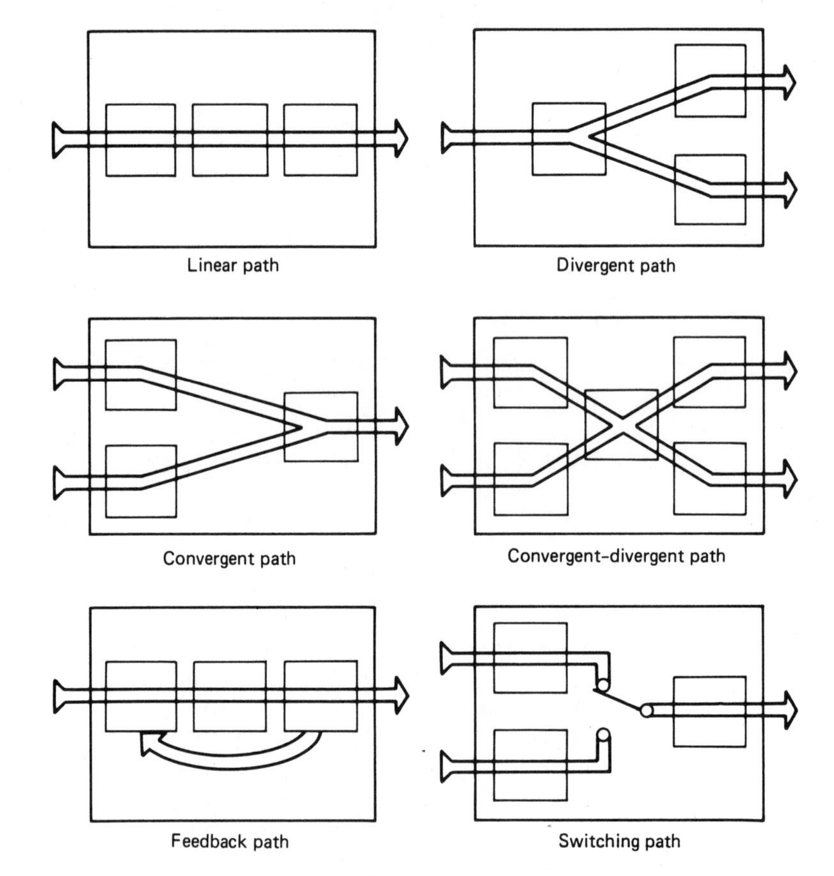

FIGURE 14-3 Basic types of signal paths.

equipment. These are

1. Linear path.
2. Divergent path.
3. Convergent path.
4. Convergent–divergent path.
5. Feedback path.
6. Switching path.

A **linear signal-flow path** is exemplified by a basic audio amplifier. A **divergent signal-flow path** is typified by a stereophonic amplifier. In a color TV receiver, the black and white and the chrome signals are combined in the color picture tube—an example of a **convergent signal-flow path**. A **convergent–divergent signal-flow path** is seen in a stereo decoder. An incoming composite stereo signal is combined with a locally generated subcarrier signal to form a reconstituted stereo signal, which is then separated

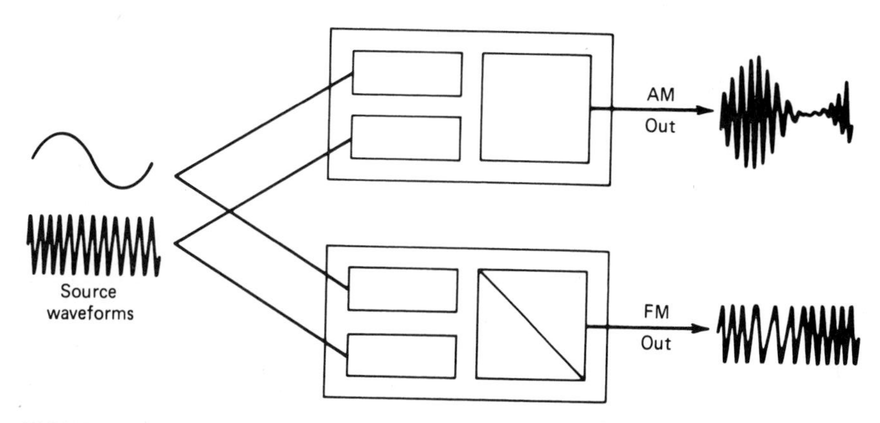

FIGURE 14-4 Amplitude modulation and frequency modulation represent basic forms of signal processing.

into left and right stereo signals. A **feedback path** is typified by an audio amplifier with a negative feedback loop. Finally, a **switching signal path** is encountered in any form of digital circuitry.

Effective signal-tracing procedures often also require that the operator recognize the type of signal processing that is being used in a particular situation. Amplitude modulation (AM) and frequency modulation (FM) represent basic forms of signal processing, as depicted in Fig. 14-4. An oscilloscope provides the most useful indication of modular circuit action. Modulation involves nonlinear circuit action, whereas amplification involves linear circuit action. AM and FM are distinguished by the type of nonlinear action that is used.

14-6 SIGNAL-INJECTION PROCEDURES

Signal injection is another of the basic troubleshooting procedures carried out with an oscilloscope. This method is used to measure stage gain or section gain, to determine where a signal may be blocked along a signal path, to check the transient response of a stage or of a section, and to determine the frequency response of a stage or of a section. With reference to Fig. 14-5, a radio-frequency (RF) sweep and marker signal is injected at the input of the mixer stage, and an oscilloscope is connected at the input of the FM detector stage to check the frequency response of the intermediate-frequency (IF) strip. Or, if the oscilloscope is connected at the output of the FM detector stage, the frequency response of the ratio detector or discriminator is displayed.

If an audio section is to be checked, a signal from an audio generator is injected at a selected point, and an oscilloscope is connected at the output of the stage or section under test. Similarly, a modulated RF signal from an RF generator can be used to inject a signal at a chosen point, and the output

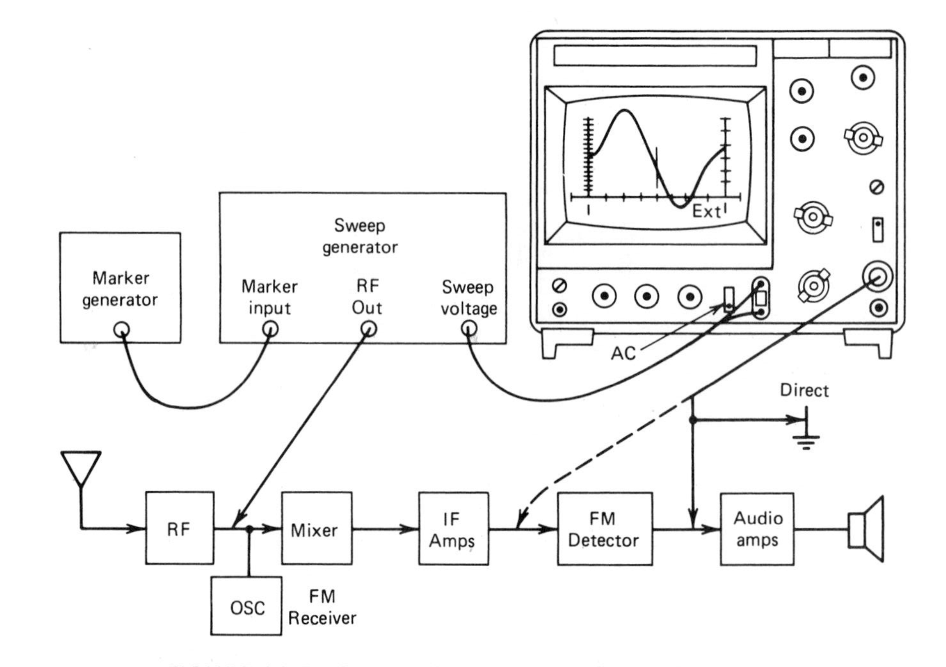

FIGURE 14-5 Sweep alignment setup for an FM receiver.

stage is checked with the oscilloscope. If the frequency of the output signal exceeds the frequency capability of the oscilloscope, use a demodulator probe; otherwise a low-capacitance probe must be used with the oscilloscope.

14-7 SIGNAL TRACING VERSUS SIGNAL INJECTION

Both signal-tracing and signal-injection methods are often used in solid-state and digital troubleshooting. In signal tracing the test probe is moved from test point to test point while the signal is applied at a fixed point. Oscilloscopes, multimeters, and loudspeakers are used as test instruments.

Signal injection or **signal substitution** is accomplished by injecting an artificial signal (from a signal generator, etc.) from point to point while the indicating device remains fixed at one point.

Troubleshooting often involves both signal tracing and signal injection. For example, an audio oscillator can be used to inject a signal at the input of a solid-state audio system, an oscilloscope to observe the waveform at each stage. Or a pulse generator can introduce pulses into a digital system as an oscilloscope displays waveforms or levels of various gates, flip-flops, and so on.

14-8 WAVEFORM MEASUREMENTS

Waveform measurements are of amplitude, frequency, and phase. Amplitude measurements are generally made in peak-to-peak voltage units (Fig. 14-6). To measure the peak voltages in pulse waveforms (or other complex waveforms), the technician selects the ac input function of the oscilloscopes. The zero-volt axis (the resting level of the horizontal trace) is noted. Then the signal voltage is applied to the ac input terminal of the oscilloscope. The positive portion of the waveform is displayed above the zero-volt axis, and the negative portion of the waveform is displayed below the zero-volt axis.

Frequency measurements are made to the best advantage with an oscilloscope that has triggered sweeps and a calibrated time base. The period of a displayed waveform is found by counting the number of horizontal divisions occupied by one cycle and by noting the setting of the Time/Div. control. The frequency is the reciprocal of this product (period in seconds).

Phase measurements can be made in several ways. The simplest method is to use a dual-trace oscilloscope. With this instrument the phase shift between the input and the output of an amplifier can be checked. If a single-trace oscilloscope is to be used for measuring phase angles, the input signal voltage to the unit under test can be applied to the vertical input channel of the oscilloscope, and the output signal voltage from the unit can be applied to the horizontal channel. A Lissajous pattern whose characteristics provide the measurement of phase differences between input and output signals is displayed.

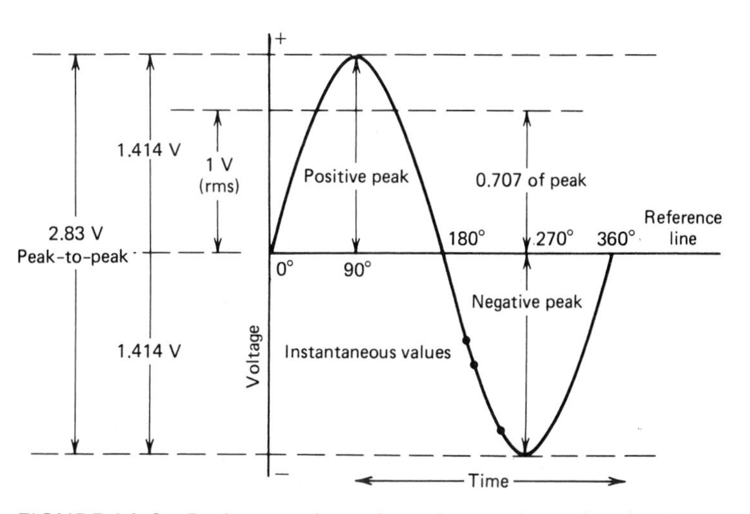

FIGURE 14-6 Peak-to-peak, peak, and rms values of a sine wave.

14-9 DISTORTION ANALYSIS

There are four basic types of distortion.

1. Amplitude distortion.
2. Frequency distortion.
3. Crossover and stretching distortion.
4. Harmonic distortion.

A general check of amplitude distortion can be made as shown in Fig. 14-7. The output from a sine-wave generator is applied to the input of the

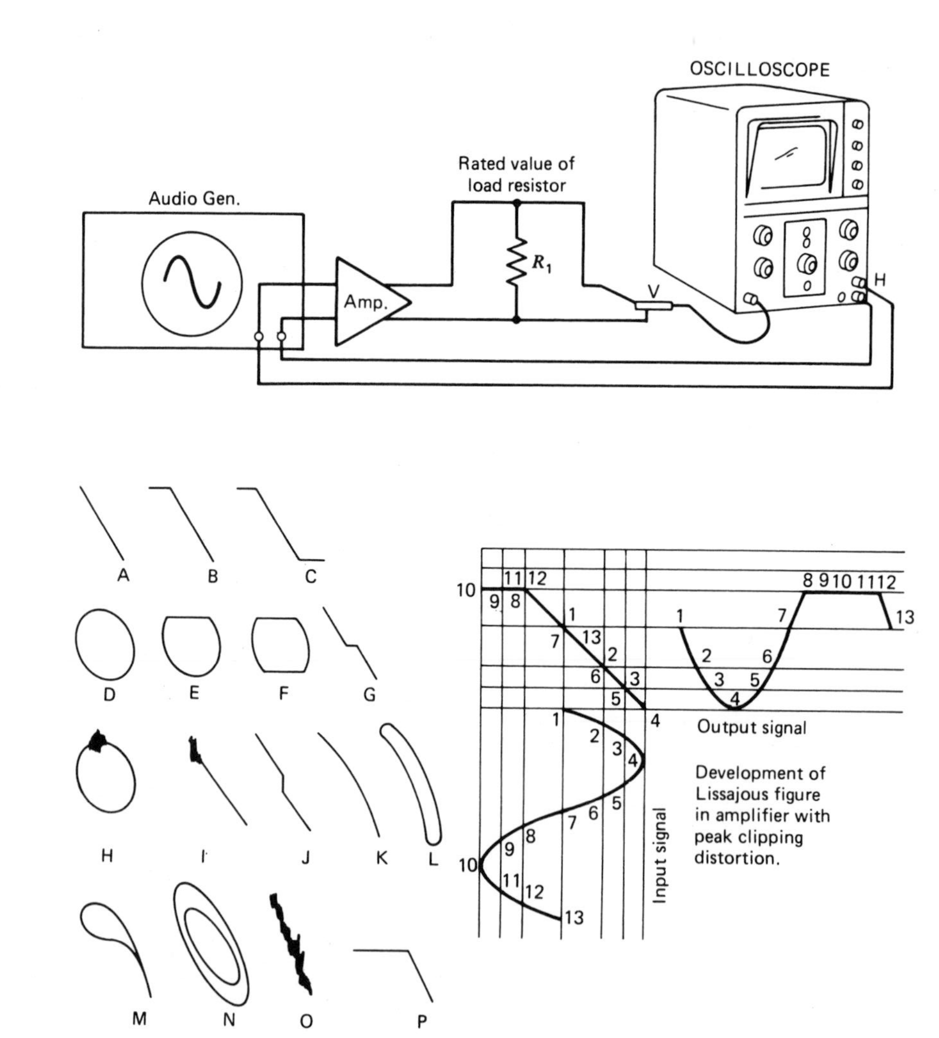

FIGURE 14-7 Distortion indicated by Lissajous figures.

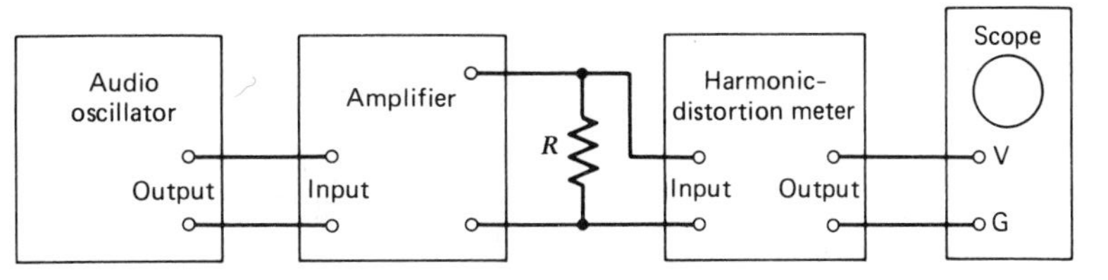

FIGURE 14-8 Test setup to measure harmonic distortion.

amplifier under test. The input signal voltage is fed to the horizontal amplifier of an oscilloscope, and the output signal voltage is fed to the vertical amplifier. A Lissajous figure is displayed on the oscilloscope screen. The pattern of the Lissajous will disclose the presence of any substantial amount of **distortion**.

Distortion values in precentages can be calculated from Lissajous patterns to determine whether substantial distortion is present and what type of distortion it is. Distortion values in percentages are measured with a harmonic distortion analyzer, as depicted in Fig. 14-8. A harmonic distortion analyzer

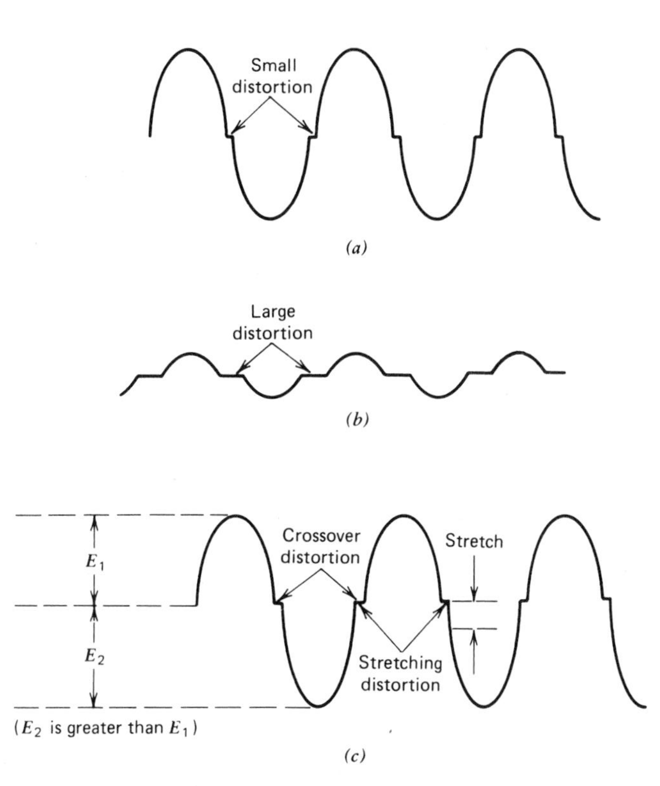

FIGURE 14-9 Examples of crossover and stretching distortion. (*a*) Crossover distortion at a high-output level. (*b*) Crossover distortion at a low-output level. (*c*) Combined crossover and stretching distortion amplitudes.

filters out the test frequency from the output signal. However, many harmonics of the test frequency are passed, and their rms value is indicated on the meter scale. Distortion analyzers are discussed in more detail in Chapter 15.

Crossover and stretching are distortions that have their maximum percentage at a low-output level. Examples of crossover and stretching distortions are shown in Fig. 14-9. These forms of distortion point to incorrect bias voltages in push–pull stages. When a conventional class B amplifier operates with zero bias on its transistors, low-level clipping results. This low-level clipping becomes apparent as crossover distortion in the output waveform. On the other hand, if too much forward bias is applied to the transistors, a dc

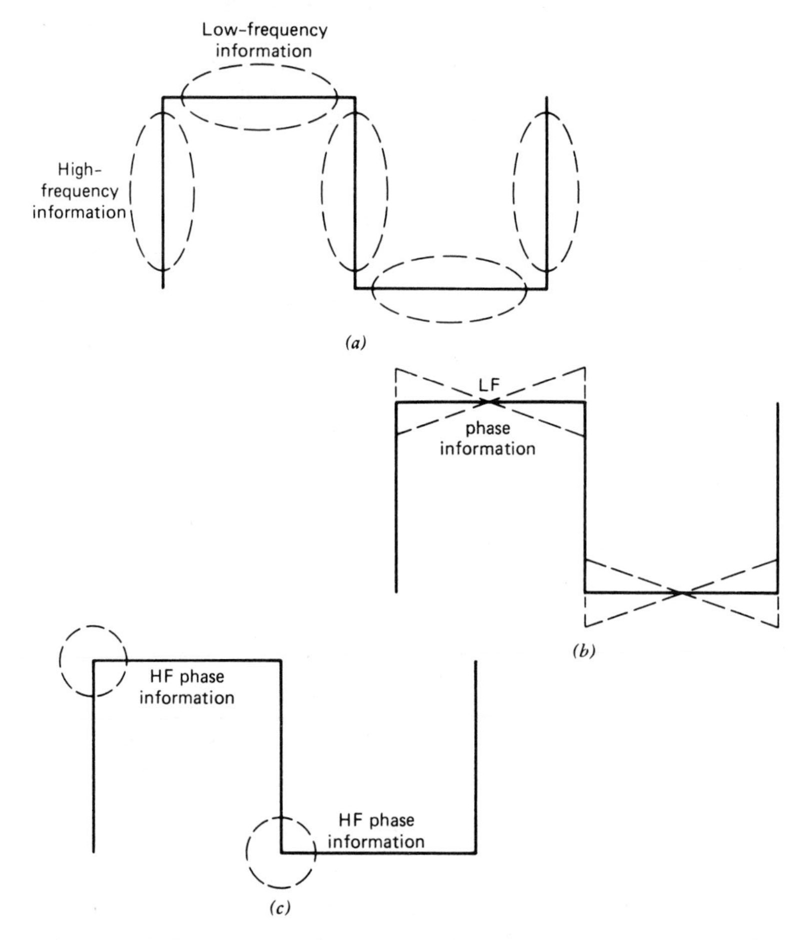

FIGURE 14-10 (*a*) High-frequency information appears in the leading and trailing edges, and low-frequency information appears along the top and bottom excursions. (*b*) Low-frequency phase information appears as tilt along the top and bottom. (*c*) High-frequency phase information appears at diagonal corners.

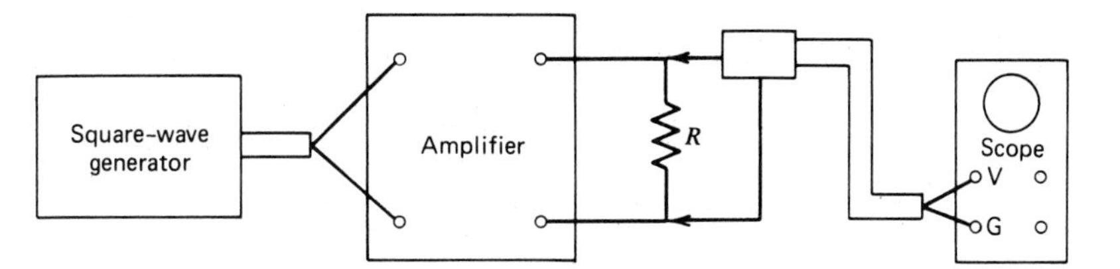

FIGURE 14-11 Setup for square-wave tests of the amplifier response. (Courtesy B&K Precision, Division of Dynascan Corporation.)

component is introduced into the output; this dc component becomes apparent as stretching distortion.

Harmonic distortion is the combined effect of a fundamental frequency and harmonic frequencies. The electronic troubleshooter is concerned primarily with the presence of harmonics and secondarily with their amplitudes. Troubleshooting is also sometimes facilitated by recognition of harmonic order—the occurrence of lower harmonics, of higher harmonics, or of a combination of the two. Accordingly, analysis of harmonic distortion is facilitated by the use of a specialized type of oscilloscope, a laboratory-quality distortion analyzer, or a spectrum analyzer.

Next, we consider the analysis of distortion by square-wave tests. With reference to Fig. 14-10, high-frequency information is contained in the leading and trailing edges of a reproduced square wave. Low-frequency information is contained along the top and bottom excursions of the square wave. Phase information at low-test frequencies is given by tilt (or lack of it) along the top and bottom excursions. Phase information at high-test excursions is given by cornering details in the reproduced square wave. An amplifier is tested for square-wave response as shown in Fig. 14-11. Basic types of square-wave distortion are given in Fig. 14-12.

14-10 HALF-SPLIT TECHNIQUE

The half-split technique is used primarily in a linear signal path. This technique is based on the idea of the simultaneous elimination of the maximum number of circuits with any test. In testing, the most logical place to make the first test is at a convenient point halfway between the good and bad locations.

Figure 14-13 is a simplified block diagram of a receiver. It shows the linear signal path of the received signal through the receiver. If test point 8 is the bad output, then the next phase of troubleshooting is to isolate the trouble to one of the circuit groups.

With signal tracing of this receiver, an amplitude-modulated (AM) signal is introduced at test point 7. An oscilloscope is then connected to monitor the waveform at various test points.

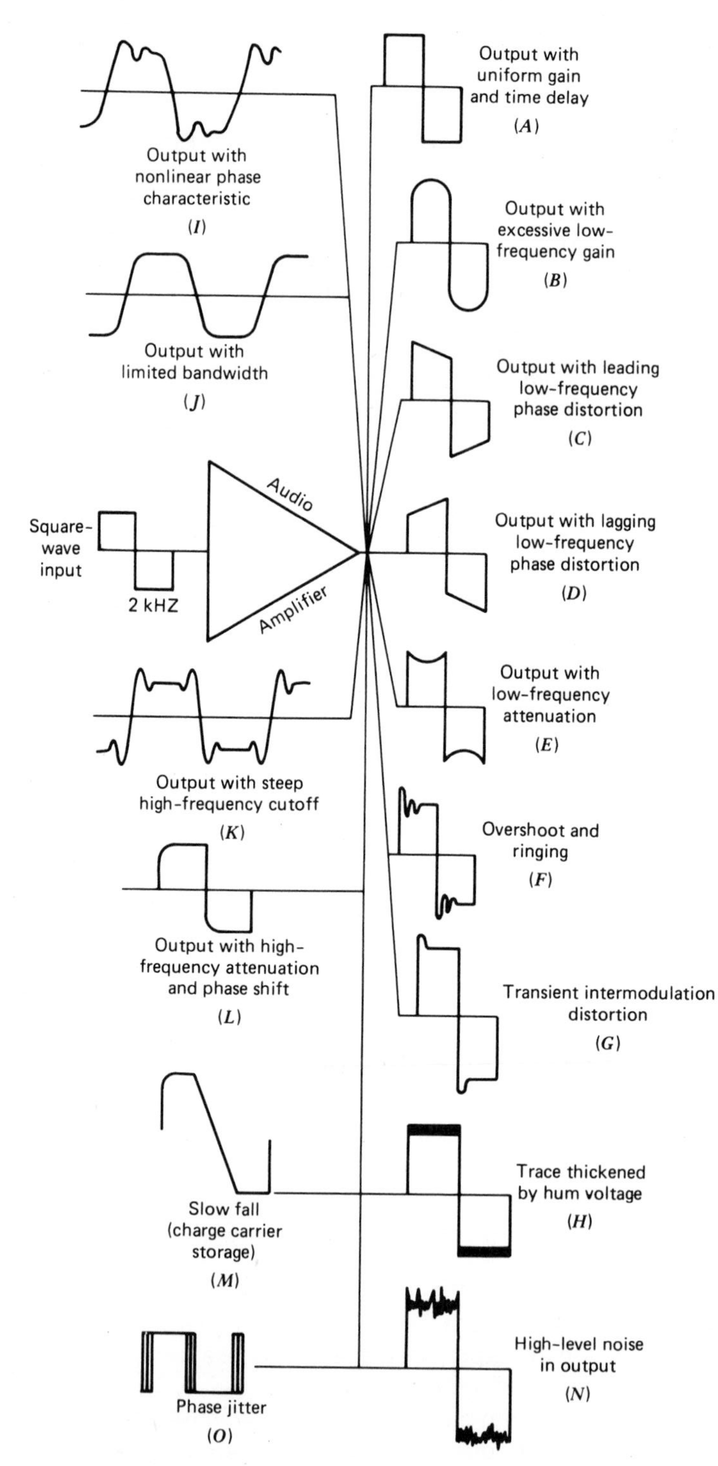

FIGURE 14-12 Basic types of square-wave distortion.

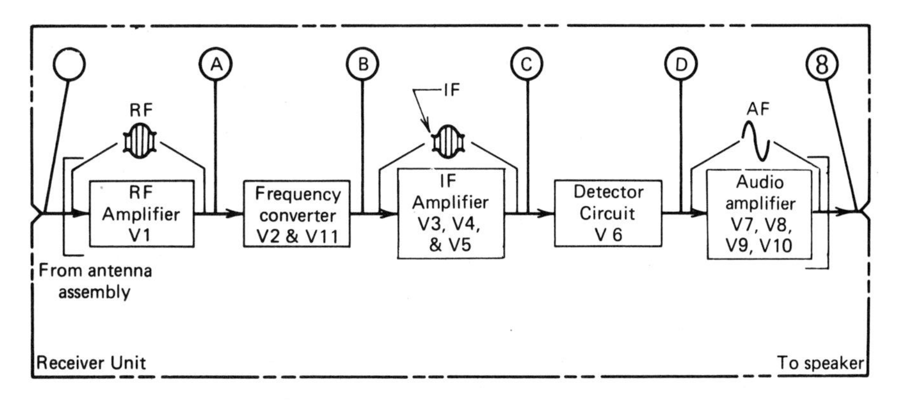

FIGURE 14-13 Simplified servicing block diagram of a receiver unit illustrating the linear path of a signal received through circuit groups.

With signal injection, signals of the right sort are injected at test points A, B, C, and D. The receiver response is analyzed on the loudspeaker. The signal injected at point A would be an AM signal. The injected signals at points B and C are modulated intermediate-frequency (IF) signals. The signal for D is an AF tone.

14-11 LOCATING A SPECIFIC TROUBLE USING THE SENSES

Troubleshooting consists of testing various branches of the defective circuit to find the specific trouble. After the problem is isolated to a circuit, we can first perform a preliminary inspection using the senses, especially those of sight, smell, hearing, and touch. For example, burned resistors can be detected by visual observation or by smell. Charred capacitors, inductors, and transformers can also be spotted. Touch can locate overheated transistors, and the sense of hearing can detect hum or high-voltage arcing (e.g., between wires) caused by overloaded or overheated transformers. These sensory procedures are more frequently referred to as visual inspection. When defective components are located by means of visual inspection, further inspection should be performed before replacing components. The real cause of trouble may be another component or situation, for example, a short. So always check for other possible sources of trouble after locating a defective component.

14-12 TESTING TO LOCATE A FAULTY COMPONENT

Solid-state circuits are analyzed by testing to locate faulty components. These devices are a convenient means of evaluating the operation of the circuit (through waveform, voltage, and resistance measurements).

The first step in solid-state circuit testing is to analyze the output waveform. If an abnormal or distorted waveform is noted, then the waveform is analyzed to determine its amplitude, duration, phase, and shape. This evaluation helps the troubleshooter to deduce in which branch (collector, emitter, or base) of the circuit the trouble may have occurred.

In-circuit testers may be used to test transistors and diodes in circuit. These testers are usually quite good for transistors used at lower frequencies. However, most in-circuit transistor testers will not show the high frequency or the switching characteristics of transistors. The same is true of out-of-circuit transistor and diode testers. For example, it is quite possible for a transistor to perform well as an audio amplifier but be hopelessly inadequate as a high-speed switching device required in digital equipment.

14-13 VOLTAGE MEASUREMENTS

After the troubleshooter has analyzed waveforms and performed in-circuit test of transistors, diodes, and so on, voltage measurements are the next logical step. These measurements are made with the circuit properly biased but with no applied signal. Manuals or data sheets are important in specifying voltages at the various test points. It is a standard practice first to check the element having the highest voltage, usually the collector. Then the terminals having lesser voltage should be checked in descending order, that is, the emitter and the base.

Generally, symptoms and output signals are the most important factors to consider. If no output signal is produced, the troubleshooter should expect a fairly large variation of voltage in the trouble area. Trouble resulting from circuits being out of tolerance causes only a slight change in circuit voltage.

14-14 RESISTANCE MEASUREMENTS

Unlike voltage measurements that are made with the equipment turned on, resistance measurements are made with all power turned off. Resistance measurement detects short circuits, open circuits, forward and reverse bias functions, and so on. A safety feature that will protect the ohmmeter is to ensure that all filter capacitors are discharged.

The shunting effects of other components in parallel can produce erroneous resistance readings. In these instances one terminal of the component being tested should be disconnected, and the resistance measured will be for this component only.

14-15 USING SCHEMATIC DIAGRAMS

In troubleshooting, the actual fault can eventually be traced to one or more of the circuit components—transistors, diodes, transformers, and so on. After

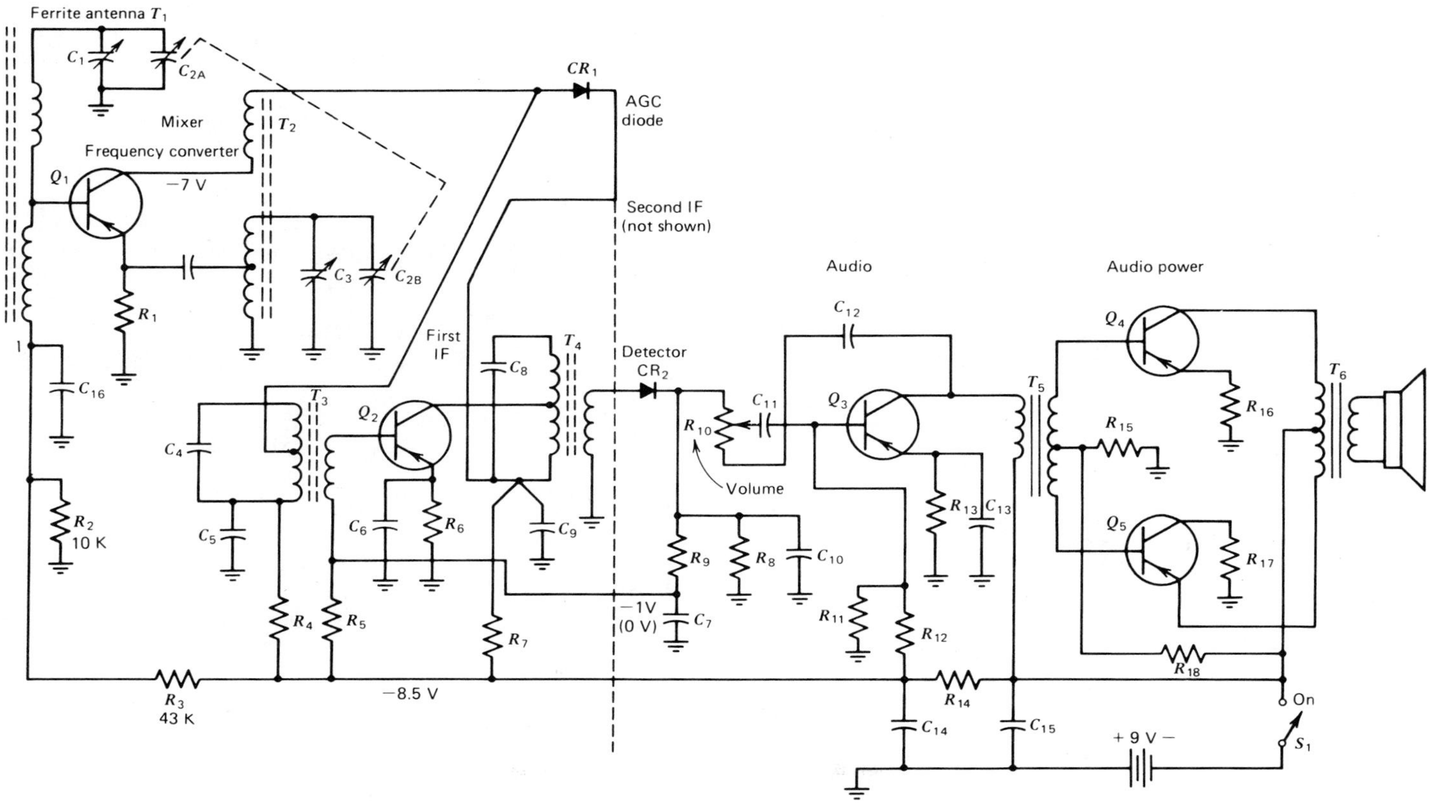

FIGURE 14-14 Typical broadcast band solid-state (transistor portable) schematic diagram.

finding the faulty circuit, the troubleshooter must locate the particular component that is causing the trouble. This requires being able to read a schematic wiring diagram. These diagrams provide the final picture of the electronic equipment.

Figure 14-14 shows the schematic diagram of a solid-state radio receiver. This receiver unit (a portable broadcast radio) differs considerably from a servicing block diagram.

14-16 DIGITAL TROUBLESHOOTING

Troubleshooting digital circuitry is fundamentally testing for normal responses from gates and flip-flops, and checking for normal clock operation. A basic test approach is to apply an input pulse to a gate under suspicion and to observe whether an output pulse is produced. Special test equipment is available for making this type of test. Most digital circuitry employs integrated circuits; each IC usually contains numerous gates, and many contain flip-flops.

A logic probe, depicted in Fig. 14-15, is the most important single tool used in digital-circuitry troubleshooting. It is employed to trace logic levels

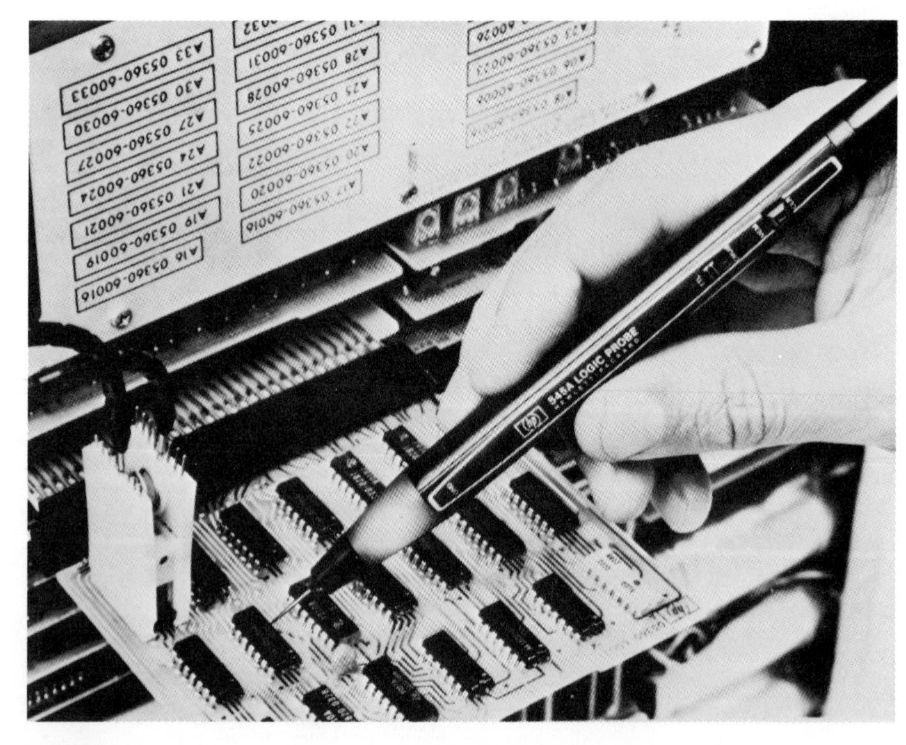

FIGURE 14-15 Logic probe used to troubleshoot digital circuitry. (Courtesy Hewlett-Packard Company.)

and pulses, through integrated circuitry to determine whether the point under test is logic high, logic low, bad level, open-circuited, or pulsing. This probe has preset logic thresholds of 2.0 and 8.0 V, which correspond to the high and low states of (TTL) transistor–transistor logic, and diode transistor logic (DTL) circuitry. When the probe is touched to a high-level point, a bright band of light appears around the probe tip. When it is touched to a low-level point, the light goes out. Open circuits or voltages in the "bad level" region between the preset thresholds produce illumination at half brilliance. Single pulses of 50 msec or less are made visibly readable by stretching to 50 msec. The lamp flashes on or blinks off, depending on the polarity of the pulse train of 50 MHz. Repetition rates cause the lamp to blink off and on at a rate of 10 Hz.

Figure 14-16 shows a logic probe in use in a circuit. The circuit can first be operated normally to detect key signals such as clock, reset, and shift pulses. Next, the circuit under test can be operated with one pulse applied and checked at this time. In this type of troubleshooting procedure we can use a logic pulser as shown in Fig. 14-17. This logic pulser is essentially a single-shot pulse generator with enough drive capability to override the

FIGURE 14-16 Logic probe in use. (Courtesy Hewlett-Packard Company.)

FIGURE 14-17 Logic pulser used to troubleshoot digital circuitry. (Courtesy Hewlett-Packard Company.)

output section of an IC. The pulser provides a convenient means of stepping a circuit one pulse at a time; the result is monitored with the logic probe. Figure 14-16 shows the pulser and probe in use.

Figure 14-18 shows a block diagram of the logic-probe circuitry and its response to different inputs.

The logic pulser is touched to the circuit under test, and the pulse button is pressed. In turn, all circuits connected to the node (outputs as well as inputs) are briefly driven to their opposite states.

The troubleshooter need not be concerned with whether the test node is in the high or low state. High nodes are pulsed low, and low nodes high, each time the button is pressed. The pulse operates at from 3 to 18 V dc and draws less than 35 mA.

14-17 APPLICATION OF A LOGIC CLIP

The logic clip (Fig. 14-19) clips onto TTL or DTL ICs and instantly displays the logic state of all 14 or 16 pins. A logic clip is easier to use than measuring devices in digital application, because it shows the state of each pin via an

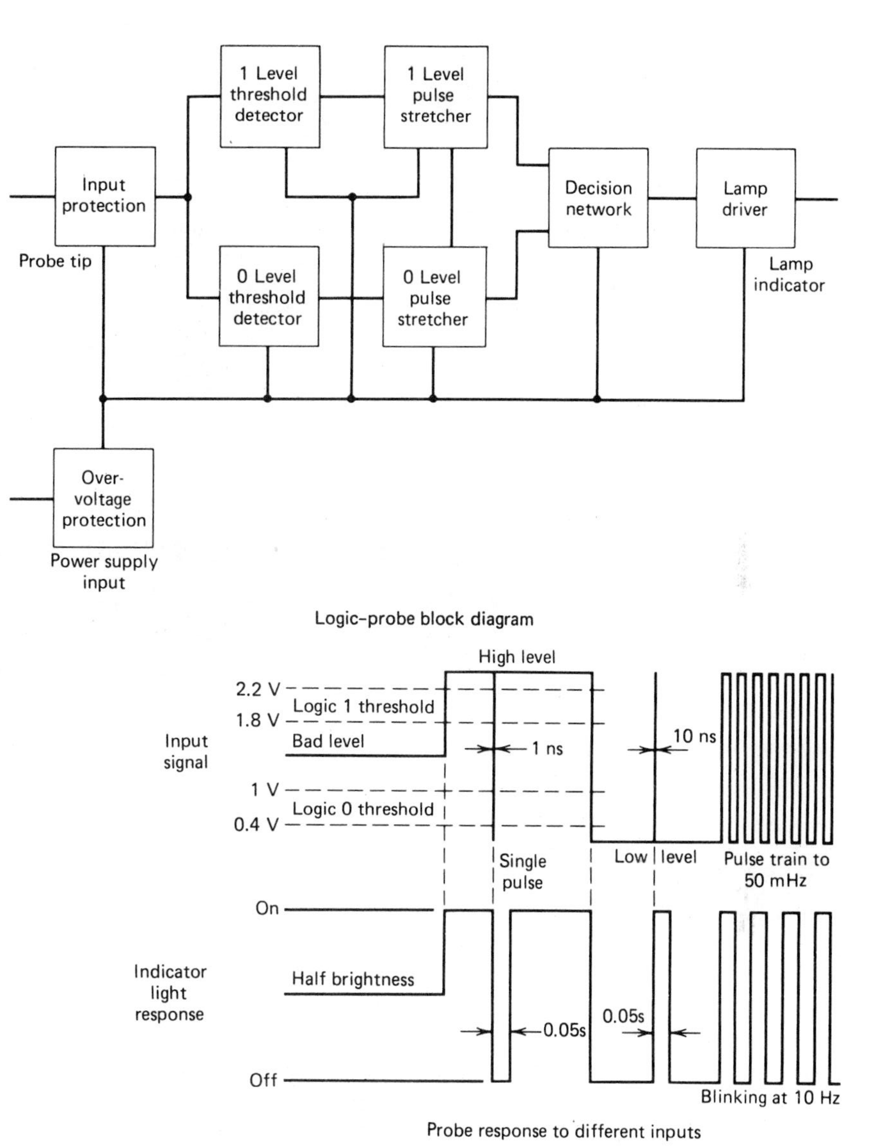

FIGURE 14-18 Block diagram of the logic-probe circuitry and its response to different inputs. (Courtesy Hewlett-Packard Company.)

individual light-emitting diode (LED). The display exhibits logic highs (on) and logic lows (off). The clip contains its own gating logic for locating the ground and the +5-V V_{cc} pins. Testing sequential circuits such as flip-flops is accomplished with the clip simultaneously monitoring all output states while the pulser applies reset pulses to the device. Faulty operation becomes immediately apparent since the IC will not go through its prescribed sequence of states.

FIGURE 14-19 Logic clip used to troubleshoot digital circuitry. (Courtesy Hewlett-Packard Company.)

14-18 APPLICATION OF A LOGIC CURRENT TRACER

A digital current tracer (Fig. 14-20) precisely locates low-impedance faults in digital circuits by "sniffing" out digital current sources or sinks. The current tracer senses the magnetic field created by current flow as small as 1 mA and up to 1 A, and the output from the sensor is amplified and turns on a light to indicate that current is flowing. To trace current, the operator merely places the probe tip next to the trace where current is known to be flowing, adjusting the sensitivities until the light just goes on. Then the tracer is moved along the main current path, following the circuit path that keeps the light on. This will lead the operator directly to the circuit that is pulling (sinking) current. Many troubleshooting problems, such as wired-AND or wired-OR configurations, can waste considerable time. A current tracer can pinpoint exactly the one faulty point on a node, even when troubleshooting multilayer boards.

In the example given in Fig. 14-21, gate U5A is short-circuited to ground, causing the node to be stuck at the low logic level and sinking all the current from U1. A current tracer quickly verifies this fault and provides a simple, clear, single-lamp indication of current activity on the node.

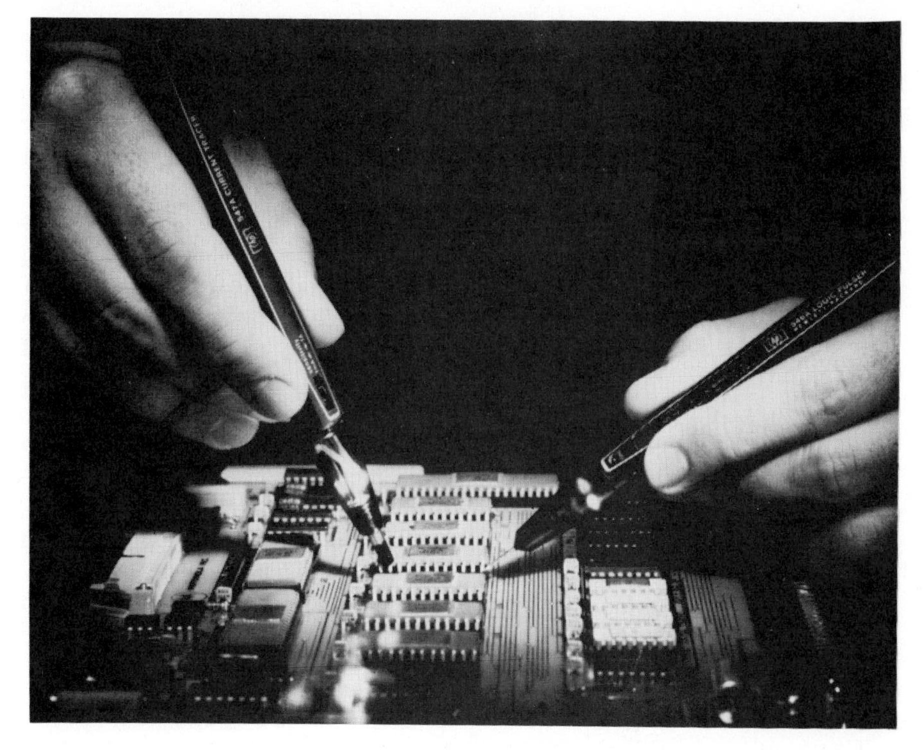

FIGURE 14-20 A digital logic current tracer. (Courtesy Hewlett-Packard Company.)

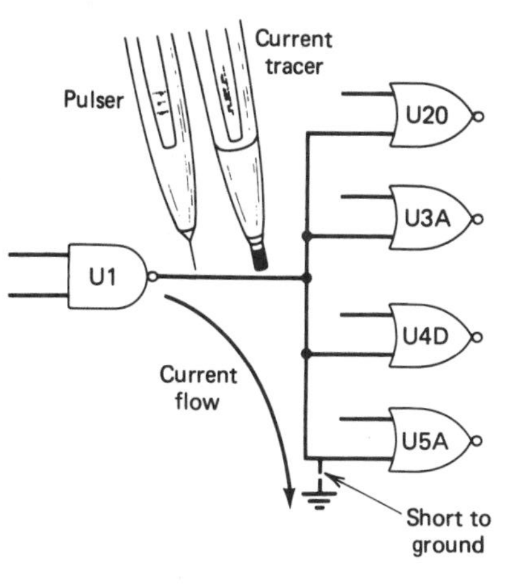

FIGURE 14-21 Location of a multiple-input fault. (Courtesy Hewlett-Packard Company.)

14-19 LOGIC COMPARATOR APPLICATION

Comparison testing consists of stimulating a known good device with the same signal that is used to stimulate the device under test, and then comparing the responses of the two components. In this manner, the technician quickly determines whether the device under test is operating properly. Such a tester, known as a logic comparator, is shown in Fig. 14-22. In its application to TTL and DTL logic families, a comparator "borrows" the input signal, causing the device under test to stimulate a known good device simultaneously. Any difference in the response of the two devices is indicated by an "on" LED corresponding to the output that failed the comparison test.

In Fig. 14-23, notice that it is a NAND gate that is under test. Recall that the output will be low only when both inputs are high. One way to test this IC is to measure the input and then decide whether the output should be high or low. Then measure it to determine whether it is. An alternate method is to connect the inputs of another NAND gate to the inputs of the suspected IC and connect a comparator (exclusive NOR) circuit to the outputs. The comparator output will be low, and the LED will illuminate if a difference ever exists between the two ICs. Since the inputs are identical, the outputs should also agree.

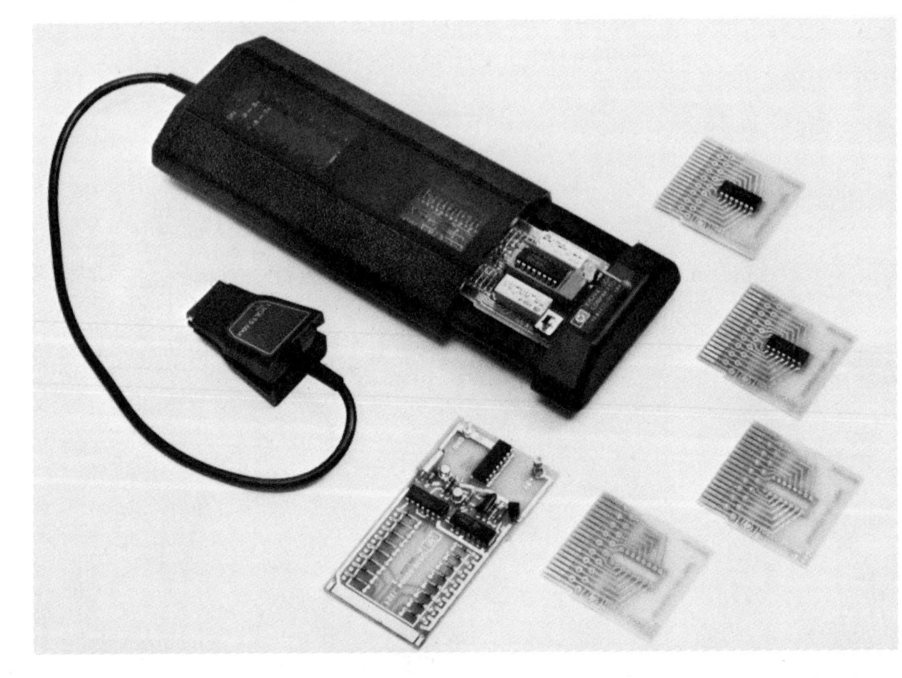

FIGURE 14-22 A logic comparator with reference cards. (Courtesy Hewlett-Packard Company.)

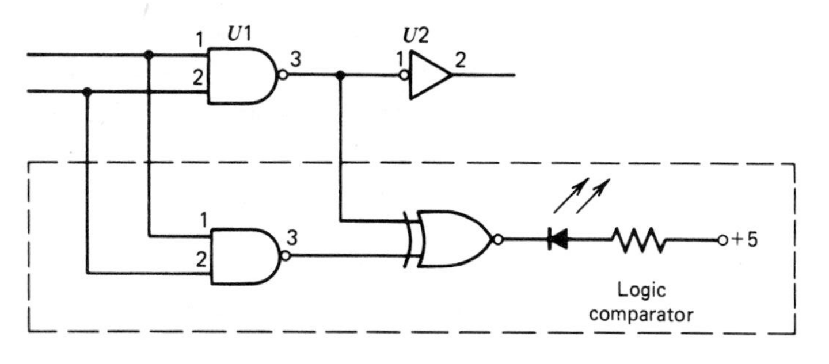

FIGURE 14-23 Comparing the operation of two identical ICs.

14-20 SUMMARY

The first step in troubleshooting any electronic equipment is to determine the symptoms of the trouble. Once these symptoms are known, an analysis of the trouble or fault may be made. Often this analysis takes only a few moments, but occasionally it may take a considerable amount of time. Time spent on a proper analysis is usually repaid by the saving of time and material in the steps that follow.

By proper analysis, the fault may be localized to a major functional unit such as RF generation and audio amplification.

The next step is to isolate the trouble or fault to a group of circuits and then to an individual circuit responsible for the fault. Careful observation of the performance of the equipment when it is turned on often helps to isolate the fault.

Location of the specific trouble within a circuit may be made in a number of ways. Component faults may be eliminated by testing or substitution. Burned or charred resistors or coils can often be spotted by visual observation or by smell. The same holds true for oil or wax-filled capacitors, chokes, and transformers. When overheated, the oil or wax in these components will expand and usually leak out or cause the case to buckle; if the oil or wax are excessively overheated, the case may explode. Overheated components can be located quickly by touching with the fingers. When power is applied, high-voltage arcing can often be heard, which can assist in locating defective high-voltage components. Thus, the senses of sight, smell, touch, and hearing can be used to locate many faulty components.

Locating trouble should begin with a close visual inspection for overheated parts or apparent damage in the wiring and component circuitry of the faulty section if this can be easily done. This inspection may lead to the faulty component. However, replacement of parts that appear to be damaged should not be made until the cause of the trouble is determined.

To locate and repair the defective circuit, proceed as follows.

1. Locate the cause of trouble by making waveform, voltage, and resistance checks.

2. Refer to previously compiled tables of normal waveforms, voltages, and resistance readings and to the schematic diagram.
3. Perform the checks at the essential points by referring to the schematic diagram.
4. Compare these readings with normal readings and recall the principles of circuit operation. Then analyze the results to determine which component is defective. For many circuits the oscilloscope will provide the only means of determining abnormal operation, especially if they are circuits in which the amplitude, the frequency, or the voltage waveshape are critical.

14-21 GLOSSARY

Block diagram: An overall representation of the functional units within the equipment as well as the signal-flow paths between them.

Convergent–divergent signal-flow path: A path formed when a single stage has multiple inputs and multiple outputs.

Convergent signal-flow path: A path in which two or more signals enter a circuit.

Distortion: An undesired change in waveform; a difference between the waveform of the output and the input of a system.

Divergent signal-flow path: A path in which two or more signal paths leave a circuit.

Feedback path: A signal path from one circuit to a point or circuit preceding it in the signal-flow sequence.

Linear signal-flow path: A series of circuits arranged so that the output of one circuit feeds the input of the following circuit.

Signal substitution: Injecting an artificial signal into a circuit to check its performance.

Signal tracing: Examining the signal at a test point with an oscilloscope.

Switching signal path: A path in which a selector switch provides a different signal path for each switch position.

14-22 REVIEW QUESTIONS

After studying the material in this chapter, try answering these questions to test your knowledge of the subject matter. If you cannot answer any particular question, look back in the text for the answer.

1. Describe briefly the generalized troubleshooting procedure.
2. What is a quick check?
3. What is a functional block diagram?
4. What are the six basic types of signal-flow paths?

5. What are the following?
   (a) Signal injection.
   (b) Signal tracing.
   (c) Half-split technique.

6. What are four basic types of distortion?

7. What are the differences between the following?
   (a) Logic probe.
   (b) Logic pulser.
   (c) Logic clip.
   (d) Logic current tracer.
   (e) Logic comparator.

14-23 PROBLEMS

14-1 Assume that a logic pulser clamps the output LOW for 0.3 μsec at approximately 0.8 V and then drives HIGH for 0.3 μsec at approximately 4.5 V. Also assume that a logic probe responds logic HIGH: "Bright": (2.0 ± 0.2 V); logic LOW: "Off": (0.8 ± 0.3 V); between HIGH and LOW: "Dim." Given the circuits in Fig. 14-24, draw the scope waveforms when the logic pulser is pulsed.

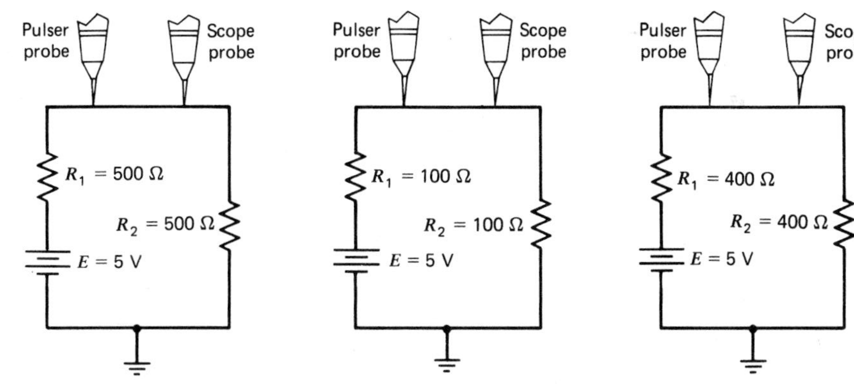

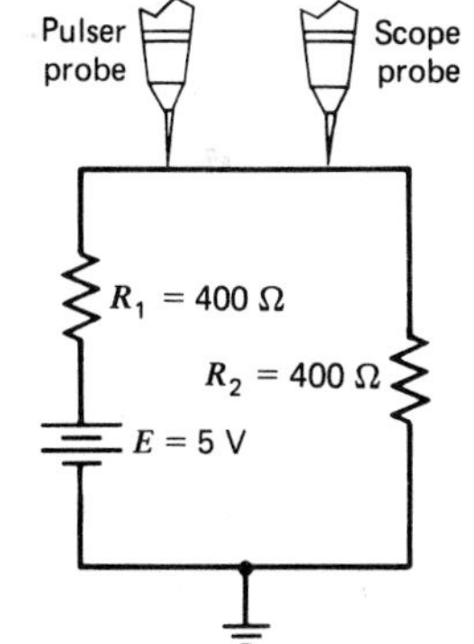

FIGURE 14-24 Circuit for Problem 14-1.

14-2 Assume that a logic pulser generates a 0.3-μsec negative pulse at 0.8 V, and then a 0.3-μsec positive pulse at 2.2-V amplitude.

   (a) A logic pulser, probe, and scope are connected across a 3-ohm resistor. Draw the scope waveform when the pulser is pulsed in the circuit of Fig. 14-25.

   (b) Draw the scope waveforms when the pulser is pulsed in the circuit of Fig. 14-26.

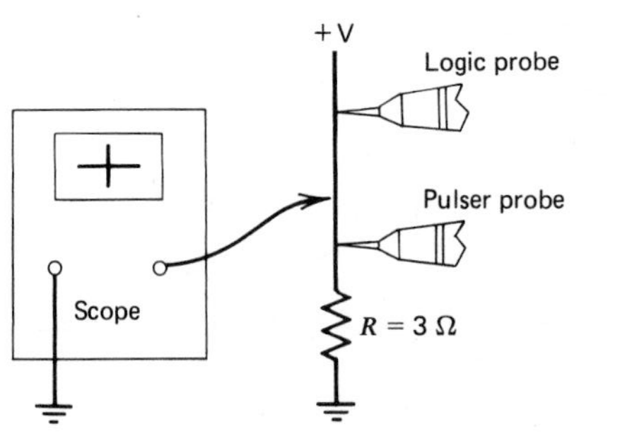

FIGURE 14-25 Circuit for Problem 14-2.

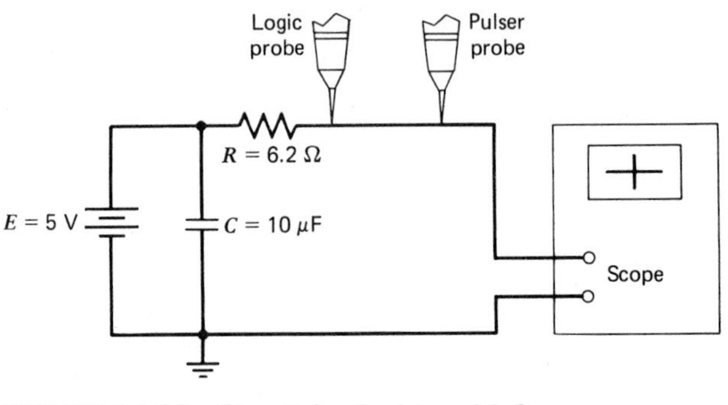

FIGURE 14-26 Circuit for Problem 14-2.

14-3 Refer to the circuit of Fig. 14-27. The probe light blinks from ON to OFF to ON when the pulser is pulsed. What is the condition of gate 1?

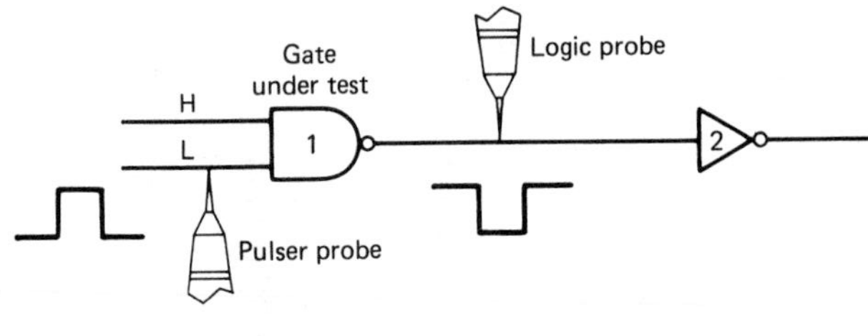

FIGURE 14-27 Circuit for Problem 14-3.

14-4 See the circuit of Fig. 14-28. The probe light remains ON when the pulser is pulsed. What is the condition of gate 2?

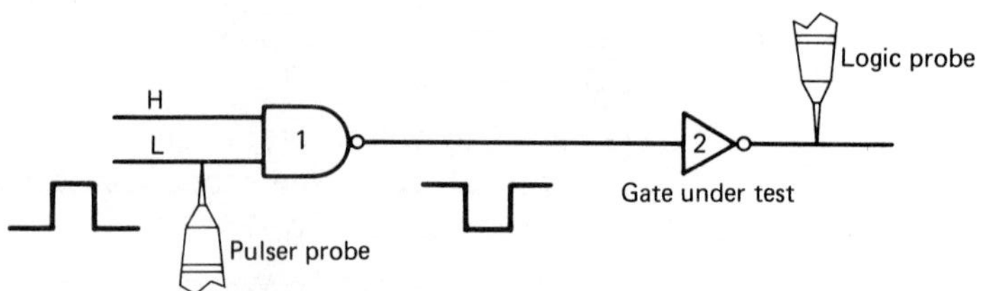

FIGURE 14-28 Circuit for Problem 14-4.

14-5 See the circuit of Fig. 14-29. What is the state of the pulsed line after the pulser is pulsed for the following conditions?

(a) The probe light remains ON.

(b) The probe light remains OFF.

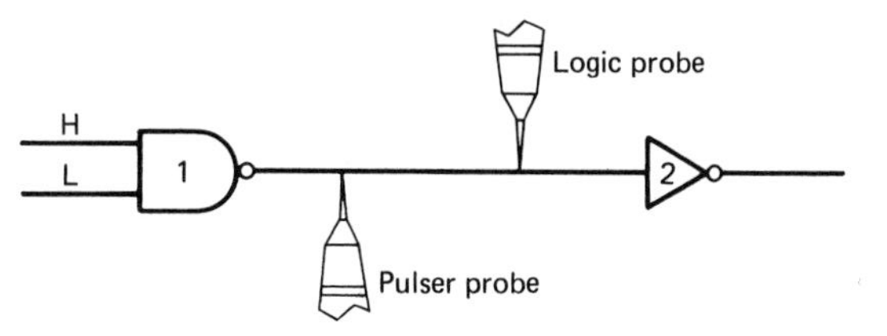

FIGURE 14-29 Circuit for Problem 14-5.

14-6 What is the most basic piece of test equipment used in electronic servicing?

14-7 What are three quick checks that can be done in troubleshooting a piece of equipment?

14-8 What is a disadvantage of DVMs compared to EVMs when used for troubleshooting?

14-9 What are three types of troubleshooting procedures?

14-10 What is the easiest way to make phase measurements with the instruments discussed in this chapter?

14-11 When measuring resistors in circuit, what should be done, in addition to turning off the power supply, to protect the ohmmeter?

14-12 What parameters of interest do waveform measurements involve?

14-13 A logic pulser is essentially what kind of circuit?

14-14 How are open circuits indicated with a logic probe?

14-24 LABORATORY EXPERIMENTS

Experiments E33 and E34, which are located at the end of the text, apply the theory that has been presented in Chapter 14. The purpose of the experiments is to provide hands-on experience to reinforce the theory.

Experiment E33 requires an AM radio receiver which will be available in many, but not all, electronic laboratories. If a receiver is not available, it may be possible to substitute another type of circuit to troubleshoot. Experiment E34 requires a logic probe and a logic pulser. As with the receiver, these will be available in most, but not all laboratories. If they are not available, the experiment can be modified to use an LED to check logic levels.

The contents of the laboratory report to be submitted by each student are included at the end of each experimental procedure.

CHAPTER 15 Signal Analyzers

15-1 Instructional Objectives

This chapter discusses instruments used for frequency-domain analysis and presents applications for each. After completing Chapter 15 you should be able to

1. List four instruments that are used for frequency-domain analysis and their primary function.
2. Define the term *spectrum analysis*.
3. Describe the cause of harmonic distortion.
4. Describe the purpose of the rejection filter in a harmonic distortion analyzer.
5. Define the term *harmonic distortion*.
6. List three other names by which wave analyzers are known.
7. Indicate the meaning of the term *non-real-time analyzer*.
8. Indicate the meaning of the term *real-time analyzer*.
9. Describe the characteristics of the bandpass filter of a wave analyzer.
10. List three applications for which Fourier analyzers are better suited than spectrum analyzers.

15-2 INTRODUCTION

In the first 14 chapters we discussed measurement techniques in the time domain, that is, measurement of parameters that vary with time. Electrical signals contain a great deal of interesting and valuable information in the frequency domain as well. Analysis of signals in the **frequency domain** is called **spectrum analysis**, which is defined as the study of the distribution of a signal's energy as a function of frequency. This analysis provides both electrical and physical system information which is very useful in performance testing of both mechanical and electrical systems. This chapter discusses the basic theory and applications of the principal instruments used for frequency-domain analysis: distortion analyzers, wave analyzers, spectrum analyzers, and Fourier analyzers. Each of these instruments quantifies the magnitude of the signal of interest through a specific bandwidth, but each measurement technique is different as will be seen in the discussion that follows.

15-3 DISTORTION ANALYZERS

Applying a sinusoidal signal to the input of an ideal linear amplifier will produce a sinusoidal output waveform. However, in most cases the output waveform is not an exact replica of the input signal because of various types of **distortion**. The extent to which the output waveform of an amplifier differs from the waveform at the input is a measure of the distortion introduced by the inherent nonlinear characteristics of active devices such as bipolar or field-effect transistors or by passive circuit components. The amount of distortion can be measured with a **distortion analyzer**.

When an amplifier is not operating in a linear fashion, the output signal will be distorted. Distortion caused by nonlinear operation is called amplitude distortion or **harmonic distortion**. It can be shown mathematically that an amplitude-distorted sine wave is made up of pure sine-wave components including the fundamental frequency f of the input signal and harmonic multiples of the fundamental frequency, $2f$, $3f$, $4f, \ldots$, and so on. When harmonics are present in considerable amount, their presence can be observed with an oscilloscope. The waveform displayed will either have unequal positive and negative peak values or will exhibit a change in shape. In either case, the oscilloscope will provide a qualitative check of harmonic distortion. However, the distortion must be fairly severe (around 10%) to be noted by an untrained observer. In addition, most testing situations require a better quantitative measure of harmonic distortion. Harmonic distortion can be quantitatively measured very accurately with a harmonic distortion analyzer, which is generally referred to simply as a distortion analyzer.

A block diagram for a fundamental-suppression harmonic analyzer is shown in Fig. 15-1. When the instrument is used, switch S_1 is set to the "set

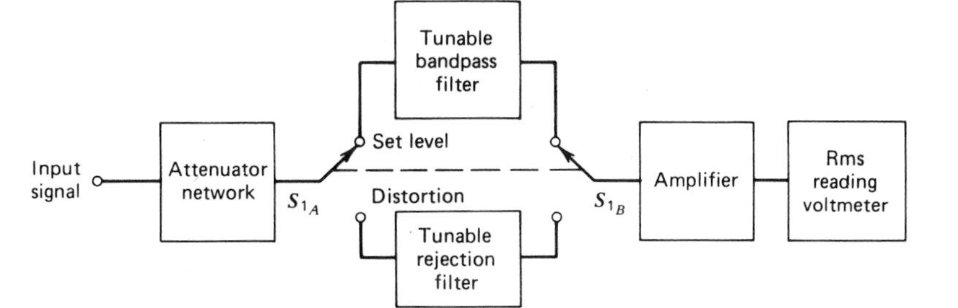

FIGURE 15-1 Block diagram of a distortion analyzer.

level" position, the bandpass filter is adjusted to the fundamental frequency, and the attenuator network is adjusted to obtain a full-scale voltmeter reading. Switch S_1 is then set to the "distortion" position, the rejection filter is turned to the fundamental frequency, and the attenuator is adjusted for a maximum reading on the voltmeter.

The total harmonic distortion (THD), which is frequently expressed as a percentage, is defined as the ratio of the rms value of all the harmonics to the rms value of the fundamental, or

$$\text{THD} = \frac{\sqrt{\Sigma(\text{harmonics})^2}}{\text{fundamental}} \tag{15-1}$$

This defining equation is somewhat inconvenient from the standpoint of measurement. An alternative working equation expresses total harmonic distortion as the ratio of the rms value of all the harmonics to the rms value of the total signal including distortion. That is,

$$\text{THD} = \frac{\sqrt{\Sigma(\text{harmonics})^2}}{\sqrt{(\text{fundamental})^2 + \Sigma(\text{harmonics})^2}} \tag{15-2}$$

On the basis of the assumption that any distortion caused by the components within the analyzer itself or by the oscillator signal are small enough to be neglected, Eq. 15-2 can be expressed as

$$\text{THD} = \frac{\sqrt{E_2^2 + E_3^2 + \cdots + E_n^2}}{E_f} \tag{15-3}$$

where

THD = the total harmonic distortion
E_f = the amplitude of the fundamental frequency including the harmonics
E_2, E_3, E_n = the amplitude of the individual harmonics

EXAMPLE 15-1 Compute the total harmonic distortion of a signal that contains a fundamental signal with an rms value of 10 V, a second harmonic with an rms value of 3 V, a third harmonic with an rms value of 1.5 V, and a fourth harmonic with an rms value of 0.6 V.

Solution Using Eq. 15-3, we compute the total harmonic distortion as

$$\text{THD} = \frac{\sqrt{3^2 + 1.5^2 + 0.6^2}}{10}$$

$$= \frac{\sqrt{11.6}}{10} = 34.07\%$$

A typical laboratory-quality distortion analyzer is shown in Fig. 15-2. The instrument shown, a Hewlett-Packard Model 334A, is capable of measuring

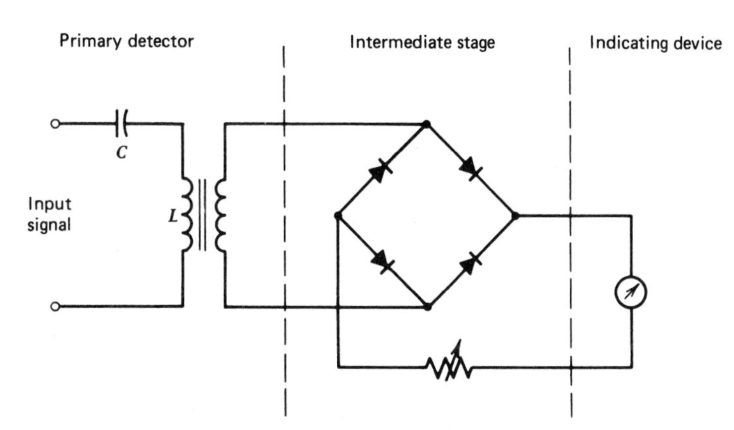

FIGURE 15-2 Laboratory-quality distortion analyzer. (Courtesy Hewlett-Packard Company.)

total distortion as small as 0.1% of full scale at any frequency between 5 Hz and 600 kHz. Harmonics up to 3 MHz can be measured.

15-4 WAVE ANALYZERS

Harmonic distortion analyzers measure the total harmonic content in waveforms. It is frequently desirable to measure the amplitude of each harmonic individually. This is the simplest form of analysis in the frequency domain and can be performed with a set of tuned filters and a voltmeter. Such analyzes have various names, including frequency-selective voltmeters, carrier frequency voltmeters, selective level meters, and wave analyzers. Any of these names is quite descriptive of the instrument's primary function and mode of operation.

A very basic **wave analyzer** is shown in Fig. 15-3. The primary detector

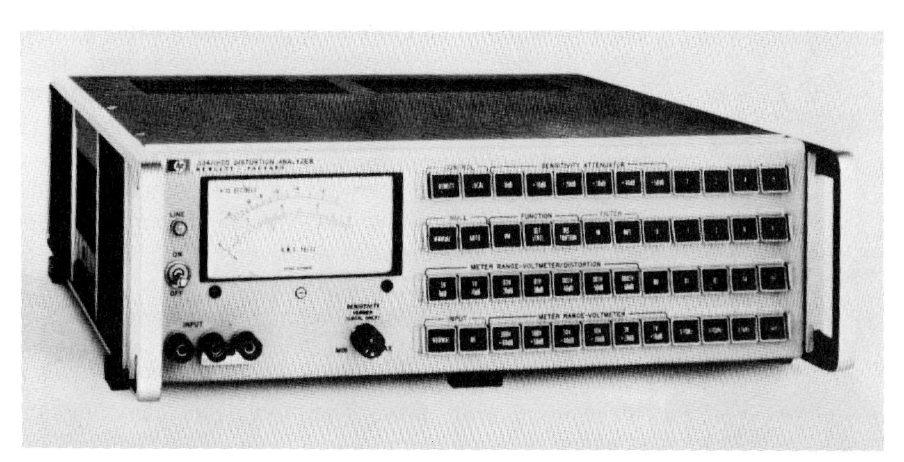

FIGURE 15-3 Basic wave analyzer circuit.

is a simple *LC* circuit which is adjusted for resonance at the frequency of the particular harmonic component to be measured. The intermediate stage is a full-wave rectifier, and the indicating device may be a simple dc voltmeter that has been calibrated to read the peak value of a sinusoidal input voltage. Since the *LC* filter in Fig. 15-3 passes only the frequency to which it is tuned and provides a high attenuation to all other frequencies, many tuned filters connected to the indicating device through a selector switch would be required for a useful wave analyzer.

Since wave analyzers sample successive portions of the frequency spectrum through a movable "window," as shown in Fig. 15-4, they are called **non-real-time analyzers**. However, if the signal being sampled is a periodic waveform, its energy distribution as a function of frequency does not change with time. Therefore, this sampling technique is completely satisfactory.

Rather than using a set of tuned filters, the heterodyne wave analyzer shown in Fig. 15-5 uses a single, tunable, narrow-bandwidth filter, which may be regarded as the window through which a small portion of the frequency spectrum is examined at any one time. In this system, the signal from the internal, variable-frequency oscillator will **heterodyne** with the input signal to produce output signals having frequencies equal to the sum and difference of the oscillator frequency f_o and the input frequency f_i. In a

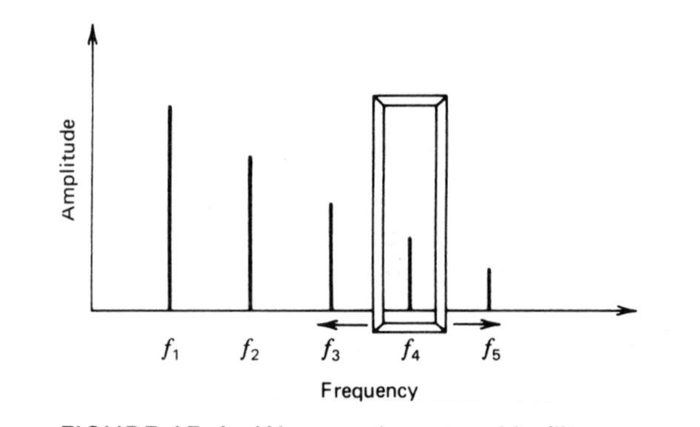

FIGURE 15-4 Wave analyzer tunable filter or "window."

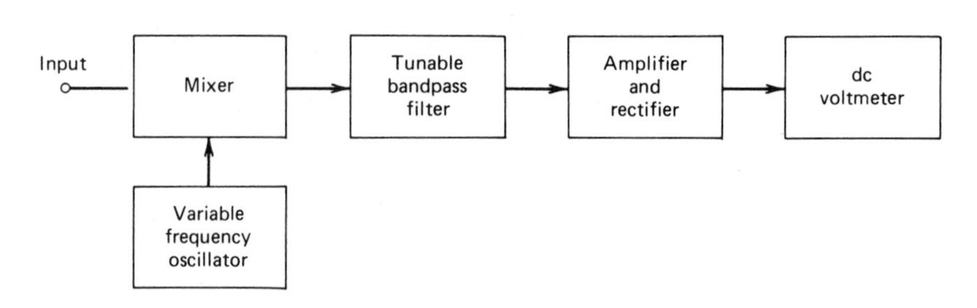

FIGURE 15-5 Heterodyne-type wave analyzer.

typical heterodyne wave analyzer, the bandpass filter is tuned to a frequency higher than the maximum oscillator frequency. Therefore, the "sum frequency" signal expressed as

$$f_s = f_o + f_i$$

is passed by the filter to the amplifier.

As the frequency of the oscillator is decreased from its maximum frequency, a point will be reached where $f_o + f_i$ is within the band of frequencies that the bandpass filter will pass. The signal out of the filter is amplified and rectified. The indicated quantity is amplified and rectified. The indicated quantity is then proportional to the peak amplitude of the fundamental component of the input signal. As the frequency of the oscillator is further decreased, the second harmonic and higher harmonics will be indicated.

The bandwidth of the filter is very narrow, typically about 1% of the frequency of interest. The attenuation characteristics of a typical commercial audio-frequency analyzer is shown in Fig. 15-6. As can be seen, at $0.5f$ and at $2f$, attenuation is approximately 75 dB. The bandwidth of a heterodyne wave analyzer is usually constant. This can make analysis very difficult, if not impossible, in applications in which the frequency of the waveform being analyzed does not remain constant during the time required for a complete analysis, which is generally several seconds. For example, the bandwidth of an audio-frequency analyzer may be on the order of 10 Hz. A 2% change in the fundamental frequency of a 1-kHz signal will shift the frequency of the fifth harmonic by 100 Hz, which is well outside the bandwidth of the instrument.

EXAMPLE 15.2

A wave analyzer has a fixed bandwidth of 4 Hz. By what percentage can a 60-Hz signal change without disrupting measurement of the fourth harmonic with the instrument?

Solution

The maximum frequency shift at any harmonic is one-half the bandwidth or 2 Hz. A frequency shift of 0.5 Hz at the fundamental frequency will cause a

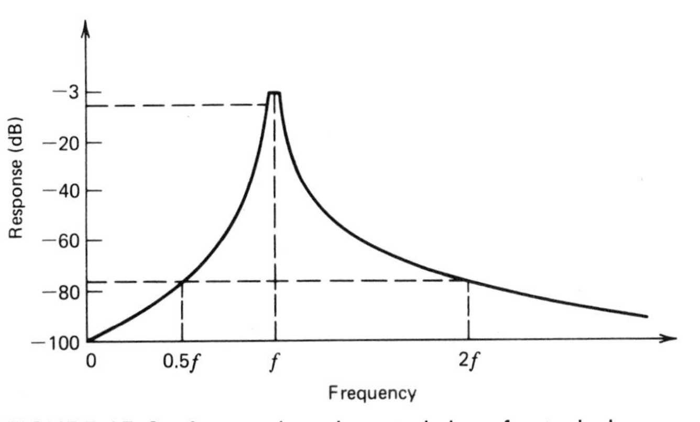

FIGURE 15-6 Attenuation characteristics of a typical wave analyzer.

2-Hz frequency shift at the fourth harmonic. The percent change in frequency is

$$\text{Percent change} = \frac{0.5\ \text{Hz}}{60\ \text{Hz}} \times 100 = 0.833\%$$

The principal applications of wave analyzers are

- Amplitude measurement of a single component of a complex waveform.
- Amplitude measurement in the presence of noise and interfering signals.
- Measurement of signal energy within a well-defined bandwidth.

15-5 SPECTRUM ANALYZERS

The problems associated with non-real-time analysis in the frequency domain can be eliminated by using a **spectrum analyzer**. A spectrum analyzer is a **real-time analyzer**, which means that it simultaneously displays the amplitude of all the signals in the frequency range of the analyzer.

Spectrum analyzers, like wave analyzers, provide information about the voltage or energy of a signal as a function of frequency. Unlike wave analyzers, spectrum analyzers provide a graphical display on a CRT. A block diagram of an audio spectrum analyzer is shown in Fig. 15-7.

The real-time, or multichannel, analyzer is basically a set of stagger-tuned bandpass filters connected through an electronic scan switch to a CRT. The composite amplitude of the signal within each filter's bandwidth is displayed as a function of the overall frequency range of the filter. Therefore, the frequency range of the instrument is limited by the number of filters and their bandwidth. The electronic switch sequentially connects the filter outputs to the CRT. Horizontal deflection is obtained from the scan generator, which has

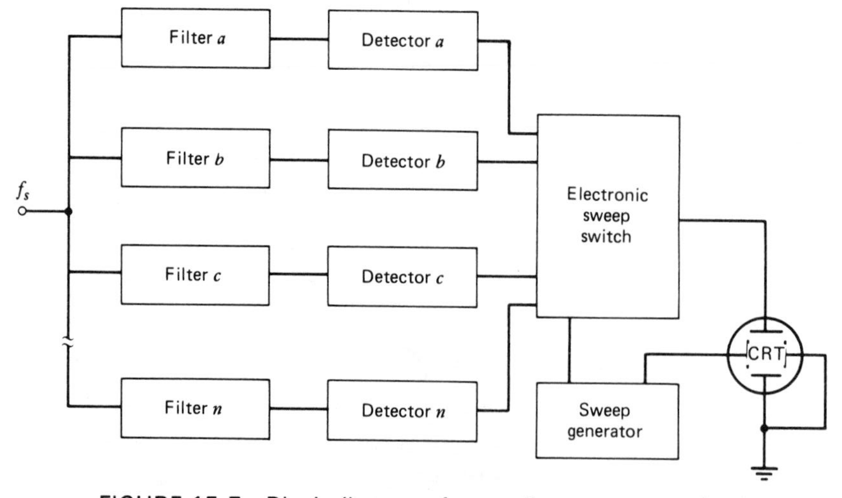

FIGURE 15-7 Block diagram of an audio spectrum analyzer.

a sawtooth output that is synchronized with the electronic switch. Such analyzers are usually restricted to audio-frequency applications and may employ as many as 32 filters. The bandwidth of each filter is generally made very narrow for good resolution.

The relationship between a time-domain presentation on the CRT of an oscilloscope and a frequency-domain presentation on the CRT of a spectrum analyzer is shown in the three-dimensional drawing in Fig. 15-8. Figure 15-8*a* shows a fundamental frequency f_1 and its second harmonic $2f_1$. An oscilloscope used to display the signal in the time–amplitude domain would display only one waveform—the composite of $f_1 + 2f_1$ as shown in Fig. 15-8*b*. A spectrum analyzer used to display the components of the composite signal in the frequency–amplitude domain would clearly display the amplitude of both the fundamental frequency f_1 and its second harmonic $2f_1$ as shown in Fig. 15-8*c*.

Spectrum analyzers are used to obtain a wide variety of information from various kinds of signals, including the following.

- Spectral purity of continuous-wave (CW) signals.
- Percentage of modulation of amplitude-modulated (AM) signals.
- Deviation of frequency-modulated (FM) signals.
- Noise such as impulse and random noise.
- Filter frequency response.

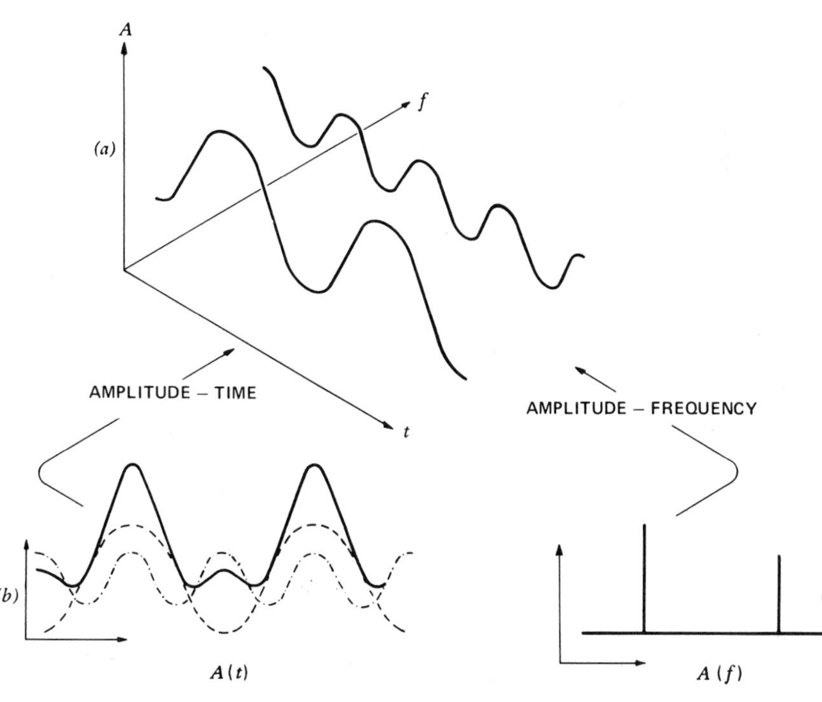

FIGURE 15-8 Three-dimensional relationship between time, frequency, and amplitude. (Courtesy Hewlett-Packard, Company.)

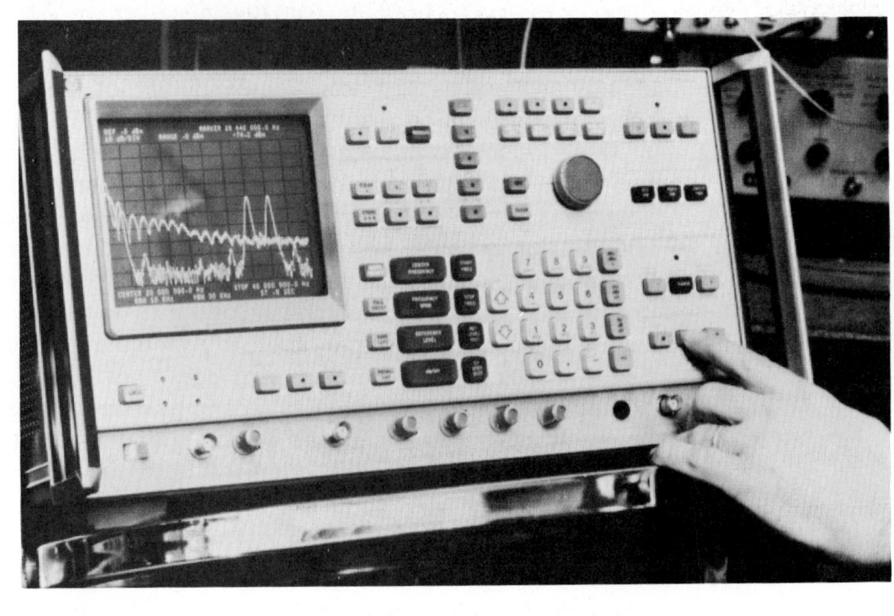

FIGURE 15-9 **Laboratory-quality spectrum analyzer. (Courtesy Hewlett-Packard Company.)**

A laboratory-quality spectrum analyzer, the Model 3585A spectrum analyzer manufactured by Hewlett-Packard, is shown in Fig. 15-9. The frequency range of the instrument is from 20 Hz to 40 MHz. Bandwidth resolution is variable from 3 Hz to 30 kHz.

15-6 FOURIER ANALYZERS

Fourier analyzers use digital signal-processing techniques to provide measurements that go beyond the capabilities of spectrum analyzers. Some of their capabilities are measurement of very low-frequency or very closely spaced signals, measurement of random signals obscured by noise, and measurement of shared properties or relationships of two or more signals.

A **Fourier analyzer** is based on the calculation of the *discrete* Fourier transform using an algorithm called the *fast* Fourier transform. This algorithm calculates the amplitude and phase of each signal component from a set of time-domain samples of the input signal.

A basic block diagram of a Fourier analyzer is shown in Fig. 15-10. The signal applied to the instrument is filtered to remove out-of-band components. The signal that comes out of the low-pass filter is applied to the A/C converter, which samples and digitizes it at regular time intervals until a full set of samples called a *time record* has been collected. The microprocessor then performs the desired series of computations on the time data to obtain the frequency-domain results. These results, which are stored in memory, can

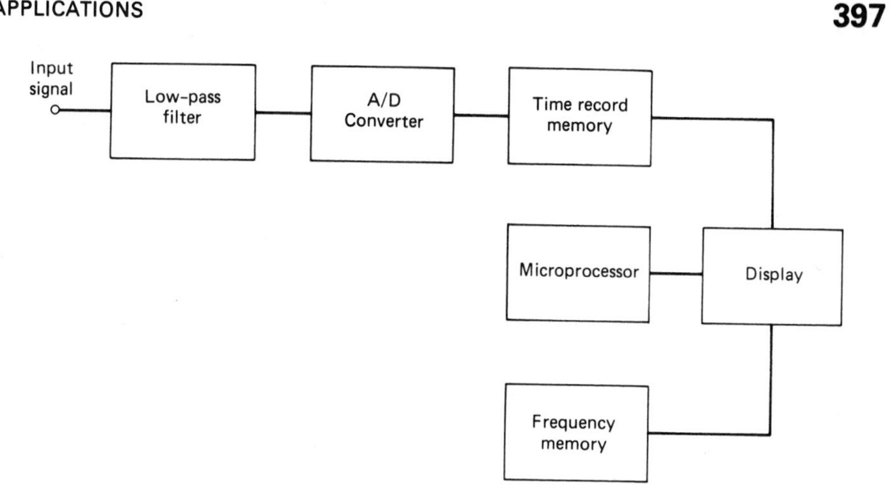

FIGURE 15-10 Basic block diagram of a Fourier analyzer.

be displayed on a CRT, or they can be recorded permanently with a recorder or plotter.

Since Fourier analyzers are primarily digital instruments, interfacing to a computer or other digital systems is relatively straightforward. Remote programming and transferring data via the IEEE 488 bus or other interfacing systems can considerably expand the range of possible applications for the instrument. By virtue of the digital nature of the instrument, a Fourier analyzer will provide a high degree of accuracy, stability, and repeatability in spectrum analysis.

15-7 APPLICATIONS

The applications for the different types of analyzers discussed in this chapter are numerous and varied. These instruments find applications in mechanical measurements as well as in electronic testing and communications.

Some of the applications in the mechanical area are to measure levels of noise and vibration. When used in conjunction with a transducer, spectrum analyzers or Fourier analyzers are employed to examine vibration signals from automobiles, airplanes, space vehicles, bridges, and other mechanical systems. This provides information on mechanical integrity, unbalance, and bearing and gear wear.

In the general field of electronic testing related to troubleshooting and quality control, signal analyzers find many applications such as identifying and measuring signals caused by nonlinear effects in the process of amplifying, filtering, or mixing signals. One of the principal applications for distortion analyzers is to measure the distortion caused by an amplifier. A test setup for measuring total harmonic distortion is shown in Fig. 15-11.

An audio oscillator with a low level of distortion serves as the source of a sinusoidal waveform. With the amplifier properly loaded, the sinusoidal

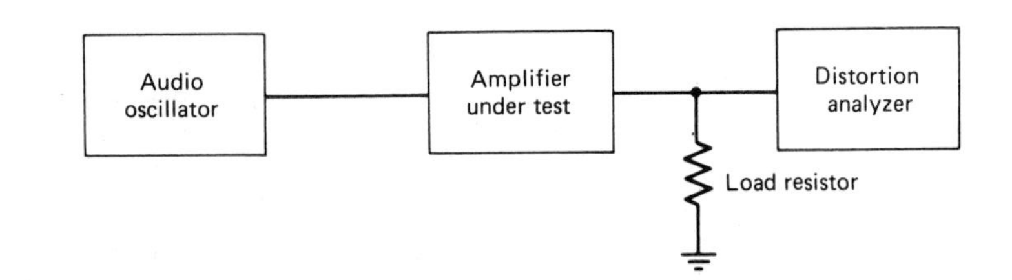

FIGURE 15-11 Test setup to measure the total harmonic distortion of an amplifier.

waveform is applied to the amplifier. The output of the amplifier is applied directly to the distortion analyzer which measures the total harmonic distortion.

In the field of microwave communications, in which pulsed oscillators are widely used, spectrum analyzers are an important tool. They also find wide application in analyzing the performance of AM and FM transmitters.

Spectrum analyzers and Fourier analyzers are widely used in applications requiring very low frequencies in the fields of biomedical electronics, geological surveying, and oceanography. They are also used in analyzing air and water pollution.

Another very important application of spectrum analyzers is the measurement of **intermodulation distortion**. This phenomenon occurs when two or more signals are applied to the input of a nonlinear circuit such as an amplifier, particularly a power amplifier. This problem is particularly troublesome in the reproduction of music.

If these signals are applied to a completely linear circuit, each passes through the circuit unaffected by the other. However, if there is nonlinearity in the circuit, heterodyning of the signals occurs. Limiting our discussion to two signals, we find that heterodyning occurs because the lower-frequency signal tends to modulate the higher-frequency signal. If f_1 and f_2 are the fundamental frequencies of the input signals, the output spectrum may contain any or all of the frequencies shown in Fig. 15-12, as well as other harmonics.

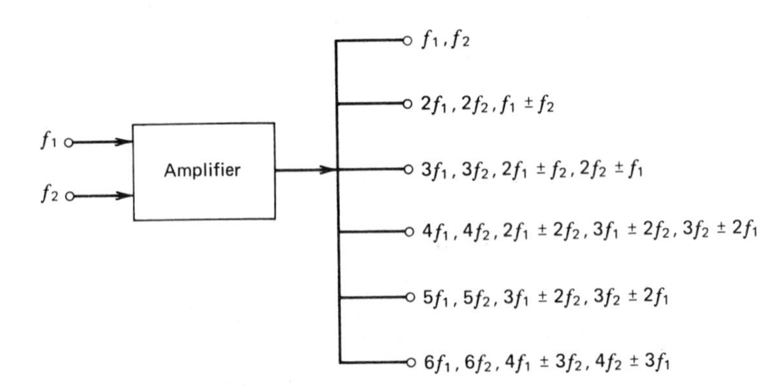

FIGURE 15-12 Some of the harmonics of f_1 and f_2 produced by amplifier nonlinearity.

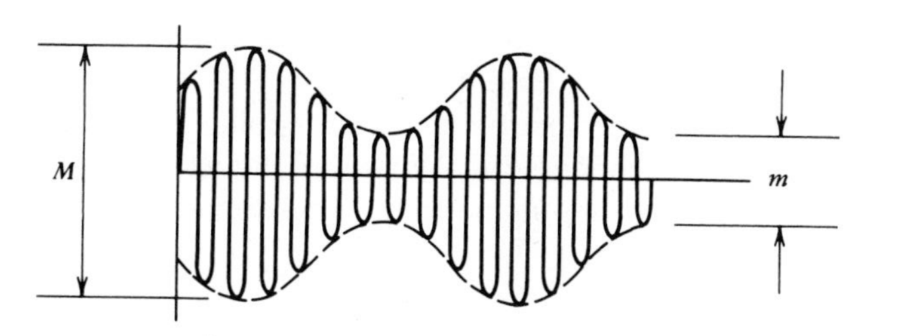

FIGURE 15-13 Amplitude-modulated waveform produced by intermodulation distortion.

If the nonlinearity of the circuit is significant, the modulation of the higher-frequency signal by the lower-frequency signal will produce the familiar amplitude modulation waveform as shown in Fig. 15-13. The percentage of intermodulation distortion is computed as

$$\text{IMD} = \frac{M - m}{M - m} \times 100\%$$

where

IMD = the intermodulation distortion expressed as a percentage
M = the peak-to-peak modulated signal
m = the minimum value of the modulated waveform

The spectrum analyzer can be used to measure the intermodulation distortion, as shown in the circuit in Fig. 15-14. The frequency of the audio oscillator is generally set to 6 kHz.

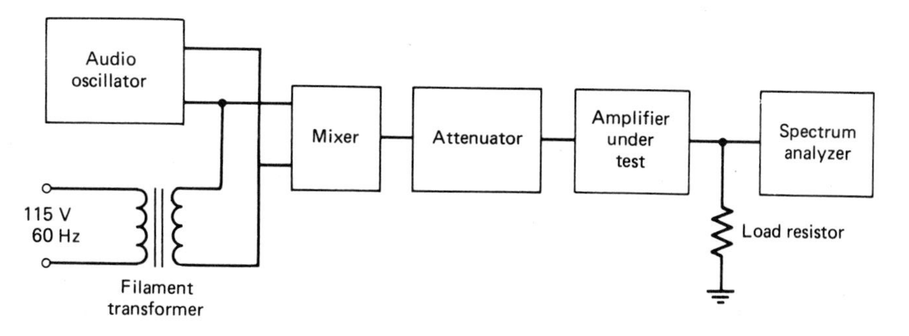

FIGURE 15-14 Using the spectrum analyzer to measure intermodulation distortion.

15-8 SUMMARY

The frequency domain contains a great deal of valuable information that can be obtained by using one of the types of analyzers discussed in this chapter.

Although the instruments are somewhat similar, the measurment technique of each is different.

Each instrument finds application in many areas of science, engineering, and technology. Whatever the application of the instrument, we can deduce much about the frequency-domain behavior of the system under test.

15-9 GLOSSARY

Distortion: An undesired change in a waveform. The principal types of distortion are frequency, phase, transient, intermodulation, and harmonic distortion.

Distortion analyzer: An instrument for measuring the amount of harmonic distortion. Also called a harmonic distortion analyzer.

Fourier analyzer: A digital instrument capable of signal analysis in the frequency domain that cannot be performed by traditional methods.

Frequency domain: A graphical representation of signal amplitude as a function of frequency.

Harmonic distortion: The ratio of the rms value of all higher harmonics to the rms value of the fundamental frequency.

Heterodyne: To mix two or more signals which produce *sum* and *difference* frequencies.

Intermodulation distortion: Created when two or more signals, or one signal containing two or more frequencies, are applied to a nonlinear circuit, and the lower-frequency signal tends to modulate the higher-frequency signal.

Real-time analyzer: A term used to describe a spectrum analyzer that is tuned to the entire spectrum at once: thus, it responds to changes in signals as they occur.

Spectrum analysis: A study of the energy distribution of a signal as a function of frequency.

Spectrum analyzer: An instrument designed to present graphically, usually on a CRT, the energy distribution of a signal as a function of frequency.

Wave analyzer: A tunable voltmeter used to measure the various frequency components of a signal.

15-10 REVIEW QUESTIONS

The following questions should be answered after a thorough study of the chapter. The purpose of the questions is to determine the reader's comprehension of the material.

1. What is the primary difference between a harmonic distortion analyzer and a wave analyzer?

2. What is the primary difference between a wave analyzer and a spectrum analyzer?
3. What are two causes of harmonic distortion?
4. What determines the size of the window in a wave analyzer?
5. What does the term *real-time analyzer* mean?
6. If the bandpass filter of a wave analyzer is adjusted to a particular frequency f, approximately how many decibels down is a signal of frequency $0.5f$?
7. What are three other names used for wave analyzers?
8. What are three applications of spectrum analyzers?

15-11 PROBLEMS

15-1 A sinusoidal signal is applied to a circuit that introduces the third, fifth, and seventh harmonics. If the input signal has a period of 20 μsec, what frequencies are present in the output signal?

15-2 A sinusoidal signal is applied to a circuit that introduces the second, fourth, and sixth harmonics. If the ampitude of the primary signal has an rms value of 5 V and the second, fourth, and sixth harmonics have rms values of 1 V, 0.4 V, and 0.1 V, respectively, what is the total harmonic distortion?

15-3 A wave analyzer is used to determine the harmonics present in a signal. Only the fifth harmonic was found to be present, and its amplitude was 0.4 V_{rms}. If the fundamental frequency had an amplitude of 4 V_{rms}, what was the total harmonic distortion?

15-4 A wave analyzer is being used to observe the fifth harmonic of a 400-Hz signal. What must its bandwidth be to continue to observe the harmonic if the fundamental frequency changes by 0.2%?

15-5 The total harmonic content of a signal is 20% and contains only the fundamental, first, third, and fifth harmonics. If the rms values of the third, fifth, and seventh harmonics are 2 V, 0.8 V, and 0.16 V, respectively, what is the rms value of the fundamental frequency?

15-6 What is the intermodulation distortion if the peak-to-peak value of a modulated signal is 3 V and the minimum value of the modulated waveform is 2 V?

15-7 The distortion caused by the third harmonic is found to be 3.33% by using a wave analyzer. The total harmonic distortion when measured with a distortion analyzer is found to be 3.5%. If the rms value of the fundamental is 18 V and if only the third and fifth harmonics are present, what is the rms value of the fifth harmonic?

15-8 Using a spectrum analyzer, we observe a signal to contain the third, fifth, and seventh harmonics in addition to the fundamental. If the third harmonic equals 0.2 times the fundamental, the fifth harmonic equals 0.3 times the third harmonic, and the seventh harmonic equals 0.24 times the fifth harmonic, what is the total harmonic distortion?

15-9 Write a one-page paper on the spectrum analyzer.

15-10 Write a one-page paper on the Fourier analyzer.

15-12 LABORATORY EXPERIMENTS

Experiments E35 and E36 apply the theory that has been presented in Chapter 15. The purpose of the experiments is to provide hands-on experience to reinforce the theory.

Both experiments use somewhat specialized equipment in the form of a distortion analyzer and a wave analyzer. If these pieces of equipment are not available, you will not be able to perform the experiments. It may be possible to borrow these items from one of the major equipment manufacturers and to perform the experiment as a demonstration.

The contents of the laboratory report to be submitted by each student are included at the end of each experimental procedure.

CHAPTER 16 Fiber Optics in Instrumentation

16-1 Instructional Objectives

The purpose of this chapter is to familiarize the reader with the principles and applications of fiber optics in the field of instrumentation. The chapter discusses the basic theory of optical fibers in a general way, and then looks at how optical fiber sensors work and at the applications for these devices. After completing the chapter you should be able to

1. Describe optical fibers.
2. List two ways that optical fibers are classifed.
3. List and describe the acceptance angle for optical fibers.
4. Describe three types of optical fibers.
5. Determine the numerical aperture for a given optical fiber.
6. List factors that affect the propagation of light through optical fibers.
7. List the advantages of optical fiber sensors over conventional sensors.
8. Describe intrinsic and extrinsic sensors.
9. Give the meaning of the term *mode* as it applies to fiber optics.
10. Describe the two basic types of fiber optic sensors.

16-2 INTRODUCTION

The instrumentation field witnessed significant developments in the technologies of fiber optic sensors during the 1980s. Instruments using optical fibers to sense various physical parameters are already on the market, and considerable research related to new applications for fiber optic sensors is being conducted in several countries around the world. By simply altering the characteristics of light in response to various stimuli, fiber optic sensors provide great accuracy and sensitivity and a broad dynamic range.

In addition to the accuracy, precision, sensitivity, and dynamic range provided by optical fibers, sensor designers are turning to them to sense parameters in hostile environments, to reduce the hazards of shock in sensing applications, and to reduce the costs of remote sensing. Fiber optic sensors appear destined to occupy a position of leadership in the growing family of sensing technologies. This chapter discusses optical fibers and their applications in sensing physical parameters and the advantages associated with their use.

16-3 OPTICAL FIBERS

Optical fibers are thin, flexible threads of transparent glass or plastic that can carry visible light. Optical fibers consist of two concentric layers called the **core** and the **cladding**, as shown in Fig. 16-1. The core and the cladding, along with the surrounding protective jacket, also shown in Fig. 16-1, make up the fiber optic cable.

Optical fibers are classified in two ways.

1. By the material of which they are made.
2. By the refractive index of the core and the number of modes by which the fiber propagates light.

Three different combinations of materials are used to construct optical fibers. These combinations are

1. Glass core and glass cladding.
2. Glass core and plastic cladding.
3. Plastic core and plastic cladding.

Optical fibers are also classified as three major types according to their refractive indexes.

1. Step-index fiber.
2. Graded-index fiber.
3. Single-mode fiber.

With **step-index** and **graded-index fibers**, the core transmits light waves that follow a nonlinear path. The cladding has a lower **index of refraction** than the core. Light waves that strike the core–cladding boundary at an angle

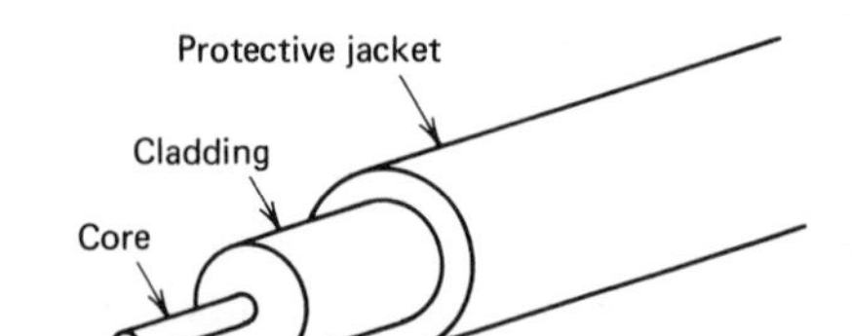

FIGURE 16-1 Fiber optic cable.

greater than the **critical angle** are therefore reflected back into the core. The critical angle is defined as the angle at which total internal reflection occurs. Since the path that a light wave follows as it propagates the core of an optical fiber is a function of its angle of incidence, the angle at which it strikes the core–cladding boundary, it follows quite naturally that there are many paths through the core. These paths are more rigorously called **modes**. A mode is a mathematical and physical concept indicating how electromagnetic waves are propagated through various media. More specifically, a mode is an allowed solution to Maxwell's equations which are the building blocks for the entire field of electromagnetism. However, for our purposes, a mode is simply a path that a light wave can follow as it travels down the core of an optical fiber. The number of modes for a fiber ranges from 1 to more than 100,000.

At this point we can make a clear distinction between optical fibers used in communications and optical fibers used in sensors. Optical fibers used in sensors are almost always **single-mode fibers**. This means that there is one path for a light wave to follow as it passes through the fiber—a straight line along the longitudinal axis of the fiber. The optical paths for the different classifications of optical fibers are shown in Fig. 16-2. Single-mode fibers are designed so that they have only one path for a light wave to travel. This is accomplished by making their core diameter exceedingly small, typically from 5 to 10 μm. This is very close to the wavelength of the propagated light wave.

In work with **multiple-mode fibers**, one important parameter is the angle over which a fiber will accept light directed toward it. This angle is

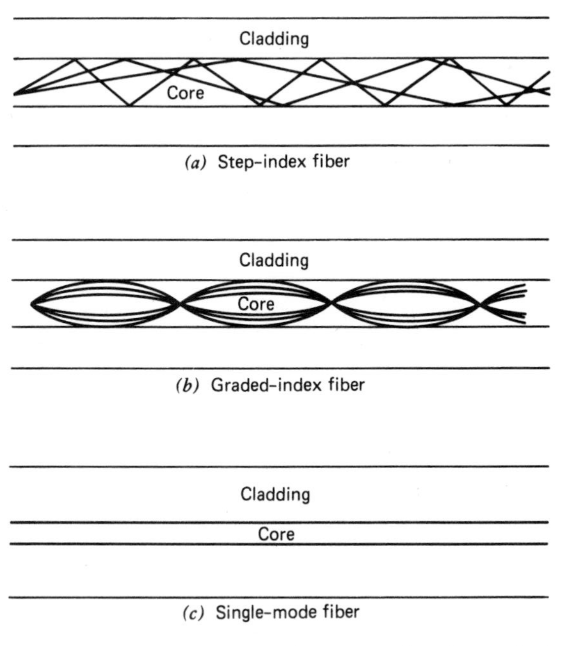

FIGURE 16-2 Modes of propagation in optical fibers.

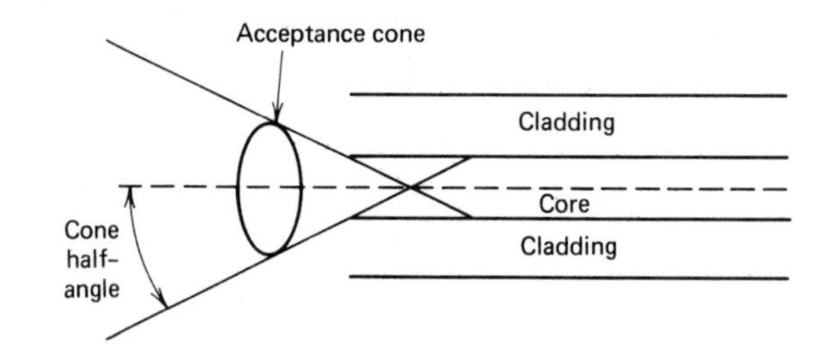

FIGURE 16-3 Acceptance cone and half-angle of an optical fiber.

called the **acceptance angle** and is defined as the half-angle of the core within which incident light is totally internally reflected by the fiber core. The acceptance cone of this angle is shown in Fig. 16-3. The "light-gathering ability" of a fiber is related to its acceptance angle and is expressed as the **numerical aperture** (NA) of the fiber. Quantitatively,

$$\text{NA} = \sin \theta$$

or

$$\theta = \arcsin \text{NA}$$

where

NA = the numerical aperture
θ = the acceptance angle

Manufacturers do not normally specify the value for the numerical aperture for their single-mode fibers since light in single-mode fibers is not reflected or refracted and, therefore, does not exit the fiber at an angle through the cladding. Typical values for the numerical aperture for multimode fibers range from 0.2 to 0.6. Although not normally specified, the numerical aperture for single-mode fibers is on the order of 0.03 to 0.1. The low numerical value of this aperture implies a small acceptance angle which, in turn, dictates the use of a laser source with single-mode fibers.

EXAMPLE 16-1

Determine the acceptance angle for a single-mode fiber with a numerical aperture of 0.096.

Solution

The acceptance angle is computed as

$$\theta = \sin^{-1} \text{NA}$$

$$= \sin^{-1} 0.096$$

$$= 5.5^\circ$$

Several factors affect the propagation of light through an optical fiber sensor.

1. The coherence of the light source.
2. The size of the fiber.
3. The composition of the fiber.
4. The numerical aperture of the source and the fiber,
5. The amount of light injected into the fiber.

The **injection laser diode** provides us with a coherent light source. When it is used with single-mode fibers, the coherence of the light is maintained along the fiber. This coherence is very important because it permits interferometric comparison of the paths taken by two beams of light, which is a technique used to make very sensitive measurements with many optical fiber sensors.

16-4 ADVANTAGES OF FIBER OPTIC SENSORS

Optical fiber sensors offer some attractive advantages when compared with conventional sensors. One advantage is compatibility with fiber optic communications systems. For example, if measurement data are to be transmitted via a fiber optic transmission system, optical sensors are inherently compatible with such systems. In addition, since optical fibers do not conduct electric current, fiber sensors are very suitable for use in explosive environments and in high-voltage equipment.

Another significant advantage of fiber optic sensors is their immunity to inductive interference and radiated signals. Optical fibers do not pick up radio-frequency interference, electromagnetic interference, or interference caused by lightning, nearby electric motors, relays, and other sources of electrical noise. Therefore, the signal obtained with a fiber optic sensor is essentially noise-free. As optical fiber sensor technology develops, it appears that very sensitive and accurate, yet simple and inexpensive, optical fiber sensors will become available.

16-5 FIBER OPTIC SENSORS

There are two basic types of fiber optic sensors: pure fiber sensors and remote optical sensors. Remote optical sensors use fibers only to carry light to a separate device that responds to the light stimuli, whereas pure fiber sensors depend on environmentally induced changes in light as it travels through a fiber. Our interest in this chapter centers on pure fiber sensors.

In most pure fiber sensors, one of two things happens to the light in response to some external effect. The light can leak from the core into the cladding where it is eventually absorbed, or it can be forced to travel a different distance than light traveling an alternate optical path. This causes a phase difference to exist between the two light waves; this difference can be detected by combining the two light waves by superposition and observing the resulting light and dark pattern.

Optical sensors use a single-mode fiber which is a low-loss fiber with a very small core and, therefore, a very small angle of acceptance of light. A basic sensor schematic is shown in Fig. 16-4. The sensing mechanism illustrated in the figure causes a change in some characteristic of the optical signal received at the detector. The sensing mechanism may, in fact, be a characteristic of the fiber itself, or it may be a device external to the fiber. If the light signal received by the detector changes as a result of some change in the fiber itself, then the sensor is called an *intensity* or *intrinsic* sensor. If the light signal at the detector changes as a result of some external transducer-type device, then the sensor is called an *interferometric* or *extrinsic* sensor.

A wide range of optical sensors are on the market. Table 16-1 is a partial list of physical parameters for whose measurement optical fiber sensors are at present available.

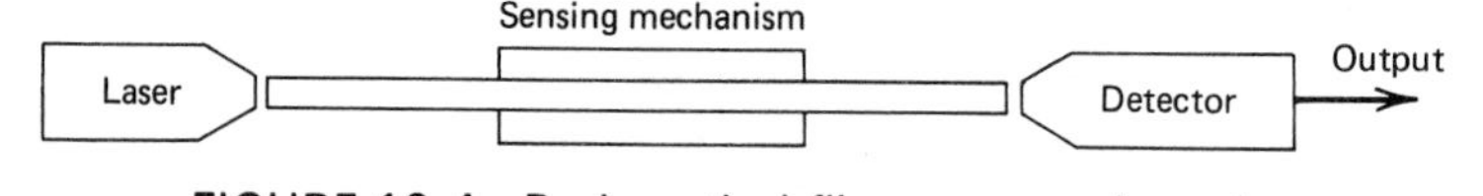

FIGURE 16-4 Basic optical fiber sensor schematic.

TABLE 16-1
Physical Parameters and the Types of Optical Sensors Currently Available for Their Measurement

	Type Sensor	
Physical Parameter	Intensity	Interferometric
Pressure	X	
Sound	X	X
Temperature	X	X
Level	X	X
Rotation		X
Acceleration	X	
Displacement		X
Strain	X	X
Electric current	X	
Voltage		X
Electric fields	X	
Magnetic fields	X	
Acoustics		X
Color		X
Pulse rate		X
Chemical pH		X

16-6 MEASURING PHYSICAL PARAMETERS WITH FIBER OPTIC SENSORS

The following sections describe how various physical parameters can be measured with fiber optic sensors.

16-6.1 Temperature

Optical fibers can be used to measure temperature since temperature induces changes in the refractive index of the fiber. The core and the cladding of optical fibers have different indices of refraction because of differences in their composition. The refractive index is a function of both composition and temperature, and it changes at different rates in response to a temperature change. A higher temperature causes the critical angle of the fiber to change slightly, which in turn changes the amount of light lost through leakage in the cladding, as shown in Fig. 16-5. Total internal reflection occurs at angles of incidence greater than the critical angle. At angles of incidence less than the critical angle, some light is lost to refraction.

16-6.2 Sound

Sound level can be measured with either an intensity or an interferometric type of optical sensor. The following paragraphs describe how sound level is measured with an interferometer-type sensor such as the one shown in Fig. 16-6. As can be seen in the figure, the laser beam is split by the beam splitter.

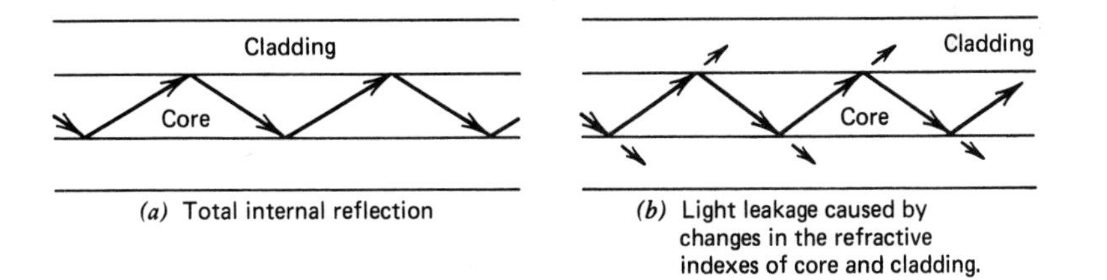

(*a*) Total internal reflection

(*b*) Light leakage caused by changes in the refractive indexes of core and cladding.

FIGURE 16-5 Optical fiber temperature sensor.

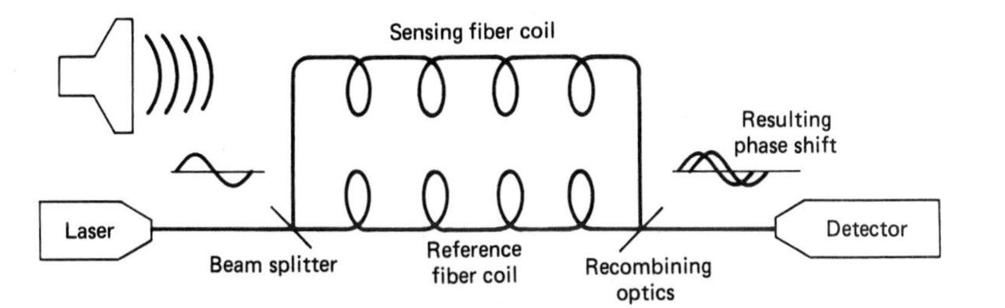

FIGURE 16-6 Interferometer-type optical fiber sound sensor.

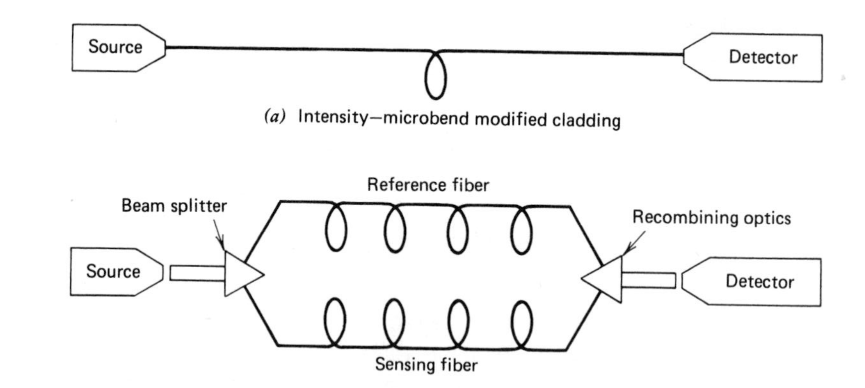

FIGURE 16-7 Fiber optic sensors. (*a*) Intensity type. (*b*) Interferometric type.

Part of the beam passes through the reference coil, which is enclosed in a stable environment, and the remainder passes through the sensing coil. Sound waves distort the sensing fiber, which causes its optical length to change, thus altering the relative phases of the light in each fiber. The sound level is a function of the phase shift, which will be seen in the interference pattern produced when the two beams are recombined.

Figure 16-7 shows in somewhat more detail the principle of operation of intensity and interferometric fiber sensors. With intensity sensors, the measured variable causes a change in the intensity of light propagating in the fiber. On the other hand, with interferometric sensors, the measured variable causes a phase change of the light propagating in the fiber. This beam is then compared with the reference beam as shown in Fig. 16-7, and the phase displacement owing to the influence of the sensed parameter causes a change in the intensity of the interference pattern as shown in Fig. 16-8.

16-6.3 Level Sensing

Another application of fiber optic sensors is in level sensing. The principle of sensing liquid levels using fiber sensor technology is quite straightforward. If

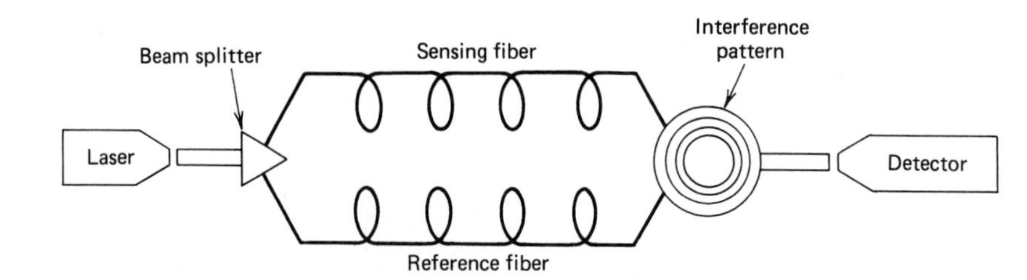

FIGURE 16-8 Change in the intensity of the interference pattern caused by phase displacement.

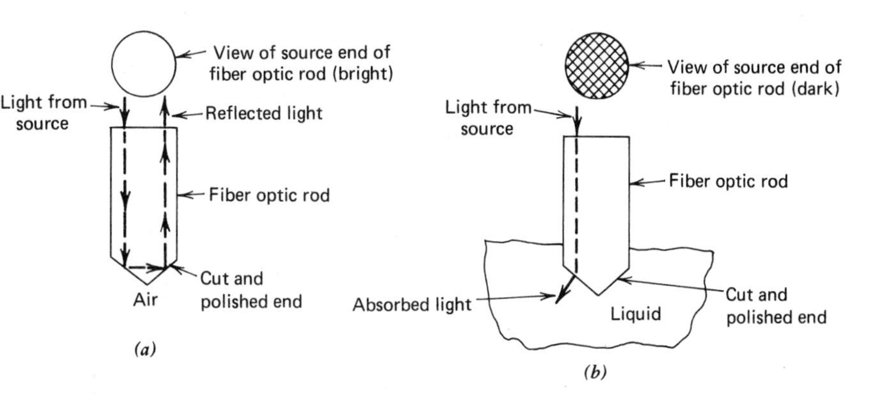

FIGURE 16-9 Optical fiber level sensor. (*a*) In air. (*b*) In liquid.

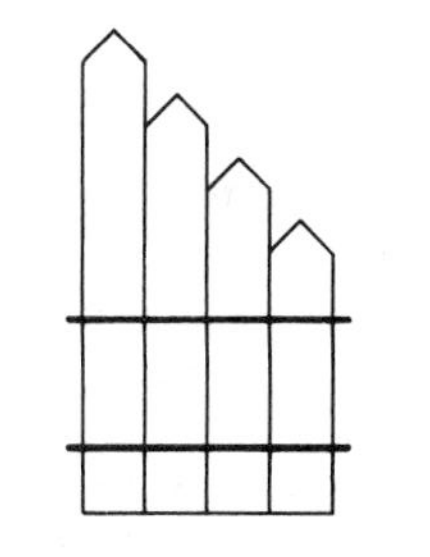

FIGURE 16-10 Optical fiber bundle for multilevel sensing.

one end of a fiber optic rod or cable is cut and polished properly to form a prism and the prism is in air, light transmitted from the other end will be reflected back with little loss. Under these conditions the end near the light source will appear bright as shown in Fig. 16-9*a*.

If the polished end of the rod is now submerged in a liquid with a higher refractive index than that of the fiber rod, most of the light transmitted through the rod will be absorbed by the liquid rather than being reflected. (Fig. 16-9*b*).

Multilevel sensing can be achieved by bundling fiber optic rods of different lengths together, as shown in Fig. 16-10.

16-7 SUMMARY

Optical fibers are thin, flexible threads of transparent glass or plastic that carry visible light. Optical fibers consist of two concentric layers, the inner layer called the *core* and the outer layer called the *cladding*. The core and the cladding may both be made of glass or of plastic, or the core may be glass with a plastic cladding.

Light waves can be transmitted along an optical fiber if the refractive index of the cladding is lower than that of the core. Light losses are affected by the angle at which light waves strike the boundary. If the light waves strike the boundary at an angle less than the critical angle, some pass into the cladding and are lost.

Optical fibers are manufactured as step-index, graded-index, or single-mode fibers. Almost all optical fibers used in fiber optic sensors are single-mode fibers because they have very low losses and because they permit interferometric measurement techniques.

The two basic types of fiber optic sensors are pure fiber sensors and remote optical sensors. Pure fiber sensors, which respond to environmentally induced change in the light level that passes through the fiber, have been our primary interest in this chapter.

16-8 GLOSSARY

Acceptance angle: The half-angle of the cone within which incident light is totally and internally reflected by the fiber core. It is equal to the arcsin NA.

Cladding: The outer concentric layer that surrounds the fiber core and has a lower index of refraction than that of the core.

Core: The central, light-carrying part of an optical fiber. It has an index of refraction higher than that of the surrounding cladding.

Critical angle: The angle of incidence of light, which has traveled through a denser medium to the interface between the denser and less dense medium, at which all the light is reflected along the interface.

Graded-index fiber: An optical fiber whose core has a nonuniform index of refraction. The core is composed of concentric rings of glass whose refractive indices decrease from the center axis. The purpose is to reduce modal dispersion and thereby increase fiber bandwidth.

Index of refraction: The ratio of the velocity of light in free space to the velocity of light in a given material, symbolized by n.

Injection laser diode: A semiconductor diode that spontaneously emits light from the p-n junction when forward current is applied. Acts as a light-emitting diode below threshold current with lasing occurring above threshold current, the current level at which light becomes coherent.

Mode: In guided-wave propagation such as through a waveguide or optical fiber; a distribution of electromagnetic energy that satisfies Maxwell's equations and boundary conditions. Loosely defined, a possible path followed by light rays.

Multiple-mode fiber: A type of optical fiber that supports more than one propagating mode.

Numerical aperture: The light-gathering ability of a fiber, defining the maximum angle to the fiber axis at which light will be accepted and propagated through the fiber. $\mathrm{NA} = \sin \theta$, where θ is the acceptance angle. NA is also used to describe the angular spread of light from a central axis, as in exiting a fiber, emitting from a source, or entering a detector.

Single-mode fiber: An optical fiber that supports only one mode of light propagation above the cutoff frequency.

16-9 REVIEW QUESTIONS

The following questions should be answered after a thorough study of the chapter. The purpose of the questions is to determine your comprehension of the material.

1. What three combinations of materials are used to construct optical fibers?
2. List the three types of optical fibers in use at present, and indicate which type is most suitable for use in instrumentation.
3. What does the term *acceptance angle* mean relative to fiber optics?
4. What five factors affect the propagation of light through optical fibers?
5. What is the difference between intrinsic and extrinsic sensors?
6. Why is it necessary to use a laser source with single-mode fibers?
7. What are the two basic types of fiber optic sensors?

16-10 PROBLEMS

16-1 What three types of optical fibers are in use at present, and which of the three is most suitable for use in instrumentation. Why?

16-2 Define the term *acceptance angle* as it applies to fiber optics.

16-3 Define the term *numerical aperture* in both quantitative and qualitative terms.

16-4 What is the typical range of values for the numerical aperture for single-mode fibers, and what are the implications of this range?

16-5 Compute the acceptance angle of a single-mode fiber with a numerical aperture of 0.06.

16-6 Describe the two basic types of fiber optic sensors.

16-7 Describe the difference between intrinsic and extrinsic sensors.

CHAPTER 17 Data Acquisition

17-1 Instructional Objectives

This chapter discusses automatic data acquisition systems, with primary emphasis on their subsystems. After completing the chapter you should be able to

1. Draw a block diagram of a basic data acquisition system.
2. Describe a "real-time" data acquisition system.
3. List the primary functions that transducers perform in data acquisition systems.
4. List and describe the types of signal conditioning discussed in the chapter.
5. Describe the operation of the following types of circuits:
   (a) Multiplexers.
   (b) Sample and hold circuits.
   (c) Digital-to-analog converters.
   (d) Analog-to-digital converters.
6. Make calculations related to filter attenuation.
7. Make calculations related to the linearization of signals.
8. Make calculations related to D/A and A/D converters.
9. Define the term *actuator* as it is used in data acquisition.

17-2 INTRODUCTION

Data acquisition is the collection of information. Usually, industrial data acquisition systems are "**real-time**" systems. Such systems are characterized by their ability to acquire data or perform a control task within an acceptable time window. The duration of the time window depends on how quickly the system must respond, which is a function of the speed and accuracy requirements for a given application.

The hardware that is used to acquire data is called a **data acquisition** system. Such systems function as an interface between the real world of

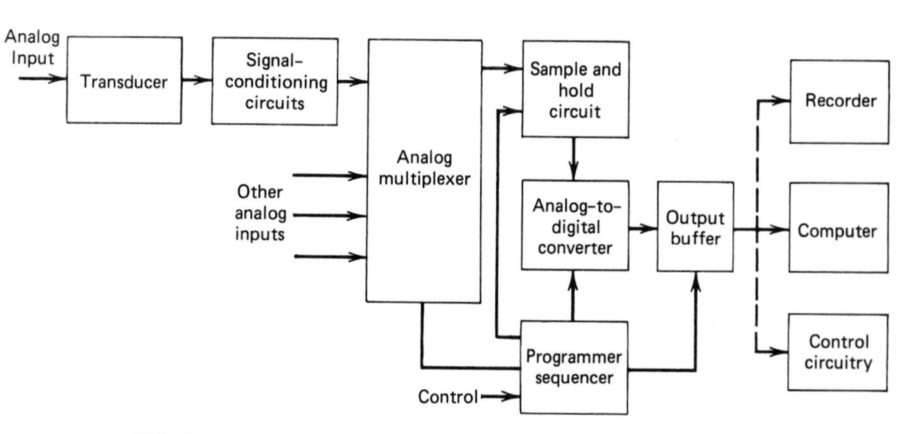

FIGURE 17-1 **Block diagram of a data acquisition system.**

physical parameters, which are *analog*, and the world of computers and digital signals. The block diagram of Fig. 17-1 shows the elements most commonly used in an industrial data acquisition system. Most of the remainder of this chapter discusses the subsystems of the system shown in Fig. 17-1.

17-3 TRANSDUCERS

Data collected by acquisition systems are usually analog data from some nonelectrical physical parameter. The purpose of the transducer is to convert the nonelectrical signal to an equivalent electrical signal. A block diagram for a transducer is shown in Fig. 17-2. The primary functions of the transducer in a data acquisition system are

1. To *sense* the presence, magnitude, change in, and frequency of the *measurand.*
2. To *provide* an electrical output that furnishes accurate quantitative *data* about the measurand.

Transducers are often classified according to the electrical principle involved in their operation. There are

1. Passive transducers that require an external source of power.
2. Self-generating transducers that require an external power source.

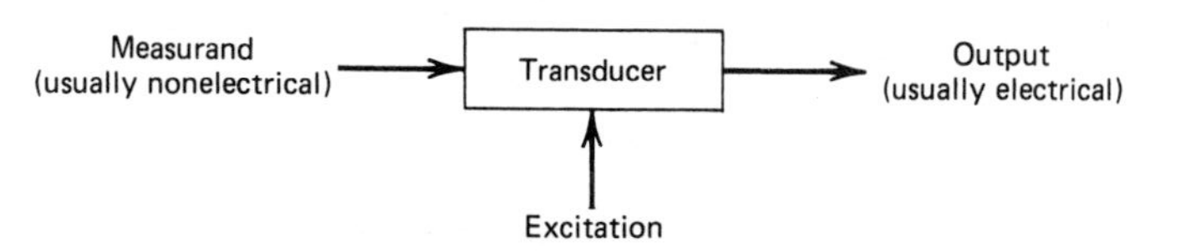

FIGURE 17-2 **Transducer block diagram.**

17-4 SIGNAL-CONDITIONING CIRCUITS

A major concern in transducer design is simply identifying some material that changes in some way owing to a change in a physical parameter being sensed. The amplitude and linearity of the transducer output signal are usually not first-order concerns in transducer design, but they are of considerable concern in data acquisition systems. Therefore, signal conditioning to increase amplitude and improve linearity is usually necessary in data acquisition systems.

The term *signal conditioning* as used here means to make any necessary changes on the analog input signal before applying the signal to the analog-to-digital converter (ADC). Some of the most frequently performed types of signal conditioning are

- Buffering.
- Filtering.
- Signal-level change.
- Signal conversion.
- Linearization.

17-4.1 Buffering

The **buffer**, which is probably the most basic and straightforward signal-conditioning circuit, can be used to provide impedance translation between the signal source and the circuitry to which the source is supplying a signal. Any *integrated-circuit* (*IC*) *operational amplifier* (*op-amp*) used in the non-inverting mode and connected for unity gain, as shown in Fig. 17-3, serves as a good buffer. However, some op-amps are designed specifically to be used as buffers.

17-4.2 Filtering

Industrial environments in which data acquisition systems are often placed tend to introduce spurious interference signals into the acquisition system.

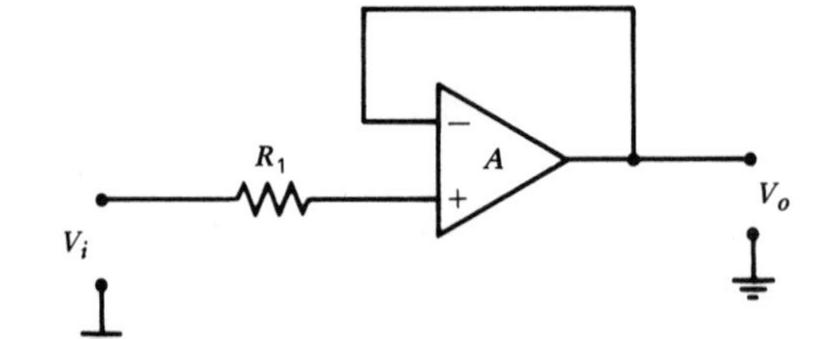

FIGURE 17-3 Op-amp configured to serve as a buffer.

These unwanted signals are categorized as *noise* and are often due to 60-Hz or 400-Hz power line interference or to transients caused by inductive load changes such as starting motors. Such interference can cause appreciable error in the input signal.

Often such interference can be significantly reduced by proper filtering. A **filter** is a circuit that passes a certain band of frequencies while attenuating the signals of other frequencies. Filters can be constructed as passive filters using only resistors, inductors, and capacitors, or as active filters using op-amps with gain and feedback. Filter performance is generally described in terms of output voltage to input voltage (V_o/V_i) at different frequencies and is generally expressed on a logarithmic scale using decibels as

$$\text{dB} = 20 \log(V_o/V_i) \tag{17-1}$$

As mentioned earlier, the signal source for many industrial data acquisition systems is a transducer. The maximum bandwidth of most transducers is about 10 Hz. Therefore, filtering noise from these transducer signals is relatively easy with a *low-pass* filter.

EXAMPLE 17-1

A low-pass filter is to reject signals above 30 Hz. How many decibels down is a 540-mV, 60-Hz input signal if its amplitude at the output of the filter is 4 mV?

Solution

The reduction, as expressed by Eq. 17-1, is

$$\text{dB} = 20 \log\left(\frac{4 \text{ mV}}{540 \text{ mV}}\right)$$

$$= -42.6$$

17-4.3 Signal-Level Change

Perhaps the most frequently performed type of signal conditioning is a level or amplitude change. A change in signal level may require either attenuation or amplification of the input signal. Attenuation is generally accomplished with a resistive voltage divider network, whereas amplification requires an active device such as a transistor or an op-amp. It is common practice to use both attenuation and amplification for signal-conditioning purposes in many electronic test instruments, such as electronic multimeters and oscilloscopes, in order to provide multiple ranges.

17-4.4 Signal Conversion

Signal conditioning often requires converting the variation of one electrical parameter into a proportional variation of another parameter. For example, many transducers provide an indication of a change in some sensed variable by a change in their resistance. This resistance change is converted to a proportional voltage or current by using the transducer in either a bridge circuit or an amplifier circuit.

Bridge circuits provide a convenient way of converting a resistance change to a proportional voltage change. Bridge circuits are passive networks that are used to measure impedances by comparison of potentials. When V_A equals V_B in Fig. 17-4, the bridge is said to be balanced. When the bridge is balanced, no current flows through the detector. Therefore, a balanced bridge indicates a null condition. When a transducer is used in a bridge circuit, its resistance is generally equal to that of each of the other bridge resistors at some desired condition, such as at a desired temperature, pressure, or light level. This produces a null condition. If the physical parameter to which the transducer is sensitive changes in value, the bridge becomes unbalanced. This causes a potential difference to exist between points A and B in Fig. 17-4. Thus, the conversion from resistance to a proportional voltage is achieved. Resistance-to-voltage conversion can also be accomplished using an amplifier whose output voltage varies with a resistance change, as shown in Fig. 17-5.

If the temperature-sensitive resistor has a positive temperature coefficient

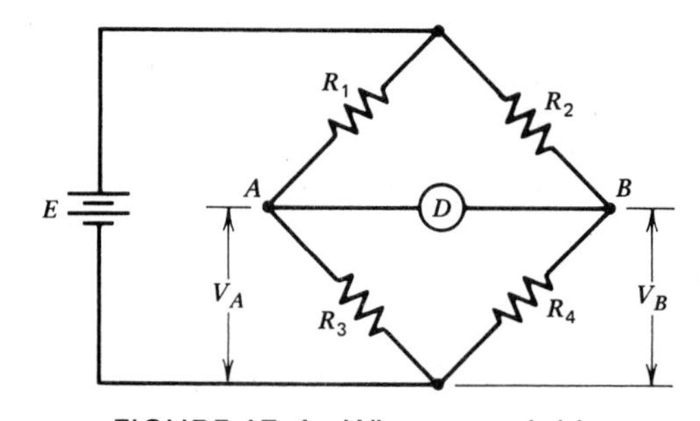

FIGURE 17-4 Wheatstone bridge.

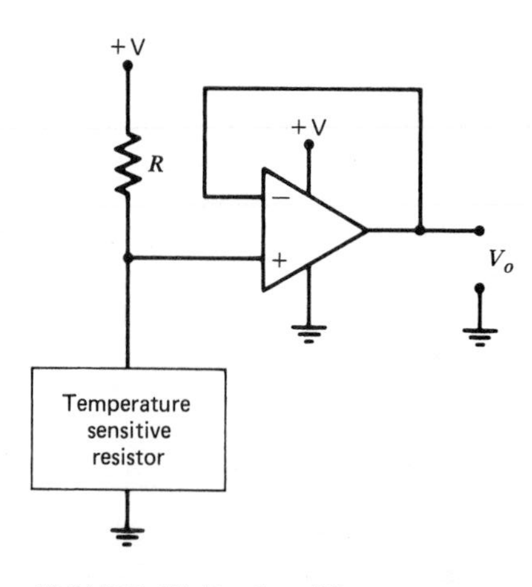

FIGURE 17-5 Amplifier to convert resistance change to voltage change.

of resistance, the output voltage will increase as temperature increases. Typically, the resistance change will be nonlinear and will require **linearization**, which is the topic discussed in the next section.

17-4.5 Linearization

As mentioned earlier, linearity is a secondary concern in transducer design but is very important in data acquisition systems. Even devices whose outputs are reasonably linear may require additional linearization when precise measurements of dynamic parameters are required.

Transducer output signals may be linearized by using an amplifier with gain that is a function of its input voltage, thereby providing a linear output. For example, the output signal of a transducer may vary exponentially with respect to some dynamic variable, such as temperature, as shown in Fig. 17-6. The output voltage produced by the transducer can be expressed as

$$V = V_o \varepsilon^{\theta T} \tag{17-2}$$

where

V = transducer output voltage at temperature T
V_o = transducer output voltage at reference temperature
ε = base of the natural logarithm system (2.718)
θ = exponential constant
T = temperature in degrees Celsius

This voltage can be linearized by using an amplifier whose output varies inversely with the input voltage according to the natural logarithmic expression

$$V_{out} = K \ln V_{in} \tag{17-3}$$

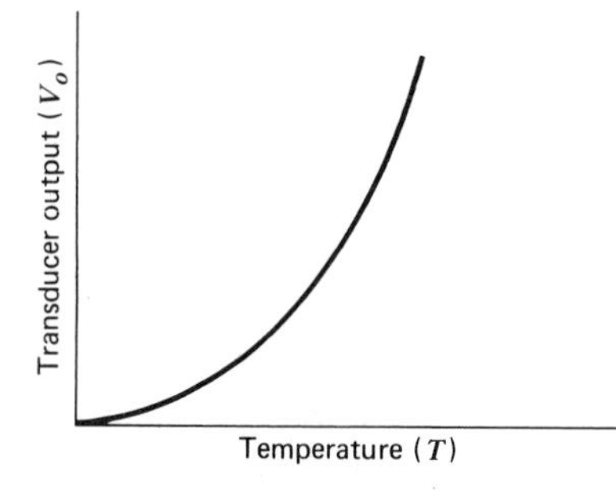

FIGURE 17-6 Transfer curves for a transducer showing the nonlinear relationship between temperature and output voltage.

where

V_{out} = amplifier output voltage
V_{in} = amplifier input voltage
K = calibration constant

Substituting Eq. 17-2 into Eq. 17-3 yields

$$V_{out} = K \ln V_o \varepsilon^{\theta T}$$
$$= K(\ln V_o + \ln \varepsilon^{\theta T})$$

Therefore, we can say that

$$V_{out} = K \ln V_o + K\theta T \quad (17\text{-}4)$$

Equation 17-4 is a linear equation. Therefore, we can see that the amplifier output is linear with respect to temperature, but with a scale factor $K\theta$ and an offset $K \ln V_o$ as shown in Fig. 17-7.

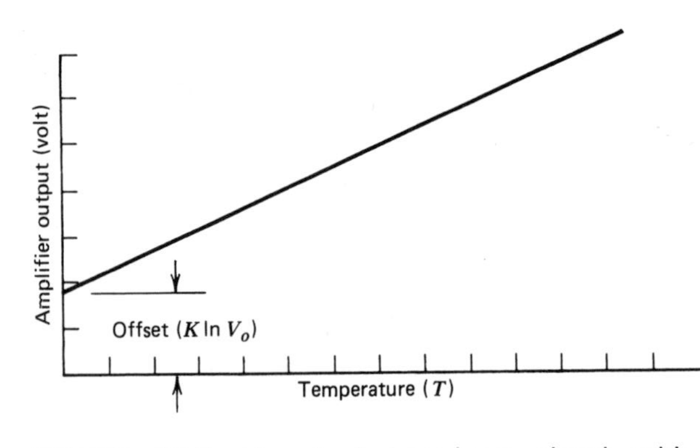

FIGURE 17-7 Linearized transducer signal with offset.

17-5 MULTIPLEXERS

A **multiplexer** is an electronic circuit with two or more input terminals and one output terminal. If a multiplexer is used in a data acquisition system, only one A/D converter is required and is used in a time-share mode as shown in Fig. 17-8. In contrast, without the multiplexer one A/D converter is required for each analog input. Since the A/D converter often represents the most expensive subsystem in a data acquisition system, multiplexing several analog input signals and using a single A/D converter represents a significant cost savings.

As shown in Fig. 17-8, an *analog multiplexer* is basically a set of parallel electromechanical or electronic switches connected to a common output line. The switches close sequentially, or nonsequentially if desired, with only one

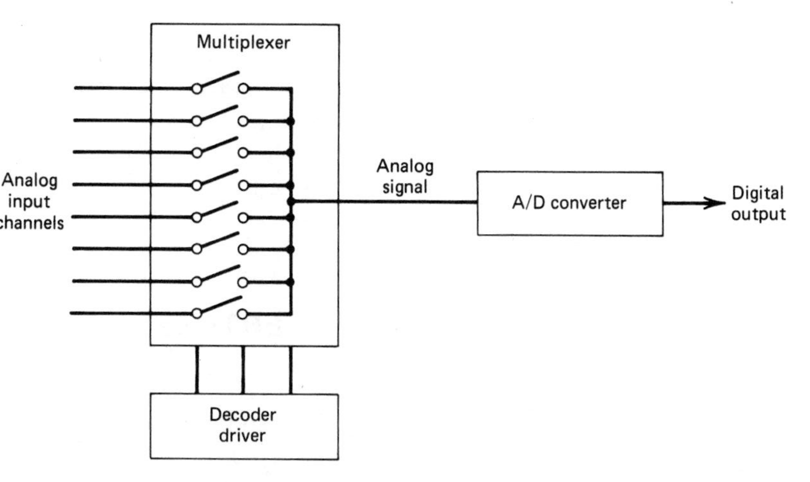

FIGURE 17-8 **Analog multiplexer.**

switch closed at a time and with switching controlled by the decoder driver. Although there are applications for both electromechanical and electronic multiplexers, most data acquisition systems use multiplexers with electronic switching devices such as junction field-effect transistors (JFETs) or complementary metal oxide semiconductor (CMOS) transmission gates.

17-6 SAMPLE AND HOLD CIRCUITS

Sample and hold circuits are essentially voltage–memory circuits that store an input voltage on a high-quality capacitor. The function of a sample and hold circuit is to take a short-duration sample, or "snapshot," of a rapidly changing input signal and hold that sample for accurate analog-to-digital conversion. A basic sample and hold circuit is shown in Fig. 17-9. Amplifier A_1 is a buffer amplifier with high input impedance to reduce loading of the previous stage and low output impedance to permit very rapid charging of the hold capacitor C.

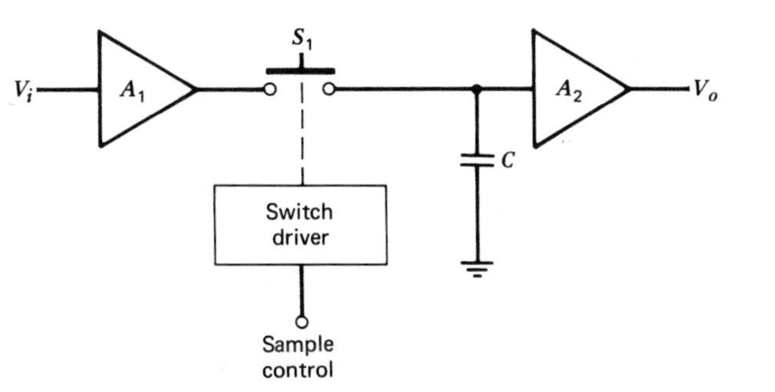

FIGURE 17-9 **Basic sample and hold circuit.**

Switch S_1 must be a fast-acting, high-quality analog device such as a CMOS transmission gate. Capacitor C is a device with low leakage and low dielectric absorption characteristics. Capacitors made of polystyrene, polypropylene, or Teflon are suitable. Amplifier A_2 is an output amplifier that acts as a buffer for the voltage on the hold capacitor. Therefore, it must have an extremely low input current. This usually dictates the use of an FET amplifier.

17-7 DIGITAL-TO-ANALOG CONVERTERS

Our primary concern in this chapter is analog-to-digital converters, since they represent one of the subsystems of a data acquisition system. However, it is often desirable to begin by discussing the **digital-to-analog (D/A) converter**, since it is frequently a subcircuit of an analog-to-digital converter. The two primary types of D/A converters are

1. Voltage-summing amplifier.
2. Binary-weighted resistor network.

17-7.1 Voltage-Summing Amplifier

The voltage-summing amplifier shown in Fig. 17-10 functions as a digital-to-analog conversion circuit. Basically, the circuit converts n digital voltage levels into an equivalent analog output voltage. The input voltages V_1 through V_4 are equal in amplitude but carry a "weight" associated with the position of the bits in a binary number. The weight of the bits in a binary number increases as we move left from the binary point:

$$2_4 \quad 2_3 \quad 2_2 \quad 2_1 \quad 2_0$$

Since the amplitudes of the digital inputs are all the same, we must compute the value of the resistors to account for the "weight" of the bits. These are

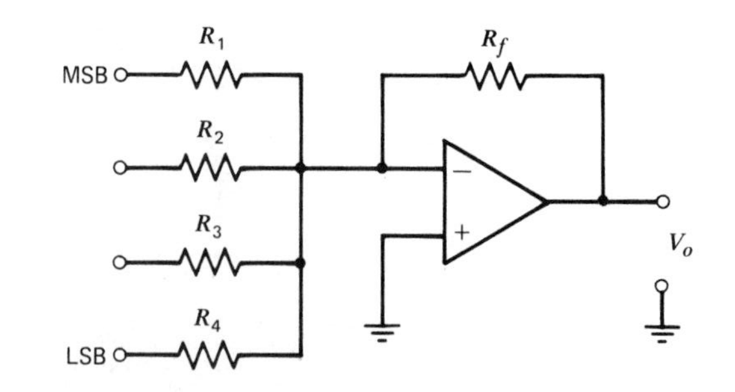

FIGURE 17-10 Voltage-summing amplifier.

computed as

$$R_1 = \frac{R_f}{2^0}$$

$$R_2 = \frac{R_f}{2^1}$$

$$R_3 = \frac{R_f}{2^2}$$

$$R_4 = \frac{R_f}{2^3}$$

The output voltage of the summing amplifier is computed as

$$V_o = -\left(V_1 \frac{R_f}{R_1} + V_2 \frac{R_f}{R_2} + V_3 \frac{R_f}{R_3} + V_4 \frac{R_f}{R_4}\right) \tag{17-5}$$

Substituting the resistance relationships into Eq. 17-5 yields

$$V_o = -(V_1 + 2V_2 + 4V_3 + 8V_4) \tag{17-6}$$

As binary inputs, V_1 represents the least significant bit (LSB) and V_4 represents the most significant bit (MSB). We can see in Eq. 17-6 that the MSB of a 4-bit binary number contributes eight times as much toward V_o as the LSB when both are at the same positive voltage representing a logic 1.

EXAMPLE 17-2

Determine the output voltage of the circuit of Fig. 17-11 if the following input voltages are applied.

$$V_1 = 1 \text{ V}$$

$$V_2 = 1 \text{ V}$$

$$V_3 = 0 \text{ V}$$

$$V_4 = 1 \text{ V}$$

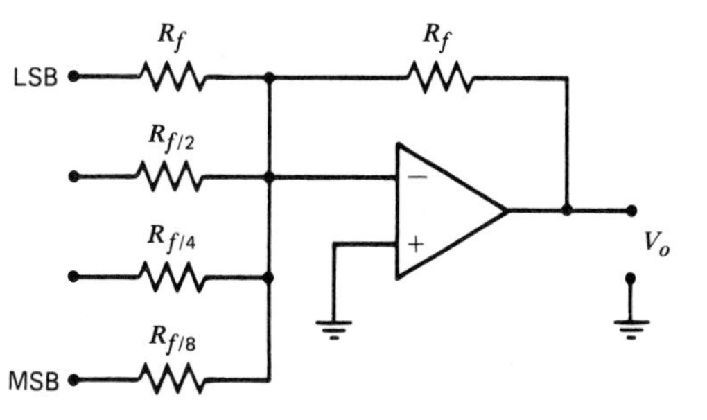

FIGURE 17-11 Voltage-summing amplifier for Example 17-2

Solution

The output voltage is computed as

$$\begin{aligned} V_o &= -[(1\text{ V})(1) + (1\text{ V})(2) + (0\text{ V})(4) + (1\text{ V})(8)] \\ &= -(1\text{ V} + 2\text{ V} + 0\text{ V} + 8\text{ V}) \\ &= -11\text{ V} \end{aligned}$$

The smallest change at the output is due to the LSB changing from logic 0 to logic 1 or vice versa. In Example 17-1 this would represent a change of 1 V at the output. This change is called the *resolution* of the converter and is expressed quantitatively as

$$\text{Res} = \frac{V_o}{2^n - 1} \tag{17-7}$$

where

V_o = maximum output voltage
n = number of binary bits

EXAMPLE 17-3

What is the resolution of a D/A converter with eight input bits and a maximum output voltage of 12.8 V?

Solution

The resolution is

$$\begin{aligned} \text{Res} &= \frac{V_o}{2^n - 1} \\ &= \frac{12.8\text{ V}}{2^8 - 1} = \frac{12.8\text{ V}}{255} = 50\text{ mV} \end{aligned}$$

Summing amplifier-type D/A converters have two rather significant disadvantages.

1. Several different values of precision resistors are required.
2. Each binary input sees a different load since each input resistor is different in value.

Both of these disadvantages can be overcome by using $R-2R$ ladder D/A converters, which are discussed in the next section.

17-7.2 D/A Conversion Using $R-2R$ Ladders

The $R-2R$ ladder-type D/A converter is a resistive network that uses only two different values of resistance, thus overcoming the first disadvantage of the voltage-summing D/A converter. An $R-2R$ ladder network designed for a 4-bit binary input is shown in Fig. 17-12. The analog output of the circuit is a properly weighted sum of the binary inputs.

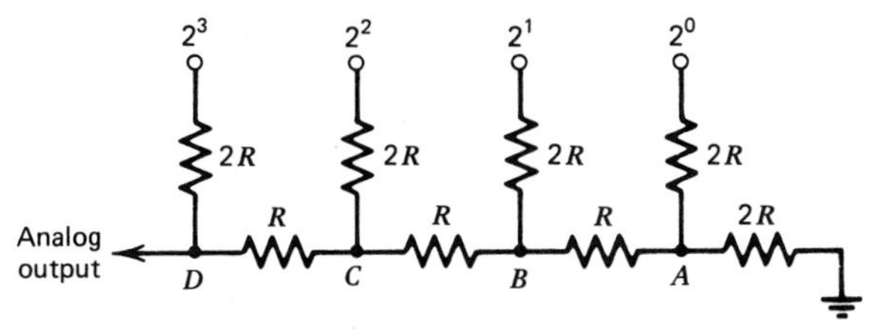

FIGURE 17-12 Four-bit $R-2R$ ladder network.

The second disadvantage of voltage-summing D/A converters is the different load that the network presents to each binary bit. The $R-2R$ converter also eliminates this problem. Analysis of the network, by use of Thévenin's theorem or other analytical tools, shows that the resistance from any node to ground or to an input terminal is $2R$. This is true whatever the number of digital inputs and whether the digital inputs are high or low.

The relation between the binary inputs and the analog output for a 10-bit $R-2R$ binary ladder network is shown in Table 17-1. Here V is the amplitude of the digital input voltage. The contribution by each bit listed in Table 17-1 holds regardless of the number of binary input bits. That is, the most significant bit (MSB) always contributes V/2 for an n-bit ladder network regardless of the value of n.

TABLE 17-1
$R-2R$ Binary Ladder Output Voltage

Bit	Output Voltage
MSB	V/2
2nd MSB	V/4
3rd MSB	V/8
4th MSB	V/16
5th MSB	V/32
6th MSB	V/64
7th MSB	V/128
8th MSB	V/256
9th MSB	V/512
LSB	V/1024

EXAMPLE 17-4

Determine the analog output voltage for a 6-bit $R-2R$ ladder network when the binary input is 101010 if 0 V corresponds to logic 0 and 5 V corresponds to logic 1.

Solution

The output voltage is the sum of the contributions of the voltages for inputs at logic 1. This is computed as

$$V_o = \frac{V}{2} + \frac{V}{8} + \frac{V}{32}$$

$$= (5\text{ V})\left(\frac{16}{32}\right) + (5\text{ V})\left(\frac{4}{32}\right) + (5\text{ V})\left(\frac{1}{32}\right)$$

$$= \frac{5\text{ V} \times 16 + 5\text{ V} \times 4 + 5\text{ V} \times 1}{32} = 3.281\text{ V}$$

17-8 ANALOG-TO-DIGITAL CONVERSION

The process of converting analog signals to an equivalent signal in digital form is called analog-to-digital (A/D) conversion. As we have indicated earlier, most physical parameters that we encounter in industry are *analog*, or continuous, in nature. The purpose of the analog-to-digital converter in a data acquisition system is to convert the analog input signal to an equivalent digital signal.

There are several methods by which conversion from analog to digital form may be accomplished. The methods vary in accuracy, have different conversion rates and costs, and vary in their susceptibility to noise. We will now examine four of the techniques used to achieve A/D conversion: these are the voltage-to-frequency, simultaneous, ramp, and successive-approximation conversions.

17-8.1 Voltage-to-Frequency Conversion

Voltage-to-frequency conversion, as its name implies, is a conversion technique used to convert an analog input voltage to a periodic waveform with a frequency that is directly proportional to the input voltage. Voltage-to-frequency conversion can be achieved by using a very linear voltage-controlled oscillator (VCO) as shown in Fig. 17-13.

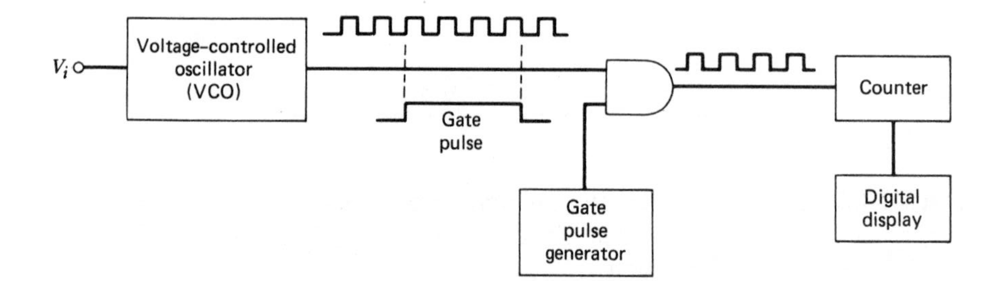

FIGURE 17-13 Block diagram of a basic voltage-to-frequency converter.

The VCO must be designed so that the relation between the output frequency and the input voltage is a constant. Quantitatively, we can express this as

$$k = \frac{f}{V_i}$$

or

$$V_i = \frac{f}{k}$$

As can be seen in Fig. 17-13, the output signal of the VCO is applied to one input terminal of a two-input AND gate. The signal applied to the other input of the AND gate is a "gating pulse." During the time that both signals are present at the AND gate inputs, the output of the AND gate is identical to the VCO output.

EXAMPLE 17-5

The constant of proportionality, k, which expresses the ratio of output frequency to input voltage for the VCO in Fig. 17-13, is 100. If 260 pulses pass through the AND gate and into the counter during a 0.1-sec gating pulse, what is the amplitude of V_i?

Solution

The output frequency of the VCO is

$$f = \frac{\text{Pulses}}{\text{Gate duration}}$$

$$= \frac{260 \text{ pulses}}{0.1 \text{ sec}} = 2600 \text{ pulses/sec}$$

The analog input voltage is

$$V_i = \frac{f}{k}$$

$$= \frac{f}{100} = \frac{2600 \text{ pps}}{100} = 26 \text{ V}$$

17-8.2 Simultaneous Method of A/D Conversion

The simultaneous, or parallel, method of A/D conversion is the fastest method of conversion available to us. However, it requires the use of a number of comparator circuits. As a general rule, the number of comparators required is $2^n - 1$ where n is the number of bits in the digital output. For example, two digital bits allow us to define four ranges of an analog input voltage because four combinations of 1s and 0s are possible with two digital bits. According to the expression $2^n - 1$, we need three comparators to accomplish the conversion. Table 17-2 shows the relation between the

TABLE 17-2
Analog Input versus Digital Output for a 2-Bit Simultaneous A/D Converter

Analog Input	Digital Output
0 to V/4	0 0
V/4 to V/2	0 1
V/2 to 3V/4	1 0
3V/4 to V	1 1

analog input voltage and the digital outputs, and Fig. 17-14 shows the three comparators in the logic circuit of a basic 2-bit simultaneous A/D converter.

Although simultaneous A/D converters are capable of very rapid conversion rates, as the number of bits in the digital output increases, the number of comparators that are required quickly becomes prohibitive. For example, an 8-bit converter requires 255 comparators, which limits the use of simultaneous converters to only those applications for which speed is of utmost importance.

17-8.3 Ramp-Type A/D Converter

We can eliminate the need for many comparators in the converter by using a *ramp*-type converter. The ramp-type A/D converter is a slightly more sophisticated and accurate (has better resolution) type of converter than the simultaneous converter and requires only one comparator. A block diagram for this type of converter is shown in Fig. 17-15. It is possible to use only one

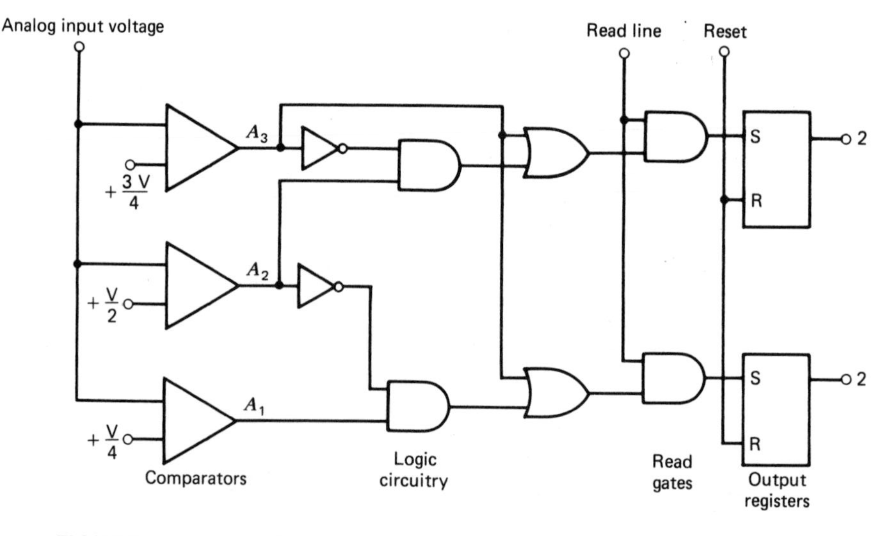

FIGURE 17-14 Logic diagram of a 2-bit simultaneous A/D converver.

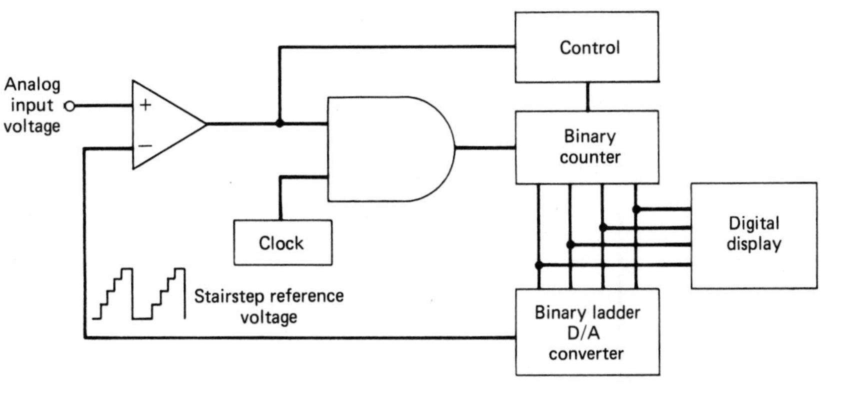

FIGURE 17-15 Ramp-type A/D converter.

comparator in the ramp-type converter because this kind of converter produces a stairstep reference voltage to which the analog input voltage is compared.

We can analyze the operation of the circuit by considering the counter to be reset and the output of the D/A converter to be zero. If the analog input voltage begins to increase, the comparator output will go high when the analog voltage exceeds the reference voltage. The high output state of the comparator enables the AND gate. Therefore, the counter begins to store pulses from the clock. As the counter advances through its binary states, it produces the stairstep reference voltage at the output of the D/A converter. When the stairstep voltage exceeds the amplitude of the analog voltage, the comparator output is switched low, thus disabling the AND gate which cuts off clock pulses to the counter. The binary number stored in the counter is displayed at the output. This display, of course, represents the amplitude of the analog voltage.

Figure 17-16 shows a conversion sequence for a 4-bit converter. As can be seen, the counter starts at zero and counts up to the point at which the

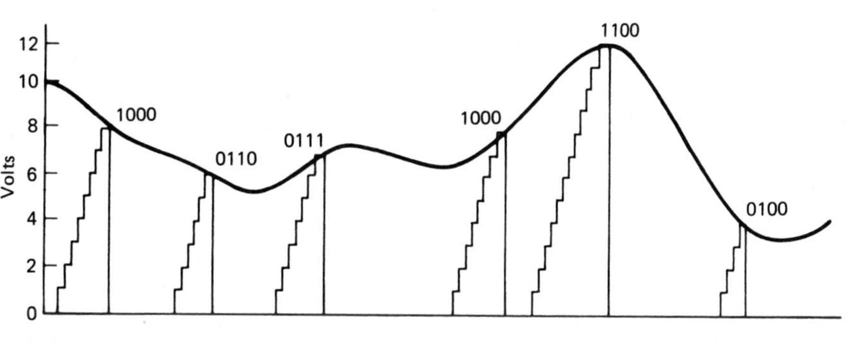

FIGURE 17-16 Graphical example of the conversion process of a ramp-type A/D converter.

stairstep reference voltage is equal to the analog input voltage. Therefore, the conversion time is directly related to the amplitude of the analog voltage.

EXAMPLE 17-6

Determine the following for a 10-bit ramp-type A/D converter driven by a 1-MHz clock.

(a) The maximum conversion time.
(b) The average conversion time.
(c) The conversion rate corresponding to maximum conversion time.

Solution

(a) For a 10-bit converter the number of count states is given as

$$\text{Count states} = 2^{10} = 1024$$

The period of the clock signal is 1 μsec. The counter advances are one count state per clock pulse. The maximum conversion time is computed as

$$\text{Maximum conversion time} = (1024 \text{ counts})\left(\frac{1\ \mu\text{sec}}{\text{count}}\right) = 1024\ \mu\text{sec}$$

(b) The average conversion time is one-half the maximum conversion time since the staircase increases linearly. Therefore,

$$\text{Average conversion time} = \frac{1024\ \mu\text{sec}}{2} = 512\ \mu\text{sec}$$

(c) The rate of conversion at maximum conversion time is determined by the greatest time required for conversion and is computed as

$$\text{Conversion rate} = \frac{1}{\text{Maximum conversion time}}$$
$$= \frac{1}{1024 \times 10^{-6}\ \text{sec}}$$
$$= 976 \text{ conversions per second}$$

17-8.4 Successive-Approximation A/D Converter

The successive-approximation technique is one of the most widely used methods of A/D conversion primarily because of its short, as well as constant, conversion time. A basic block diagram for a 4-bit successive-approximation A/D converter is shown in Fig. 17-17. As can be seen, it consists of a voltage comparator, a D/A converter, a successive-approximation (SA) register, and a clock.

In going through a conversion cycle, the system starts by enabling the bits of the D/A converter one bit at a time, starting with MSB. As each bit is enabled, its amplitude is compared to the analog voltage, V_i, by the voltage comparator. The comparator then produces an output that indicates whether the analog voltage is greater or less in amplitude than the output of the D/A converter. If the output of the D/A converter is greater than the analog voltage, the MSB is reset to zero since it will not be required in the digital representation of the analog input. If the D/A converter output is less than the analog input, the MSB is retained in the register.

The system makes this comparison with each bit, starting with the MSB, then the next MSB, and so on. As each bit of the D/A converter is compared, those that contribute to the digital representation of the analog input are stored in the register, and those bits not required are reset.

EXAMPLE 17-7

A 12-V input signal is applied to the 4-bit successive-approximation A/D converter shown in Fig. 17-17. There is a linear relationship between the digital input and the analog output of the D/A converter, that is, $2^0 = 1$ V, $2^1 = 2$ V, $2^2 = 4$ V, $2^3 = 8$ V. What bits are stored in the SA register to represent the 12-V analog input voltage?

Solution

The same relationship that exists between the digital input and analog output for the D/A converter exists between the analog input and digital output for the A/D converter. Therefore, a 12-V analog input voltage equals 1100 digital, so the two most significant bits generated by the D/A converter are stored by the SA register.

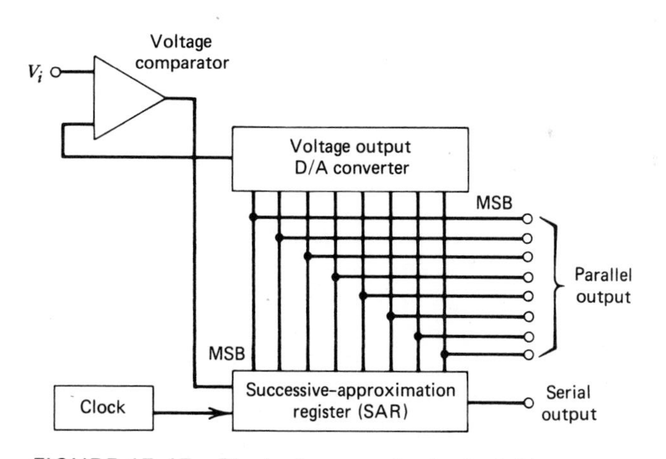

FIGURE 17-17 **Block diagram of a basic 4-bit successive-approximation A/D converter.**

17-9 DATA ACQUISITION SYSTEMS AND COMPUTERIZED CONTROL

Data are often acquired and stored for future use. However, our interest in data acquisition systems centers primarily on their use in industrial control. The circuitry discussed in this chapter permits us to acquire analog data from

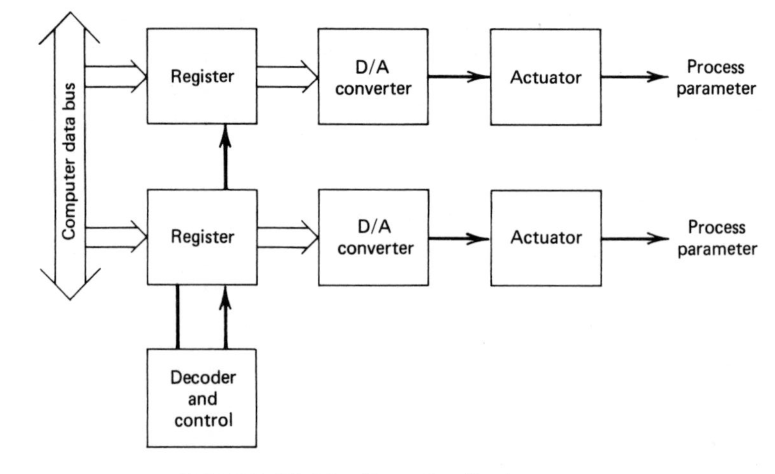

FIGURE 17-18 Data distribution system.

one or more sources and to convert the data to an equivalent digital signal recognizable to a computer. For control of industrial processes, the computer output signals must be converted back to analog signals by digital-to-analog converters and applied to some sort of control circuitry. Electronic components that are used for control purposes are often called actuators, which means "to put into motion." The portion of the total computerized industrial system from the computer to the device or process being controlled is called a data distribution system and is shown in block diagram form in Fig. 17-18.

17-10 SUMMARY

Data acquisition systems, which are used to collect information, provide an interface between the world of physical parameters, which are analog in their behavior, and the world of computers and digital signals.

The analog input signal is converted to an electrical signal by a transducer. Once the signal is in electrical form, all further processing, such as signal conditioning and A/D conversion, is done by electronic circuits. The five major subsystems of a data acquisition system are the transducer, the signal-conditioning circuits, the multiplexer, the sample and hold circuit, and the A/D converter.

17-11 GLOSSARY

Buffer: A circuit or device that isolates one electrical circuit from another.

Data acquisition: The process by which events in the real world are translated to machine-readable signals.

Digital-to-analog converter: A unit or device that converts a digital signal to a proportional analog voltage or current.

Filter: An electrical or electronic circuit or device that passes a certain range of frequencies while effectively blocking frequencies outside the range.

Linearization: The process of correcting the output signal of a device such as a transducer to make the output signal change linearly with respect to an input signal.

Multiplexer: A device used to select one of a number of inputs and switch its information to a single output.

Real time: The actual time during which physical events occur.

Sample and hold: A circuit used in conjunction with an analog-to-digital converter to capture and retain a signal so that it may be converted by the A/D converter.

17-12 REVIEW QUESTIONS

The following questions should be answered after a thorough study of the chapter. The purpose of the questions is to determine the reader's comprehension of the material.

1. What are the five major subsystems of a data acquisition system?
2. What does the term *real time* mean when applied to a data acquisition system?
3. What is a "time window," and what determines whether it is of sufficient duration?
4. Why is signal conditioning usually necessary in data acquisition systems in which the signal source is a transducer?
5. In what quantitative terms is filter performance generally described?
6. What is the primary benefit of using a multiplexer in a system in which information is taken from several analog sources?
7. What is the function of a sample and hold circuit in a data acquisition system?
8. What are the two primary types of digital-to-analog counters?
9. Does either type of digital-to-analog converter have advantages with respect to the other? If so, what are they?
10. What does the term *actuator* mean in the context of data acquisition?

17-13 PROBLEMS

17-1 What is the input voltage to a filter if its output voltage is 5 mV and the input signal experiences a 50-dB reduction?

17-2 The output voltage of a transducer is 2.8 mV at a reference temperature of 0°C. If the exponential constant for the transducer equals 6.4×10^{-3}/°C, what is the transducer output voltage at 300°C?

17-3 The transducer described in Problem 17-2 is connected to the input terminals of a logarithmic amplifier. Plot the output of the amplifier if it is properly designed to linearize the transducer voltage over a temperature range from 0°C to 375°C. The calibration constant for the amplifier is 0.8 V. Treat V_o as a dimensionless number.

17-4 Compute the output voltage for the voltage-summing D/A converter shown in Fig. 17-19 if the binary number 1010 is applied to the circuit and logic 1 equals 1 V and logic 0 equals 0 V.

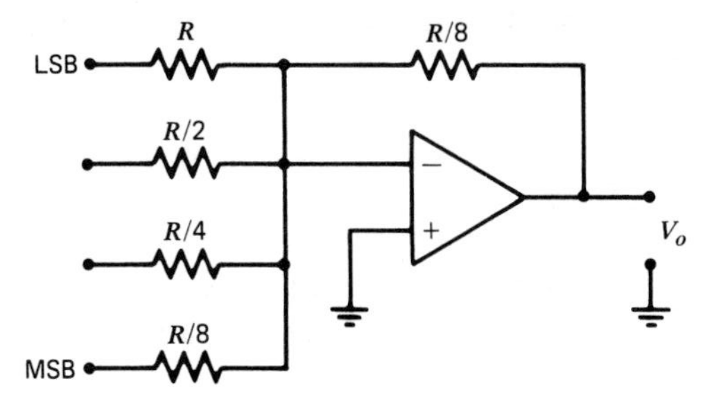

FIGURE 17-19 Circuit for Problem 17-4.

17-5 How many input bits are required to obtain a resolution of 15 mV if the maximum output voltage of a voltage summing D/A converter is 15.36 V?

17-6 If the constant of proportionality for a VCO, used in a voltage-to-frequency type A/D converter, is 200 and if 4400 pulses are stored in the counter during a 0.1-sec gating pulse, what is the amplitude of V_i?

17-7 How many comparators are required in a 6-bit simultaneous A/D converter?

17-8 What is the maximum conversion time with a 16-bit ramp-type A/D converter driven by a 4-MHz clock?

17-9 What is the major advantage of a successive-approximation A/D converter when compared to a ramp-type A/D converter?

17-14 LABORATORY EXPERIMENTS

Experiment E37 applies the theory that has been presented in Chapter 17. The purpose of the experiment is to provide hands-on experience to reinforce the theory.

The experiment uses standard components and equipment that are readily available in any electronics laboratory. The contents of the laboratory report to be submitted by each student are included at the end of the experimental procedure.

Laboratory Experiments

LABORATORY REPORT WRITING GUIDE

As part of performing an experiment in the laboratory, each student is required to submit a technical laboratory report. There are many different formats for laboratory reports. The following comments are intended to serve only as a guide. Your instructor will provide more specific instructions.

The Informal Report

The informal or "short-form" report generally emphasizes the data section. The report is normally due at the beginning of the laboratory period one week after the experiment is performed. The contents of a typical informal report, in outline form, are as follows.

A. Cover Page

1. Your name.
2. Lab partner's name.
3. Course number and section.
4. Experiment number and title.
5. Date the experiment was performed.

B. Data Section

1. A schematic of the experimental circuit. The drawing should be done neatly using a straight edge and should be drawn on quadrille paper.
2. A list of the equipment used indicated by the manufacturer's name, model number, and serial number.
3. A neat, accurate, and complete data table.
4. Graphs drawn on graph paper, accurately plotted, properly labeled, and titled.
5. Sample calculations.

C. Analysis Section
The analysis section should contain a concise technical discussion of your data. Compare your data with expected results and discuss the probable causes of any errors noted. Depending on the format you are to follow, conclusions may be in the same paragraph as the analysis or there may be a separate paragraph for conclusions.

The Formal Report

The formal or "long-form" report includes all that is part of an informal report as well as several additional sections. Frequently the instructor will ask that the formal report be submitted in a suitable laboratory report folder. It is normally due at the beginning of the laboratory period one week after the experiment is performed. The following outline is intended to serve only as a guide in formal report preparation.

A. Title Page or Folder Cover

1. Your name.
2. Lab partner's name.
3. Course number and section.
4. Experiment number and title.
5. Date the experiment was performed.

B. Introduction Section
The introductory section generally consists of a statement of the objective of the experiment.

C. Theory Section
The theory section should consist of a discussion of pertinent theory.

D. Method of Investigation Section
The section describing the experimental investigation usually includes the following.

1. A brief outline of the experimental procedure.
2. A neat schematic drawing of the experiment circuit.

E. Equipment List
All equipment used should be listed by the manufacturer's name, model number, and serial number.

F. Data Section
The data section should include a neat, accurate, and complete data table showing all measured and computed values.

G. Sample Calculations Section

The calculation section consists of a sample of each type of calculation made in the experiment.

H. Analysis Section

The analysis section should contain a discussion of the following.

1. A comparison of the experimental data and expected results.

2. A discussion of the probable causes of any errors noted.

I. Rough Data Section

The section giving rough data should include the actual rough experimental data recorded in the laboratory. The data should be submitted exactly as they were recorded when performing the experiment.

Preparation of a quality laboratory report is a time-consuming task, one that is by no means easy. However, the time spent in learning to communicate technical information well, in writing, is time well spent. This will become obvious to you when you finish school and enter the "world of work."

A possible benefit of quality laboratory reports that you may not have considered is related to job interviewing. Since few students have work experience to list on an employment application, you might consider providing a copy of two or three of your best lab reports to the interviewer as an indication of the quality of your work.

SAMPLE EXPERIMENT: THE LINEAR OPERATIONAL AMPLIFIER

Objective

To investigate the properties of the linear operational amplifier operating in the noninverting mode.

Discussion

Linear operational amplifiers are of fundamental importance in signal-processing circuitry and are used in a wide variety of instrumentation. Linear operational amplifiers or "op-amps" may be discrete-component units. However, most op-amps available today are monolithic integrated-circuit packages.

Op-amps are direct-coupled amplifiers having a high-voltage gain, wide bandwidth, high input impedance, and low output impedance.

The three basic op-amp configurations are the inverting, noninverting, and differential amplifiers. Each configuration has characteristics that makes it particularly suitable for certain applications. Two very desirable characteristics of the noninverting configuration, which is the operating mode of interest in

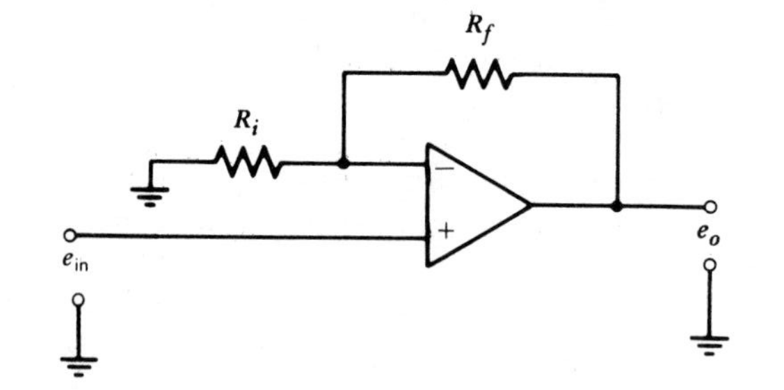

FIGURE UE-1 Noninverting op-amp.

this experiment, are the in-phase relationship between the input and output signals and the very high input impedance.

An op-amp connected in the noninverting configuration is shown in Fig. UE-1. The input signal is applied to the positive (+) input terminal. A signal applied to the positive terminal is amplified and appears at the output in phase with the input signal as shown. The voltage gain of a noninverting op-amp is computed as

$$A_v = \frac{R_f + R_i}{R_i}$$

where

A_v = voltage gain
R_f = feedback resistor
R_i = input resistor

Apparatus

- 1 ±15-V power supply.
- 1 741 op-amp IC.
- 1 signal generator.
- 1 oscilloscope.
- 3 resistors, 10 kΩ, 100 kΩ, 1 MΩ.
- 1 proto board.

Experimental Circuit

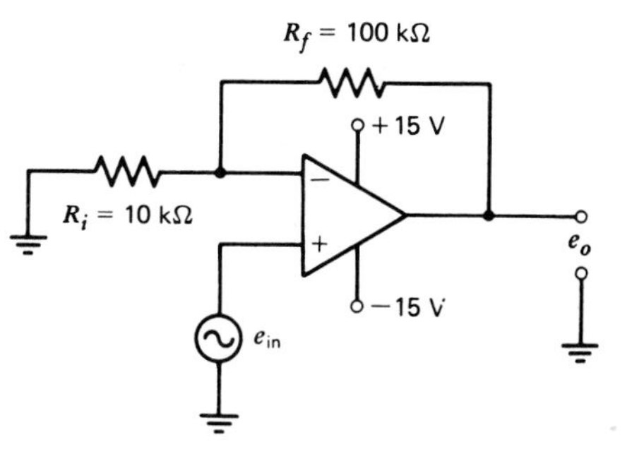

FIGURE UE-2 The experimental circuit—a noninverting op-amp.

Procedure

1. Connect the experimental circuit shown.
2. Set the input signal to 0.1 V_{p-p} at a frequency of 50 Hz.
3. Measure and record the output voltage.
4. Set the frequency to each value of frequency shown in the data table and record the output voltage. Make sure the input signal remains constant at 0.1 V throughout the experiment.
5. Calculate and record the gain for each frequency setting.
6. Plot a frequency response curve for the amplifier.
7. Replace R_f with the 1-MΩ resistor and repeat steps 2 through 6.

Laboratory Report

Your laboratory report should include the following as a minimum.

- Title page.
- Schematic diagram.
- Data table.
- List of equipment.
- Graphs.
- Analysis of results.

Data Table

	$R_f = 100\ \text{k}\Omega$		$R_f = 1\ \text{M}\Omega$	
Frequency	Output	Gain	Output	Gain
50 Hz				
100				
150				
200				
400				
600				
800				
1 kHz				
2				
4				
6				
8				
10				
15				
20				
30				
40				
50				
60				
70				
80				
90				
100 kHz				

ELECTRONICS TECHNOLOGY

Course No. 2244

Sec. No. 3

Experiment No. 8

THE LINEAR OPERATIONAL AMPLIFIER

(Sample Laboratory Report)

Name: A. D. Casey

Lab Partner: R.R. Smith

Date: ______________

Schematic Diagram

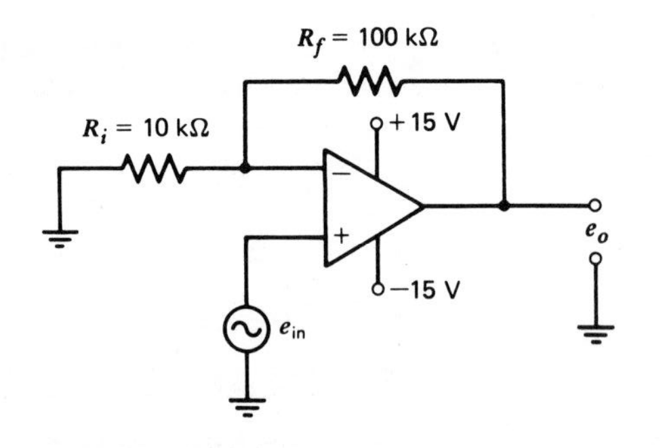

FIGURE UE-3 Op-amp connected on a noninverting amplifier.

Equipment List

- 1 ±15-V power supply, PASCO, Model 8000, SN 4051.
- 1 IC operational amplifier, LM 741.
- 1 signal generator, Hewlett-Packard, Model 3311A, SN 38716.
- 1 oscilloscope, Tektronix, Model 7633, SN 12,938.

Data Table

	R_f = 100 kΩ		R_f = 1 MΩ	
Frequency	Output	Gain	Output	Gain
50	0.97	9.7	4.5	45
100	1.0	10	6.4	64
150	1.01	10.1	7.4	74
200	1.05	10.5	8.4	84
400	1.06	10.6	9.5	95
600	1.06	10.6	9.7	97
800	1.04	10.4	9.8	98
1 kHz	1.1	11.0	10.1	101
2 kHz	1.1	11.0	9.8	98
4 kHz	1.05	10.5	8.4	84
6 kHz	1.06	10.6	7.4	74
8 kHz	1.1	11.0	6.6	66
10 kHz	1.03	10.3	6.0	60
15	1.01	10.1	4.8	48
20	0.97	9.7	3.9	39
30	0.76	7.6	2.5	25
40	0.59	5.9	1.7	17
50	0.41	4.1	1.1	11
60	0.26	2.6	0.8	8
70	0.19	1.9	0.5	5
80	0.10	1.0	0.3	3
90	0.04	0.4	0.1	1
100 kHz	0.01	0.1	0.0	0

Sample Calculations

$R_f = 100\ \text{k}\Omega$ $\qquad\qquad$ $R_f = 1\ \text{M}\Omega$

$$A_f = \frac{R_f + R_i}{R_i} = \frac{100\ \text{k}\Omega + 10\ \text{k}\Omega}{10\ \text{k}\Omega} = 11 \qquad A_f = \frac{R_f + R_i}{R_i} = \frac{1\ \text{M}\Omega + 10\ \text{k}\Omega}{10\ \text{k}\Omega} = 101$$

Analysis of Results

The results of the experiment agree well with the expected results. The gain equation given in the discussion indicates a maximum voltage gain of 11 when the feedback resistor has a value of 100 kΩ and a maximum gain of 101 when the feedback resistor equals 1 MΩ. As can be seen in the data table, an output voltage of 1.1 V, which is a gain of 11, was observed at three different frequencies with R_f equal to 100 kΩ. With R_f equal to 100 kΩ, the gain varied by less than 10% from 150 Hz to 15 kHz.

With the feedback resistor R_f equal to 1 MΩ, a maximum output voltage of 10.1 V was measured. This represents a gain of 101, which agrees with theory. However, this gain was observed only at 1 kHz. At both higher and lower frequencies the gain decreased fairly rapidly, as can be seen in Fig. UE-4. The difference between the two curves shown in the figure is probably due to the effects of feedback.

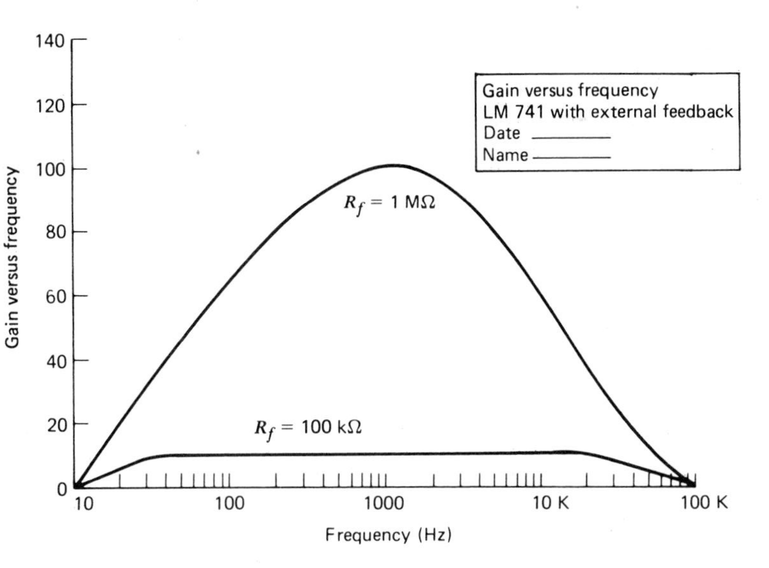

FIGURE UE-4 Frequency response curve.

EXPERIMENT E1: ERROR IN EXPERIMENTAL DATA

Objective

To investigate sources of error in measurements and to apply the equations provided to analyze the error.

Discussion

Error is defined as the **deviation** of a reading (or set of readings) from the **expected** value of the measure variable. When we make measurements, some error is inevitable because no measurement can yield the exact value of any quantity. There are several sources of error in any experimental data. The primary concerns about analyzing experimental data are the sources of error and the extent to which the error has affected the validity of the data.

This experiment consists of two parts. Part 1 is concerned with error introduced by component tolerances. Although there are other sources of error in your data, such as instrument errors or observational errors, the experimental data will be treated as though the only source of error is due to variations in resistance values owing to manufacturing tolerances.

Part 2 covers errors in reading, which may be classified as **gross error** or **observational error**. The instructor will provide a single resistor of unknown value, R_x. Each student, in turn, measures and records on a piece of paper his or her value of the resistor. All readings should be made with the **same** ohmmeter. Students should **not** disclose their measurement to their classmates until the instructor or a designated student has recorded all measurements on the blackboard.

Equations

The following equations are used in performing this experiment.

1. $R_{ave} = \dfrac{R_1 + R_2 + \cdots + R_n}{n}$

2. Range of error $= \dfrac{(R_{max} - R_{ave}) + (R_{ave} - R_{min})}{2}$

3. Percent of error $= \dfrac{R_{ave} - R_x}{R_x} \times 100\%$

where

R_x = actual value of the resistor
R_{ave} = average value of the resistor

4. $R_b = R_a \times \dfrac{E_o}{E_{in} - E_o}$. See Fig. E1-1.

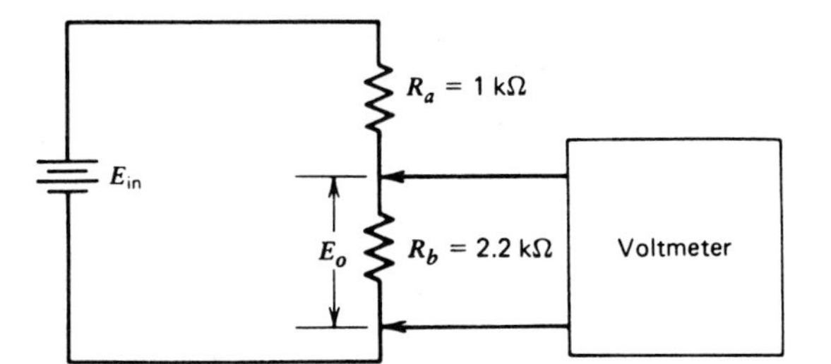

FIGURE E1-1 Circuit for Experiment E1.

Apparatus

- 1 dc power supply.
- 1 high-input-impedance voltmeter.
- 10 composition resistors, 2.2 kΩ.
- 1 composition resistor, 1 kΩ.

Procedure

Part A—Error Caused by Component Tolerance
(*Perform this part immediately.*)

1. Set up the experimental circuit shown in Fig. E1-1.
2. Connect the voltmeter across R_b and adjust the input voltage until the output is exactly 10 V. Record the value of E_{in} in the data table provided. Do not change E_{in} throughout the remainder of the experiment.
3. Replace the decade resistor with one of the 2.2 kΩ composition resistors and again record the value of E_o in the data table provided.
4. Place each of the remaining 2.2-kΩ resistors in the circuit, one at a time, and record the value of E_o for each resistor.
5. Use equation 4 to compute the value of R_b for each of the ten resistors.
6. Compute the average value of R_b. Enter in the data table.
7. Compute the range of errors in the values for R_b. Enter in the data table.
8. Compute the percentage of error for the average value of R_b, computed in step 6, against the color-coded value of the resistors. Enter in the data table.

Part B—Error in Reading
(*Perform whenever R_x is available.*)

1. Each student measures and records the value of the resistor provided by the instructor, using the meter provided by the instructor.

Data Table for Experiment E1

Component Tolerance Error, E1A			Error in Reading, E1B	
E_o	R_b Values	E_{in}	Resistance Values	Average Resistance
		Average R_b		
				Range of errors
		Range of errors		
				Percentage of error
		Percentage of error		
				Actual value of R

2. The instructor or a designated student will record the student's readings on the blackboard, after all readings have been made. The students will enter these values in the data table.
3. Compute the average of the readings from step 2, using equation 1. Enter in the data table.
4. Compute the range of errors in the readings from step 2, using equation 2. Enter in the data table.
5. Compute the percentage of error of the average resistance computed in step 3 compared to the actual value of the resistor as measured by the instructor on a resistance bridge. Enter in the data table.

Report

Your laboratory report should include the following as a minimum.

- Title page.
- Data table.
- List of equipment including serial numbers.
- Analysis of results.

EXPERIMENT E2: BASIC STATISTICAL SAMPLING

Objective

To permit the student to observe the value of statistical analysis in such applications as sampling large quantities for variation and in examining applications as sampling experimental data.

Discussion

A statistical analysis is frequently performed on samples of very large quantities to determine the probable variation in values of the entire lot. The percentage of the entire lot, which will fall within a specific range of values, can be predicted quite accurately from the statistical analysis of the sample. For example, a company may purchase a large quantity of some component, such as resistors, and decide to use the resistors in an application requiring a high degree of accuracy—if a statistical analysis on a sample of resistors shows that a very high percentage of the resistors are likely to fall within some predetermined range of values.

Under ideal conditions, a very large number of measurements will provide a distribution of readings, with the greatest number of readings approximately equal to the actual value. On either side of the actual value, the frequency of readings will decrease, producing an approximately normal distribution curve, as shown in Fig. E2-1.

In this experiment you are to perform a statistical analysis on three samples of different sizes. Using your data, you are to attempt to draw a conclusion related to the size of the sample in such techniques. The small size of the total lot and the samples may make it difficult to draw a completely valid conclusion. However, the laboratory experience will still prove valuable to you.

Equations

The following equations are used in conducting this experiment.

Mean: $$\bar{X} = \frac{x_1 + x_2 + x_3 + \cdots + x_n}{n} = \frac{\Sigma X_i}{n}$$

where

$\bar{X}$ = mean
X_i = readings taken
n = number of readings

Deviation: $$d_1 = x_1 - \bar{x}$$

$$d_2 = x_2 - \bar{x}$$

$$d_n = x_n - \bar{x}$$

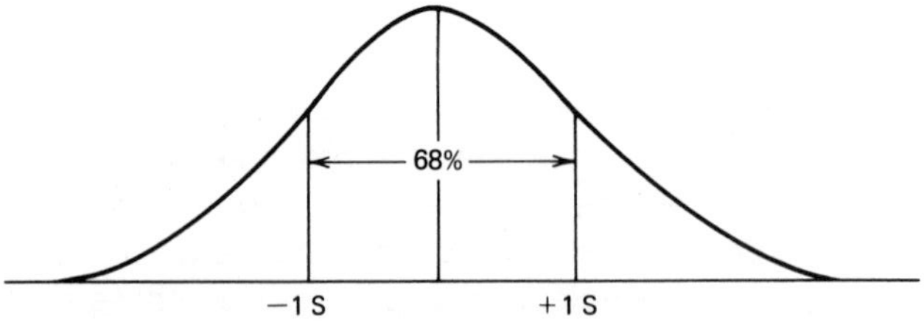

68%
−1 S
+1 S

F'

FIGURE E2-1 Normal distribution curve.

Average deviation: $D = \dfrac{|d_1| + |d_2| + \cdots + |d_n|}{n}$

Standard deviation: $S = \sqrt{\dfrac{d_1^2 + d_2^2 + d_3^2 + \cdots + d_n^2}{n}}$

Apparatus

- 30 composition resistors of same color-coded value.
- 1 multimeter (preferably a digital multimeter).

Procedure

1. Measure and record the value of each resistor.
2. Select eight resistors at random from the total lot and measure and record their values in Table E2-1 (Sample 1).
3. Mix all the resistors together and select, at random, any 12 resistors. Your selection may or may not include resistors from the previous sample. Measure and record the values of the 12 resistors in Table E2-1 (Sample 2).
4. Mix all the resistors together and select, at random, a sample of 16 resistors and measure and record their values in Table E2-1 (Sample 3).
5. On one sheet of graph paper, plot the value of each resistor in the total lot and make a bar graph, or histogram, as shown in Fig. E2-2, in which each block represents a resistor having that value of resistance.
6. Divide a second sheet of graph paper three ways vertically. Plot a histogram for each of the three samples of resistors.
7. Connect the maximum points of each histogram by a smooth curve. If the numbers of resistors in the samples were much larger, this would give an

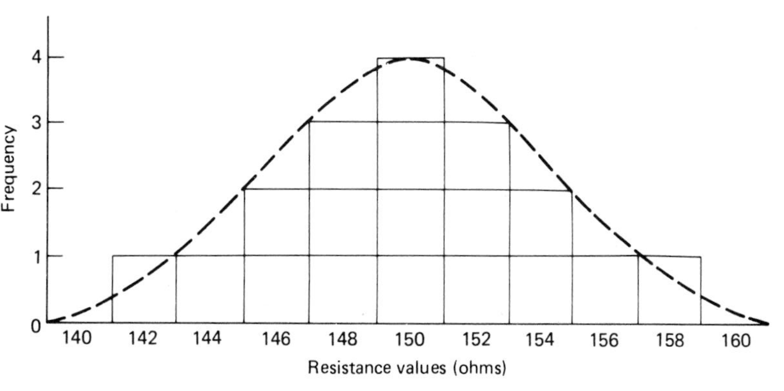

FIGURE E2-2 Distribution of resistance values around the expected value of 150 Ω.

approximate normal distribution curve such as the one shown in Fig. E2-1. However, your curves may be skewed because samples are small.

8. Compute and record in Table E2-1 the average deviation D for the entire lot and for each of the three samples.

9. Compute and record Table E2-1 the standard deviation S for the entire lot and for each of the three samples.

10. Record in Table E2-2 which sample (1, 2, or 3) most nearly describes the total lot with regard to average deviation and standard deviation.

Report

Your laboratory report should include the following as a minimum.

- Title page.
- Histograms.
- Data table.
- Calculations.
- Analysis of results.

Data Tables for Experiment E2

TABLE E2-1

Resistance Value	Sample 1	Sample 2
	Sample 1 D	
	Sample 2 D	Sample 3
	Sample 3 D	
	Sample 1 S	
	Sample 2 S	
	Sample 3 S	

TABLE E2-2

Best Sample
Second Best Sample
Worst Sample

EXPERIMENT E3: BASIC VOLTMETER DESIGN

Objective

To design and build a basic dc voltmeter accurate to within 5% of an accepted standard.

Discussion

Voltmeters are one of the most frequently used pieces of test equipment in the electronics industry. The basic dc voltmeter uses a current-sensitive meter movement with the meter face calibrated in voltage units. A typical laboratory-quality voltmeter uses a 50-μA meter movement. For this meter movement to be used as a voltmeter, it is necessary to put a resistor called a multiplier in series with the meter movement. Design a voltmeter as described in Section 2-6.

Apparatus

- 1 variable dc power supply.
- 2 potentiometers, 25 kΩ and 100 kΩ.
- 1 panel meter, 0 to 100 μA.
- 1 EVM or digital multimeter.
- Selected composition resistors, as required.

Procedure

Determine the internal resistance of the 100 μA meter movement by each of the following three methods.

1. *Variable-resistor method* (Fig. E3-1).
   - (a) Set the 25-kΩ potentiometer to *maximum* and the voltage source to *minimum*.
   - (b) Slowly increase the voltage to 1 V.
   - (c) Slowly decrease the resistance of the potentiometer until a full-scale reading is obtained on the meter.
   - (d) Carefully disconnect the potentiometer and measure the value of the potentiometer resistance (using a digital multimeter) that was in the circuit. Record this as R_1 in Data Table E3-1.
   - (e) Reconnect the circuit and slowly vary the potentiometer until the meter reads exactly half-scale. Record this as R_2 in Table E3-1.
   - (f) Compute the internal resistance of the meter movement by the equation $R_{m1} = R_2 - 2R_1$.

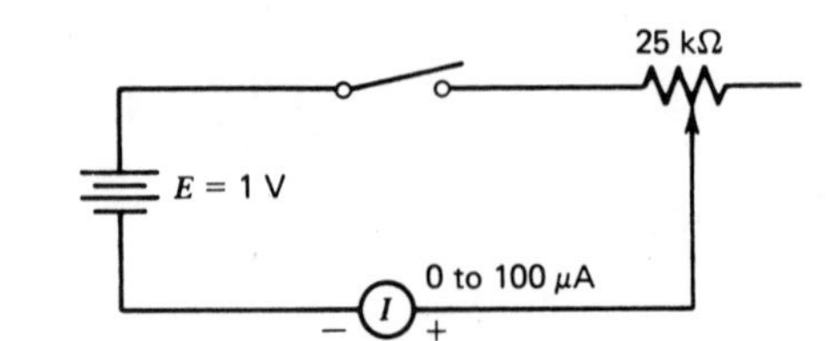

FIGURE E3-1 Variable-resistor method.

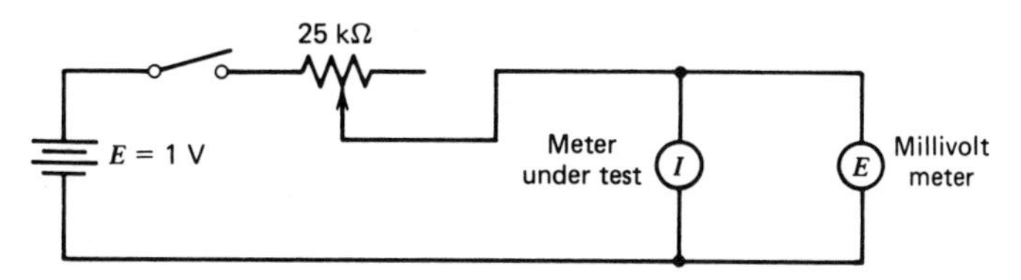

FIGURE E3-2 Potentiometer method.

2. *Potentiometer method* (Fig. E3-2).
   (a) Set the 25-kΩ potentiometer to *maximum* and the voltage source to *minimum*.
   (b) Slowly increase the voltage to 1 V.
   (c) Slowly decrease the resistance of the potentiometer until a full-scale reading is obtained on the meter under test.
   (d) Compute the internal resistance, R_{m2}, of the meter under test using the millivolt and microampmeter readings. Record the results in the data table.
3. *Shunt resistor method* (Fig. E3-3).
   (a) Connect the circuit shown and vary the 100-kΩ potentiometer for a full-scale reading on the meter under test.
   (b) Close switch S_1 and vary the 25-kΩ potentiometer until the meter under test reads exactly one-half scale.
   (c) Remove the 25-kΩ potentiometer from the circuit and measure the potentiometer shunting the meter under test, using a digital multimeter. Record this as R_{m3} in Table E3-1.
4. Calculate the average internal resistance of the meter under test as

$$R_{m_{\text{ave}}} = \frac{R_{m1} + R_{m2} + R_{m3}}{3}$$

5. Using the test meter movement, design a voltmeter having the following five ranges.
   (a) 0 to 1 V
   (b) 0 to 2 V
   (c) 0 to 5 V
   (d) 0 to 10 V
   (e) 0 to 20 V

   Use the average meter resistance from step 4 as R_m in the calculations for multipliers.

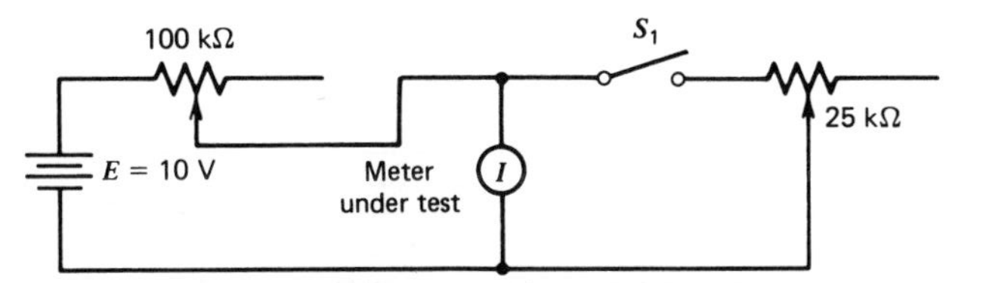

FIGURE E3-3 Shunt resistor method.

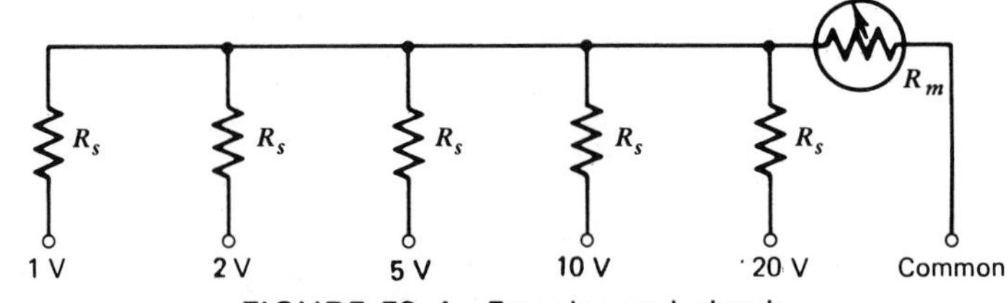

FIGURE E3-4 Experimental circuit.

6. Construct the experimental circuit shown in Fig. E3-4. Use fixed composition resistors for the multipliers.
7. Apply voltage to each range of the experimental circuit. Adjust the voltage for full-scale deflection on the test meter, and record these readings as E_2 in Table E3-2. In addition, measure the applied voltage on each range with the EVM or digital meter and record the readings as E_1 in the data table.
8. Calculate and record the percent of error of each scale on your meter by the following equation:

$$\text{Percent error} = \frac{E_1 - E_2}{E_1} \times 100\%$$

where

E_1 = laboratory voltmeter reading of voltage source
E_2 = your voltmeter reading

Report

Your laboratory report should contain the following as a minimum.

- Title page.
- Circuit diagrams.
- Data table.
- Calculations.
- Analysis of results.

Data Tables for Experiment E3

TABLE E3-1
Meter Movement Data

Variable Resistor Method	Potentiometer Method	Shunt Resistor Method	Internal Resistance of Meter
R_1	E (mV)		
R_2	I (μA)		
R_{m1}	R_{m2}	R_{m3}	$R_{m\,ave}$

TABLE E3-2
Voltmeter Data

Volts Scale	Multiplier Resistor	E_1	E_2	Percent Error
0–1				
0–5				
0–10				
0–20				
0–50				

EXPERIMENT E4: MULTIMETERS

Objective

To design, fabricate, troubleshoot, and calibrate a functional basic multimeter circuit.

Discussion

Since the ammeter, the voltmeter, and the ohmmeter all use the basic d'Arsonval meter movement as an indicating device, it is reasonable that, with the proper switching arrangement, the three circuits might be contained in the same case, making use of the same meter movement. This is indeed possible—the result is known as a volt–ohm–milliammeter (VOM) or a multimeter. For general-purpose and service-type measurements this multipurpose instrument is excellent. A typical commercial multimeter is capable of ac and dc voltage measurements from less than 1 V to 1000 V or greater, dc measurements from approximately 50 μA to 10 A, and resistance measurements from approximately 1 Ω to several megohms.

This experiment enables design, fabrication, and calibration of a basic multimeter circuit. An additional aspect of the experiment is troubleshooting practice. Troubleshooting is basic to the technician's job. Therefore, the experiment will not be considered complete until the VOM is functional and accurate within 3% on all settings.

Apparatus

- 1 meter movement, 0 to 100 μA.
- 1 variable power supply.
- 1 EVM or digital multimeter or *RLC* bridge.

- 3 switches (spdt).
- 4 potentiometers, 25 kΩ, 100 kΩ, 500 kΩ, 1 MΩ.
- Selected composition resistors.

Procedure

1. Calculate the value of the multimeter resistors necessary to construct a voltmeter with scales of 2, 20, and 50 V using the 0- to 100-μA meter (R_{in} = 1825 Ω or the value determined in Experiment E3).
2. Using the resistance, inductance, capacitance *RLC* bridge or digital multimeter, set a potentiometer to each of the three resistances calculated in step 1.
3. Connect your voltmeter, taking care not to change potentiometer settings.
4. Calibrate your voltmeter on each range by connecting the circuit of Fig. E4-1. Use the EVM or digital multimeter as the standard.
5. Disconnect the potentiometers and measure their resistances again. Calculate the percent of error between this resistance value and the values calculated in step 1.
6. Reconnect your voltmeter circuit, as done in step 3.
7. Calculate the value of shunt resistors necessary for current ranges of 10 and 100 mA.
8. Construct the two-range milliammeter. (Leave the voltmeter assembled.)
9. Check the accuracy of your meter by connecting the circuit of Fig. E4-2. Use the digital meter as your standard.

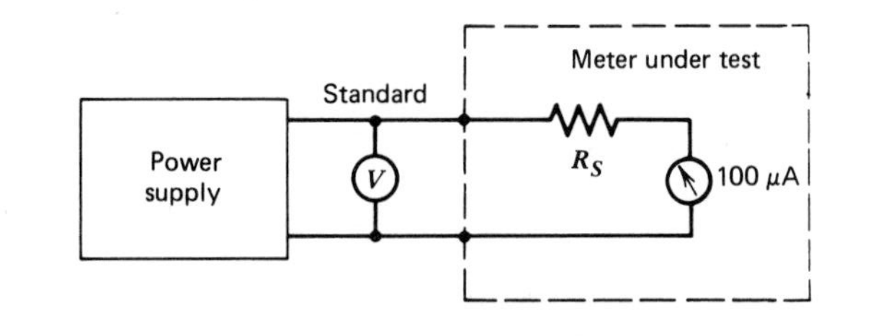

FIGURE E4-1 Calibration circuit.

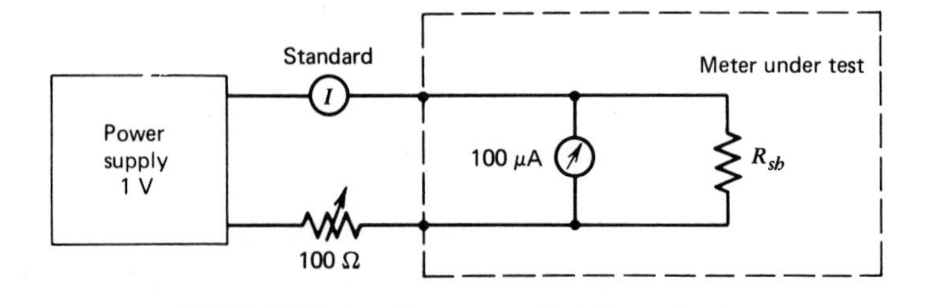

FIGURE E4-2 Ammeter calibration circuit.

10. Adjust the value of the shunt resistor for exactly full-scale deflection on the meter under test. Measure and record this value of resistance.

11. Calculate the percentage of error between the measured and calculated value of shunt resistance.

12. Leaving your voltmeter and milliammeter circuits connected, construct the ohmmeter circuit of Fig. E4-3.

13. One way to connect the total VOM circuit is shown in Fig. E4-4.

14. Connect the leads of your ohmmeter together and vary the 20-k**Ω** potentiometer for full-scale deflection. Measure and record the resistance at R_{zero} setting.

15. Determine the internal resistance of your meter (including R_{zero}) by connecting a 20-k**Ω** potentiometer between $X-Y$ and varying it to get a half-scale reading. Remove the potentiometer and measure its resistance.

16. Place a potentiometer of at least 2-M**Ω** resistance between $X-Y$ and set it to maximum resistance. Slowly decrease the resistance until you can detect the up-scale movements of the pointer.

17. Remove the potentiometer and measure its resistance. This is the maximum possible resistance that can be measured with your ohmmeter.

18. Reconnect the potentiometer and demonstrate your total instrument to the instructor.

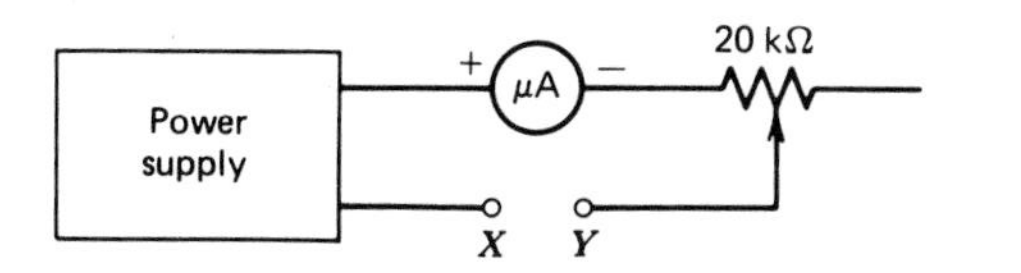

FIGURE E4-3 Ohmmeter circuit.

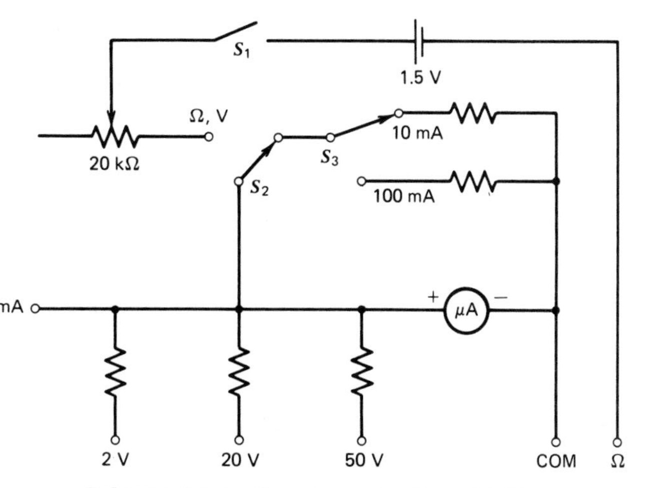

FIGURE E4-4 Complete circuit for VOM.

Troubleshooting

The following is intended to assist you in troubleshooting your circuit.

Step 1. Check the wiring of your circuit for correctness ______________.

Step 2. Check the value of your resistors for correctness ______________.

Step 3. Using a commercial ohmmeter, make the following measurements.

Terminal		Terminal	Your measurement	Correct
mA	to	2 V	______________	18.2 kΩ
mA	to	20 V	______________	198.2 kΩ
mA	to	50 V	______________	498.2 kΩ
2 V	to	20 V	______________	216.4 kΩ
2 V	to	50 V	______________	516.4 kΩ
20 V	to	50 V	______________	696.4 kΩ

Step 4. Remove the meter movement from the circuit and make the following measurements with switch S_2 in the milliampere position. With S_3 at the 10-mA setting:

Terminal		Terminal	Your measurement	Correct
mA	to	common	______________	18.2 Ω

With the S_3 at the 100-mA setting:

Terminal		Terminal	Your measurement	Correct
mA	to	common	______________	1.82 Ω

Step 5. Set switch S_2 to the ohms, volts position. With an ohmmeter, measure the resistance from the milliampere terminal to the common terminal of switch S_1.

Terminal		Terminal	Your measurement	Correct
mA	to	S_1	______________	13.2 kΩ

Step 6. Analyze your data and make any corrections you feel are necessary to permit your multimeter to work properly.

Step 7. Write a brief summary of your analysis from step 6 and state what changes you made to your circuit.

Step 8. Reconnect your meter movement and recheck the operation of your circuit.

Step 9. If your circuit still does not work properly, consult the instructor.

Report

Your laboratory report should include the following as a minimum.

- Title page.
- Data table.
- Calculations.
- Schematics.
- Analysis of results.

Data Table for Experiment E4

Voltmeter				Milliammeter				Ohmmeter		
	Multiplier Resistor				Shunt Resistor					
Scale (volts)	(ohms) calculated	(ohms) measured	% error	Scale (mA)	(ohms) calculated	(ohms) measured	% error	R_{zero} Setting	Internal Resistance	R_{max}
2				10						
20				100				X	X	X
50				X	X	X	X	X	X	X

EXPERIMENT E5: AC VOLTMETERS

Objective

To design, fabricate, and calibrate a basic ac voltmeter circuit using both half-wave and full-wave rectification.

Discussion

Several types of meters are available with which to measure alternating current or voltage. By far the most widely used is the rectifier-type meter. This instrument can be made by using the basic dc meter movement in conjunction with a rectifier. The rectifier may be a half-wave or full-wave rectifier, using a vacuum tube or semiconductor diodes, or a copper oxide instrument rectifier.

This experiment is done with half-wave ac voltmeters, which use the dc meter movements and semiconductor rectifying diodes.

Apparatus

- 1 10- to 100-μA dc meter movement.
- 5 semiconductor diodes.
- 1 variac.

- 1 oscilloscope with probe.
- Selected composition resistors.
- 1 isolation transformer.

Procedure

1. Calculate the ac sensitivity of the 100-μA meter movement.
2. Calculate the value of R_s for the following rms ac voltage ranges using half-wave rectification: 10 V, 25 V, 50 V, 60 Hz.
3. Construct the half-wave ac voltmeter circuit shown in Fig. E5-1. Use fixed-value resistors for the multiplier resistors.
4. Apply the rms voltages given in step 2 to your ac voltmeters and adjust the voltage for full-scale deflection on each range.
5. With a full-scale reading on the 25-V rms range, connect the oscilloscope across the resistor R_s and sketch the waveform.
6. Remove the resistors and measure and record their actual resistance.
7. Calculate the value of R_s for the following rms ac voltage ranges using full-wave rectification: 10 V, 25 V, 50 V, 60 Hz.
8. Construct the full-wave ac voltmeter circuit shown in Fig. E5-2.

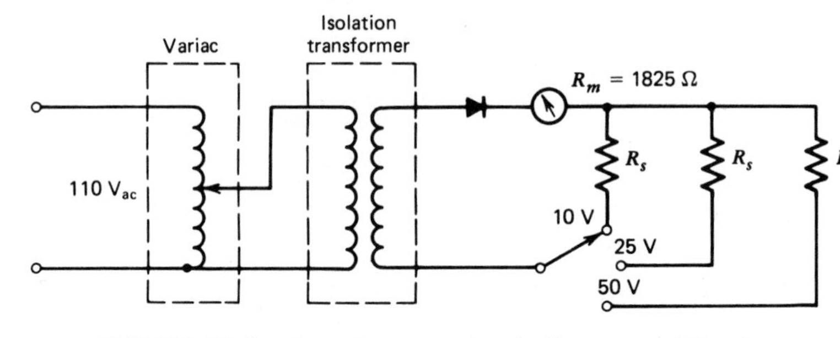

FIGURE E5-1 Ac voltmeter using half-wave rectification.

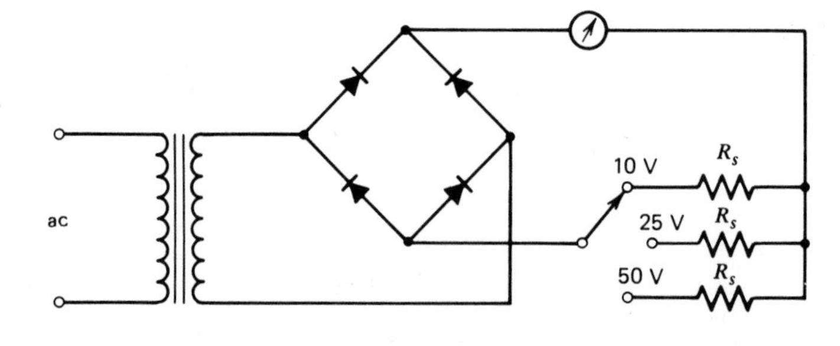

FIGURE E5-2 Ac voltmeter using full-wave rectification.

9. Apply the rms voltages given in step 7 to your ac voltmeter, and adjust the voltage for the full-scale deflection on each range.
10. With a full-scale reading on the 25-V rms range, connect the oscilloscope across the resistor R_s and sketch the waveform.
11. Remove the resistors and measure and record their actual resistance.
12. Compute the percentage of error between the actual and computed values of R_s for each range of the half-wave and full-wave rectifier-type ac voltmeter.

Report

Your laboratory report should include the following as a minimum.

- Title page.
- Circuit diagrams.
- Data table.
- Calculations.
- Analysis of results.

Data Table for Experiment E5

Half-Wave Rectifier				Full-Wave Rectifier			
Ac Sensitivity				Ac Sensitivity			
Multiplier Resistor, R_s				Multiplier Resistor, R_s			
Range (volts)	Calculated	Measured	Percent Error	Range (volts)	Calculated	Measured	Percent Error
				10			
				25			
				50			
Waveform				Waveform			

EXPERIMENT E6: FREQUENCY RESPONSE OF AC VOLTMETERS

Objective

To investigate the frequency response of general-purpose ac voltmeters.

Discussion

General-purpose ac voltmeters have a flat frequency response curve over a limited range of frequencies. The meter response is inaccurate above or below this range of frequencies. This inaccuracy is due to the frequency characteristics of components such as rectifier diodes, instrument-type rectifiers, capacitors, and wire-wound resistors in the voltmeter circuit. The major effect is generally attributed to the capacitance associated with the rectifying elements in ac voltmeters that use a d'Arsonval meter movement. The other meter movements discussed in Chapter 3 are affected in varying degrees by the frequency of the ac signal being measured.

Apparatus

- 3 ac voltmeters with different types of meter movements.
- 1 cathode-ray oscilloscope.
- 1 sine-wave generator.
- 1 resistor, 1 kΩ.

Procedure

1. Connect the experimental circuit shown in Fig. E6-1. Use any one of the three meter movements.

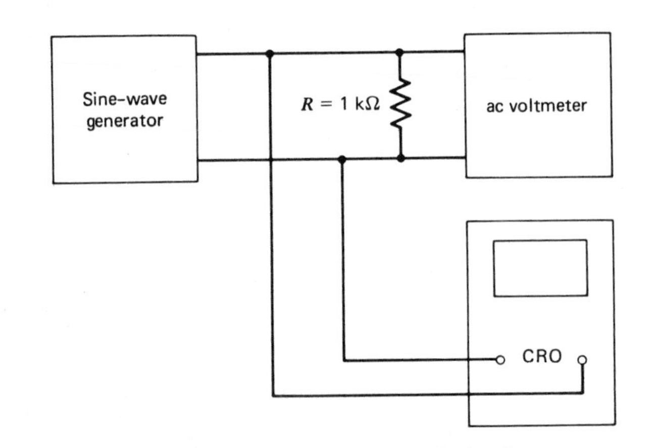

FIGURE E6-1 **Experimental circuit.**

2. Use the CRO to set the sine-wave generator output to 2 V_{rms} at 1 kHz.
3. Decrease the generator frequency until the voltmeter reading decreases to 1.4 V_{rms}.
4. Increase the generator frequency until the voltmeter reading again decreases to 1.4 V_{rms}.
5. Set the generator to each of the frequencies shown in the data table and record the voltmeter reading. Be sure the amplitude of the generator output remains constant at 2 V_{rms}.
6. Repeat steps 1 through 5 with the two remaining meter movements.
7. Plot the frequency response curve for each meter movement.

Report

Your laboratory report should include the following as a minimum.

- Title page.
- Data table.
- List of equipment including serial numbers.
- Schematic diagram.
- Graphs.
- Analysis of results.

Data Table for Experiment E6

Frequency (hertz)	Oscilloscope (rms volts)	Meter 1 (rms volts)	Meter 2 (rms volts)	Meter 3 (rms volts)
25				
50				
100				
500				
1,000				
10,000				
50,000				
100,000				
500,000				
1,000,000				
2,000,000				
5,000,000				

EXPERIMENT E7: VOLTMETER CALIBRATION USING THE POTENTIOMETER

Objective

To teach the student to use the potentiometer to check the calibration of a voltmeter.

Discussion

The potentiometer is frequently used in calibrating test instruments. The Cenco, Model 83338 potentiometer, which may be used in this experiment, is shown in Fig. E7-1. This potentiometer can be used directly to measure voltages between 0 and 2 V. A precision voltage divider network, sometimes called a volt box or a calibration ladder, is needed to measure higher voltages.

Before this potentiometer is used, it must first be calibrated or standardized against an internal mecury cell that has a potential of 1.3562 V. After the potentiometer is standardized, it can then be used in the circuit shown in Fig. E7-2 to record calibration data. Although this discussion is of a specific type

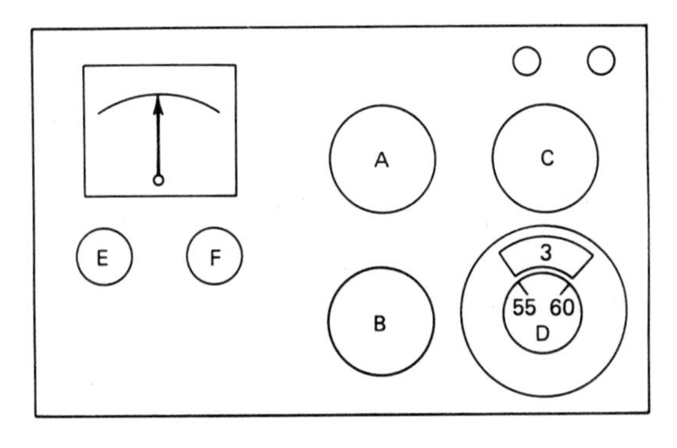

FIGURE E7-1 Basic potentiometer.

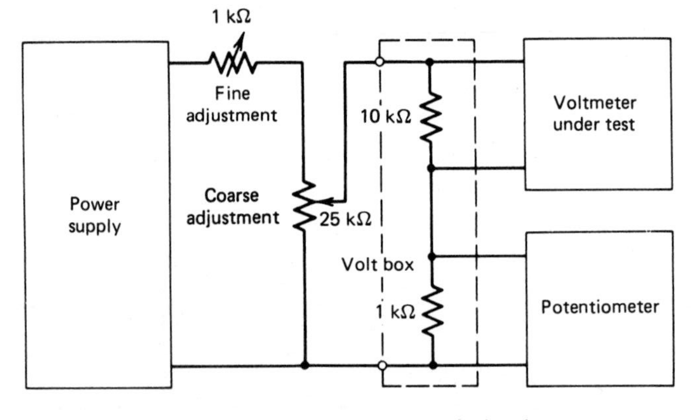

FIGURE E7-2 Experimental circuit.

of potentiometer, any basic potentiometer operates in a similar manner and may be used without difficulty.

Apparatus

- 1 potentiometer.
- 1 power supply.
- 1 dc voltmeter.
- 1 1-kΩ potentiometer.
- 1 25-kΩ potentiometer.
- 1 10-kΩ, 0.5-W resistor.
- 1 1-kΩ, 0.5-W resistor.

Procedure

1. Calibrate the potentiometer against the internal calibration source.
2. Connect the calibration circuit shown in Fig. E7-2.
3. Set up the potentiometer to measure voltage following the procedure set forth by the instructor or by the manufacturer of the instrument.
4. Set the power supply to approximately 3 V and adjust the course and fine adjustments until the voltmeter reads 0.2 V. Measure the actual voltage with the potentiometer.
5. Record the voltmeter and potentiometer readings in the data table.
6. Obtain the following voltmeter readings; then measure the actual voltage with the potentiometer.

	Voltmeter Reading	Potentiometer Reading
(a)	0.4 V	
(b)	0.5 V	
(c)	0.6 V	
(d)	0.7 V	
(e)	0.8 V	

7. Set the voltmeter on the 3-V range.
8. Set the power supply to approximately 5 V and adjust the fine and coarse adjustments until the voltmeter reads 0.5 V. Measure the actual voltage with the potentiometer.
9. Record the voltmeter and potentiometer readings in the data table.

10. Obtain the following voltmeter readings; then measure the actual voltage with the potentiometer.

	Voltmeter Reading	Potentiometer Reading
(a)	1.0 V	
(b)	1.5 V	
(c)	2.0 V	
(d)	2.5 V	
(e)	3.0 V	

11. Set the voltmeter on the 10-V range.

12. Set the power supply to approximately 12 V and adjust the fine and coarse adjustment until the voltmeter reads 2.0 V. Measure the actual voltage with the potentiometer.

13. Record the voltmeter and potentiometer readings in the data table.

14. Obtain the following voltmeter readings; then measure the actual voltage with the potentiomer.

	Voltmeter Reading	Potentiometer Reading
(a)	3.0 V	
(b)	6.0 V	
(c)	8.0 V	
(d)	10.0 V	

15. Determine the correction voltage by subtracting the voltmeter reading from the potentiometer reading.

16. Plot the calibration curve like the one shown in Fig. E7-3 for each of the

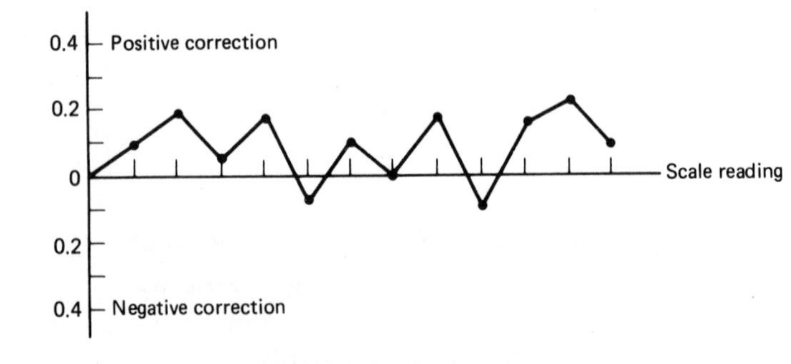

FIGURE E7-3 Calibration curve.

three ranges checked. Use an $8\frac{1}{2} \times 11$-in. sheet of graph paper divided into three equal sections. Identify the instrument checked by its serial number.

Report

Your laboratory report should include the following as a minimum.

- Title page.
- Data table.
- List of apparatus including serial numbers.
- Calculations.
- Curves or graphs.
- Analysis of results.

Data Table for Experiment E7

One-Volt Range		
Voltmeter Reading (Scale Reading)	Potentiometer Reading (True Reading)	Correction
0.0		
0.2		
0.4		
0.6		
0.8		
1.0		
Three-Volt Range		
Voltmeter Reading (Scale Reading)	Potentiometer Reading (True Reading)	Correction
0.5	0.0	0.0
1.0		
1.5		
2.0		
2.5		
3.0		
Ten-Volt Range		
Voltmeter Reading (Scale Reading)	Potentiometer Reading (True Reading)	Correction
2.0		
4.0		
6.0		
8.0		
10.0		

EXPERIMENT E8: BASIC REFERENCE VOLTAGE SOURCE

Objective

To fabricate and test a basic reference voltage source.

Discussion

A zener diode is a semiconductor device designed to operate in a reverse-biased mode. When sufficient reverse-biased voltage is applied to a zener diode, breakdown occurs. This breakdown voltage is called the **zener voltage**. If we continue to increase the bias voltage, additional current flows but the zener voltage remains essentially constant.

This characteristic makes zener diodes very useful in many applications related to voltage regulation. In this experiment we wish to fabricate and evaluate a basic reference voltage source that utilizes the zener voltage phenomenon.

Apparatus

- 1 dc power supply.
- 1 zener diode, 5- to 10-V zener voltage, 1 W.
- 1 multimeter.
- 1 decade resistance box.
- 2 current meters.
- 1 composition resistor, 1 kΩ, 0.5 W.

Procedure

1. Connect the experimental circuit shown in Fig. E8-1. Before turning on the power supply, make sure that resistor R is adjusted to 1 kΩ.
2. Adjust the power supply to 12 V.
3. Connect a voltmeter across the load resistor and record the load voltage V_L, the total current I_T, and the zener current I_Z.
4. Adjust the resistor R in 100-Ω steps from 1 kΩ to 100 Ω and record V_L, I_T, and I_Z at each setting.
5. Set the decade resistor to 500 Ω and interchange the decade resistor and the composition resistor.
6. With the decade resistor connected as the load resistor, connect the voltmeter across the load resistor. Measure and record V_L, I_T, and I_Z.

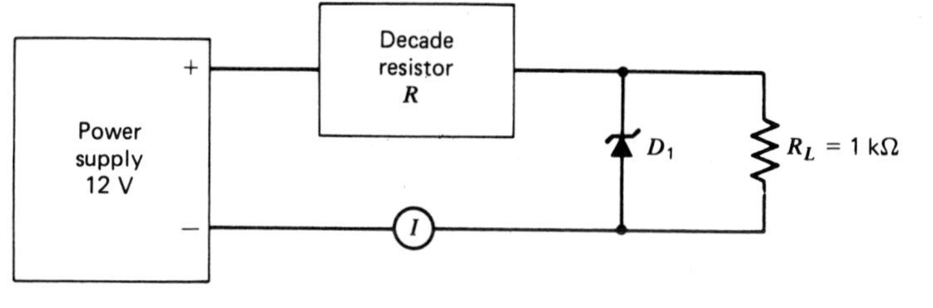

FIGURE E8-1 Experimental circuit.

7. Increase the value of the load resistor in 500-Ω steps to maximum value of 5 kΩ and record V_L, I_T, and I_Z at each step.
8. Set the decade resistor back to 1 kΩ.
9. Starting at 12 V, increase the supply voltage in 2-V steps to a maximum value of 25 V. Record V_L, I_T, and I_Z at each step.
10. Plot four separate graphs, one of R versus V_L, one of R_L versus V_L, one of E versus V_L, and one of I_Z versus V_L from data Table E8-1.
11. Compute the percentage by which R changes versus the percentage by which V_L changes, and similarly, R_L versus V_L, and E versus V_L. Use Eq. E8-1 to compute the percentage of change.

$$\text{Percent change} = \frac{|\text{Initial value} - \text{Final value}|}{\text{Initial value}} \times 100\% \qquad \text{(E8-1)}$$

Report

Your laboratory report should include the following as a minimum.

- Title page.
- List of apparatus.
- Data table.
- Graphs.
- Schematic.
- Analysis of results.

Data Tables for Experiment E8

TABLE E8-1

R versus V_L			
R	V_L	I_T	I_Z
1000 Ω			
900 Ω			
800 Ω			
700 Ω			
600 Ω			
500 Ω			
400 Ω			
300 Ω			
200 Ω			
100 Ω			

TABLE E8-2

R_L versus V_L			
R_L	V_L	I_T	I_Z
0.5 kΩ			
1.0 kΩ			
1.5 kΩ			
2.0 kΩ			
2.5 kΩ			
3.0 kΩ			
3.5 kΩ			
4.0 kΩ			
4.5 kΩ			
5.0 kΩ			

TABLE E8-3

E versus V_L			
E	V_L	I_T	I_Z
12 V			
14 V			
16 V			
18 V			
20 V			
22 V			
24 V			
26 V			

EXPERIMENT E9: THÉVENIN'S THEOREM

Objective

To develop an appreciation of the power of Thévenin's theorem as an analytical tool in solving electronic problems by analyzing several circuits and their Thévenin's equivalent circuit.

Discussion

Someone once said, "Happiness in electronics is being able to solve problems using Thévenin's theorem." There is considerable truth in this statement. Thévenin's theorem is perhaps the most powerful analytical tool available in electronics. The ability to use this tool, once mastered, will prove to be of great value. It is often difficult to see that an original circuit and Thévenin's equivalent circuit are equivalent circuits when looking back from a load. The purpose of this experiment is to show that a circuit connected to a load can be replaced by Thévenin's equivalent circuit without changing the value of the current through the load.

Apparatus

- 1 EVM or digital multimeter.
- 1 dc power supply.
- 1 meter movement, 100 μA.
- 1 meter movement, 10 mA.
- 1 decade resistance box.
- Selected resistors.

Procedure

1. Connect the circuit of Fig. E9-1 and record the current through the meter movement.
2. Compute Thévenin's equivalent voltage and resistance for the circuit to the left of points A and B in the circuit of step 1.
3. Connect Thévenin's equivalent circuit and record the current through the meter movement in Fig. E9-2.
4. Repeat steps 1, 2, and 3 for the circuit of Fig. E9-3.
5. Construct the circuit shown in Fig. E9-4.
6. Measure and record the current through the 10-mA meter in Fig. E9-4.
7. Compute Thévenin's equivalent circuit to the left of points A and B for the circuit of step 5.

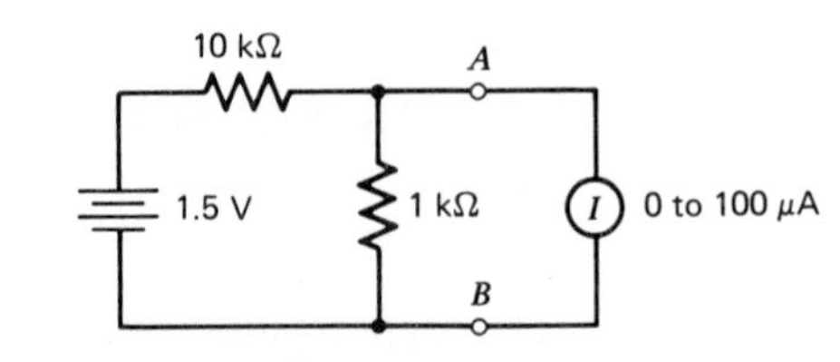

FIGURE E9-1 Circuit for step 1—simple series–parallel circuit.

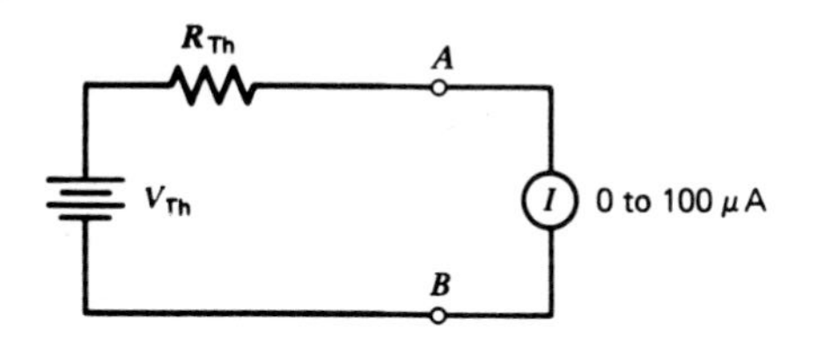

FIGURE E9-2 Thévenin's equivalent circuit for the circuit of Fig. E9-1.

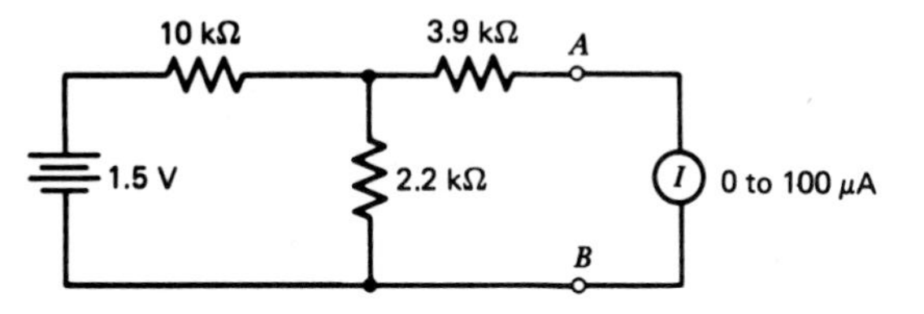

FIGURE E9-3 Circuit for step 4—series–parallel circuit.

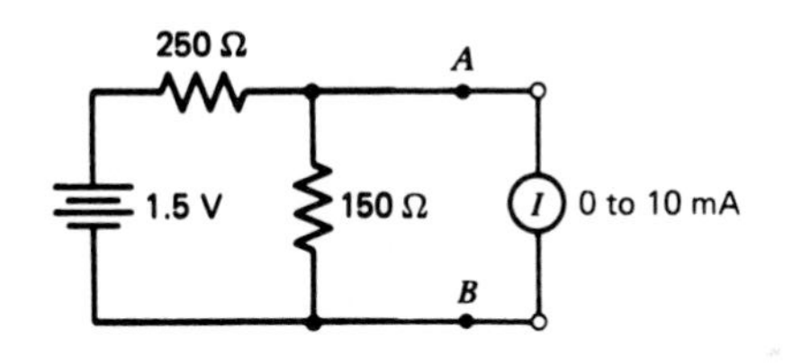

FIGURE E9-4 Series–parallel circuit for step 5.

8. Construct Thévenin's equivalent circuit, connect the 10-mA meter as the load, and determine the internal resistance of the meter.
9. Repeat steps 1, 2, and 3 for the circuit in Fig. E9-5.
10. Construct the circuit shown in Fig. E9-6.
11. Measure and record the open-circuit voltage, V_{oc}, between points A and B for the circuit in step 10 (V_{oc} is Thévenin's equivalent voltages.)

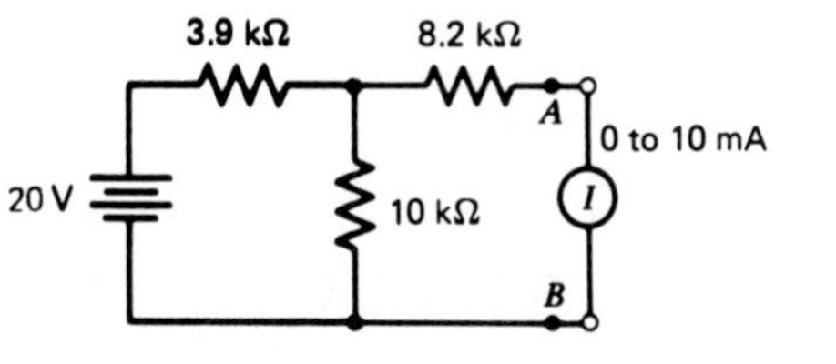

FIGURE E9-5 Series–parallel circuit for step 9.

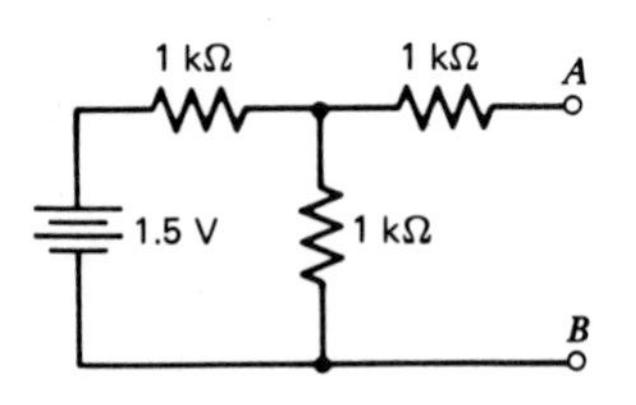

FIGURE E9-6 Circuit with open output terminals to be Thévenized.

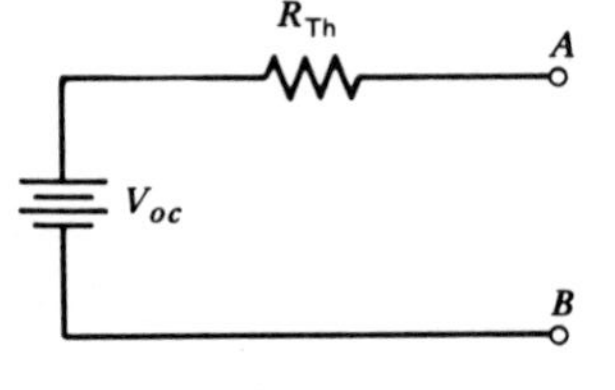

FIGURE E9-7 Thévenin's equivalent circuit for the circuit of Fig. E9-6.

12. Measure and record the short-circuit current, I_{sc}, by connecting the 10-mA meter movement between points A and B.

13. Compute Thévenin's equivalent resistance by the formula

$$R_{Th} = \frac{V_{oc}}{I_{sc}}$$

and construct the circuit of Fig. E9-7.

14. Connect the 10-mA meter movement between points A and B and record the current through the meter.

15. Compute the percentage of error between the current in the original circuits and the current in Thévenin's equivalent circuit for each of the five circuits.

Report

Your laboratory report should include the following as a minimum.

- Title page.
- Schematics of original circuits and equivalent circuits.
- Data table.
- Calculations.
- Analysis of results.

Data Table for Experiment E9

	I_{meas}	I_{Th}	Percent Error	E_{Th}	R_{Th}	R_{in}
C_1						X
C_2						X
C_3						
C_4						X
C_5						X

EXPERIMENT E10: THE WHEATSTONE BRIDGE

Objective

To be able to analyze a basic Wheatstone bridge in a balanced condition and to apply Thévenin's theorem to an unbalanced bridge.

Discussion

The basic Wheatstone bridge has been used extensively since the earliest days of electricity. It is still widely used in a large number of null-type instruments. A null reading is obtained on a Wheatstone bridge by a comparison of the voltage drops in the passive resistance arms of the bridge. When the equation $R_1 R_4 = R_2 R_3$ for the circuit of Fig. E10-1 is satisfied, the bridge is balanced and a "null" or zero reading is obtained on the detector. The primary use of the basic Wheatstone bridge is to measure the resistance value of unknown resistors.

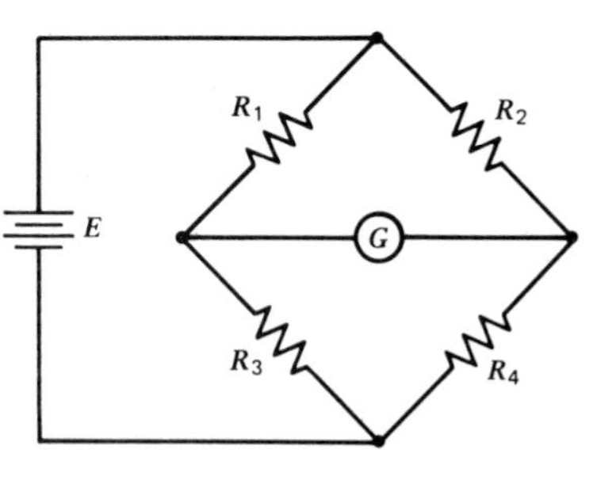

FIGURE E10-1 Basic Wheatstone bridge.

Apparatus

- 1 dc power supply.
- 1 panel meter or galvanometer, 0 to 100 μA.
- 1 EVM or digital multimeter.
- 4 composition resistors, 2.2 k$\Omega \pm$ 5%.
- 1 decade resistor.
- Assorted composition resistor.

Procedure

1. Measure and record in the Table E10-1 the exact value of the resistors color-coded as 2.2 kΩ.
2. Connect the circuit shown in Fig. E10-2.
3. Apply 5 V to the circuit and record the galvanometer current and the current through each resistor.
4. Disconnect the galvanometer from the circuit of Fig. E10-2 and compute Thévenin's equivalent circuit looking back into the bridge circuit from the output. Record V_{Th} in the data table.
5. Construct Thévenin's equivalent circuit as calculated in step 4.
6. Connect the galvanometer to the output terminals of Thévenin's equivalent circuit and measure and record the current.
7. Using the data from step 6, compute and record in Table E10-1 the internal resistance of the galvanometer.
8. Apply the following approximations to your original circuit of Fig. E10-2:

$$V_{Th} \equiv \frac{E}{4R} \Delta R$$

$$Z_{Th} \equiv R$$

The term ΔR is the cumulative variation of each resistor from its marked value. If the actual value of a resistor is above the marked value, the

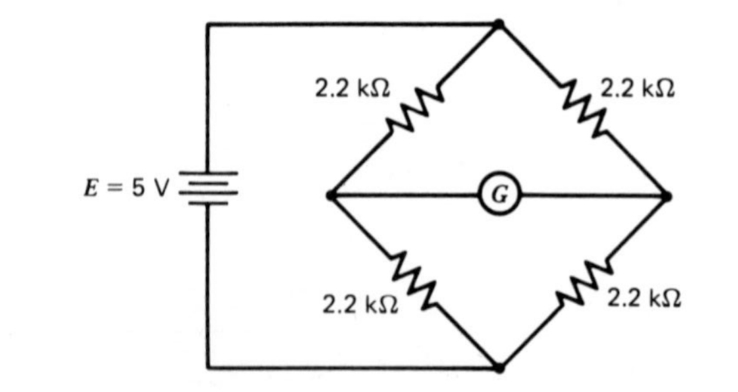

FIGURE E10-2 Circuit for step 2.

difference is assigned a plus. If the actual value is below the marked value, the difference is assigned a negative. Take the algebraic difference for the two resistors in each arm, which is always a negative from a zero or a positive. Now take the algebraic difference for the total of each arm. This result is ΔR. Record in Table E10-2.

To see how ΔR is computed, consider the following example.

EXAMPLE E10-1 If each resistor in the Wheatstone bridge in Fig. E10-3 is marked as 100 **Ω**, what is ΔR?

Solution

$$R_{\text{left}} = +10\,\Omega - (-10\,\Omega) = +20\,\Omega$$

$$R_{\text{right}} = +5\,\Omega - 0\,\Omega = +5\,\Omega$$

$$\Delta R = R_{\text{left}} - R_{\text{light}} = +20\,\Omega - (+5\,\Omega) = +15\,\Omega$$

9. Construct Thévenin's equivalent circuit using the approximate values of step 8.

10. Connect the galvanometer to the output of Thévenin's equivalent circuit of step 9 and measure and record in Table E10-2 the current through the galvanometer.

11. Compute and record the percentage of error between the galvanometer currents of steps 3 and 6.

12. Compute and record the percentage of error between the galvanometer currents of steps 3 and 10.

13. Compute and record in Table E10-2 the percentage of error between the V_{Th}'s of steps 4 and 8.

14. Compute and record the percentage of error between the R_{Th}'s of steps 4 and 8.

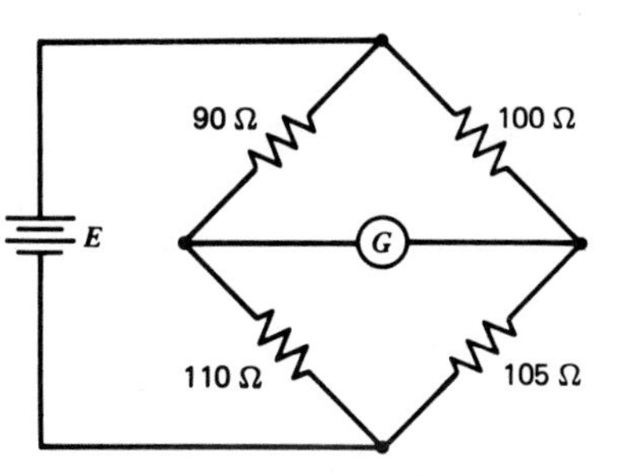

FIGURE E10-3 **Circuit for Example E10-1.**

Report

Your laboratory report should include the following as a minimum.

- Title page.
- Circuit diagrams.
- Data table.
- Calculations.
- Analysis of results.

Data Tables for Experiment E10

TABLE E10-1

Resistor Values	Current (step 3)	Thévenin's Equivalent Circuit (steps 4 and 6)	Galvanometer Resistance (step 7)
R_1	I_1	V_{Th}	R_g
R_2	I_2	R_{Th}	
R_3	I_3		
R_4	I_4		
	I_g	I_g	

TABLE E10-2

Approximate Thévenin's Equivalent Circuit (steps 8 and 10)	Percent Error	
ΔR	I_g (step 11)	
R_{Th}	I_g (step 12)	
V_{Th}	V_{Th} (step 13)	
I_g	R_{Th} (step 14)	

EXPERIMENT E11: BRIDGE-CONTROLLED CIRCUITS

Objective

To investigate the behavior of a basic bridge-controlled circuit.

Discussion

Although the Wheatstone bridge is widely used in electrical measuring instruments, it also finds wide application in both manually operated and automatic control systems. Control is obtained by varying the resistance in one arm of the bridge to produce an error voltage that drives an active device such as a tube or transistor.

Apparatus

- 1 dc power supply.
- 1 resistor, 100 **Ω**, 5 W.
- 1 decade resistance box.
- 1 panel meter, 0 to 100 μA.
- 1 panel meter, 0 to 50 mA.
- 1 *n-p-n* transistor, $\beta \approx 100$, $I_c \approx 100$ mA.
- 1 EVM or digital multimeter.
- 3 resistors, 500 **Ω**.

Procedure

1. Construct the experimental circuit shown in Fig. E11-1.
2. Adjust the decade resistor R_d until the bridge is balanced. Record the setting of the R_d in the data table.
3. Increase the resistance of the decade resistor in 20-**Ω** increments up to 700 **Ω**. Record R_d, I_b, I_c and V_{a-b} in the data table for each 20-**Ω** increment.
4. Plot the following graphs, each on a separate sheet of graph paper with resistance plotted along the horizontal axis.
   (a) Resistance change (Δr) versus error voltage.
   (b) Δr versus I_b.
   (c) Δr versus I_c.

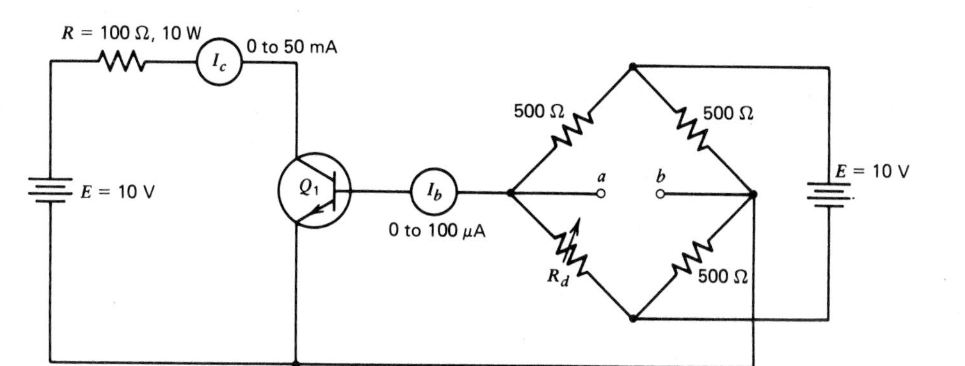

FIGURE E11-1 Basic bridge-controlled circuit.

Report

Your laboratory report should include the following as a minimum.

- Title page.
- Data table.
- List of apparatus.
- Graphs.
- Schematic diagram.
- Analysis of results, including a description of how the circuit can be made more sensitive to resistance changes.

Data Table for Experiment E11

	R_d at Balance		
R_d Setting	V_{a-b}	I_b	I_c
520 Ω			
540 Ω			
560 Ω			
580 Ω			
600 Ω			
620 Ω			
640 Ω			
660 Ω			
680 Ω			
700 Ω			

EXPERIMENT E12: THE MAXWELL BRIDGE

Objective

To be able to apply basic ac circuit theory to the Maxwell bridge.

Discussion

The ac bridge, a natural outgrowth of the dc bridge, consists in its basic form of four bridge arms, a source of excitation, and a null detector. The source of excitation is an ac signal at the desired frequency. The detector may be a set of headphones, an ac voltmeter, an oscilloscope, or another device capable of responding to alternating currents.

The general form of an ac bridge is given in Fig. E12-1. The four bridge arms Z_1, Z_2, Z_3, and Z_4 are shown as unspecified impedances. The bridge is said to be balanced when the detector response is zero. One or more of the bridge arms are varied to balance the bridge so that a null response is obtained. The condition for bridge balance requires that the potential difference from A to C be zero. This happens when the voltage drop from B to A equals the voltage drop from B to C, in both magnitude and phase. In complex notation we can write

$$I_1 Z_1 = I_2 Z_2 \tag{E12-1}$$

We can also write

$$I_3 Z_3 = I_4 Z_4 \tag{E12-2}$$

and

$$I_1 Z_3 = I_2 Z_4 \tag{E12-3}$$

since

$$I_1 = I_3$$

and

$$I_2 = I_4$$

Dividing Eq. E12-1 by Eq. E12-3 yields

$$\frac{Z_1}{Z_3} = \frac{Z_2}{Z_4} \tag{E12-4}$$

$$Z_1 Z_4 = Z_2 Z_3 \tag{E12-5}$$

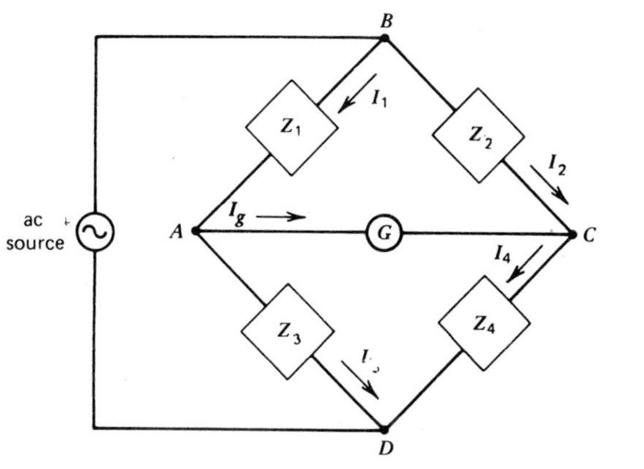

FIGURE E12-1 General ac bridge.

If the impedances are written in polar form, $Z\underline{/\theta}$, where Z represents the magnitude and θ the phase angle of the complex impedance, Eq. E12-5 can be rewritten as

$$(Z_1\underline{/\theta_1})(Z_4\underline{/\theta_4}) = (Z_2\underline{/\theta_2})(Z_3\underline{/\theta_3}) \tag{E12-6}$$

To multiply these complex numbers we multiply the magnitudes and add the phase angles. Equation E12-6 can be rewritten as

$$Z_1Z_4(\theta_1+\theta_4) = Z_2Z_3(\theta_2+\theta_3) \tag{E12-7}$$

Equation E12-7 shows that two conditions must be met simultaneously when balancing an ac bridge. The first condition is that the magnitude of the impedances satisfy the relationship

$$Z_1Z_4 = Z_2Z_3 \tag{E12-8}$$

The second condition requires that the phase angles of the impedances satisfy the relationship

$$\theta_1+\theta_4=\theta_2+\theta_3$$

This expression states that the sums of the phase angles of the opposite arms must be equal.

The Maxwell bridge shown in Fig. E12-2 measures an unknown inductance in terms of a known capacitance. Observing the bridge, we can see that

$$Z_1 = \frac{(R_1)(-j\omega C_1)}{R_1 - j\omega C_1}$$

$$Z_2 = R_2$$

$$Z_3 = R_3$$

$$Z_x = R_x + j\omega L_x$$

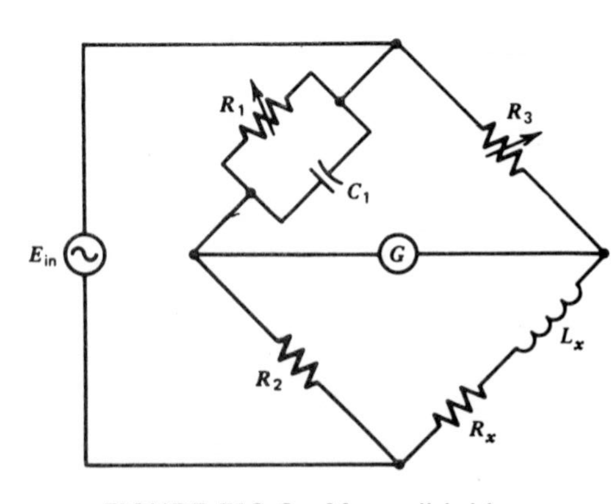

FIGURE E12-2 Maxwell bridge.

Substituting these expressions into Eq. E12-8 and separating the real and imaginary terms yields

$$R_x = \frac{R_2 R_3}{R_1} \tag{E12-9}$$

and

$$L_x = R_2 R_3 C_1 \tag{E12-10}$$

Both Eqs. E12-9 and E12-10 must be satisfied for the bridge to be balanced.

Apparatus

- 1 sine-wave generator.
- 1 ac voltmeter.
- 1 vector impedance meter.
- 1 resistor, 1 kΩ.
- 2 potentiometers, 10 kΩ, 100 kΩ.
- 1 inductor, 2.5 mH.
- 1 capacitor, 0.001 μF.

***Note*:** The values of the components for this experiment were determined experimentally and may need to be changed if the inductor is not very similar to the one used in the experimental setup.

Procedure

1. Set up the experimental circuit shown in Fig. E12-3.
2. Balance the bridge by varying both R_1 and R_3.

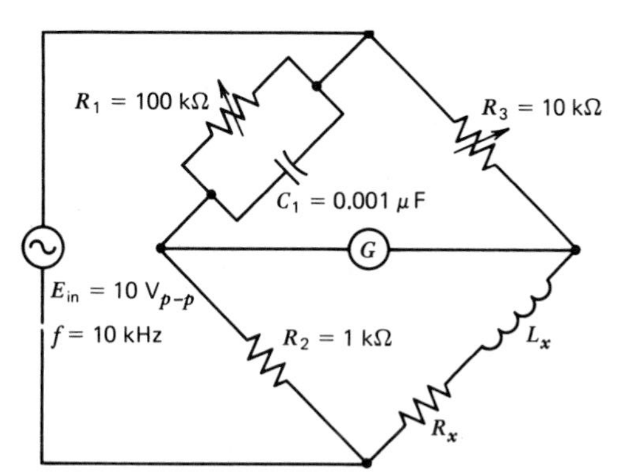

FIGURE E12-3 Experimental circuit.

3. Remove R_1 and R_3 from the circuit and measure and record their values.
4. Compute and record the inductance L_x of the inductor.
5. Compute and record the value of the dc resistance L_x of the inductor.
6. Compute and record the Q of the inductor.
7. Compute and record the phase angle θ of the inductor.
8. Compute the total impedance of the inductor using the expression $Z = \sqrt{R_x^2 + X_L^2} \underline{/\tan^{-1(X_L/R_x)}}$.
9. Measure and record the total impedance Z_x and the phase angle of θ of the inductor with the vector impedance meter.
10. Measure and record the dc resistance R_x of the inductor with an ohmmeter.
11. Compute and record the total impedance of the inductor using the expression $Z_x = Z_2 Z_3 / Z_1$.
12. Compute the percentage of error between the measured and computed values of R_x.
13. Compute the percentage of error between Z_x computed in step 11 and measured Z_x.
14. Compute the percentage of error between the two computed values of Z_x. Treat the value of Z_x from step 11 as the correct value.

Report

Your laboratory report should include the following as a minimum.

- Title page
- Schematic diagram.
- Data table.
- Calculations.
- Analysis of results.

Data Table for Experiment E12

Step	Parameter	Computed	Measured	Percent Error
3	R_1	X		X
3	R_3	X		X
4	L_x		X	X
5	R_x		X	X
6	Q		X	X
7	θ		X	X
8	Z_x		X	X
9	Z_x	X		X
9	θ	X		X
10	R_x	X		X
11	Z_x		X	X
12	R_x	X	X	
13	Z_x	X	X	
14	Z_x	X	X	

EXPERIMENT E13: THE SCHERING BRIDGE

Objective

For basic ac circuit theory to be applied to the Schering bridge, measured and computed values of the unknown should agree within 5% tolerance.

Discussion

The Schering bridge (Fig. E13-1) is one of the most important ac bridges and is used extensively to measure capacitors. This bridge is useful for measuring insulating properties, that is, for phase angles that are very nearly 90°. The capacitors in the bridge have very small electric fields. Thus, the insulating material to be tested can easily be kept out of any strong fields. The balance equations are obtained in the usual manner, and by substituting the corresponding values in the general equation, we obtain

$$Z_x = \frac{Z_2 Z_3}{Z_1}$$

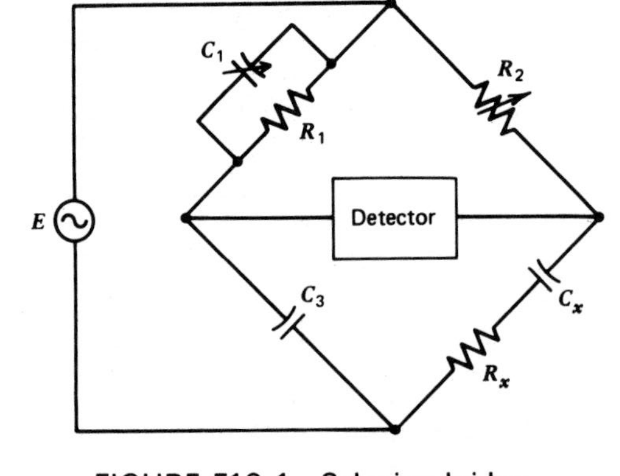

FIGURE E13-1 Schering bridge.

Expanding and equating the real terms and the imaginary terms, we find that

$$R_x = R_2 \frac{C_1}{C_3} \tag{E13-1}$$

$$C_x = C_3 \frac{R_1}{R_2} \tag{E13-2}$$

As can be seen from the circuit diagram, the two variables chosen for the balance adjustment are capacitor C_1 and resistor R_2.

Apparatus

- 1 sine-wave generator.
- 1 1-kΩ resistor.
- 1 25-kΩ potentiometer.
- 1 0.1-μF capacitor.
- 1 0.01-μF capacitor.
- 1 capacitor substitution box.

Procedure

1. Balance the bridge by varying C_1 and R_2 of Fig. E13-2.
2. Remove C_1 and R_2 from the circuit and measure their values.
3. Compute and record the values of the capacitance C_x and the resistance R_x.
4. Compute and record the total impedance of the capacitor C_x from the equation

$$Z_x = \sqrt{R_x^2 + X_C^2}$$

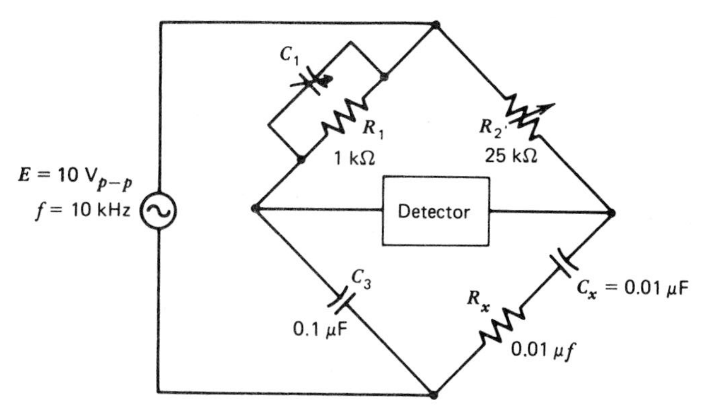

FIGURE E13-2 Experimental circuit.

5. Compute and record Z'_x using $Z'_x = Z_2 Z_3 / Z_1$.
6. Measure and record the capacitance of the capacitor C'_x on a vector impedance meter.
7. Compute and record R'_x from the measured phase angle on the vector impedance meter.
8. Compute and record the percentage of error between C_x computed and C'_x measured.
9. Compute and record the percentage of error between Z_x computed and Z'_x computed.
10. Compute and record the percentage of error between R_x computed and R'_x computed.

Report

Your laboratory report should include the following as a minimum.

- Title page.
- List of apparatus.
- Schematic diagram.
- Data table.
- Calculations.
- Analysis of results.

Data Table for Experiment E13

	Computed	Measured	Percent Error
C_1			
R_1			
C_x			
C'_x			
Z_x			
Z'_x			
R_x			
R'_x			

EXPERIMENT E14: ELECTRONIC VOLTMETERS

Objective

To analyze the dc voltmeter with a direct-coupled amplifier.

Discussion

The dc electronic voltmeter represents a straightforward application of electronics to measuring instruments. The instrument usually consists of one or more dc amplifier stages and a dc meter movement. The dc amplifiers used in voltmeters can be classified into two groups: (1) direct-coupled dc amplifiers; and (2) chopper-type dc amplifiers. Since direct-coupled dc amplifiers are quite economical, they are frequently found in lower-priced voltmeters. Figure E14-1 shows a schematic diagram of a basic dc voltmeter with an FET input direct-coupled dc amplifier. The dc input voltage is applied to the input alternator, which is a calibrated front panel control marked range. This input voltage divider permits a maximum voltage of 0.5 V to be applied to the gate of the *n*-channel FET without causing distortion. The FET is connected as a source follower and is directly coupled to an *n-p-n* transistor Q_2 that is operating as an emitter follower.

Transistor Q_2 is one arm of a bridge circuit whose remaining arms consist of the 10-kΩ emitter resistor for Q_2, the 2.5-kΩ potentiometer, and the 2.2-kΩ resistor. Bridge balance is achieved by adjusting the zero potentiometer. Full-scale calibration is achieved by adjusting the 10-kΩ potentiometer marked calibration, which is in series with the 50-μA meter movement. The input impedance of the voltmeter is 10 MΩ, which is sufficiently high to ignore any loading effects by the instrument on the circuit under test.

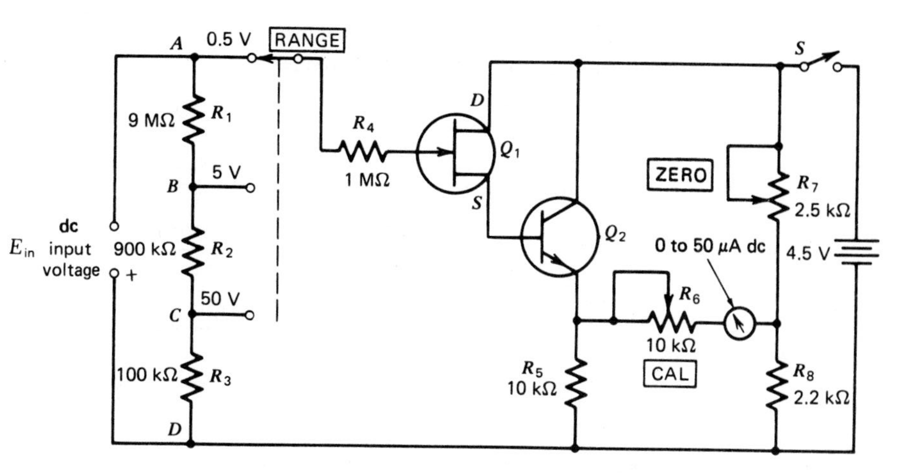

FIGURE E14-1 Basic dc voltmeter circuit with FET input.

The circuit shown in Fig. E14-1 is analogous to the Wheatstone bridge circuit shown in Fig. E14-2 where Z_2 represents the Q_1 to Q_2 source follower. When $E_{in} = 0$ V (Fig. E14-2), the bridge is balanced; therefore,

$$\frac{V_a}{V_b} = \frac{V_c}{V_d} \tag{E14-1}$$

By design, $V_b = V_d$; therefore,

$$I_g = \frac{V_b - V_d}{R_m} = 0 \tag{E14-2}$$

When $E_{in} = 0.5$ V, the bridge is unbalanced, so that $V_b > V_d$ and

$$I_g = \frac{V_b - V_d}{R_m} = I_{fs} \tag{E14-3}$$

where I_{fs} is the full-scale deflection current of the meter movement.

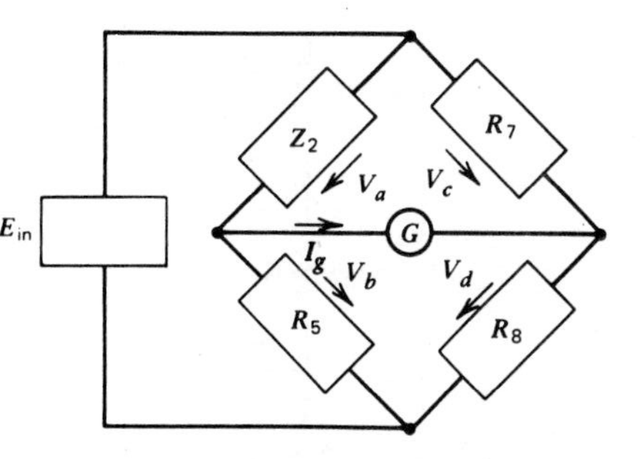

FIGURE E14-2 Wheatstone bridge representation of the voltmeter circuit in Fig. E14-1.

The input alternator network can be designed for three ranges with the total input impedance expressed as

$$R_{in} = R_1 + R_2 + R_3 \qquad \text{(E14-4)}$$

The value of the individual resistors can be computed with the following equations:

$$R_1 = \frac{E_{in} - \text{Range 2}}{E_{in}} R_{in} \qquad \text{(E14-5)}$$

$$R_3 = \frac{(\text{Range 3})(R_{in})}{E_{in}} \qquad \text{(E14-6)}$$

$$R_2 = R_{in} - (R_1 + R_3) \qquad \text{(E14-7)}$$

Apparatus

- 1 dc power supply.
- 1 meter movement, 0 to 50 μA.
- 1 *n-p-n* transistor, 2N 6004.
- 1 JFET, EGC 132.
- 5 resistors, 2.2 kΩ, 100 kΩ, 900 kΩ, 1 MΩ, 9 MΩ.
- 2 potentiometers, 2.5 kΩ, 10 kΩ.

Procedure

1. Set up the experimental circuit shown in Fig. E14-3 with $E_{in} = 0$ V.
2. Adjust resistor R_7 until the bridge is balanced.
3. Set resistor R_6 to its maximum value.

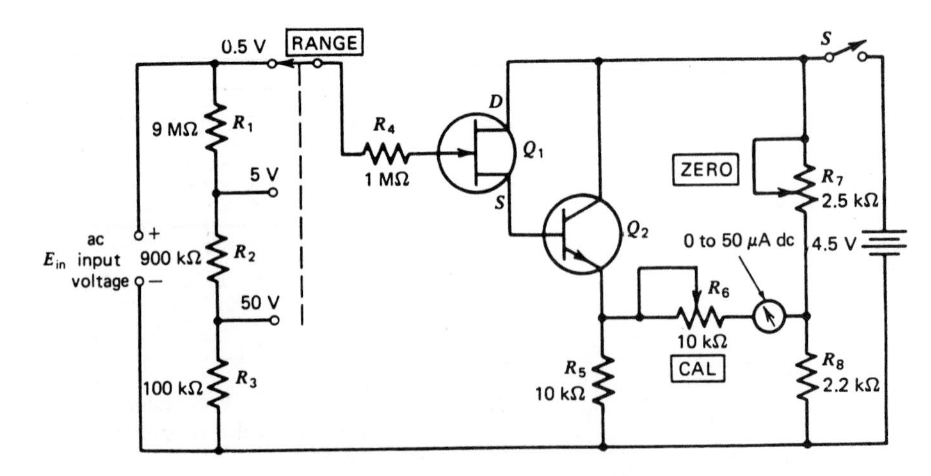

FIGURE E14-3 Experimental circuit.

4. Apply an input voltage of 0.5 V.
5. Adjust resistor R_6 until the meter current is at full-scale deflection.
6. Remove R_6 and readjust R_7, if necessary, so that the meter reads zero.
7. Set the range switch to the 5-V range and increase E_{in} to 5 V.
8. Record the percentage of full-scale deflection caused by the 5-V input.
9. Reduce the input voltage in 0.5-V increments and record the percentage of full-scale deflection.
10. Plot a graph of input voltage versus the percentage of full-scale deflection.

Report

Your laboratory report should include the following as a minimum.

- Title page.
- Data table.
- Schematic diagram.
- Sample calculations.
- List of apparatus.
- Graph.
- Analysis of results.

Data Table for Experiment E14

E_{in}	Percent Full-Scale Deflection
5.0	
4.5	
4.0	
3.5	
3.0	
2.5	
2.0	
1.5	
1.0	
0.5	
0.0	

EXPERIMENT E15: BALANCED-BRIDGE DC AMPLIFIER

Objective

To familiarize the student with the balanced-bridge dc amplifier with input attenuator and indicating meter.

Discussion

One of the most versatile general-purpose shop instruments is the solid-state electronic multimeter or VOM. Figure E15-1 shows the schematic diagram of a balanced-bridge dc amplifier using field-effect transistors or FETs. This is also known as a source-follower type of EVM. For identical FETs it is clear that the circuit is balanced. If the dc input is a positive input voltage, current flows through the meter in the direction indicated. The maximum voltage that can be applied to the gate of Q_1 is determined by the operating range of the FET and is usually on the order of a few volts. The range of input voltages can easily be extended by an input attenuator or range switch.

In this experiment the circuit will first be balanced by the zero adjust control for true null indication. Then with a dc input the calibration control is

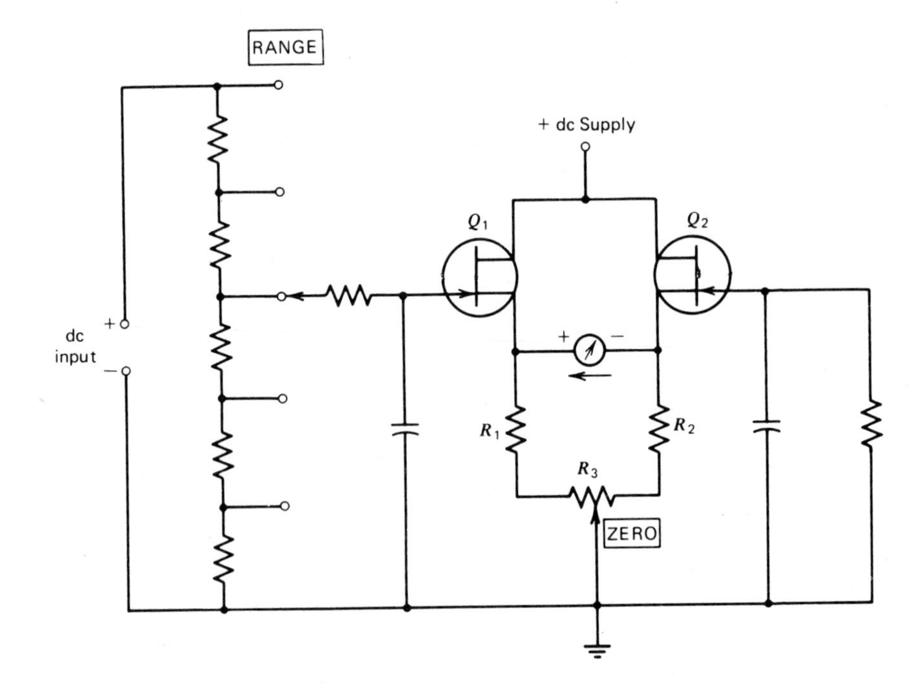

FIGURE E15-1 Balanced-bridge dc amplifier with input attenuator and indicating meter.

adjusted to cause full-scale meter deflection. The settings of an input attenuator will cause the meter readings to be directly proportional.

Apparatus

- 1 meter movement, 0 to 100 μA.
- 1 variable power supply.
- 1 VOM.
- 2 JFET ECG 132 or equivalent.
- 1 100-**Ω** potentiometer.
- 2 35-k**Ω** resistor.
- 1 100-k**Ω** potentiometer.
- 2 1-M**Ω** resistor.
- 1 6-M**Ω** resistor.
- 1 1.2-M**Ω** resistor.
- 1 800-k**Ω** resistor.

Procedure

1. Assemble the circuit shown in Figure E15-2. For zero signal input, adjust R_3 until there is no meter current. Set R_4 at its highest position.

2. Apply −3 V inputs and adjust R_4 until the meter reads maximum.

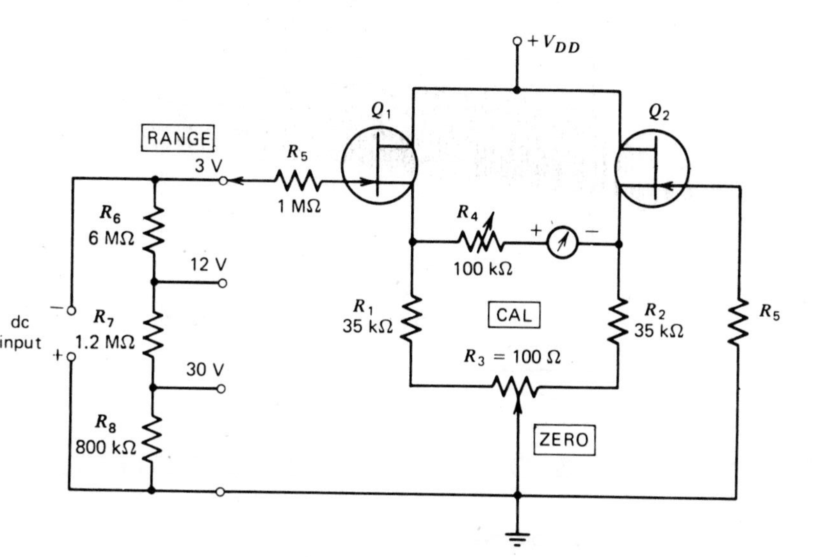

FIGURE E15-2 Experimental circuit.

3. Decrease the input voltage to −2.5 V and record the meter reading in the data table.
4. Complete the data table.
5. Plot a graph of voltage versus current.
6. Set the range switch to "12 V" and apply an input voltage of 12 V. What is the meter reading? ____________
7. Decrease input voltage by one-half. What is the meter reading? ____________
8. Repeat steps 6 and 7 for 30 V.

Report

Your laboratory report should include the following as a minimum.

- Title page.
- List of apparatus.
- Schematic diagram.
- Data table.
- Graph.
- Analysis of results.

Data Table for Experiment E15

Input Voltage (volts)	Meter Current (μA)
−3.0	
−2.5	
−2.0	
−1.5	
−1.0	
−0.5	
0	

EXPERIMENT E16: OSCILLOSCOPE OPERATION

Objective

To identify the operating controls of a triggered oscilloscope and to adjust the controls properly to observe an ac or dc voltage waveform.

Discussion

The cathode-ray oscilloscope (CRO) is the most versatile instrument in electronics. The oscilloscope is used to evaluate a circuit through visual analysis of the waveform. The waveform to be analyzed appears on the face of the cathode-ray tube (CRT).

For the purposes of this manual, oscilloscopes will be classified either as general-purpose service (nontriggered) or as a laboratory (triggered) type. A service-type oscilloscope is employed for viewing ac waveforms at test points in a circuit and for measuring the peak-to-peak amplitude of these waveforms. Other applications include frequency and phase measurements, signal tracing, null indications, alignments, and so on.

The laboratory type is used for more sophisticated measurements, such as observation of low- and high-frequency waveforms, phase measurement of time, and timing relationships. It should be noted that the laboratory oscilloscope may be used in any application for which the general-purpose service type is used. In these oscilloscope experiments we will be concerned only with the laboratory-type oscilloscopes.

Figure E16-1 shows an elementary block diagram of an oscilloscope. Signal waveforms are applied to the vertical input of the oscilloscope and the vertical deflection plates via the vertical amplifiers. Since the oscilloscope must handle a wide range of signal voltage amplitudes, a vertical attenuator acts as a range selector.

An oscilloscope automatically graphs a time-varying voltage. That is, it displays the instantaneous amplitude of an ac voltage waveform versus time. Triggered oscilloscopes with two traces are in common use. By means of an electronic switching arrangement, two traces are developed on the screen of

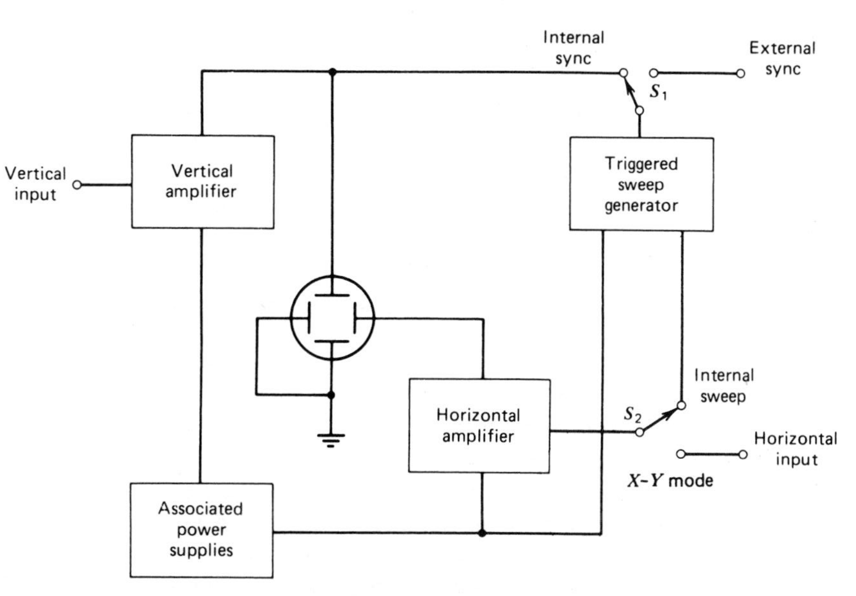

FIGURE E16-1 Elementary block diagram of an oscilloscope.

the oscilloscope. Dual-trace oscilloscopes make it possible to observe simultaneously two time-related waveforms at different points on the electronic circuit.

The operational controls of most common types of oscilloscopes have the following functions.

1. The *Brilliance* control regulates the level of intensity of the light trace on the CRT.
2. The *Focus* control is adjusted in conjunction with the Brilliance control for the sharpest trace on the screen.
3. *Astigmatism* is another beam-focusing control that operates in conjunction with the Focus control to regulate the distinctiveness of the trace.
4. *Horizontal and vertical positioning or centering* controls are trace-positioning controls. They are adjusted so that the trace is positioned or centered both vertically and horizontally on the screen. In front of the CRT screen is an etched faceplate called the graticule. The etchings appear in the form of horizontal and vertical graph lines. Calibration markings are usually placed on the center vertical and horizontal lines on this faceplate.
5. *Volts/Div* (also called *Volts/Cm*) consists of two concentric controls that act as attenuators of the vertical input signal waveform (which is to be viewed on the screen). The inner Volts/Div control marked *Variable* is continuously variable for adjusting the height (vertical amplitude). When it is positioned completely clockwise (CW), it is *calibrated* for making peak-to-peak voltage measurements of the vertical input signal. The outer of the two concentric vertical attenuators is the Volts/Div switch control. This is a rotary switch with the positions to which it can be set marked in Volts/Div values. When the Variable control is set to its calibrated position, the setting of the Volts/Div control determines the voltage that is equivalent to every division of vertical input deflection on the screen.
6. *Time/Div* (also called *Time/Cm*) consists of two concentric controls that affect the timing of the sweep or time-base generator. The inner control marked Variable is continuously adjustable over each range of Time/Div control. When positioned completely clockwise (CW), it is *calibrated* for making time measurements on the horizontal axis. When the Variable control is in its calibrated position, the settings of the Time/Div control determine the time it takes the trace to move horizontally across one division of the graticule.

The *triggering controls* usually have several controls associated with a simple calibrated time base. Thus, one oscilloscope has the following.

1. *Level control*. In the automatic position the trigger circuit is free-running and on each cycle triggers the sweep generator. Hence, a trace always appears on the screen. The oscilloscope is frequently used in this mode of operation. When the oscilloscope is not in the automatic mode, triggering depends on some external signal, and the setting of the level control

determines the stability or synchronization of the sweep. In the nonautomatic mode there will be *no trace* on the screen if there is no triggering signal.

2. The *Stability* control adjusts the sensitivity of the sweep generator.
3. The *Slope* (±) switch is marked + and −, and its setting determines whether triggering of the sweep is affected by the positive or negative portion of the triggering signal.
4. The *Coupling* (ac or dc) control selects the manner in which trigger coupling is achieved. For very low input frequency, direct current should be selected.
5. The *Source* (INT and EXT) controls may be external or internal. This enables the sweep to be triggered either internally, from the vertical amplifier, or externally.

Note: The controls described may have other names depending on the manufacturer and the oscilloscope model.

Apparatus

Triggered-type oscilloscope.

Procedure

1. List each manual control, switch on your oscilloscope, and state its function in the data table. Also include the input jacks.
2. Turn on the oscilloscope. Wait 1 to 2 minutes until the oscilloscope is warmed up.
3. If a trace does not appear on the screen, check and set the triggering switch on Auto(matic).
4. If there is still no trace, turn the Brilliance or Intensity controls completely clockwise.
5. If there is still no trace, then adjust the horizontal and vertical position controls until a trace does appear.
6. Adjust the Focus, Astigmatism, and Brightness controls for a clear, sharp trace. Center the beam or trace vertically and horizontally.
7. Set the triggering mode controls to normal and automatic.
8. Set the triggering Slope to +.
9. Set the triggering Coupling to alternating current.
10. Set the triggering Source to INT.
11. Calibrate Variable vertical attenuators (Volts/Div) by setting fully clockwise.
12. Calibrate Variable Time/Div control by setting fully clockwise.
13. Set the Stability control for free-running by adjusting fully clockwise.

Report

Your laboratory report should include the following as a minimum.

- Title page.
- Data table.
- Description of the oscilloscope including
  Manufacturer.
  Model.
  Maximum sensitivity and bandwidth operating modes (alternate, chopped, etc.)
- Analysis of results.

Data Table for Experiment E16

Manual Controls and Switches and Their Functions

Control or Switch	Function

EXPERIMENT E17: BASIC OSCILLOSCOPE MEASUREMENTS

Objective

To make dc and ac voltage measurements and frequency measurements with the oscilloscope.

Discussion

In general, the oscilloscope has three principal functions.

1. To measure ac voltages.
2. To measure time periods.
3. To view ac wave shapes.

The height of the voltage waveform displayed on the oscilloscope screen is directly proportional to the peak-to-peak amplitude of the voltage. Thus, for the same setting of Volts/Div, or vertical gain control, a 100-V signal will have twice the height of a 50-V signal.

The method just described is also used to measure direct current. The ac–dc switch is placed in the dc position, and the probe is connected to the point in the circuit where the dc voltage is located. The ground lead of the oscilloscope is connected to the ground of the circuit. The number of divisions the trace rises above or below the zero trace setting is a measure of the positive or negative dc voltage.

Before ac voltages can be accurately measured, the vertical deflection must be calibrated. If we now apply an unknown value of an ac voltage to the vertical input and observe a deflection, we can conclude that the unknown voltage has a peak-to-peak voltage of

$$\begin{aligned} E_{p\text{-}p} &= \text{Deflection sensitivity} \times \text{Deflection} \\ &= \text{Volts/Div} \times \text{Number of divisions} \end{aligned}$$

In using the oscilloscope for time measurements, we must calibrate the horizontal deflection as a function of time. If a signal of unknown duration is now observed to produce a complete cycle, the period of the signal is

$$\begin{aligned} T &= \text{Horizontal deflection factor} \times \text{Horizontal sweep per cycle} \\ &= \text{Time/Div} \times \text{Number of horizontal divisions} \end{aligned}$$

For example, a signal has a vertical deflection of 4 cm and a horizontal sweep per cycle of 3.33 cm. The Volts/Div control is set on 10 V, and the Time/Div control is set on 5 msec. Then

$$\begin{aligned} E_{p\text{-}p} &= \text{Volts/Div} \times \text{Deflection} \\ &= 10\ \text{V/cm} \times 4\ \text{cm} = 40\ \text{V} \end{aligned}$$

$$\begin{aligned} T &= \text{Time/Div} \times \text{Horizontal sweep per cycle} \\ &= 5\ \text{msec/cm} \times 3.33\ \text{cm} = 16.66\ \text{msec} \end{aligned}$$

The frequency is found from the period to be

$$f = \frac{1}{T} = \frac{1}{16.66\ \text{msec}} = 60\ \text{Hz}$$

Apparatus

- 1 power supply: dc and ac at line frequency.
- 1 oscilloscope.
- 1 5.1-kΩ resistor.
- 1 10-kΩ resistor.
- 1 15-kΩ resistor.

Procedure

1. Set up Fig. E17-1 and note the peak-to-peak voltage from *A* to *G*, with a voltmeter.
2. Use the oscilloscope measure and record the peak-to-peak voltage from *A* to *G*, *A* to *B*, *B* to *C*, and *C* to *G*.
3. Compute and record in the data table the respective peak-to-peak voltages.
4. Compute and record the percentage of difference between measured and computed values.
5. Measure and record the period of the sinusoidal waveform from *A* to *G*.
6. Compute and record the frequency.
7. Compute and record the percentage of difference between the measured frequency and the line frequency (60 Hz).
8. Replace the ac input at *AG* with a dc power supply set at 18 V.
9. Using the oscilloscope measure, record the dc voltages across *AG*, *AB*, *BC*, and *CG*.
10. Compute and record the dc voltages across *AG*, *AB*, *BC*, and *CG*.
11. Compute and record the percentage of difference between measured and computed values.

Report

Your laboratory report should include the following as a minimum.

- Title page.
- List of apparatus.
- Schematic diagram.
- Data table.
- Calculations.
- Analysis of results.

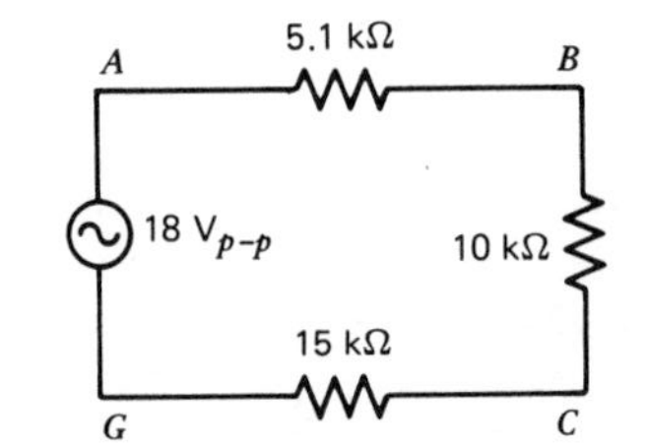

FIGURE E17-1 Ac voltage across divider network.

Data Table for Experiment E17

Test Points	AC Voltage (volts peak to peak)		Percentage Difference	DC Voltage (volts)		Percentage Difference
	Measured	Computed		Measured	Computed	
A to G		18			18	
A to B						
B to C						
C to G						
Frequency						
Period						

EXPERIMENT E18: MEASURING FREQUENCY AND PHASE SHIFT WITH THE OSCILLOSCOPE

Objective

To measure frequency and phase differences using Lissajous patterns.

Discussion

Frequency Measurements Using Lissajous Patterns

Oscilloscopes are used to measure the frequency of ac signals by the method of Lissajous patterns. With this method the sweep switch is set to Ext. Sweep (thus disabling the internal sweep of the oscilloscope). On some dual-trace oscilloscopes this is activated by an $X-Y$ where channel 1 provides a vertical shift and channel 2 a horizontal shift. A signal from an accurately calibrated generator is applied to the horizontal input of the oscilloscope. The signal whose frequency F_x we wish to measure is applied to the vertical input. The frequency F_H of the standard signal generator is then varied manually until a stable pattern appears on the screen. This is called a Lissajous pattern. Several Lissajous patterns are shown in Fig. E18-1. If a horizontal line and a vertical line are drawn tangent to the figure, the number of points of horizontal (T_H) and vertical (T_V) tangency may be obtained. The relationship between the known frequency F_H and the unknown frequency F_x is given by the equation

$$F_x = F_H \frac{T_H}{T_V} \tag{E18-1}$$

Thus, in Fig. E18-1*d*, $T_H = 3$, $T_V = 2$, and $F_x = F_H \times 3/2$. In each case we see that the pattern depends on the phase relationship between F_x and F_H.

A pattern called a ring Lissajous pattern may be used at higher-frequency

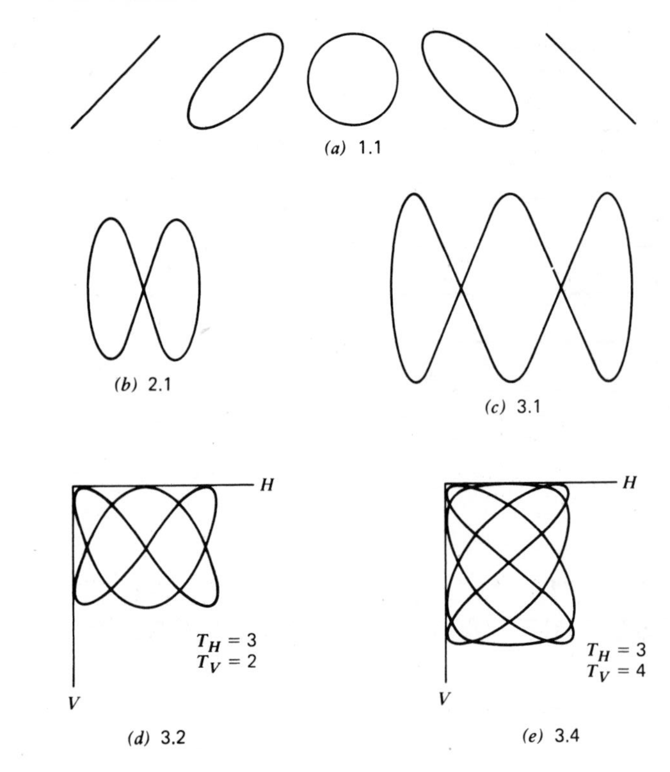

FIGURE E18-1 Characteristic Lissajous patterns.

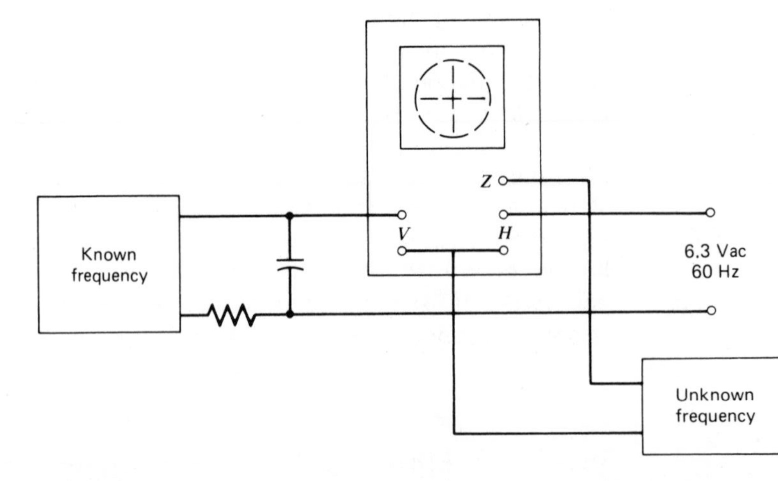

FIGURE E18-2 Measuring an unknown frequency with a ring Lissajous pattern.

ratios if the oscilloscope has a Z-axis input. This signal is used to modulate the beam or to "blank out" the beam at a rate proportional to the ratio of known to unknown frequencies. The necessary circuitry and the resulting pattern are shown in Fig. E18-2. If the unknown signal in Fig. E18-2 was not connected to the Z input, a circle would be observed. This means that the "known frequency" at the vertical input is equal to the frequency at the horizontal input, or 60 Hz. This is desirable but not absolutely necessary. The pattern in Fig. E18-2 shows a 8:1 ratio of unknown frequency to known frequency. Therefore, the unknown frequency is 420 Hz. The 8:1 ratio is determined by counting the dashes in the circle.

Phase Measurements Using Lissajous Patterns

The phase relationships between two sine waves of the same frequency may also be determined by Lissajous patterns. Thus, one of the signals is connected to the vertical and the other to the horizontal input of the oscilloscope. The function switch is set to Ext. Sweep (or the $X-Y$ control is activated). If the signals have the same phase, the resulting waveform will be a straight line slanting from the left to the right. The slant angle is a function of the amplitude of the two signals, of the phase relationship between the input and output of the oscilloscope's vertical and horizontal amplifiers, and of the gain of the amplifiers. In Fig. E18-1*a*, as the phase angle increases from 45° to 90°, the pattern changes from an ellipse to a circle. As the phase angle increases from 90° to 180°, the ellipse slants in the opposite direction, decreasing from an ellipse to a straight line at 135°. For phase shifts between 180° and 360°, the Lissajous patterns are repeated in reverse order. Thus, the pattern for 225° is the same for 135°, that for 270° is the same for 90°, and so forth.

Figure E18-3 shows how the phase angle θ may be calculated. The pattern

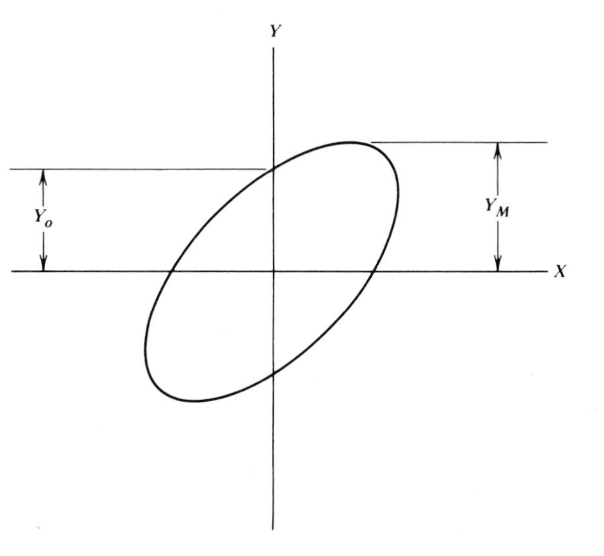

FIGURE E18-3 Phase angle measurements.

must be centered on the $X-Y$ axis of the oscilloscope. Y_M is the distance from the X-axis to the maximum positive point on the ellipse, and Y_o is the distance from the X-axis to the Y-intercept. The phase angle θ may be found from the equation

$$\theta = \sin^{-1} \frac{Y_o}{Y_M} \tag{E18-2}$$

Apparatus

- 1 oscilloscope.
- 1 AF sine-wave generator.
- 1 transformer 100:1, 6.3 V.
- 1 30-mH inductor.
- 1 1-kΩ resistor.

Procedure

Frequency Measurements

***Note*:** The line-isolated 6.3-V filament source of a power supply may be used instead of the transformer.

1. Connect the circuit of Fig. E18-4.

***Note*:** An oscilloscope with a line sweep position automatically applies a line-derived sinusoidal voltage to the horizontal input. For this type of oscilloscope an external 6.3-V sine wave is not required.

2. Set the oscilloscope sweep control to Ext. Sweep or activate the $X-Y$ mode.

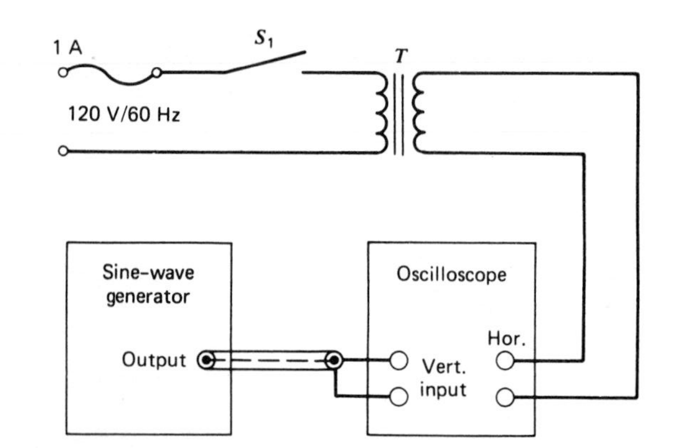

FIGURE E18-4 Circuit for checking frequency calibration.

3. Apply the output of an AF sine-wave generator to the vertical input of the oscilloscope. Adjust the frequency of the generator until a circular pattern appears. This pattern may vary from a circle to an ellipse to a straight line. Draw the pattern in Table E18-1. Record the generator frequency F_V as shown on the dial. Using Eq. E18-1, compute and record the frequency of the signal generator.
4. Vary the frequency of the generator until a pattern appears according to the next tangency pattern given in Table E18-1. Again record this pattern, the dial frequency, and the computed frequency.
5. Repeat step 4 for each T_H/T_V ratios shown in Table E18-1. Compute and record the percentage difference for each set of frequencies.
6. Connect the circuit shown in Fig. E18-5 to make frequency measurements with a ring Lissajous pattern.
7. Vary the unknown frequency until a pattern of dashes is observed on the CRT. Sketch the pattern and determine and record in Table E18-2 the unknown frequency for five different patterns.

Phase Measurements

8. Connect the circuit of Fig. E18-6.
9. Set the frequency of the signal generator to 50 Hz.

***Note*:** If you have difficulty in obtaining sufficient horizontal deflection, reverse the positions of the vertical and horizontal input leads.

10. Center the resulting pattern vertically and horizontally.

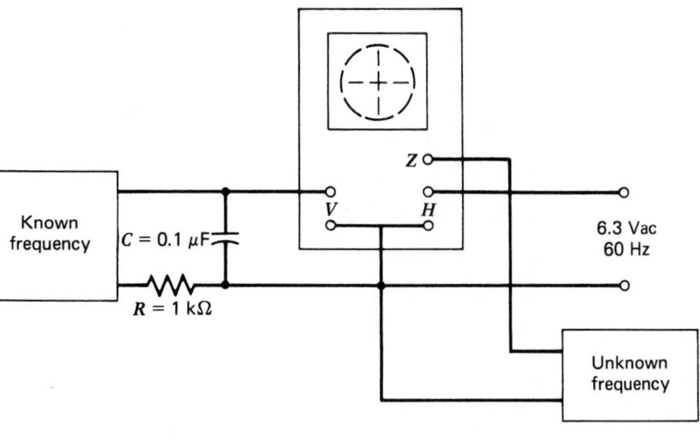

FIGURE E18-5 Circuit for making frequency measurements with a ring Lissajous pattern.

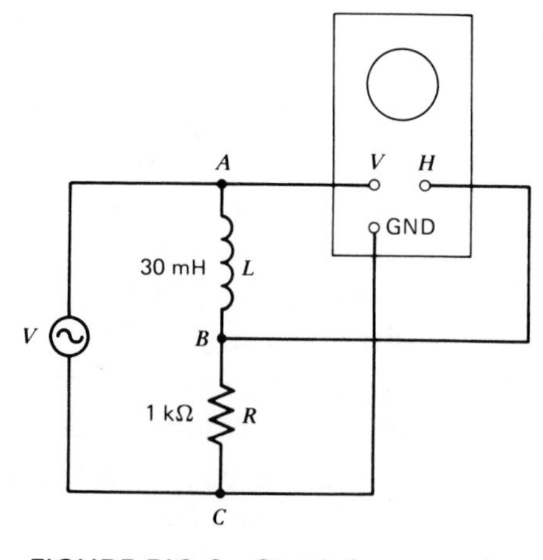

FIGURE E18-6 **Circuit for measuring the phase angle between *v* and *i* in an inductive circuit.**

11. Draw the resulting pattern. Record the value of Y_o and Y_M in Table E18-3.

12. Calculate and record the phase angle θ.

13. Calculate and record the phase angle θ' from the equation

$$\theta' = \tan^{-1} X_L / R$$

14. Compute and record the percentage difference for the phase angle.

15. Set the signal generator frequency to 10 kHz.

16. Repeat steps 8 through 12.

Report

Your laboratory report should include the following as a minimum.

- Title page.
- List of apparatus.
- Schematic diagram.
- Data table.
- Calculations.
- Analysis of results.

Data Tables for Experiment E18: Lissajous-Pattern Frequency Measurements

TABLE E18-1

Step	Lissajous Patterns	Number of Points of Tangency		Generator Frequency (F_V)		Percentage Difference
		T_H	T_V	Dial Setting	Computed	
3						
4		1	1			
5		1	2			
		1	3			
		2	3			
		2	1			
		3	1			
		3	2			

TABLE E18-2

Selected Frequencies	Ring Pattern	Unknown Frequency
f_{x1}		
f_{x2}		
f_{x3}		
f_{x4}		
f_{x5}		

TABLE E18-3

Generator Frequency	Lissajous Patterns	Y_o	Y_M	Phase Angle		Percentage Difference
				θ	θ'	
50 kHz						
10 kHz						

EXPERIMENT E19: OSCILLOSCOPE PROBE CALIBRATION

Objective

To make external sync phasing observations and to calibrate a low-capacitance oscilloscope probe.

Discussion

Some characteristics of an oscilloscope that determine its usefulness for a specific application are as follows.

Frequency Response of the Vertical Amplifiers

The bandwidth of the vertical amplifier determines the range of signal frequencies that may be viewed. Thus, a narrowband scope (i.e., one whose vertical amplifiers are "flat" to 500 kHz) cannot be used for viewing a high-frequency (say, 4.5 MHz) sine wave, whereas a wideband scope can. Checking nonsinusoidal waveforms with a narrowband scope may permit distorted presentation of these waveforms.

Input Impedance

The input impedance of an oscilloscope will determine to what extent the instrument will load the circuit. This characteristic is usually expressed in terms of a resistance in parallel with a capacitance. Since this impedance is frequency-dependent, an oscilloscope will tend to load high-frequency circuits more than low-frequency circuits. Special low-capacitance probes that increase the input impedance of a scope may be employed in measuring high-impedance circuits. Other characteristics include vertical sensitivity, sweep-frequency range, and so on.

Oscilloscope Probes

Low-capacitance probe. A low-capacitance probe reduces the loading effects of an oscilloscope by increasing the input impedance of the scope. Figure E19-1 shows that a low-capacitance probe is essentially a large resistance R in parallel with a small-valued trimmer capacitor C. The total impedance of the probe and scope is additive since the impedance of the probe is in series with the input impedance of the scope. A low-capacitance probe reduces the amplitude of the signal applied to the input of the vertical amplifier. The probe factor determines the reduction factor.

Demodulation probe. A demodulation probe is also known as an RF probe or detector probe. It acts as an AM demodulator and is used in signal-tracing an AM RF carrier in high frequency RF or IF circuits. Figure E19-2 is a schematic diagram of a demodulation probe. It rectifies and filters the signal and then applies the envelope of modulation to the vertical input of the oscilloscope.

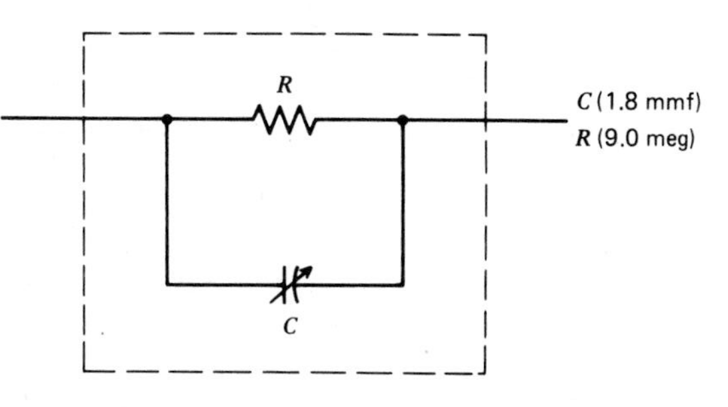

FIGURE E19-1 Low-capacitance probe.

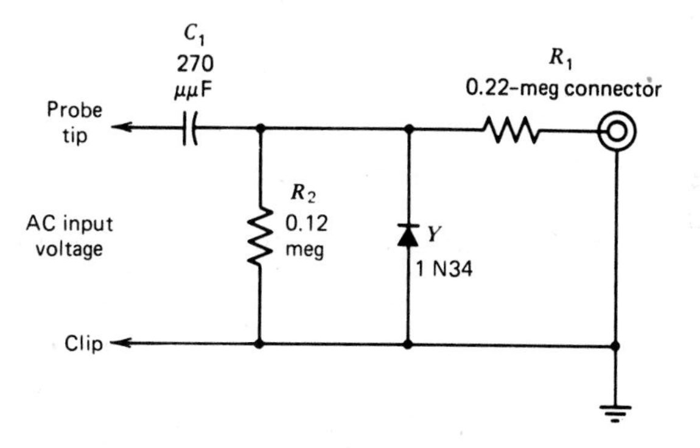

FIGURE E19-2 Demodulation probe.

Apparatus

- 1 filament transformer with a 6.3-V center-tapped secondary.
- 1 oscilloscope with a direct and low-capacitance probe.
- 2 500-kΩ potentiometers
- 2 0.1-μF capacitors.

Procedure

External Sync Phasing

1. Connect the circuit of Fig. E19-3. Connect point *A* to the external sync jack of the oscilloscope. Turn the trigger selector to EXT (or depress the EXT TRIG button).
2. Connect point *D* to the vertical input of the oscilloscope and point *B* to the ground lead.
3. Set R_2 at its midpoint and vary R_1, the sync-phasing control, over its entire range. Observe the effect on the waveform presentation seen on the oscilloscope.

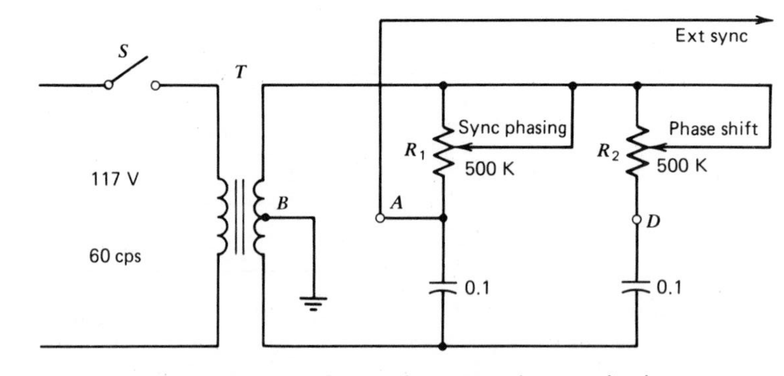

FIGURE E19-3 Circuit for external sync phasing.

4. Set R_1 in the middle of its range and vary R_2 over its entire range. What happens to the amplitude and phase?
5. Set the trigger selector to INT (or depress the EXT TRIG button).

Low-Capacitance Probe Calibration and Probe Factor

6. Connect the vertical input leads of the oscilloscope (direct probe) to the output of a square-wave generator set at 10 $V_{p\text{-}p}$, 15 kHz. Using the internal sync-phasing control, obtain two cycles as the reference waveform.
7. Replace the direct probe with a low-capacitance probe and connect to the output of a square-wave generator. You might switch the oscilloscope Volts/Div control to a more sensitive range. (Leave the generator output at 10 $V_{p\text{-}p}$.) If the low-capacitance probe is not properly matched to the oscilloscope, adjust the variable capacitor in the probe until the waveshape is the same as that originally viewed with the direct probe. The probe is now properly matched to the oscilloscope. Record the apparent voltage E_{App} in Table E19-1.
8. Compute and record the probe factor $M (M = E/E_{\text{App}})$. To obtain actual voltages, multiply all measurements made with the low-capacitance probe by M.
9. Replace the square-wave generator with a 6.3-V rms (18 peak-to-peak) voltage and repeat steps 6 through 9.

Report

Your laboratory report should include the following as a minimum.

- Title page.
- List of apparatus.
- Data table.
- Schematic diagram.
- Analysis of results.

TABLE E19-1
Low-Capacitance Probe Characteristics

Step	Probe Type	Point of Measurement	Peak-to-Peak Volts		Probe Factor
			E	E_{App}	
6	Direct	Output square-wave generator	10		
7, 8	Low capacitance		10		
9	Direct	6.3-V rms source	18		
	Low capacitance		18		

EXPERIMENT E20: CATHODE-RAY TUBE

Objective

To become familiar with the general characteristics of the electrostatic cathode-ray tube, and to observe what external conditions control focusing and beam deflection of the tube.

Discussion

The cathode-ray tube (Fig. E20-1), or CRT, is the heart of the oscilloscope. Basically, the CRT produces a sharply focused beam of electrons, accelerated to a very high velocity. This focused and accelerated beam of electrons travels from its source (the electron gun) to the front of the CRT, where it strikes the fluorescent material deposited on the inside face of the CRT (the screen) with sufficient energy to cause the screen to light up in a small spot. The electron beam passes between a set of vertical deflection plates and then a set of horizontal deflection plates. Voltages applied to the vertical deflection plates

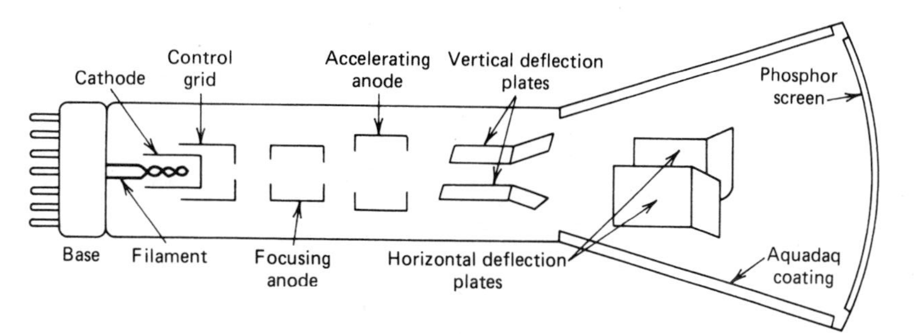

FIGURE E20-1 Cathode-ray tube.

can move the beam in the vertical plane. Voltages applied to the horizontal deflection plates can move the beam in the horizontal plane. These movements are independent of one another, so that the CRT spot can be positioned anywhere on the screen by the simultaneous application of appropriate vertical and horizontal voltage inputs.

The amount of beam deflection on the screen is governed by a number of factors. The two most important factors are the voltage on the deflection plates and the voltage on the accelerating anode. The length of the field and the distance between the plates also influence the beam deflection. The higher the anode potential, the greater the velocity of the electrons as they pass through the region of the deflection plates. As a result, higher deflecting potentials are required at higher values of accelerator anode potentials.

Apparatus

- 2 dc power supplies (high voltage).
- 1 CRT on protective board (any type similar to 3KP1).
- 1 VTVM or EVM.
- 1 sawtooth generator (optional section).

Procedure

1. Refer to the manufacturer's specifications and record the type and number of the CRT used in this experiment. Record the horizontal and vertical deflection sensitivity listed.
2. Insert the proper pin numbers for the schematic of Fig. E20-2, which

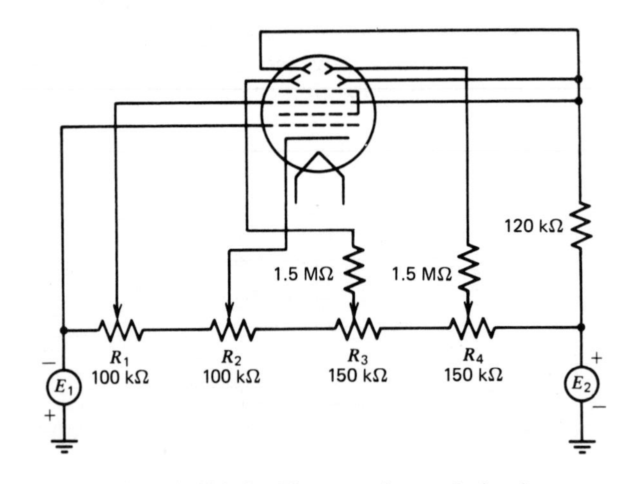

FIGURE E20-2 The experimental circuit.

shows a CRT with voltages connected, onto the pin diagram in Fig. E20-3, which is part of the data table.

3. Assemble the circuit of Fig. E20-2.
4. Apply filament power to the CRT. Set E_1 and E_2 in series for approximately 1000 V. (**Caution**: Use care in handling this circuit. The applied voltage is dangerous and could cause injury.)
5. Adjust R_1, R_2, R_3, and R_4 until a clearly defined dot appears near the center of the screen. Vary each control slightly and identify the name of each: intensity, focus, vertical, and horizontal.
6. Determine the deflection factor for the tube. Vary the voltage of the vertical controls and measure the voltage change and the amount of beam deflection. Record the deflection factor: $S'_V = \Delta_V / d_V$. Repeat for the horizontal control.
7. Connect a sawtooth generator to the horizontal deflection plates and a sine-wave generator to the vertical deflection plates to observe oscilloscope action. Record the wave shape. (**Caution**: Apply voltages through 0.1-μF capacitors, at 1000 working volts.)

Report

Your laboratory report should include the following as a minimum.

- Title page.
- List of apparatus.
- Schematic diagram.
- Data table.
- Calculations
- Analysis of results.

Data Table for Experiment E20

1. Type ______________ No. ______________

Vert. Sen. = ______________ Horiz. Sen. = ______________

2. Identify the pin numbers and element terms (grid, cathode, etc.) in Fig. E20-3.

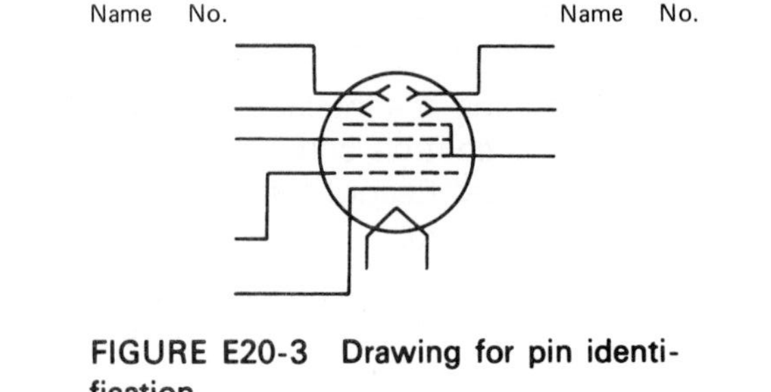

FIGURE E20-3 Drawing for pin identification.

3. S'_V ______________ S'_H ______________

EXPERIMENT E21: SWEEP GENERATOR

Objective

To permit the student to design a basic frequency generator that will produce a sawtooth waveform meeting given specifications.

Discussion

The sweep generator of an oscilloscope produces a sawtooth waveform to sweep the electron beam from left to right across the face of the CRT. To

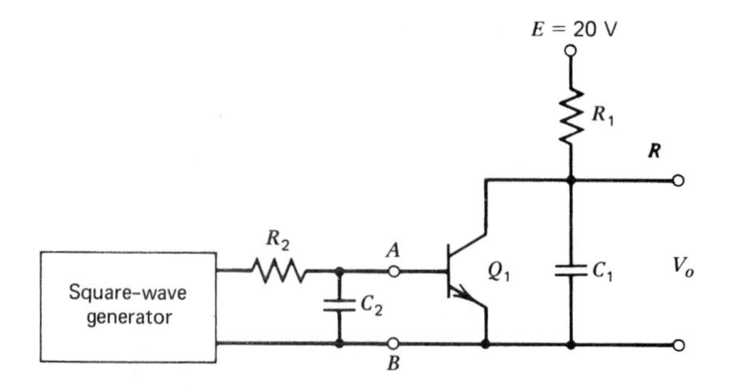

FIGURE E21-1 Experimental circuit.

reproduce the input signal accurately, the beam must move equal distances per unit of time, which requires a linear sawtooth. In this experiment capacitor C_1 charges through R_1 and the variable resistor R and discharges through the transistor shown in Fig. E21-1 to produce a sawtooth.

Apparatus

- 1 dc power supply.
- 1 square-wave generator.
- 1 oscilloscope.
- 1 transistor 2N2219 or similar.
- 1 100-kΩ potentiometer.
- 1 51-kΩ resistor.
- 1 resistor, value to be determined.
- 2 capacitors, values to be determined.

Procedure

1. Set up the experimental circuit shown in Fig. E21-1.

2. You are to design a sawtooth generator that will allow you to display one cycle of a 2.5-kHz sine wave on a CRT screen that is 10 cm wide.

The signal to turn the transistor on periodically is to be supplied by the square-wave generator. The square wave is to be differientiated by the C_2R_2 combination. The R_2C_2 time constant should be about 5% to 10% of the period of a 2.5-kHz square wave. Use a value of about 1 to 3 kΩ for R_2 and solve for C_2. You should also connect a diode between points A and B to shunt the negative spike of the differentiated signal to ground.

The capacitor should discharge to 0.05 V on retrace. Solve for a value for C_1 by assuming a resistance value of the transistor, when it is on, of 50 Ω and a retrace time of no more than 6 μsec. With a value for C_1 you can now solve for the resistance that is necessary to display the 2.5-kHz sine wave.

The capacitor should charge to no more than 2.5 V, and there should be no more than 10% nonlinearity associated with the ramp. Display the sawtooth on the oscilloscope, and verify that the sweep time is correct to display one cycle of a 2.5-kHz sine wave and that the retrace time is no more than 6 μsec.

Report

Your laboratory report should include the following as a minimum.

- Title page.
- List of apparatus.

- Schematic diagram.
- Data table.
- Calculations.
- Analysis of results.

Data Table for Experiment E21

Parameter	Value
C_1	
C_2	
R_1	
t_3	
t_r	
V_c	
% Nonlinearity	

EXPERIMENT E22: OSCILLOSCOPE DELAYED-SWEEP MEASUREMENTS

Objective

To introduce the student to the use of an oscilloscope with delayed-sweep capabilities for measuring rise time.

Discussion

Oscilloscopes are versatile, widely used instruments for measuring time-varying signals. They are able to determine the frequency, period, and amplitude of waveforms as well as to measure such pulse parameters as rise and fall time, duty cycle, and pulse width. Their versatility is increased by incorporating a delayed-sweep feature, which is the topic of this experiment.

Apparatus

- 1 dual-trace oscilloscope with delayed-sweep features.[1]
- 1 square-wave generator.
- 1 1-kΩ resistor.
- 3 capacitors, 0.01 μF, 0.001 μF, 100 pF.

[1]This experiment is written specifically for the Tektronix 2215 oscilloscope but should be suitable for use with similar dual-trace oscilloscopes.

Procedure

1. Connect the experimental circuit shown in Fig. E22-1. The *RC* circuit across the output of the square-wave generator will reduce the rise time on the leading edge of the square wave appreciably, thus making the measurement easier for you.
2. Connect the channel 1 probe across the capacitor and display the waveform. The rise time for the waveform can be measured by setting the following front panel controls:

   - Vertical mode to CH 1.
   - Trigger mode to NORM.
   - Trigger slope to NEGATIVE.
   - Trigger source to INT.
   - Horizontal mode to A.

   After a stable display is observed, switch the horizontal model to ALT. Use the B delay time position dial to move the intensified zone to a point just before the first positive transition. Set the B trigger level into the "B runs after delay" region (blue area) to obtain a second trace on the CRT. Then use the A/B Sweep Sep to separate the traces.
3. After you have obtained two traces, change the Time/Sec control several times to see that both traces are the display of the CH 1 input.
4. Pull out the Sec/Div knob and rotate it clockwise to change the trace B sweep to a higher sweep speed; adjust the B delay time position dial to position trace B toward the left side of the screen.
5. Switch the horizontal mode to B and position the waveform to measure the rise time. Record all pertinent data.
6. Set the frequency of the square-wave generator to 10 kHz, replace the capacitor with the 0.001-μF capacitor, and repeat steps 2 through 5.
7. Set the frequency of the square-wave generator to 100 kHz, replace the capacitor with the 100-pF capacitor, and repeat steps 2 through 5.

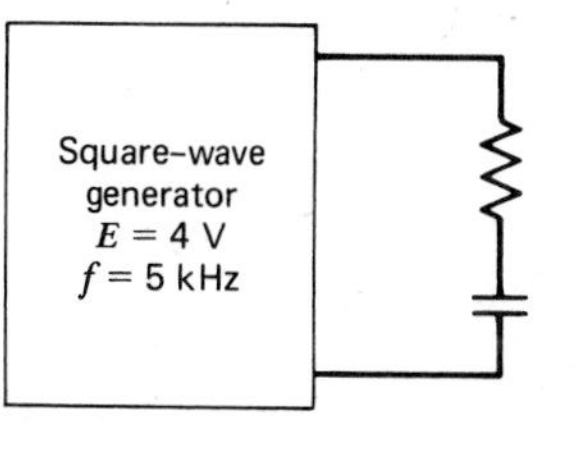

FIGURE E22-1 Experimental circuit.

Report

Your laboratory report should include the following as a minimum.

- Title page.
- Data table, waveforms, or both.
- List of equipment, including serial numbers.
- Schematic diagram.
- Calculations.
- Analysis of results.

Data Table for Experiment E22

Parameter Values	Rise Time	Waveforms
$C = 0.01\ \mu F$ $f = 1$ kHz		Channel A ________ Channel B ________
$C = 0.001\ \mu F$ $f = 10$ kHz		Channel A ________ Channel B ________
$C = 100$ pF $f = 100$ kHz		Channel A ________ Channel B ________

EXPERIMENT E23: THE STRIP-CHART RECORDER

Objective

To familiarize the student with the operation and possible applications of the strip-chart recorder by using it to record waveforms in various kinds of circuits.

Discussion

A strip-chart recorder serves two of the basic functions of electronic instrumentation—indicating and recording. Most electronic recording instruments used in industry are self-balancing, null-type instruments operating on a comparison principal. The input signal is compared to a reference voltage by an error detector. The difference is amplified and is then used to drive a servomotor to which a recording pen is connected.

In this experiment the strip-chart recorder is used to provide waveforms that may be employed to analyze the operation of several RC circuits.

Apparatus

- 1 strip-chart recorder.
- 2 dc power supplies.
- 2 dc relays.
- 1 toggle switch.
- 2 1000-μF, 25-V capacitors.
- 1 2.2-kΩ resistor.
- 1 100-kΩ resistor.
- 1 560-Ω resistor.
- 1 diode.
- 1 2-μF, 25-V capacitor.

Procedure

1. Set up the strip-chart recorder for 23-cm deflection with 10 V applied.
2. Determine the chart speed on the low-speed setting and the high-speed setting.
3. Set up the circuit shown in Fig. E23-1.
4. Record the charge waveform across the capacitor C in Fig. E23-1.
5. With the capacitor charged to 10 V, open switch S_1 and record the discharge waveform across the capacitor.
6. Reconnect the recorder across the 100-kΩ resistor in Fig. E23-1.
7. Close switch S_1 and record the waveform across the 100-kΩ resistor.
8. Connect the circuit shown in Fig. E23-2.
9. Set the output of power supply No. 1 to 10 V. Be sure that 10 V will not overdrive the recorder.
10. With the recorder running, slowly increase the output of power supply No. 2 until the relay contacts close.
11. Connect the circuit shown in Fig. E23-3. Set the power supplies to the proper voltages and record the waveform across the 2.2-kΩ resistor.

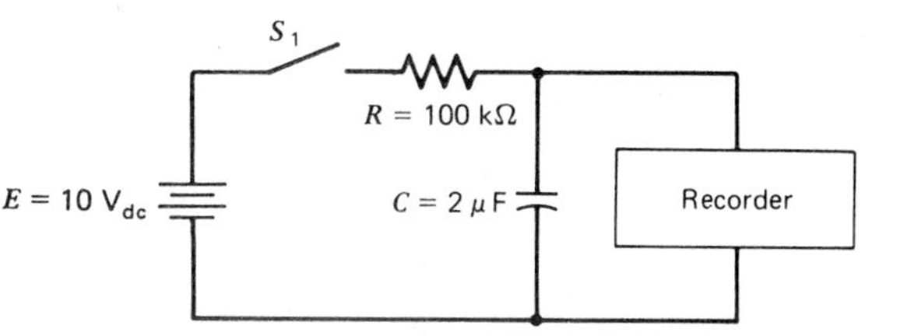

FIGURE E23-1 **Basic *RC* circuit with recorder for recording waveforms.**

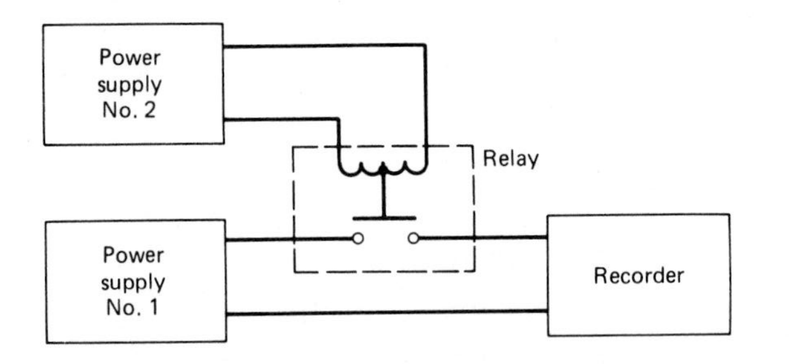

FIGURE E23-2 Circuit for observing relay operation.

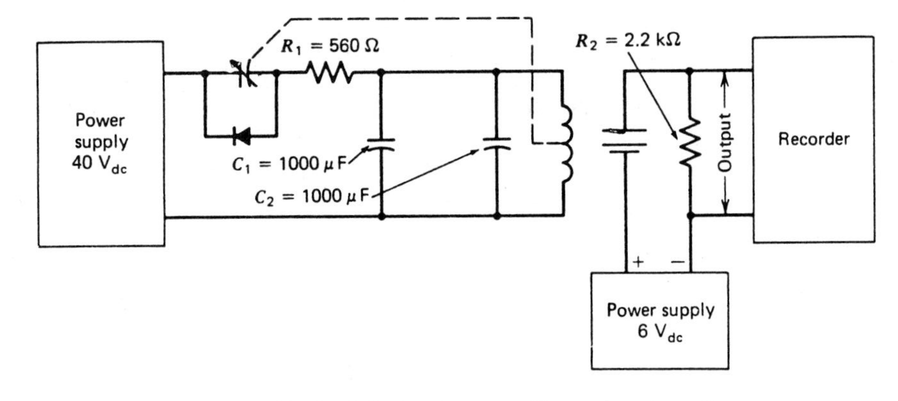

FIGURE E23-3 *RC* circuit controlling relay operation.

12. Reconnect the recorder across the capacitors and record the waveform.

13. Reconnect the recorder across the 560-**Ω** resistor and record the waveform.

Calculations and Circuit Analysis

1. Compute the time constant of the circuit of Fig. E23-1 using the formula $T = RC$.

2. Using the capacitor discharge curve from step 5, determine the output impedance of the recorder. (Read the time constant from the curve and compute R using the formula $T = RC$.)

3. Using the data from step 10, analyze the relay contact bounce. (Was there contact bounce, how severe, etc.?)

4. Using the data from step 11 to 13, describe the operation of the circuit in Fig. E23-3.

Report

Your laboratory report should include the following as a minimum.

- Title page.
- Data table.
- Graphs or waveforms.
- Calculations.
- Analysis of results.

Data Table for Experiment E23

Data for Strip-Chart Recorder				
Chart speed, low				
Chart speed, high				
Time constant, computed				
Time constant, graph				
Recorder output impedance				
Contact bounce noted	Yes		No	
Type circuit shown in Fig. E23-3				
Waveforms				
Across capacitor, step 4				
Across capacitor, step 5				
Across resistor, step 7				
Power supply output, step 10				
Across 2.2-kΩ resistor, step 11				
Across capacitor, step 12				
Across resistor, step 13				

EXPERIMENT E24: THE *X–Y* RECORDER

Objective

To gain experience in using the *X–Y* recorder to record the change in one parameter as it varies with respect to a second changing parameter, other than time.

Discussion

The *X–Y* recorder has many applications in electricity and electronics. This experiment deals with one basic application—that of recording the *E–I* curve for a *p-n* junction diode. In order to obtain the classical power function curve, the voltage that deflects the recorder in the *X*-plane must increase linearly with respect to time. A very low-frequency function generator can be used to obtain the required ramp voltage. However, in this experiment a simple *RC* circuit with a very long time constant will be used. During the first 10% to 15% of the charge curve, the voltage across the capacitor increases quite linearly with respect to time. Therefore, this voltage can be used for deflection in the *X*-plane. This linearly increasing voltage is also applied to the diode, as shown in Fig. E24-1, which causes a nonlinear increase in current and provides us with the classical *E–I* curve.

Apparatus

- 1 *X–Y* recorder.
- 1 dc power supply.
- 1 resistor; value to be computed.
- 1 capacitor, 500 μF.
- 1 resistor, 10 kΩ.
- 1 semiconductor diode.

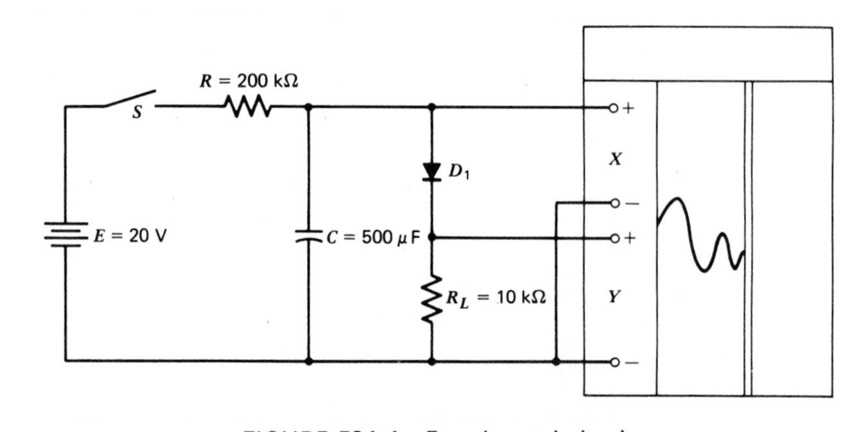

FIGURE E24-1 Experimental circuit.

Procedure

1. Compute the value of the resistor R if the power supply is set to 20 V and the voltage across the 500-μF capacitor is to be 3 V after 16 sec. Use the equation

$$V_C = E(1 - \varepsilon^{-t/RC})$$

and solve for the value of R.

2. Since most laboratories will have only one or two $X-Y$ recorders, work in groups of four to ensure sufficient time to complete the experiment.
3. Set up the experimental circuit shown in Fig. E24-1 and adjust the power supply to 20 V.
4. Turn on the X-channel amplifier and set it to its least sensitive setting.
5. Close switch S and select a range on the X-channel amplifier for full-scale deflection.
6. Turn off the X-channel amplifier, turn on the Y-channel amplifier, and adjust for full-scale deflection of the Y-plane.
7. Open switch S and discharge capacitor C.
8. Place a sheet of graph paper on the recorder and turn on both channels of the instrument.
9. Close switch S and record the $E-I$ curve for the diode.
10. Repeat steps 7, 8, and 9 until each student in the group has a copy of the $E-I$ curve.

Report

Your laboratory report should include the following as a minimum.

- Title page.
- List of apparatus.
- Schematic diagram.
- Calculations.
- $E-I$ curve.
- Analysis of results.

EXPERIMENT E25: SWEEP-FREQUENCY GENERATOR

Objective

To become familiar with the sweep-frequency generator and to learn to use the instrument to obtain test data.

Discussion

In this experiment a sweep-frequency generator is to be used in conjunction with an $X-Y$ recorder to analyze a bandpass filter. Although point-by-point measurements can be taken to obtain a bandpass curve, the sweep-frequency generator makes it possible to obtain the necessary data more rapidly and conveniently and eliminates the chance of missing important information between data points.

Apparatus

- 1 sweep-frequency generator.
- 1 $X-Y$ recorder.
- 2 1-H inductors.
- 2 14-H inductors.
- 2 0.1-μF capacitors.
- 1 10-μF capacitors.
- 2 1-kΩ, $\frac{1}{4}$-W resistors.
- 1 semiconductor diode.

Procedure

1. Connect the experimental circuit shown in Fig. E25-1.
2. Set the generator to sweep through a frequency of 1 kHz.
3. Adjust the gain on the recorder for approximately three-quarters of full-scale deflection.
4. Sweep the generator through 1 kHz and record the shape of the bandpass curve.
5. Place a 1-kΩ resistor in series with the 1-H inductors.

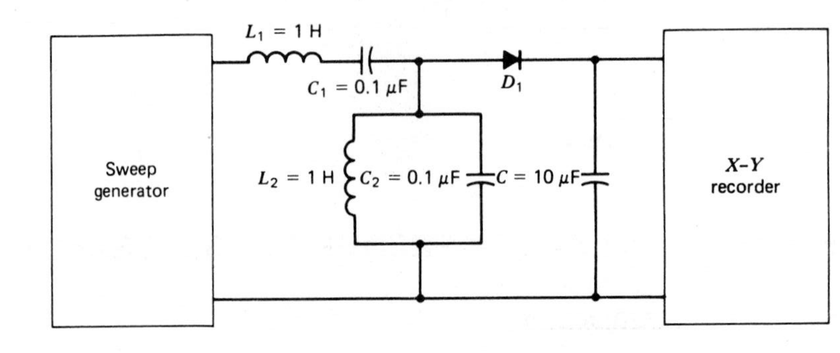

FIGURE E25-1 Experimental circuit.

6. Sweep the generator through 1 kHz and again record the shape of the bandpass curve.
7. Remove the 1-kΩ resistors, replace the 0.1-μF capacitors, and set the generator to sweep through 3 kHz.
8. Sweep the generator through 3 kHz and record the shape of the bandpass curve.
9. Place the 1-kΩ resistors in series with the inductors and repeat step 8.

Report

Your laboratory report should include the following as a minimum.

- Title page.
- Bandpass curves from the $X-Y$ recorder.
- Schematic diagram.
- List of apparatus including serial numbers.
- Analysis of results.

EXPERIMENT E26: BASIC FUNCTION GENERATOR

Objective

To design and analyze a basic function generator.

Discussion

Function generators are very important general-purpose signal sources and find wide application in production testing, instrument repair, service bench testing, and general electronic laboratory work.

In this experiment, a basic function generator is to be designed around the LM741 operational amplifier. The generator is to have a square-wave and a triangular-wave output. The square-wave output appears at the output of the voltage comparator, and the triangular waveform appears at the output of the integrator.

Apparatus

- 1 power supply, 15 V.
- 1 oscilloscope.
- 2 operational amplifiers, LM741.
- 3 resistors; values to be computed.
- 1 capacitor, 0.005 μF.

Procedure

1. Design a function generator that has the circuit configuration shown in the experimental circuit of Fig. E26-1 and that will satisfy the following specifications:

$$V_{02} = 0.5V_{cc}$$
$$f = 1 \text{ kHz}$$
$$C = 0.005\ \mu\text{F}$$
$$R = 5R_2$$
$$V_{cc} = 12 \text{ to } 15 \text{ V}$$

2. Using the values from step 1, set up the experimental circuit shown in Fig. E26-1.
3. Apply power to the circuit and observe the waveforms at V_{01} and V_{02} with the oscilloscope. Record the amplitude and frequency of both waveforms.
4. Sketch both waveforms to scale on graph paper.
5. Replace resistor R with a 250-kΩ potentiometer set to the same resistance as R.
6. Adjust the potentiometer from $0.5R$ to $2R$ and record the effect on the operation of the circuit.
7. Compute the percentage of error between specified V_{02} and measured V_{02}.
8. Compute the percentage of error between specified frequency and measured frequency.
9. Compute the percentage change in the frequency observed in step 6, using the expression

$$\text{Percent change} = \frac{f_{max} - f_{min}}{f_{min}} \times 100\%$$

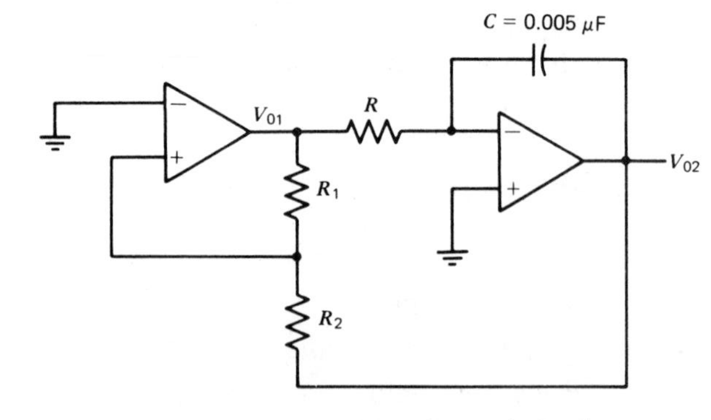

FIGURE E26-1 Experimental circuit.

Report

Your laboratory report should include the following as a minimum.

- Title page.
- Schematic diagram.
- Data table.
- Calculations.
- Waveform sketches.
- Analysis of results.

Data Table for Experiment E26

	V_{01}	V_{02}	Percent Error	Percent Change
E_p			X	X
f				
f_{min}			X	X
f_{max}			X	X
% Error	X		X	X

EXPERIMENT E27: THE THERMOCOUPLE

Objective

To study the principles and practical use of a two-metal junction as a temperature indicator.

Discussion

If two wires of dissimilar metals are connected in a loop as shown in Fig. E27-1, and the two junctions A and B are held at temperatures of θ_A and θ_B, respectively, a current will circulate around the loop.

The current is due to small emfs generated by two quite separate effects, which sum algebraically. They are called the Peltier and the Thompson effects.

1. *The Peltier effect.* An emf is generated at each of the two junctions, shown as e_A and e_B in Fig. E27-1. These emfs are dependent on the "absolute temperatures" of the junctions. If A is hotter than B, there will be a

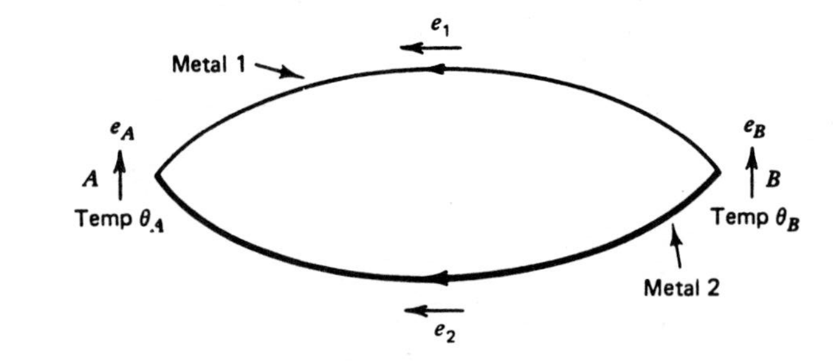

FIGURE E27-1 Thermocouple of metals 1 and 2 showing Peltier effect.

resultant emf of

$$e_A - e_B = P(\theta_A - \theta_B) \tag{E27-1}$$

where P is the Peltier coefficient.

2. *The Thompson effect.* Each wire of the loop generates a small emf, e_1 and e_2 in Fig. E27-1, simply as a result of the difference in temperatures at its ends. The emf is different for different metals. If T_1 is the Thompson coefficient for metal 1 and T_2 is the Thompson coefficient for metal 2, then

$$e_1 = T_1(\theta_A - \theta_B) \tag{E27-2}$$

$$e_2 = T_2(\theta_A - \theta_B) \tag{E27-3}$$

so that the resultant emf is

$$e_2 - e_1 = (T_2 - T_1)(\theta_A - \theta_B) \tag{E27-4}$$

Combining Eqs. E27-1 and E27-4, we get a resultant emf round the loop:

$$(e_A - e_B) + (e_2 - e_1) = (P + T_2 - T_1)(\theta_A - \theta_B)$$

In practice, the values of T_1 and T_2 are much smaller than the value of P, but for a given pair of metals they can all be lumped together into a single constant. Let us call it K. Thus,

$$E = K(\theta_A - \theta_B) \tag{E27-5}$$

Referring back to Fig. E27-1, if the total loop resistance is R, then by Kirchhoff's law for a circuit loop

$$I = E/R \tag{E27-6}$$

Now we can see that if one of the junctions, say B, is held at a known temperature and called the *cold* or *reference* junction, then by measuring the current I we can determine the temperature of the other or *hot* junction, provided we know the values of K and R.

In order to measure the current, a meter of some kind must be inserted in the loop as in Fig. E27-2. This meter is likely to have various metals making up parts of its internal construction. The law of intermediate metals states that any number of junctions may be introduced into a circuit provided they are all at the same temperature. The net emf is not altered by inserting the meter,

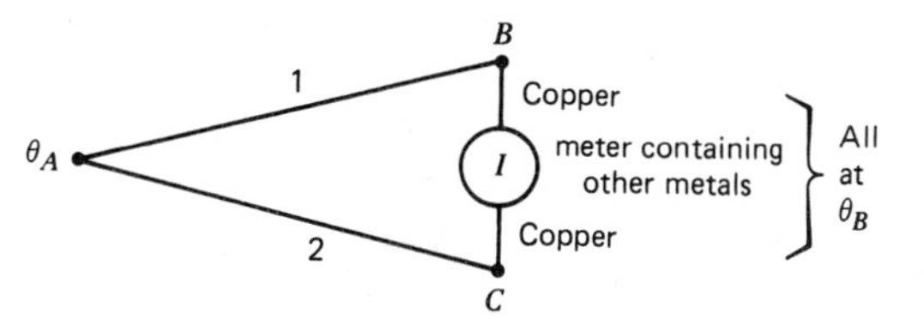

FIGURE E27-2 Meter inserted to measure thermocouple current.

provided that points B and C and all the other junctions within the meter are at the same (cold junction) temperature.

Practically, the instrument should be far enough from the hot junction that it is not affected by its temperature. In these thermocouples special compensating leads made of metals that can cancel out errors caused by junctions that are at intermediate temperatures are used. Figure E27-3 illustrates this. If the emf generated at junction 1–3 for temperature θ_x is equal and opposite to that generated at junction 2–4, inserting the compensating leads has no effect on the net emf, and the effective cold junction is still at the meter.

We saw that the current in a thermocouple loop is given by

$$I = K(\theta_x - \theta_c)/R \qquad \text{(E27-6)}$$

where

θ_x = unknown temperature
θ_c = cold junction temperature
R = total circuit resistance
K = thermocouple constant

If the measuring instrument has a very high input resistance and is voltage-sensitive, the resistance of the thermocouple and its extension leads will be negligible, and the emf measured by the meter will be the "open-circuit emf" of the thermocouple. Figure E27-4 illustrates this; the cold junction is still at the meter terminals, both of which must be at the same temperature

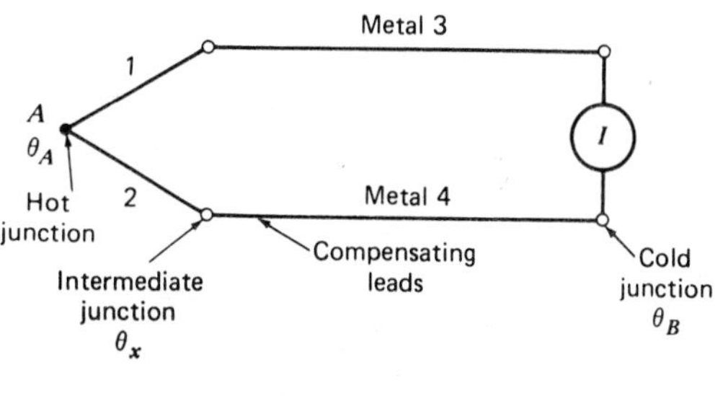

FIGURE E27-3 Thermocouple circuit with compensating leads.

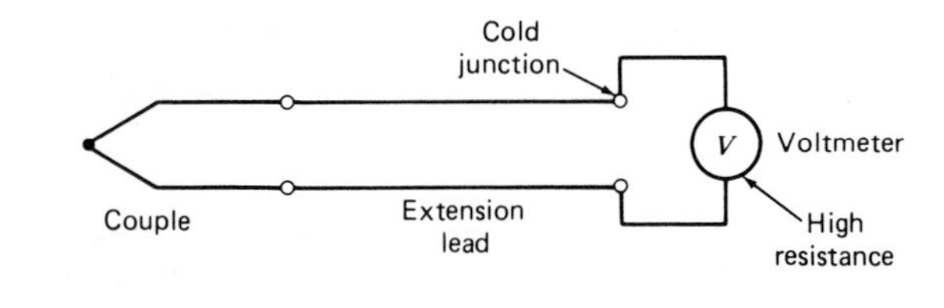

FIGURE E27-4 High-resistance voltmeter used to measure the thermocouple emf.

as before. The circuit resistance of the thermocouple assembly is very small, and the input resistance of an amplifier is large enough that virtually the whole of the emf appears at the amplifier input terminals.

Apparatus

(Equipment suggested by Feedback)

- 1 module 294B (amplifier).
- 1 heat bar.
- 2 thermometers.
- 1 calibration tank.
- 1 thermocouple transducer (copper and constantan).
- 1 voltmeter.

Procedure

1. Clip the tank assembly on the heat bar (Fig. E27-5) at notch 20 or 8.7 in. from the heater assembly.
2. Fill the calibration tank (Fig. E27-6) with water to about $1\frac{1}{2}$ in. from the top.

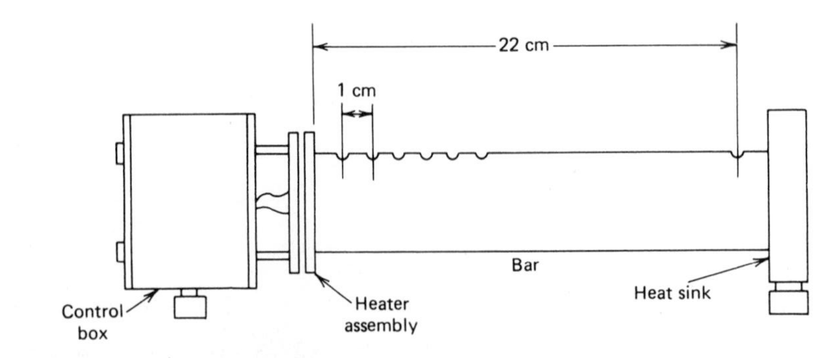

FIGURE E27-5 The heater bar assembly.

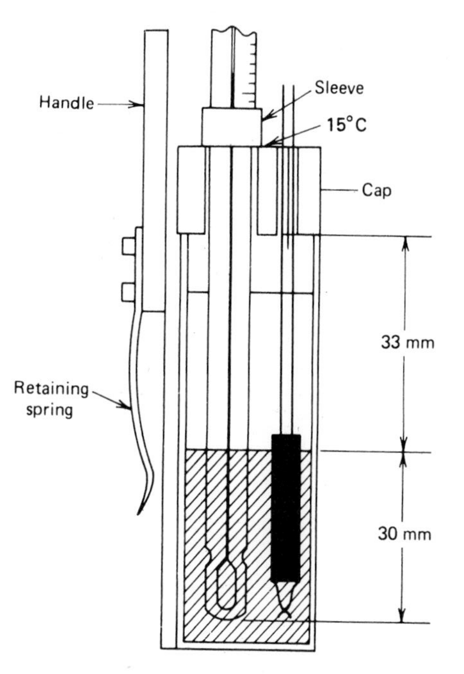

FIGURE E27-6 Calibration tank.

3. Place the thermocouple probe in the tank cap to about 1 in. from the bottom.
4. Place the thermometer in the tank cap to about $\frac{1}{4}$ in. from the bottom.
5. Connect thermocouple leads to the input of the op-amp set at a gain of 1000 (Fig. E27-7).
6. Place a second thermometer near the amplifier at the cold junction (blue lead) to measure room temperature.
7. Turn on the power to the heat bar.
8. When the temperature is steady (after about 15 minutes), record the thermometer and voltmeter readings.
9. Repeat for notches 18, 16, etc., up to the point at which 100°C is reached.
10. Look up in a book of physical data (e.g., in a table of physical and chemical constants) the emf/°C for a copper–constantan thermocouple. Compute the emf per notch and record.
11. Compare the measured emf and the computed emf.
12. Plot the tank temperature versus the measured emf and analyze the slope.

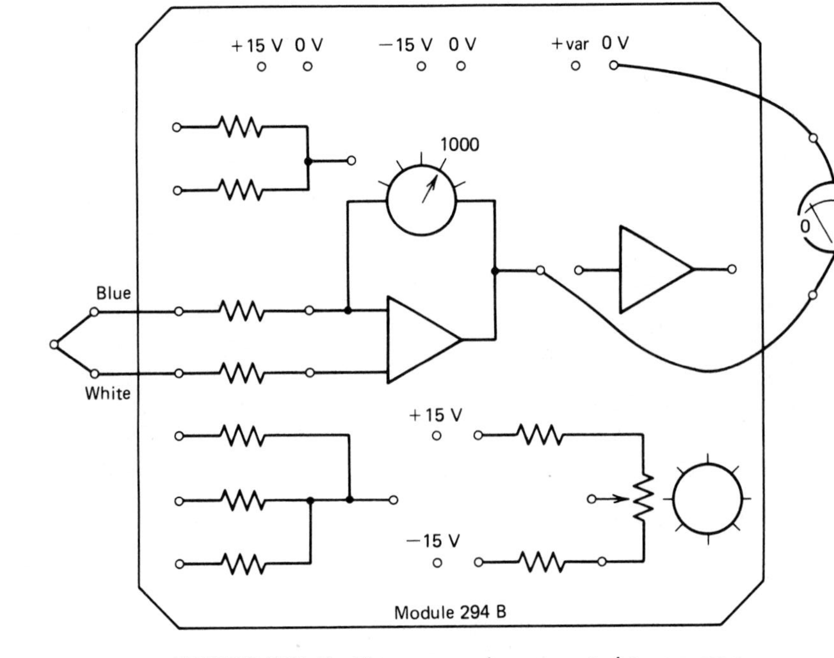

FIGURE E27-7 Thermocouple connected to op-amp.

Report

Your laboratory report should include the following as a minimum.

- Title page
- Data table.
- Experimental circuit drawing.
- Sample calculations.
- Graph.
- Analysis of results.

Data Table for Experiment E27

Notch	Tank Thermometer, °C	Room Temp., °C	Difference, °C	Meter Reading, V	Couple emf, mV
20					
18					
etc.					

EXPERIMENT E28: PHOTOELECTRIC TRANSDUCERS

Objective

To examine the characteristics and operation of photoelectric transducers.

Discussion

Electrons may acquire enough energy to escape from a metal, even at low temperatures, if the material is illuminated by light of sufficiently short wavelength.

This phenomenon is known as the photoelectric effect. Photoelectric devices are classified by the way the electric output is furnished to the circuit. Devices that emit electrons are **photoemissive**. Devices that change their resistance as a function of light intensity are **photoconductive** or **photoresistive**. Photoelectric cells of one type or another are being used in many places around the home and community. An example is the automatic eye that controls outside lights around the home.

Apparatus

- 2 VOMs.
- 1 photovoltaic cell.
- 1 photoconductive cell.
- 1 20-Ω resistor.

- 1 light source (200-W lamp).
- 1 meterstick.
- 1 sheet of black paper.
- 3 sheets of white paper.

Procedure

1. Set up the circuit with the photovoltaic cell (shown in Fig. E28-1).
2. With the light source off, place a sheet of black paper over the photovoltaic cell. Measure the voltage and record in Table E28-1.
3. Remove the paper and record the voltage for normal room light.
4. With the light source on, record the voltages for distances of 4, 3, 2, 1, and $\frac{1}{2}$ ft between the light source and the cell.
5. Place a 20-Ω resistor between points A and B of Fig. E28-1.
6. Repeat step 4.
7. Calculate the current and power in the resistor for each set of data.
8. Set up the circuit with the photoconductive cell as shown in Fig. E28-2.
9. With the light source off, place a sheet of black paper over the photoconductive cell. Measure the resistance and record in Table E28-1.
10. Remove the paper and record the resistance for normal room light.
11. With the light source on, record the resistances for distances of 4, 3, 2, 1, and $\frac{1}{2}$ ft between the light source and the cell.
12. With the light source 1 ft away, place one sheet of white paper between the light source and the cell.
13. Record the resistance in Table E28-2.
14. Repeat for two and three sheets of paper.
15. Plot graphs of
    (a) Voltage output versus illumination distance.
    (b) Resistance versus illumination distance.
    (c) Resistance versus the obstructions for the photoconductive cell.

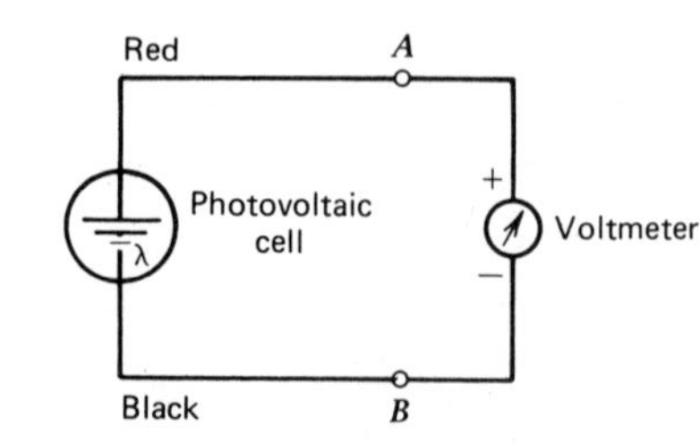

FIGURE E28-1 Photovoltaic circuit.

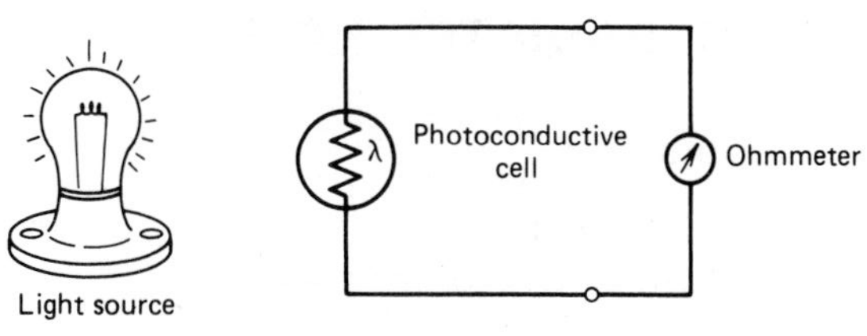

FIGURE E28-2 Photoconductive circuit.

Report

Your laboratory report should include the following as a minimum.

- Title page.
- Equipment list.
- Drawing of test circuit.
- Data table.
- Graphs.
- Analysis of results.

Data Tables for Experiment E28

TABLE E28-1

	Black Paper	Room Light	4 ft	3 ft	2 ft	1 ft	6 in.
Photovoltaic voltage with 20-Ω resistor (voltage)							
Current (mA)							
Power (mW)							
Photoconductor resistance							

TABLE E28-2

Number of sheets	1	2	3
Photoconductor resistance			

EXPERIMENT E29: RADIATED NOISE

Objective

To be able to observe how radiated noise enters a circuit and to make basic measurements of radiated noise.

Discussion

Noise is one of the few topics in electronics with which everyone in the field must deal on occasion. Radiated noise is difficult to treat quantitatively, and it is often equally difficult to eliminate. The objective of this experiment is primarily one of observation; however, you will gain valuable knowledge in the observations you make.

When we measure signals of very low amplitude, the presence of noise may cause appreciable error. Circuits designed for low-level signals are constructed to minimize noise pickup. Construction techniques may include proper grounding, shielding, or filtering.

In this experiment we wish to investigate ways in which noise may enter a circuit, as well as techniques we might use to reduce the noise level in a circuit.

Apparatus

- 1 audio oscillator or signal generator.
- 1 ac voltmeter (Hewlett-Packard 400 series or equivalent).
- 4 composition resistors, 1 MΩ and 2 MΩ.
- 1 coaxial cable.
- 1 metal sheet (approximately 1 ft square).

Procedure

1. Connect the experimental circuit shown in Fig. E29-1. Use the longest test leads you have. Make no attempt at good wiring practices since the presence of noise is desirable in this experiment.

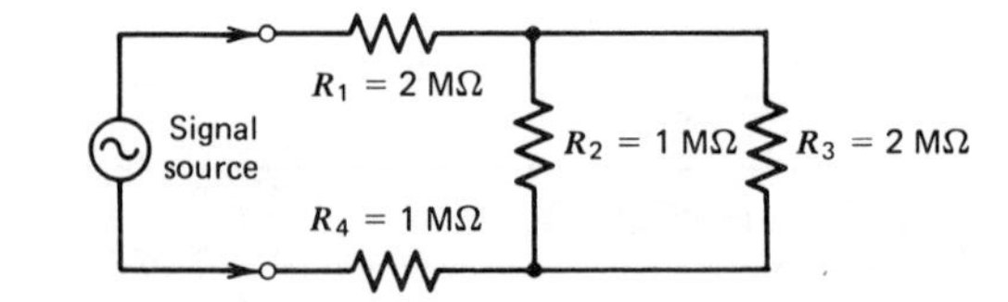

FIGURE E29-1 Experimental circuit.

2. Turn the signal source off, set the voltmeter to a range of approximately 1 V ac and connect the test leads of the voltmeter across resistor R_1. Switch the instrument to more sensitive ranges until a usable reading is obtained. (Any observable reading obtained will be a measurement of noise in the circuit.) Record your noise reading in the data table.
3. Set the voltmeter back to a range of approximately 1 V ac before touching the instrument test leads, and then reconnect the instrument across resistor R_2. Switch to more sensitive ranges until a usable reading is obtained. Record the reading in the data table.
4. Repeat step 3 for resistor R_4.
5. Remove the signal source from the circuit. Turn the instrument that provides the signal source "on" and adjust the signal to 50 mV at 10 kHz. Consider this to be a noise-free signal and record the amplitude as the "measured signal" in the data table. Reconnect the signal source to the resistance network.
6. Measure and record the voltage across resistors R_1, R_2, and R_4.
7. Repeat the measurements in step 6. As each measurement is taken, observe the effect of placing your hand near, but not touching, the resistor.
8. Carefully place the circuit on the metal sheet covered with a sheet of paper and repeat steps 6 and 7.
9. Connect the metal sheet to ground with a test lead and repeat steps 6 and 7.
10. Remove the circuit from the metal sheet. Replace the test leads used to connect the resistors with coaxial cable and repeat steps 6 and 7.

Calculations

1. Use the data from step 5 and step 2 to calculate the signal-to-noise ratio for resistors R_1, R_2, and R_4.
2. Subtract the measurement of step 5 from the measurements of steps 6 through 10. The differences represent the noise signal. Calculate the signal-to-noise ratio for steps 6 through 10 using the noises just calculated and the signal voltage from step 5.

Report

Your laboratory report should include the following as a minimum.

- Title page.
- Schematic diagram.
- List of apparatus.

- Sample calculations.
- Data table.
- Analysis of results.

Data Table for Experiment E29

	Signal Measurement	Noise Measurement			Signal Plus Noise Measurement		
		R_1	R_2	R_4	R_1	R_2	R_4
Step 2							
Step 3							
Step 4							
Step 5							
Step 6							
Step 7							
Step 8							
Step 9							
Step 10							

EXPERIMENT E30: AMPLIFIER NOISE MEASUREMENT

Objective

To be able to observe and measure noise present in an amplifier.

Discussion

Noise should be measured with the circuit under test in a well-shielded environment. The power supply for the circuit should be outside the shielded environment, and shielded wire should be used for power supply leads and output leads. It is also good practice to bypass to ground the power supply terminals with high-quality capacitors on the order of 0.01 μF in value. Circuit resistors should be low-noise carbon film resistors or good-quality metal film resistors.

If a screen room is not available for shielding the circuit under test, a metal enclosure, such as an enclosed equipment cabinet or a small industrial oven, will provide fairly good shielding. If nothing else is available, you can devise a reasonably well-shielded environment by wrapping a cardboard box with aluminum foil. Be sure to ground whatever enclosure is used.

Apparatus

- 1 ±15-V dc power supply.
- 1 operational amplifier, Fairchild μA 741, or the equivalent.
- 1 oscilloscope or ac voltmeter.
- 1 breadboard such as E&L Instruments SK-10 or a Global Specialties proto board.
- 4 resistors: one each 100 **Ω**, 50 k**Ω**, two 10 M**Ω**. (Low-noise resistors are preferred but are not absolutely necessary.)
- 1 10-k**Ω** potentiometer.
- 1 shielded enclosure.

Procedure

1. Connect the experimental circuit of Fig. E30-1. Do not shield the circuit at this time.

2. Turn on the power supply, close switches S_1 and S_2, and adjust the potentiometer R for a small indication at the output.

3. Open switches S_1 and S_2 and record the noise at the output of the amplifier.

4. Place the experimental circuit in the shielded enclosure. Be sure to ground the enclosure. Measure and record the noise at the output.

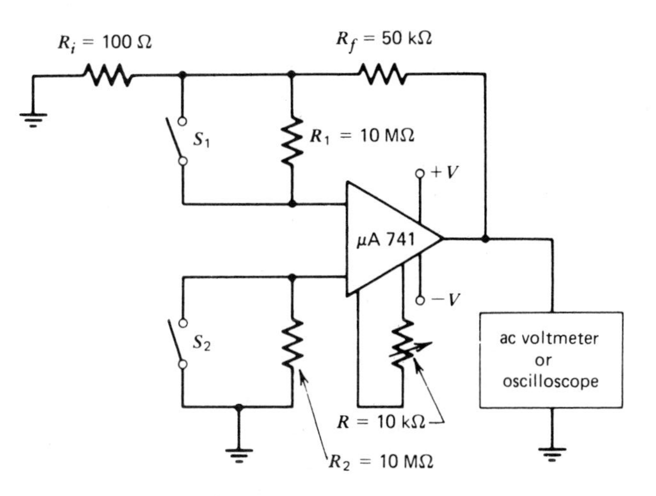

FIGURE E30-1 Experimental circuit.

5. Close switches S_1 and S_2 and measure and record the output voltage V_o.
6. The output voltage of the circuit with the switches closed is

$$V_o = e_n\left(\frac{R_f}{R_i} + 1\right)$$

Using the output voltage, V_o, measured in step 5, compute the noise voltage e_n, which is generated within the amplifier where

$$e_n = \frac{V_o}{R_f/R_i + 1}$$

Report

Your laboratory report should include the following as a minimum.

- Title page.
- Schematic diagram.
- List of apparatus.
- Data table.
- Sample calculations.
- Analysis of results.

Data Table for Experiment E30

	No Enclosure	Shielded Enclosure
e_n (S_1 and S_2 open)		
V_o (S_1 and S_2 closed)		
e_n (computed)		

EXPERIMENT E31: MEASURING PHASE SHIFT WITH AN ELECTRONIC COUNTER

Objective

To investigate using an electronic counter in the time-interval mode for the purpose of measuring phase shift.

Discussion

There are many applications for time-interval measurements made with an electronic counter. Some typical applications are measuring the propagation delay in digital or linear circuits, pulse width and rise time, cable length, and phase shift. In this experiment phase-shift measurement will be investigated.

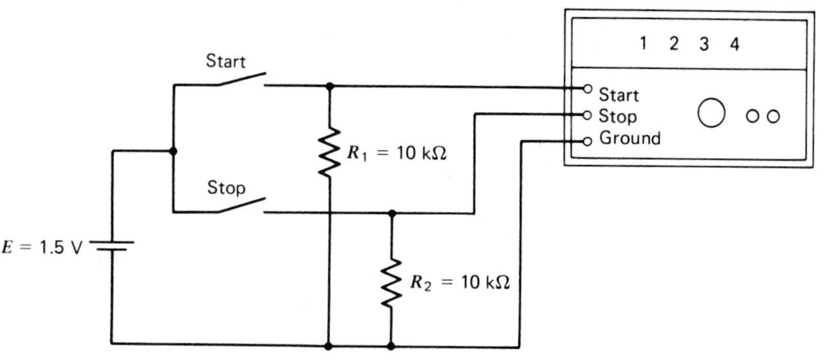

FIGURE E31-1 Simple circuit for time-interval measurement.

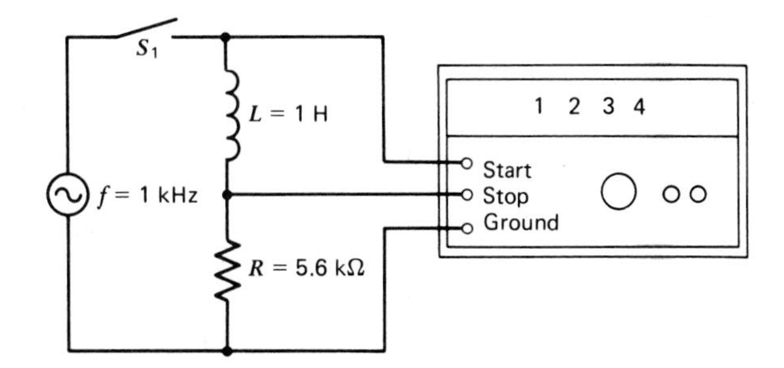

FIGURE E31-2 Experimental circuit.

The simplest system for time-interval measurement consists of an electronic counter operating in the time-interval mode, a voltage source, two resistors, and two switches connected as shown in Fig. E31-1. The counter starts counting when the start switch is closed and stops counting when the stop switch is closed.

The experimental circuit for measuring phase shift is very similar to the circuit in Fig. E31-1 and is shown in Fig. E31-2. The reference phase that starts the counter is taken at the output of the signal generator. The *RL* circuit causes the phase shift that is to be measured. The stop signal is taken between the phase network components. One necessary precaution is to make sure the counter is triggering at zero volts. Otherwise a phase shift will be indicated because of the difference in the amplitude of the start and stop signals.

Apparatus

- 1 electronic counter with a time-interval mode.
- 1 sine-wave generator.
- 1 inductor, 1 H.

- 1 resistor, 5.6 kΩ.
- 1 switch, SPDT.

Procedure

1. Calculate the phase shift for the *RL* circuit in Fig. E31-2.
2. Set up the experimental circuit shown in Fig. E31-2.
3. Set the counter for time-interval measurement.
4. Adjust the trigger level to trigger at zero volts.
5. Close switch S_1 and note the time difference in microseconds registered by the counter (t_o).
6. Calculate the phase shift in degrees by using the formula:

$$\text{Phase shift} = \frac{t_o}{T} \times 360$$

   where t_o is the time between the *a* point on the reference waveform and the same point on the shifted waveform and T is the period of the reference waveform in microseconds.
7. Calculate the percentage of error between the calculated phase shift of step 1 and the phase shift determined with the counter.

Laboratory Report

Your laboratory report should include the following as a minimum.

- Title page.
- Schematic diagram.
- List of apparatus.
- Data table.
- Calculations.
- Analysis of results.

Data Table for Experiment E31

Phase shift (Step 1)	
t_o	
T	
Phase shift (step 6)	
Percent error	

EXPERIMENT E32: VOLTAGE-TO-FREQUENCY CONVERSION USING A VOLTAGE-CONTROLLED OSCILLATOR

Objective

To investigate the use of a voltage-controlled oscillator to achieve voltage-to-frequency conversion.

Discussion

An integral part of any digital voltmeter is the circuitry necessary to convert a dc input voltage to a proportional periodic waveform suitable for application to the counting circuitry of the instrument. One way to convert the dc signal to a periodic waveform is by using a voltage-controlled oscillator (VCO). Increasing the input voltage increases the frequency of the output signal. This increases the count stored by the counter. The stored count is displayed as a voltage level on the digital readout of the instrument.

In this experiment an actual commercial frequency counter will be used to count and display the output of the VCO. However, the concept of the frequency counter is the same as that of a digital voltmeter.

Apparatus

- 1 voltmeter.
- 1 dc power supply.
- 1 electronic counter.
- 1 voltage-controlled oscillator IC, LM566.
- 1 decade resistor.
- 2 resistors, 10 kΩ.
- 2 capacitors, 0.01 μF.

Procedure

1. Construct the experimental circuit shown in Fig. E32-1.
2. Set the decade resistor to 1 kΩ and record the frequency as displayed by the electronic counter and the voltmeter reading.
3. Increase the decade resistor setting in 1-kΩ increments to 10 kΩ, and record the frequency and the voltmeter reading at each setting.
4. Replace the 0.01-μF capacitor with a 0.02-μF capacitor and repeat steps 2 and 3.
5. Plot the data from steps 2, 3, and 4 on an $8\frac{1}{2} \times 11$-in. sheet of graph paper with voltage plotted along the abscissa.

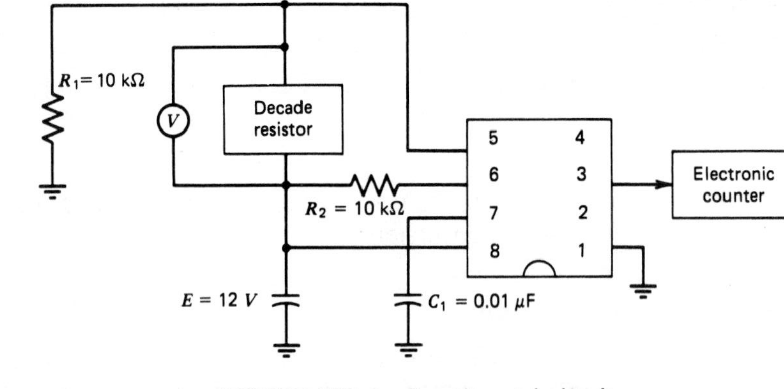

FIGURE E32-1 Experimental circuit.

Report

Your laboratory report should include the following as a minimum.

- Title page.
- Schematic diagram.
- List of apparatus.
- Graph.
- Analysis of results.

Data Table for Experiment E32

R (kΩ)	C_1		C_2	
	V	f	V	f
1				
2				
3				
4				
5				
6				
7				
8				
9				
10				

EXPERIMENT E33: RECEIVER TROUBLESHOOTING

Objective

To examine some of the techniques that can be used to locate the cause of a receiver failure.

Discussion

A superheterodyne receiver has five different signals that it processes in one way or another.

1. The modulated RF input signal.
2. The unmodulated oscillator signal.
3. The modulated IF signal.
4. The direct-current automatic gain control (AGC) signal.
5. The modulation signal.

These five signals appear at different points in the receiver circuit. Figure E33-1 shows the block diagram of a receiver and the type of signal processed by each block.

In troubleshooting a dead receiver, first check the V_{cc} supply voltage. If V_{cc} is normal, then connect an audio test signal across the volume control. Turn the volume up and listen. If the test tone is heard in the speaker and has normal volume, the audio section is working.

On the other hand, if a normal test tone cannot be heard in the speaker, there is a failure of some kind in the audio section. In this case we proceed to troubleshoot the audio amplifiers.

Failures of the audio amplifier may be divided into a number of major categories. The following major divisions cover the vast majority of troubles.

1. An amplifier with a normal input signal may have no output at all.
2. An amplifier with a normal input signal may have seriously reduced output.
3. An amplifier with a normal input signal may produce a distorted output.
4. An amplifier may oscillate at some frequency.
5. An amplifier output may contain objectionable amounts of noise or hum.

In troubleshooting an amplifier (or other electronic system), we normally make two basic assumptions.

1. The circuit was functional before the failure.
2. There is only one problem to be located in the circuit.

Let us now proceed with each of the five types of troubles.

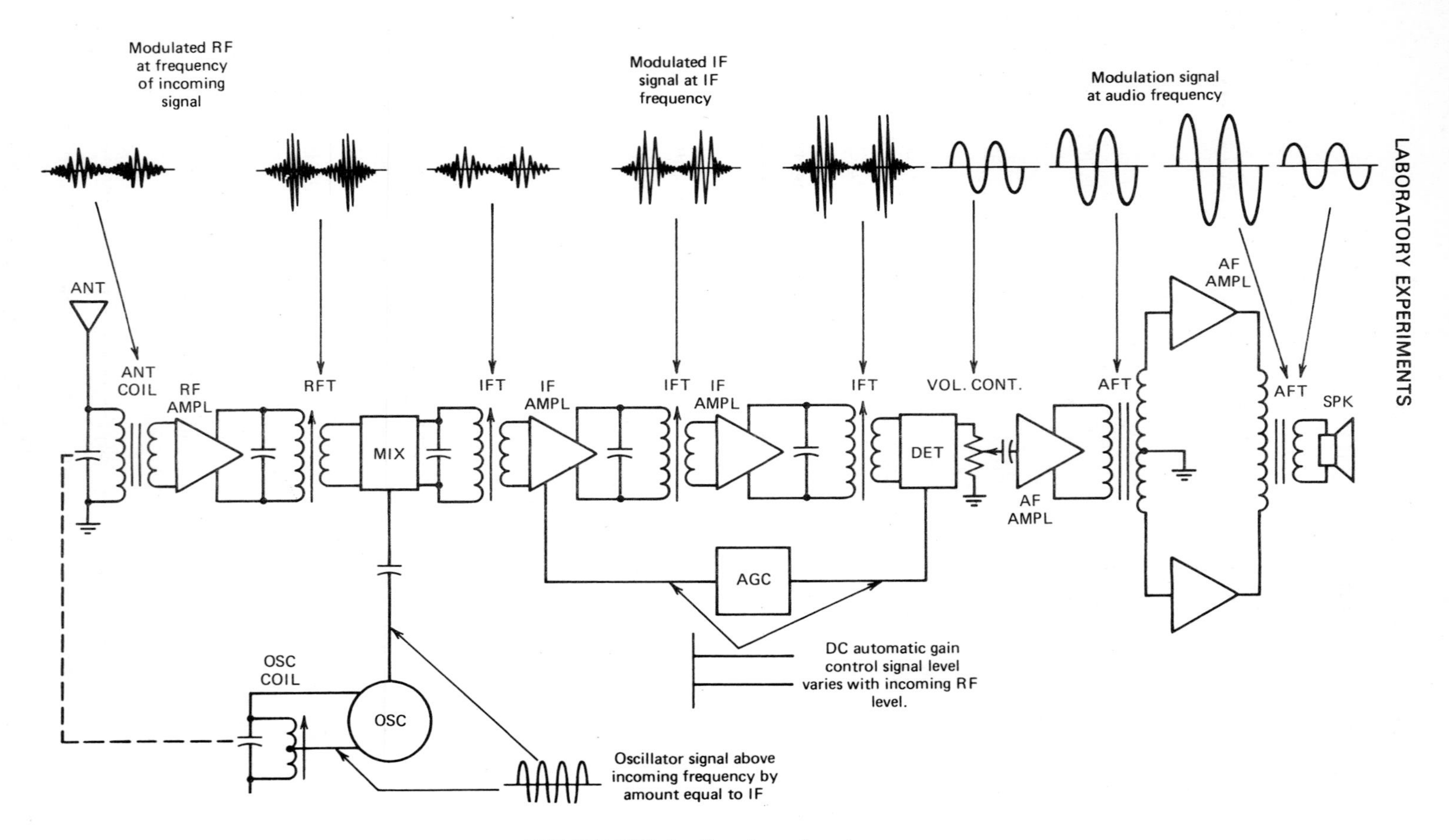

FIGURE E33-1 Receiver signals.

No Output Signal

The first step in any troubleshooting process is to check the dc operating potentials. In the example circuit of Fig. E33-2, we could measure the value of V_{cc} and the V_{CE} of each transistor. Normally, in a class *A* amplifier we expect V_{CE} to be approximately $0.5V_{cc}$. A class *B* amplifier, on the other hand, usually has a V_{CE} of about $0.9V_{cc}$.

As a result, we would expect V_{CE} to be about 6 V for the first two transistors and about 10 V for the two output transistors. If the V_{CE} value measured for one of the transistors is greatly different from the expected value, we could suspect this stage of being defective.

After checking the operating potentials of the circuit, we apply a normal signal to the input of the amplifier. We may then check the signal level at various points within the circuit. By starting at the input and working toward the output, we can locate the point at which the signal becomes lost. In the example circuit, the signal should be checked at the input terminals, the base and collector of Q_1, Q_2, Q_3, and Q_4, respectively, and the output terminals.

Reduced (or Weak) Output Signal

A seriously reduced output signal can be caused either by a weak device (a transistor with little life) or by a defective coupling network—an abnormally high reactance of a coupling capacitor or of a transformer with shorted windings.

Distorted Output

Output is most frequently distorted by a shift in the *Q*-point of one of the

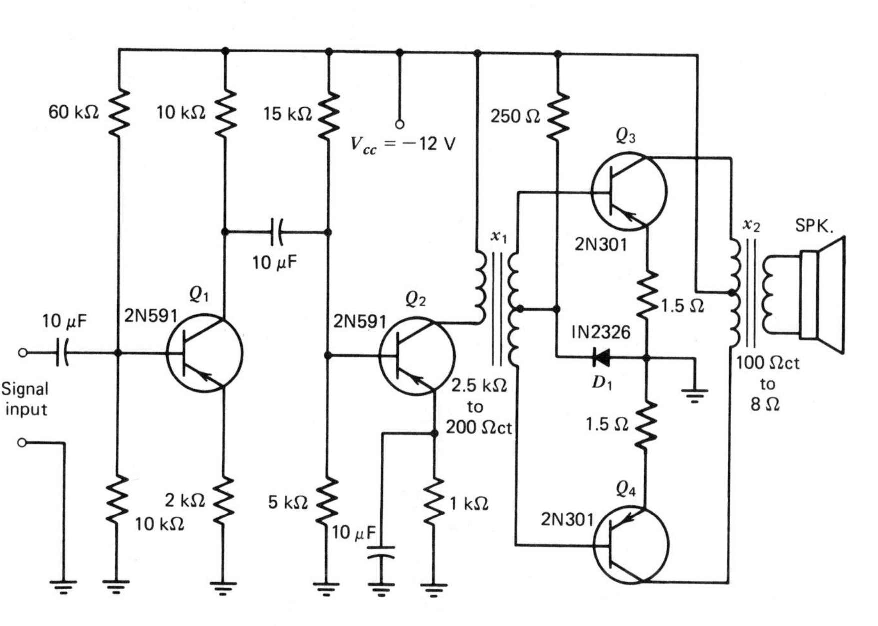

FIGURE E33-2 A typical audio section.

devices. The two most common causes of distortion are (1) defective devices (transistors, tubes, FETs) and (2) leaky coupling capacitors.

Oscillation

Oscillation is the result of regenerative feedback. Accidental regenerative feedback can occur when (1) circuit wiring is carelessly done, (2) circuit or component shielding is inadequate, (3) some component has signals from several stages applied to it, and (4) the circuit gain has sharp spikes.

Noise and Hum

Noise is randomly distributed energy usually produced by small disturbances within electronic components (resistors, transistors, capacitors, etc.). Hum is low-frequency noise and is almost always related frequencywise to the 60-Hz ac line. Hum may get into an amplifier in one of four basic ways.

1. The dc power source may be inadequately filtered, allowing a hum signal to pass through into an amplifier stage.
2. A ground loop, ground points that are at different ac levels, can induce hum into amplifier circuits.
3. Any high-impedance circuit can have hum induced into it by stray magnetic fields from transformers, power lines, and so on.
4. Vacuum tubes with ac filament voltages are particularly bad about inducing hum into amplifier circuits.

If the audio section is working normally but the receiver is still dead, we move to the IF section. Figure E33-3 shows a typical IF and detector circuit. First measure the V_{CE} of each transistor. It should be about $0.9V_{CC}$.

In the signal tracing we inject a modulated IF signal at the input of the last IF stage and note the responses in the speaker. We continue signal tracing by moving back to the first IF stage.

If the IF section is working properly but the set is still dead, the trouble must be in the RF mixer or the oscillator circuits. Application of a modulated IF or RF signal at the antenna can identify trouble in the oscillator, mixer, or RF amplifier stage.

Apparatus

- 1 AM radio receiver.
- 1 circuit diagram for the receiver.
- 1 RF signal generator.
- 1 oscilloscope.
- 1 VOM.
- 1 nonmetallic hex-type tuning tool.
- 1 nonmetallic screw-type tuning tool.

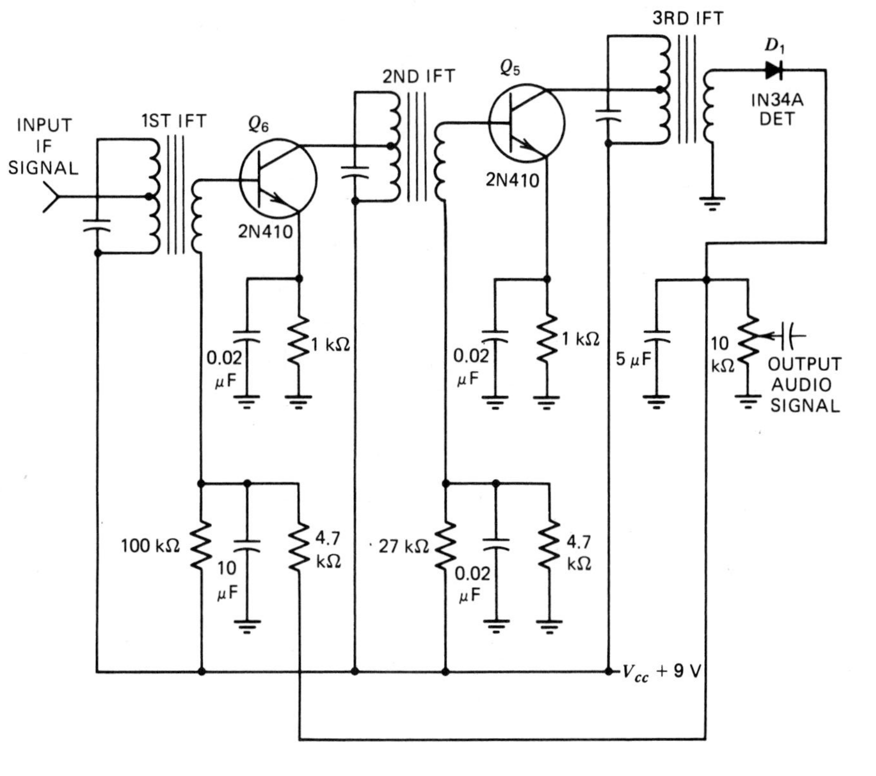

FIGURE E33-3 A typical IF section with detector.

Procedure

1. Check the general operation of the receiver to ensure that it is working.
2. Check the IF, oscillator, and the RF alignment to ensure that they are correct.
3. The instructor will now put a trouble in your receiver.
4. When your trouble has been installed, carry out the following steps.
   (a) Sectionalize the trouble to the audio, IF, or RF section. Record the tests you make, the results of the test, and your conclusion. For example,
   - Test—audio signal across volume control.
   - Results—no sound from speaker.
   - Conclusion—trouble in audio section.

   (b) Localize the trouble to a particular stage by measuring voltages or by injecting a test signal. Again record your test, results, and conclusion. For example,
   - Test—measure V_{CE} of first audio stage.
   - Results—V_{CE} is zero.
   - Conclusion—defect in first audio stage.

(c) Isolate the defect within the stage by measuring voltage and resistance. Record each test you make, the results of the test, and your conclusion. For example,

- Test—measure voltage across the emitter resistor in the first audio stage.
- Results—voltage is nearly equal to V_{CE}.
- Conclusion—transistor is shorted or saturated.
- Test—check transistor.
- Results—shorted emitter to collector.
- Conclusion—should replace transistor.

5. When you have located the defect, tell your instructor. He or she will verify your conclusion and then put another trouble in your receiver.

6. Repeat steps 4 and 5 as many times as possible during the lab period.

Report

Your laboratory report should include the following as a minimum.

- Title page.
- List of apparatus.
- Schematic diagram.
- Data table drawn by the student.
- Analysis of results.

EXPERIMENT E34: TOOLS AND TECHNIQUES FOR DIGITAL TROUBLESHOOTING

Objective

To troubleshoot a TTL circuit using logic probes, logic clips, and logic pulsers.

Discussion

Digital troubleshooting is done with various types of electronic test equipment. One of the simplest and most useful is a logic probe. Logic probes have a light that indicates the logic level on the test point being monitored. The light is on for a high and off for a low. It blinks on and off for a changing signal.

The logic pulser is essentially a single-shot pulse generator in probe form. All circuits connected to the node (outputs as well as inputs) are briefly

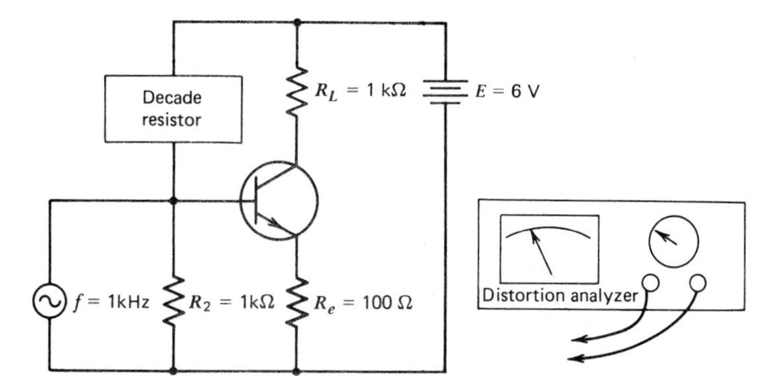

FIGURE E34-1 Sample showing a manufacturers' data sheet.

driven to their opposite state. High nodes are pulsed low, and low nodes are pulsed high each time that the button is pressed.

A sensing device that shows the state of all 16 (or 14) pins on a DIP (dual in-line package shown in Fig. E34-1) is a logic clip. It shows the state of each pin via an individual LED. The display exhibits logic highs (on) and logic lows (off).

Apparatus

- 1 power supply.
- 1 SN 7400 IC.
- 1 IC socket.
- 1 oscilloscope with storage.
- 1 HP 10525T logic probe.
- 1 HP 10526T logic pulser.
- 1 HP 10528A logic clip.
- 1 Model 320 logic probe.

Procedure

1. Connect pin 14 of SN 7400 to +5 V, pin 7 to ground.
2. Connect the logic clip to the IC socket to verify logic and pin assignments. Disconnect the logic clip.
3. Connect the logic pulser clips to +5 V and ground. Connect pin 1 to +5 V and connect the logic pulser tip to pin 2.
4. Activate the pulser and draw the oscilloscope waveforms at pin 2 (input) and pin 3 (output).
5. Connect Model 320 logic probe clips to +5 V and ground.

6. What do you observe when the probe tip is touched on pin 14, pin 7, pin 1, pins 2 and 3 when the pulser is activated?
7. Disconnect the 320 logic probe.
8. What do you observe when the HP logic probe is touched on pin 14, pin 7, pin 1, pins 2 and 3 when the pulser is activated?

Report

Your laboratory report should include the following as a minimum.

- Title page.
- List of apparatus.
- Logic diagram and pin assignment for the SN 7400 IC.
- Data table of results (drawn by the student).
- Analysis of results.

EXPERIMENT E35: DISTORTION ANALYZERS

Objective

To learn to use a commercial distortion analyzer to measure the distortion of a sinusoidal waveform.

Discussion

Distortion analyzers are used to measure the total harmonic distortion present in a sinusoidal waveform without attempting to determine which harmonics are present. In this experiment we use a commercial distortion analyzer to measure the percentage of distortion present in a sinusoidal waveform as a result of amplifying it with a single-stage transistor amplifier.

Since there are a number of commercial models of distortion analyzers, no attempt will be made to describe how to use an analyzer. Instead, you should refer to the operating manual for the particular instrument. In addition, we will not discuss the single-step amplifier since it is assumed that students in an instruments course have an understanding of basic transistor amplifiers.

Apparatus

- 1 oscilloscope.
- 1 distortion analyzer.

- 1 dc power supply.
- 1 sine-wave generator.
- 1 *n-p-n* transistor, $\beta = 50$ to 100.
- 2 resistors, 1 kΩ.
- 1 resistor, 100 Ω.
- 1 capacitor, 10 μF.

Procedure

1. Set up the experimental circuit shown in Fig. E35-1.
2. Set the decade resistor to 4 kΩ.
3. Set the signal generator to 0.3 $V_{p\text{-}p}$.
4. Use the oscilloscope to observe the waveform at the collector. If the signal is distorted, adjust the decade resistor. If this does not correct the problem, decrease the amplitude of the input signal slightly.
5. When the output of the amplifier is distorted, measure and record any distortion of the general output. Follow the procedure for using the analyzer provided by your instructor.
6. Measure and record the distortion present in the signal out of the amplifier.
7. Subtract any distortion measured in step 5 from that measured in step 6 to determine the distortion caused by the amplifier.
8. Change the value of the decade resistor by 10% and repeat steps 6 and 7.
9. Change the value of the decade resistor by an additional 10% and repeat steps 6 and 7.

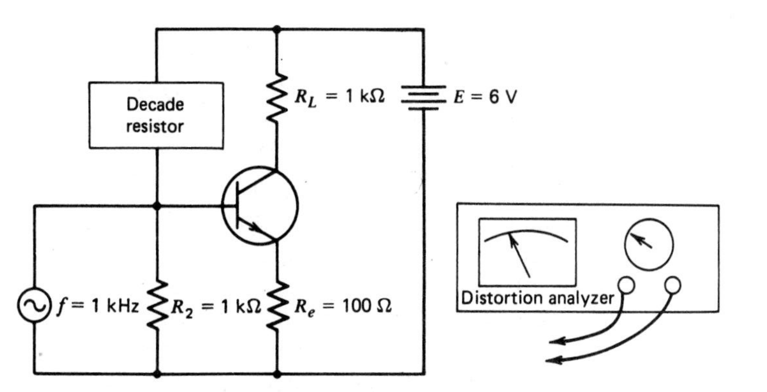

FIGURE E35-1 Experimental circuit.

Report

Your laboratory report should include the following as a minimum.

- Title page.
- List of apparatus, including serial numbers.
- Schematic diagram.
- Data table.
- Analysis of results.

Data Table for Experiment E35

Generator distortion	
Total distortion	
Amplifier distortion	
Total distortion	
Amplifier distortion	
Total distortion	
Amplifier distortion	

EXPERIMENT E36: WAVE ANALYZERS

Objective

To investigate the principle of operating a wave analyzer by fabricating and analyzing the operation of a basic wave analyzer circuit.

Discussion

Wave analyzers are used to measure the amplitude of individual harmonics present in a waveform. A wave analyzer is essentially a frequency-selective voltmeter. Frequency selectivity is achieved by switch-selectable bandpass filters that pass only a very narrow range of frequencies.

If a waveform of frequency f is to be analyzed for harmonic content, filters turned to $2f$, $3f$, $4f$, . . . , nf would be switched into the analyzer circuit. If a reading is noted on the voltmeter when a particular filter is switched into the circuit, this harmonic is present in the waveform being analyzed. For example, if the $3f$ filter is switched into the circuit when a voltmeter reading is noted, third harmonic distortion exists in the waveform.

If the value of either the inductor or the capacitor is known, the value of the other can be computed. However, since a wider selection of capacities are generally available, it is advisable to select available values of inductance and to compute the required value of capacitance using the expression

$$C = \frac{1}{4\pi^2 L f_r^2} \qquad \text{(E36-1)}$$

where

C = the value of capacitance in either the series or the parallel resonant circuit
L = the value of inductance in either the series or the parallel resonant circuit
f_r = the center frequency of the bandpass curve

Apparatus

- 1 sine-wave generator.
- 1 ac voltmeter.
- 1 dc power supply.
- 1 oscilloscope.
- 1 *n-p-n* transistor, $\beta = 50$ to 100.
- 1 decade resistor box.
- 1 resistor, 100 Ω.
- 2 resistors, 1 kΩ.
- 1 capacitor, 10 μF.
- 2 capacitors, values to be calculated.
- 2 inductors, 1 H.
- 2 switches, spdt.

Procedure

1. Compute and record the value of the capacitors required at 1.2 kHz (C_s and C_p) and at 6 kHz (C_s' and C_p').
2. Connect the experimental circuit shown in Fig. E36-1.
3. Set the decade resistor to 4-kΩ resistance.
4. Adjust the sine-wave generator output to 1.2 kHz at 0.2 V rms.
5. Use the oscilloscope to observe the waveform at the collector of the transistor.
6. If the waveform is distorted, adjust the decade resistor, decrease the amplitude of the input signal, or increase the output of the power supply slightly.
7. If an undistorted output waveform is observed, set switches S_1 and S_2 so that the capacitors C_s and C_p are connected in the circuit.
8. Note and record the ac voltmeter reading of the fundamental.

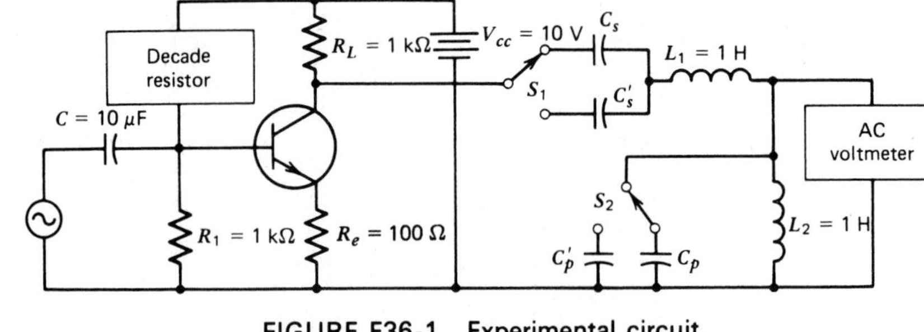

FIGURE E36-1 Experimental circuit.

9. Set the signal generator to 6 kHz. The voltmeter reading should decrease to near zero.
10. Change the position of the switches so that the capacitors C_s' and C_p' are connected in the circuit. A reading should be observed on the voltmeter.
11. Set the signal generator back to 1.2 kHz. The voltmeter reading should decrease to near zero.
12. Decrease the value of the decade resistor to 3 kΩ. A reading should now be observed on the voltmeter. This represents the fifth harmonic distortion.
13. Record the voltmeter reading in step 11. Because of component variations, the filter may not be tuned to exactly 6 kHz and a frequency adjustment may be necessary.

Report

Your laboratory report should include the following as a minimum.

- Title page.
- List of apparatus.
- Schematic diagram.
- Data table.
- Calculations.
- Analysis of results.

Data Table for Experiment E36

C_s and C_p	
C_s' and C_p'	
Amplitude of the fundamental	
Amplitude of the fifth harmonic	

EXPERIMENT E37: SIGNAL CONDITIONING FOR DATA ACQUISITION

Objective

To investigate the use of an op-amp configured as a mixer to perform signal conditioning for data acquisition purposes.

Discussion

Voltage-summing circuits find many applications in the acquisition of data. Such circuits may serve as adders, mixers, averaging circuits, or other similar functions. For example, in some industrial processes, three parameters may be monitored individually, but there might be a certain combination of the parameters that produces a desirable, or undesirable, condition. This condition can be detected with a "mixer" that combines the signal from each parameter. The circuit shown in Fig. E37-1 can be used to mix the signals.

Apparatus

- 1 741 op-amp.
- 1 dc power supply to power the 741.
- 1 dc power supply for voltage inputs.
- 1 VOM or digital multimeter
- Selected resistors.

Procedure

1. Construct the experimental circuit shown in Fig. E37-2.
2. Apply the combinations of input voltages shown in the data table and measure and record the output voltages.

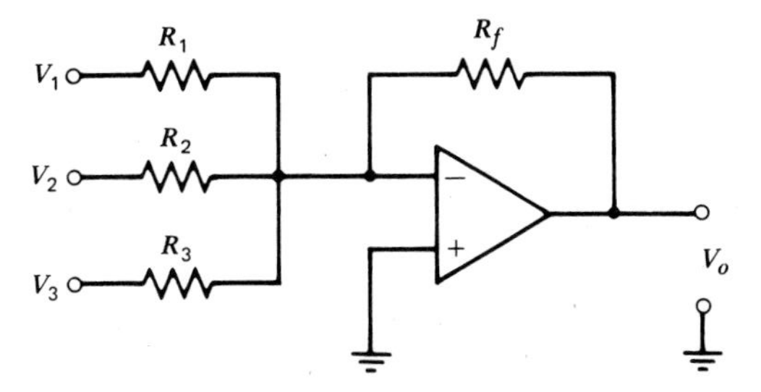

FIGURE E37-1 Inverting amplifier used as a mixer.

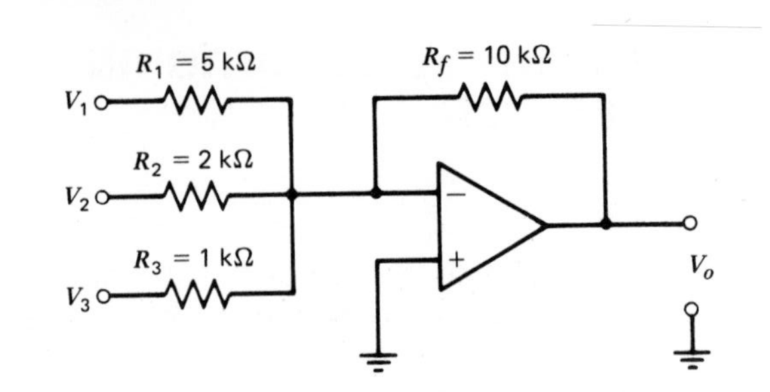

FIGURE E37-2 Experimental circuit.

3. Use the equation

$$V_o = -\left(V_1\frac{R_f}{R_1} + V_2\frac{R_f}{R_2} + V_3\frac{R_f}{R_3}\right)$$

to compute the output voltage.

4. Calculate the percentage of error between the measured and expected values of V_o for each combination.

Report

Your laboratory report should include the following as a minimum.

- Title page.
- List of apparatus.
- Schematic diagram.
- Data table.
- Calculations.
- Analysis of results.

Data Table for Experiment E37

Input (volts)			Output (volts)		
V_1	V_2	V_3	V_0(meas)	V_0(cal)	% Error
2.00	−2.0	0.20			
1.75	−1.6	0.40			
1.50	−1.2	0.60			
1.25	−0.8	0.80			
1.00	−0.6	0.50			
0.75	−0.4	0.20			
0.50	0	−.20			
0.25	0.40	−.60			
0	0.80	−.50			
−.25	1.00	−.20			
−.50	1.20	0			
−.75	1.40	0.20			
−1.0	1.60	0.40			
−1.5	1.80	0.60			
−2.0	2.00	0.80			

EXPERIMENT E38: EQUIPPING AN ELECTRONICS LABORATORY

Objective

To allow the student to consider, in terms of their value in equipping a hypothetical laboratory, the various pieces of equipment discussed in the theory portion of this book and already used in the laboratory.

Discussion

Many different kinds of electronic instruments have been discussed in this book, most of them general-purpose instruments that are used in many electronic laboratories. Others are specialized instruments used in only special applications.

When equipping a laboratory, first consider the kinds of measurements that must be made. Then consider the cost, reliability, and accuracy of the instruments and the frequency with which they will be used.

Project

Assume you have been assigned the task of equipping an electronics laboratory with a set of test equipment. The lab may be a repair and calibration lab for the company for which you work, for a company for which you would like to work, or for an educational institution.

You are to list the equipment, specifications, and cost for each item, and to justify your choice.

Select a minimum of ten pieces of equipment.

APPENDIX A

Derivation of the Equations for Computing the Value of an Unknown Resistor When Using a Kelvin Bridge

The equation for computing the value of an unknown resistor R_x can be developed in a manner similar to that for developing the balance equation for the Wheatstone bridge. When the Kelvin bridge is balanced, the potential at point A in Fig. 5-12 equals the potential at point B. The potential at point A can be expressed as

$$V_A = E\left[\frac{R_3}{R_1 + R_3}\right]$$

The potential from point B to the negative terminal of the voltage source is the sum of the voltage drops across R_b and R_x and may be written as

$$V_B = E\left[\left(\frac{R_{1c} \| R_a + R_b}{R_2 + R_x + R_{1c} \| R_a + R_b}\right)\left(\frac{R_b}{R_a + R_b}\right) + \frac{R_x}{R_2 + R_x + R_{1c} \| R_a + R_b}\right]$$

Since $V_A = V_B$ when the bridge is balanced, we can set the right side of the two expressions equal to each other as

$$\frac{R_3}{R_1+R_3}=\left(\frac{R_{1c}\|R_a+R_b}{R_2+R_x+R_{1c}\|R_a+R_b}\right)\left(\frac{R_b}{R_a+R_b}\right)\frac{R_x}{R_2+R_x+R_{1c}\|R_a+R_b}$$

Solving this equation for the unknown resistor R_x yields

$$R_x=\frac{R_2R_3}{R_1}+\frac{R_{1c}}{R_a+R_b+R_{1c}}\frac{R_3}{R_1}-\frac{R_b}{R_a} \tag{A-1}$$

Since at balance $R_a/R_b=R_1/R_3$, we can see that Eq. A-1 reduces to the balance equation for the Wheatstone bridge, which is

$$R_x=\frac{R_2R_3}{R_1}$$

APPENDIX B Derivation of Equations for Selected AC Bridges

B-1 OPPOSITE-ANGLE BRIDGE

Problem

To derive an expression for the equivalent-series inductance and resistance of an opposite-angle bridge at null.

Solution

The impedance of the arms of the bridge can be written as

$$Z_1 = R_1 \qquad Z_2 = R_2 - jX_{C2}$$

$$Z_3 = R_x + jX_{Lx} \quad Z_4 = R_4$$

Substituting these values in Eq. 6-4 gives the balance equation

$$R_1 R_4 = (R_2 - jX_{C2})(R_x + jX_{Lx})$$

$$R_1 R_4 = R_2 R_x + jR_2 X_{Lx} - jR_x X_{C2} + X_{C2} X_{Lx}$$

$$R_1 R_4 = R_2 R_x + jR_2 X_{Lx} - jR_x X_{C2} + \frac{\omega L_x}{\omega C_2}$$

By setting both the real and imaginary parts equal to zero, we have

$$R_1 R_4 = R_2 R_x + \frac{L_x}{C_2} \tag{B-1}$$

$$R_2 X_{Lx} - R_x X_{C2} = 0 \tag{B-2}$$

From Eq. 6-15, we get

$$R_2 X_{Lx} = R_x X_{C2}$$

$$\omega R_2 L_x = \frac{R_x}{\omega C_2}$$

$$L_x = \frac{R_x}{\omega^2 R_2 C_2} \tag{B-3}$$

Substituting Eq. B-3 in Eq. A-1 leads to

$$R_1 R_4 = R_2 R_x + \frac{R_x}{\omega^2 R_2 C_2^2}$$

$$R_x\left(\frac{1}{\omega^2 R_2 C_2^2} + R_2\right) = R_1 R_4$$

$$R_x\left(\frac{1 + \omega^2 R_2^2 C_2^2}{\omega^2 R_2 C_2^2}\right) = R_1 R_4$$

$$R_x = \frac{\omega^2 R_1 R_2 R_4 C_2^2}{1 + \omega^2 R_2^2 C_2^2} \tag{B-4}$$

Substituting Eq. B-4 in Eq. A2-3 leads to

$$L_x = \left(\frac{\omega^2 R_1 R_2 R_4 C_2^2}{1 + \omega^2 R_2^2 C_2^2}\right)\left(\frac{1}{\omega^2 R_2 C_2}\right)$$

$$= \frac{R_1 R_4 C_2}{1 + \omega^2 R_2^2 C_2^2} \tag{B-5}$$

B-2 WIEN BRIDGE

Problem

To derive an expression for

(a) The equivalent-parallel components.

(b) The equivalent-series components of the unknown impedance of a Wien bridge at null.

Solution

(a) Find R_2 and C_2. The impedance of the arms of the bridge can be written as

$$Z_1 = R_1 - jX_{C1} \quad Z_3 = R_3$$

$$Z_2 = \frac{1}{\frac{1}{R_2} - \frac{1}{jX_{C2}}} \quad Z_4 = R_4$$

Substituting these values in Eq. 6-4 gives

$$(R_1 - jX_{C1})R_4 = \left(\frac{1}{\frac{1}{R_2} - \frac{1}{jX_{C2}}}\right) R_3$$

$$\frac{1}{R_2} - \frac{1}{jX_{C2}} (R_1 - jX_{C1})R_4 = R_3$$

$$\frac{R_1 R_4}{R_2} - \frac{jR_4 X_{C1}}{R_2} - \frac{R_1 R_4}{jX_{C2}} + \frac{jR_4 X_{C1}}{jX_{C2}} = R_3$$

$$\frac{R_1 R_4}{R_2} - \frac{jR_4}{\omega R_2 C_1} + j\omega R_1 R_4 C_2 + \frac{R_4 C_2}{C_1} = R_3$$

Setting both the real and imaginary parts to zero leads to

$$\frac{R_1 R_4}{R_2} + \frac{R_4 C_2}{C_1} = R_3 \tag{B-6}$$

$$\omega R_1 R_4 C_2 = \frac{R_4}{\omega R_2 C_1} \tag{B-7}$$

Solving Eq. B-7 for C_2 and substituting in Eq. B-6, we get

$$C_2 = \frac{R_4}{\omega^2 R_1 R_2 R_4 C_1} = \frac{1}{\omega^2 R_1 R_2 C_1} \tag{B-8}$$

$$\frac{R_1 R_4}{R_2} + \frac{R_4}{\omega^2 R_1 R_2 C_1^2} = R_3$$

$$R_1 R_4 + \frac{R_4}{\omega^2 R_1 C_1^2} = R_2 R_3$$

$$R_2 = \frac{R_1 R_4}{R_3} + \frac{R_4}{\omega^2 R_1 R_3 C_1^2}$$

$$= \frac{R_4}{R_3}\left(R_1 + \frac{1}{\omega^2 R_1 C_1^2}\right) \tag{B-9}$$

Substituting Eq. B-9 in Eq. B-8 leads to

$$C_2 = \frac{1}{\omega^2 R_1 C_1 \left[\frac{R_4}{R_3}\left(R_1 + \frac{1}{\omega^2 R_1 C_1^2}\right)\right]}$$

$$= \frac{1}{\frac{R_4}{R_3}\left[\omega^2 R_1 C_1 \left(\frac{\omega^2 R_1^2 C_1^2 + 1}{\omega^2 R_1 C_1^2}\right)\right]}$$

$$= \frac{R_3}{R_4}\left(\frac{1}{1 + \omega^2 R_1^2 C_1^2}\right) C_1 \tag{B-10}$$

(b) Find R_1 and C_1. Solving Eq. B-7 for R_1 and substituting in Eq. B-6, we get

$$R_1 = \frac{R_4}{\omega^2 R_2 R_4 C_1 C_2} \qquad \text{(B-11)}$$

$$\frac{R_4^2}{\omega^2 R_2^2 R_4 C_1 C_2} + \frac{R_4 C_2}{C_1} = R_3$$

$$\frac{R_4}{\omega^2 R_2^2 C_2} + R_4 C_2 = R_3 C_1$$

$$C_1 = \frac{R_4 C_2}{R_3} + \frac{R_4}{\omega^2 R_2^2 R_3 C_2} = \frac{R_4}{R_3}\left(C_2 + \frac{1}{\omega^2 R_2^2 C_2}\right) \qquad \text{(B-12)}$$

Substituting Eq. B-12 in Eq. B-11 leads to

$$R_1 = \frac{R_4}{\omega^2 R_2 R_4 C_2 \left[\frac{R_4}{R_3}\left(C_2 + \frac{1}{\omega^2 R_2^2 C_2}\right)\right]}$$

$$= \frac{R_3}{R_4}\left(\frac{1}{\omega^2 R_2 C_2^2 + \frac{1}{R_2}}\right)$$

$$= \frac{R_3}{R_4}\left(\frac{1}{\frac{\omega^2 R_2^2 C_2^2 + 1}{R_2}}\right)$$

$$= \frac{R_3}{R_4}\left(\frac{R_2}{1 + \omega^2 R_2^2 C_2^2}\right) \qquad \text{(B-13)}$$

B-3 RADIO-FREQUENCY BRIDGE

Problem

To derive an expression for the equivalent-series elements of an RF bridge at null after it is rebalanced.

Solution

The impedance of the arms of the bridge can be written as

$$Z_1 = -jX_{C1} \qquad Z_2 = R_2 - jX_{C2}$$

$$Z_3 = \frac{1}{\frac{1}{R_3} - \frac{1}{jX_{C3}}} \qquad Z_4 = R_4$$

Substituting these values in Eq. 6-4, we obtain the resulting balance:

$$-jR_4X_{C1} = (R_2 - jX_{C2})\left(\frac{1}{\frac{1}{R_3} - \frac{1}{jX_{C3}}}\right)$$

$$\frac{-jR_4X_{C1}}{R_3} + \frac{jR_4X_{C1}}{jX_{C3}} = R_2 - jX_{C2}$$

$$\frac{-jR_4}{\omega R_3C_1} + \frac{R_4C_3}{C_1} = R_2 - \frac{j}{\omega C_2}$$

Setting both the real and imaginary parts to zero leads to

$$\frac{R_4C_3}{C_1} = R_2 \tag{B-14}$$

$$\frac{R_4}{\omega R_3C_1} = \frac{1}{\omega C_2} \tag{B-15}$$

From Eq. B-14 we get

$$C_3 = \frac{R_2}{R_4}C_1 \tag{B-16}$$

From Eq. B-15 we get

$$C_2 = \frac{R_3}{R_4}C_1 \tag{B-17}$$

When an unknown impedance represented by $Z_x = R_x \pm jX_x$ is inserted and the bridge is rebalanced by adjusting C_2 and C_3, the impedance of the arms of the bridge can be written as

$$Z_1 = -jX_{C1} \qquad Z_2 = R_2 + R_x + jX_x - jX_{C'_2}$$

$$Z_3 = \frac{1}{\frac{1}{R_3} - \frac{1}{jX_{C'_3}}} \qquad Z_4 = R_4$$

where C'_2 and C'_3 are adjusted values of C_2 and C_3, respectively. Substituting these values in Eq. 6-4 gives the balance equation

$$-jR_4X_{C1} = (R_2 + R_x + jX_x - jX_{C'_2})\left(\frac{1}{\frac{1}{R_3} - \frac{1}{jX_{C'_3}}}\right)$$

$$-jR_4X_{C1}\left(\frac{1}{R_3} - \frac{1}{jX_{C'_3}}\right) = R_2 + R_x + jX_x - jX_{C'_2}$$

$$\frac{-jR_4}{\omega R_3C_1} + \frac{j\omega R_4C'_3}{j\omega C_1} = R_2 + R_x + jX_x - \frac{j}{\omega C'_2}$$

Setting both the real and imaginary parts to zero leads to

$$R_x = \frac{R_4 C'_3}{C_1} - R_2 \tag{B-18}$$

$$X_x = \frac{1}{\omega C'_2} - \frac{R_4}{\omega R_3 C_1} \tag{B-19}$$

Substituting Eq. B-14 in Eq. B-18, we get

$$R_x = \frac{R_4 C'_3}{C_1} - \frac{R_4 C_3}{C_1}$$

$$= \frac{R_4}{C_1}(C'_3 - C_3) \tag{B-20}$$

Solving Eq. B-17 for C_1 and substituting in Eq. B-19 leads to

$$C_1 = \frac{R_4}{R_3} C_2$$

$$X_x = \frac{1}{\omega C'_2} - \frac{R_3 R_4}{\omega R_3 R_4 C_2}$$

$$= \frac{1}{\omega C'_2} - \frac{1}{\omega C_2}$$

$$= \frac{1}{\omega}\left(\frac{1}{C'_2} - \frac{1}{C_2}\right) \tag{B-21}$$

Answers

Chapter 1

1-1 0.04 A, 2.67%

1-3 0.992

1-5 (a) 1100 Ω, (b) 1210 Ω + 10%, (c) 990 Ω − 10%

1-7 (a) 1 MHz, (b) 0.9 MHz − 10%, (c) 1.1 MHz + 10%

1-9 −9% at R_{max} to +11.23% at R_{min}

1-11 (a) 6 V, (b) 3.43%

1-13 0.106 kΩ

Chapter 2

2-1 85 mV

2-3 25 μA

2-5 Meter A

2-7 $R_1 = 18$ kΩ, $R_2 = 80$ kΩ, $R_3 = 100$ kΩ, $R_4 = 1000$ kΩ

2-9 $R_1 = 1$ Ω, $R_2 = 9$ Ω, $R_3 = 90$ Ω, $R_4 = 900$ Ω

2-11 Only the 100-V range has less than 3% error.

2-13 6.66 V

Chapter 3

3-1 47.75 μA

3-3 $S_{ac} = 450$ Ω/V, $S_{dc} = 1$ kΩ/V, $R_s = 13.3$ kΩ

3-5 $S_{ac} = 225$ Ω/V, $S_{dc} = 1$ kΩ/V for the meter movement but 500 Ω/V for the meter, $R_s = 4075$ Ω

3-7 $S_{dc} = 20$ Ω/V, $R_s^2 = 198$ Ω

3-9 (a) Not sinusoidal, (b) sinusoidal, (c) sinusoidal

3-11 Meter B provides a more accurate reading.

Chapter 4

4-1 147.5 Ω
4-3 (a) 20 mA, (b) 132.2 Ω, (c) 28.6 mA
4-5 (a) 20 mA, (b) 69.5 Ω, (c) 20 μV
4-7 200
4-9 $R_1 = 4.5$ MΩ, $R_2 = 500$ kΩ
4-11 $R = 500$ Ω, 0.625 W
4-13 145.5 Ω
4-15 1840 Ω

Chapter 5

5-1 25 kΩ
5-3 80 Ω
5-5 0.3 Ω
5-7 ±0.6%
5-9 4.86 kΩ
5-11 64.3 μA
5-13 2,353 m

Chapter 6

6-1 $R_x = 122.6$ Ω, $C_x = 0.33$ μF
6-3 $R_x = 23.18$ kΩ, $C_x = 1.08$ μF
6-5 $R_x = 166.67$ Ω, $L_x = 0.1$ H
6-7 $R_x = 9.75$ Ω, $L_x = 0.247$ mH
6-9 $R_x = 575.9$ Ω, $C_x = 0.086$ μF
6-11 $R_x = 10$ Ω, $C_x = 15.62$ μF
6-13 $R_x = 150$ Ω, $C_x = 0.01$ μF

Chapter 7

7-1 3.57 V
7-3 6.67 MΩ, 2.33 MΩ, 0.67 MΩ, 0.23 MΩ, 66.6 kΩ, 33.3 kΩ
7-5 7.19 kΩ
7-7 25.9 kΩ
7-9 18.5 kΩ

Chapter 8

8-1 583 V

8-3 22 V

8-7 $R_1 = 6$ kΩ, $R_2 = 9$ kΩ, $R_3 = 45$ kΩ, $R_4 = 90$ kΩ, $R_5 = 450$ kΩ, $R_6 = 900$ kΩ, $R_7 = 4.5$ MΩ, $R_8 = 9$ MΩ

8-9 $f = 33.33$ kHz, $V_{p\text{-}p} = 600$ mV

8-11 5 pF

8-13 23.33 nsec

8-15 36 pF

8-17 53°

Chapter 9

9-6 0.5 Hz

9-7 75 mm/sec

Chapter 10

10-1 19.96

10-3 0.0013 μF

10-5 152 μH

10-7 11.72 kHz

10-9 937-5 kΩ

10-11 69 μsec

Chapter 11

11-1 0.3 in., 0.5 in., 0.8 in.

11-3 1.5 V

11-5 (a) 300 Ω, (b) 3 V

11-7 8×10^{-3} Ω

11-9 105.6 Ω

11-11 1×10^{-2} Ω

11-13 (a) 70.8 nF, (b) 283.3 pF

11-15 0.01 V

11-17 (a) −3.75 V, (b) 3.75 V to −2 V

11-19 200 mJ

11-21 152.55°C

11-23 4.37 mV

11-25 0.5 mA, 1 mA

11-27 (a) 3 kΩ, (b) 0.19 mA
11-29 (a) 0.9 mA, (b) 0.18 V, (c) 0.32 V
11-31 (a) 8.5 V, (b) 11.5 V, (c) 5.6 mA

Chapter 12

12-3 12.65 μV
12-5 1×10^{-19} W/Hz
12-7 11.6°K
12-9 3
12-11 0.67 mV
12-13 1000

Chapter 13

13-1 4 V
13-3 60 V
13-5 Lamps A and D
13-7 4 kHz
13-9 0.6 V

Chapter 14

14-3 Gate 1 is okay.
14-5 (a) High, (b) Low

Chapter 15

15-1 $f_f = 50$ kHz, $f_3 = 150$ kHz, $f_5 = 250$ kHz, $f_7 = 350$ kHz
15-3 10%
15-5 10.8 V
15-7 0.21 V

Chapter 16

16-5 3.44°

Chapter 17

17-1 1.58 V
17-5 10
17-7 63

INDEX

A

D

E

F

G

H

I

J

K

L

M

N

O

P

Q

R

S

T

U

V